Moral, Ethical, and Social Dilemmas in the Age of Technology:

Theories and Practice

Rocci Luppicini
University of Ottawa, Canada

Information Science
REFERENCE

Managing Director:	Lindsay Johnston
Editorial Director:	Joel Gamon
Book Production Manager:	Jennifer Yoder
Publishing Systems Analyst:	Adrienne Freeland
Assistant Acquisitions Editor:	Kayla Wolfe
Typesetter:	Deanna Jo Zombro
Cover Design:	Nick Newcomer

Published in the United States of America by
 Information Science Reference (an imprint of IGI Global)
 701 E. Chocolate Avenue
 Hershey PA 17033
 Tel: 717-533-8845
 Fax: 717-533-8661
 E-mail: cust@igi-global.com
 Web site: http://www.igi-global.com

Library of Congress Cataloging-in-Publication Data

Moral, ethical, and social dilemmas in the age of technology: theories and practice / Rocci Luppicini, editor.
    p. cm.
 Includes bibliographical references and index.
 Summary: "This book highlights the innovations and developments in the ethical features of technology in society, bringing together research in the areas of computer, engineering, and biotechnical ethics"--Provided by publisher.
  ISBN 978-1-4666-2931-8 (hbk.) -- ISBN 978-1-4666-2932-5 (ebook) -- ISBN 978-1-4666-2933-2 (print & perpetual access) 1. Technology--Moral and ethical aspects. 2. Information technology--Moral and ethical aspects. I. Luppicini, Rocci.
 BJ59.M665 2013
 147'.96--dc23
                                   2012039281

British Cataloguing in Publication Data
A Cataloguing in Publication record for this book is available from the British Library.

The views expressed in this book are those of the authors, but not necessarily of the publisher.

Table of Contents

Section 1
Historical and Theoretical Perspectives

Chapter 1
Miles Kennedy, National University of Ireland Galway, Ireland

Chapter 2
Raphael Cohen-Almagor, University of Hull, UK

Chapter 3
Marcus Schulzke, State University of New York at Albany, USA

Chapter 4
Tommaso Bertolotti, University of Pavia, Italy
Emanuele Bardone, University of Pavia, Italy
Lorenzo Magnani, University of Pavia, Italy

Chapter 5
Christopher Wareham, European School of Molecular Medicine, Italy
*   & The University of Milan, Italy*

Section 2
Ethical Dilemmas of Technoselves in a Technosociety

Section 3
Morality and Techno-Mediated Violence

Detailed Table of Contents

Section 1
Historical and Theoretical Perspectives

Chapter 1

Miles Kennedy, National University of Ireland Galway, Ireland

This paper is an attempt to establish a foundation for technoethics of IT that makes an account of the virtual environment based within the lived situation of those who work and dwell in that emerging realm. The most important phenomenon for technoethics of IT is the relationship between knowledge about information and the capacity to turn information into knowledge. This relationship is embodied in being a Master of Information Technology. To achieve mastery of information and mould it into knowledge, a useful tool-like entity, is to have power in the contemporary world. Once this situation is recognised ethical questions arise of their own volition. A selection of these questions are dealt with in the following paper, they are the questions of the distinction between information and knowledge, the central issue of virtue and virtuality, and the distinction between stealing and sharing in the virtual environment. This paper constitutes a think piece; readers who have a stake in the virtual environment and its ethical makeup are urged to ask themselves these questions and come up with others in turn.

Chapter 2

Raphael Cohen-Almagor, University of Hull, UK

This paper outlines and analyzes milestones in the history of the Internet. As technology advances, it presents new societal and ethical challenges. The early Internet was devised and implemented in American research units, universities, and telecommunication companies that had vision and interest in cutting-edge research. The Internet then entered into the commercial phase (1984-1989). It was facilitated by the upgrading of backbone links, the writing of new software programs, and the growing number of interconnected international networks. The author examines the massive expansion of the Internet into a global network during the 1990s when business and personal computers with different operating systems joined the universal network. The instant and growing success of social networking-sites that enable Netusers to share information, photos, private journals, hobbies, and personal as well as commercial interests with networks of mutual friends and colleagues is discussed.

This paper examines property relations in massively multiplayer online games (MMOGs) through the lens of John Locke's theory of property. It argues that Locke's understanding of the common must be modified to reflect the differences between the physical world that he dealt with and the virtual world that is now the site of property disputes. Once it is modified, Locke's theory provides grounds for recognizing player ownership of much of the intellectual material of virtual worlds, the goods players are responsible for creating, and the developer-created goods that players obtain through an exchange of labor or goods representing labor value.

This paper analyzes the impact of new technologies on a range of practices related to activism. The first section shows how the functioning of democratic institutions can be impaired by scarce political accountability connected with the emergence of moral hazard; the second section displays how cyberactivism can improve the transparency of political dynamics; in the last section the authors turn specifically to cyberactivism and isolate its flaws and some of the most pernicious and self-defeating effects.

Chapter 5

Artificial agents such as robots are performing increasingly significant ethical roles in society. As a result, there is a growing literature regarding their moral status with many suggesting it is justified to regard manufactured entities as having intrinsic moral worth. However, the question of whether artificial agents could have the high degree of moral status that is attributed to human persons has largely been neglected. To address this question, the author developed a respect-based account of the ethical criteria for the moral status of persons. From this account, the paper employs an empirical test that must be passed in order for artificial agents to be considered alongside persons as having the corresponding rights and duties.

Section 2
Ethical Dilemmas of Technoselves in a Technosociety

The term biometrics is derived from the Greek words: bio (life) and metrics (to measure). "Biometric technologies" are defined as automated methods of verifying or recognizing the identity of a living person based on a physiological or behavioral characteristic. Several techniques and features were used over time to recognize human beings several years before the birth of Christ. Today, this research field has become very employed in many applications such as security applications, multimedia applications and banking applications. Also, many methods have been developed to strengthen the biometric accuracy and reduce the imposture errors by using several features such as face, speech, iris, finger vein, etc. From a security purpose and economic point of view, biometrics has brought a great benefit and has become an important tool for governments and institutions. However, citizens are expressing their thorough worry, which is due to the freedom limitations and loss of privacy. This paper briefly presents some new technologies that have recently been proposed in biometrics with their levels of reliability, and discusses the different social and ethic problems that may result from the abusive use of these technologies.

Chapter 7

Daniele Cantore, University of Pavia, Italy

This paper advances an analysis of biometrics and profiling. Biometrics represents the most effective technology in order to prove someone's identity. Profiling regards the capability of collecting and organizing individuals' preferences and attitudes as consumers and costumers. Moreover, biometrics is already used in order to gather and manage biological and behavioral data and this tendency may increase in Ambient Intelligence context. Therefore, dealing with individuals' data, both biometrics and profiling have to tackle many ethical issues related to privacy on one hand and democracy on the other. After a brief Introduction, the author introduces biometrics, exploring its methodology and applications. The following section focuses on profiling both in public and private sector. The last section analyzes those issues concerning privacy and democracy, within also the Ambient Intelligence.

Chapter 8

Steven E. Stern, University of Pittsburgh at Johnstown, USA
Benjamin E. Grounds, Penn State College of Medicine, USA

Changes in technology often affect patterns of social interaction. In the current study, the authors examined how cellular telephones have made it possible for members of romantically involved couples to keep track of each other. The authors surveyed 69 undergraduates on their use of cellular telephones as well as their relationships and their level of sexual jealously. Results find that nearly a quarter of romantically involved cellular telephone users report tracking their significant other, and evidence shows that tracking behavior correlates with jealousy. Furthermore, participants frequently reported using countermeasures such as turning off their cellular telephones in order to avoid being tracked by others. In conclusion, newer communication technologies afford users to act upon protectiveness and jealousy more readily than before these technologies were available to the general public.

Chapter 9

Xue Lin, University of Ottawa, Canada
Rocci Luppicini, University of Ottawa, Canada

Technoethical inquiry deals with a variety of social, legal, cultural, economic, political, and ethical implications of new technological applications which can threaten important aspects of contemporary life and society. GhostNet is a large-scale cyber espionage network which has infiltrated important political, economic, and media institutions including embassies, foreign ministries and other government offices in 103 countries and infected at least 1,295 computers. The following case study explores the influences of GhostNet on affected organizations by critically reviewing GhostNet documentation and relevant literature on cyber espionage. The research delves into the socio-technical aspects of cyber espionage through a case study of GhostNet. Drawing on Actor Network Theory (ANT), the research examined key socio-technical relations of Ghostnet and their influence on affected organizations. Implications of these findings for the phenomenon of GhostNet are discussed in the hope of raising awareness about the importance of understanding the dynamics of socio-technical relations of cyber-espionage within organizations.

Chapter 10

Troy J. Strader, Drake University, USA

J. Royce Fichtner, Drake University, USA

Suzanne R. Clayton, Drake University, USA

Lou Ann Simpson, Drake University, USA

Employees have access to a wide range of computer-related resources at work, and often these resources are used for non-work related personal activities. In this study, the authors address the relationship between employee's utilitarian ethical orientation, the factors that create the context that influences their ethical perceptions, and their overall perceptions regarding the level of acceptability for 14 different non-work related computing activities. The authors find that time and monetary cost associated with an activity has a negative relationship to perceived acceptability. Results indicate that contextual variables, such as an employee's supervisory or non-supervisory role, opportunity, computer self-efficacy, and whether or not an organization has computer use policies, training, and monitoring, influence individual ethical perceptions. Implications and conclusions are discussed for organizations and future research.

Chapter 11

Kurt Reymers, Morrisville State College, USA

In 2008, a resident of a computerized virtual world called "Second Life" programmed and began selling a "realistic" virtual chicken. It required food and water to survive, was vulnerable to physical damage, and could reproduce. This development led to the mass adoption of chicken farms and large-scale trade in virtual chickens and eggs. Not long after the release of the virtual chickens, a number of incidents occurred which demonstrate the negotiated nature of territorial and normative boundaries. Neighbors of chicken farmers complained of slow performance of the simulation and some users began terminating the chickens, kicking or shooting them to "death." All of these virtual world phenomena, from the interactive role-playing of virtual farmers to the social, political and economic repercussions within and beyond the virtual world, can be examined with a critical focus on the ethical ramifications of virtual world conflicts. This paper views the case of the virtual chicken wars from three different ethical perspectives: as a resource dilemma, as providing an argument from moral and psychological harm, and as a case in which just war theory can be applied.

Section 3
Morality and Techno-Mediated Violence

Chapter 12

Lorenzo Magnani, University of Pavia, Italy, & Sun Yat-sen University, China

A kind of common prejudice is the one that tends to assign the attribute "violent" only to physical and possibly bloody acts – homicides, for example – or physical injuries; but linguistic, structural, and other various aspects of violence – also embedded in artifacts – have to be taken into account. The paper will deal with the so-called "technology-mediated violence" taking advantage of the illustration of the case of profiling. If production of knowledge is important and central, this is not always welcome and so people have to acknowledge that the motto introduced in the book *Morality in a Technological World* (Magnani, 2007) knowledge as a duty has various limitations. Indeed, a warning has to be formulated regarding the problem of identity and cyberprivacy. The author contends that when too much knowledge about people is incorporated in external artificial things, human beings' "visibility" can become excessive and dangerous. Two aims are in front of people to counteract this kind of technological violence, which also jeopardizes *Rechtsstaat* and constitutional democracies: preserving people against the various forms of circulation of knowledge about them and building new suitable "technoknowledge" (also to originate new "embodied" legal institutions) to reach this protective result.

Chapter 13

Cameron Shelley, University of Waterloo, Canada

The purpose of this article is to explore how the design of technology relates to the fairness of the distribution of violence in modern society. Deliberately or not, the design of technological artifacts embodies the priorities of its designers, including how violence is meted out to those affected by the design. Designers make implicit predictions about the context in which designs will perform, predictions that will not always be satisfied. Errors from failed predictions can affect people in ways that designers may not appreciate. In this article, several examples of how artifacts distribute violence are considered. The Taylor-Russell diagram is introduced as a means of representing and exploring this issue. The role of government regulation, safety, and social role in design assessment is discussed.

Chapter 14

Emanuele Bardone, University of Pavia, Italy, & Tallinn University, Estonia

This paper outlines a theoretical framework meant to help people understand the emergence of violent mediators in human cognitive niches enriched by technology. Violent mediators are external objects that mediate relations with the environment in a way facilitating – not causing – the adoption of violent behaviors. In order to cast light on the dynamics violent mediators are involved in, the author illustrates the role played by what is called unintended affordances. In doing so, the author presents three specific examples of unintended affordances as violent mediators: multitasking while driving, desultory behavior and cyberstalking. The last part of the paper presents the notion of counteractive cognitive niche as a possible, yet partial, solution to the problem concerning the emergence of unintended affordance.

Chapter 15

Jeffrey Benjamin White, KAIST, Korea

This paper reviews the complex, overlapping ideas of two prominent Italian philosophers, Lorenzo Magnani and Luciano Floridi, with the aim of facilitating the nonviolent transformation of self and world, and with a focus on information technologies in mediating this process. In Floridi's information ethics, problems of consistency arise between *self-poiesis*, *anagnorisis*, entropy, evil, and the narrative structure of the world. Solutions come from Magnani's work in distributed morality, moral mediators, moral bubbles and moral disengagement. Finally, two examples of information technology, one ancient and one new, a Socratic narrative and an information processing model of moral cognition, are offered as mediators for the nonviolent transformation of self and world respectively, while avoiding the tragic requirements inherent in Floridi's proposal.

Chapter 16

Tommaso Bertolotti, University of Pavia, Italy

Over the past years, mass media increasingly identified many aspects of social networking with those of established social practices such as gossip. This produced two main outcomes: on the one hand, social networks users were described as gossipers mainly aiming at invading their friends' and acquaintances' privacy; on the other hand the potentially violent consequences of social networking were legitimated by referring to a series of recent studies stressing the importance of gossip for the social evolution of human beings. This paper explores the differences between the two kinds of gossip-related sociability, the traditional one and the technologically structured one (where the social framework coincides with the technological one, as in social networking websites). The aim of this reflection is to add to the critical knowledge available today about the effects that transparent technologies have on everyday life, especially as far as the social implications are concerned, in order to prevent (or contrast) those "ignorance bubbles" whose outcomes can be already dramatic.

Section 4
Current Trends and Applications in Technoethics

Chapter 17

Rocci Luppicini, University of Ottawa, Canada

University degree programs in STS (Science and Technology Studies) represent a popular training ground for scholars and other professions dealing with advanced studies in science and technology. Degree programs in STS are currently offered at universities around the globe with various specializations and orientations. This study explores the nature of science and technology in Canada and the state of ethics within STS curriculum in Canada. STS degree programs offered under various titles at nine universities in Canada are examined. Findings reveal that ethical aspects of science and technology study is lacking from the core content of most Canadian academic programs in STS. Key challenges are addressed and suggestions are made on how to leverage STS programs within Canadian universities. This study advances the understanding of the developing field of STS in Canada from a technoethical perspective.

Chapter 18

Jeffrey Reiss, University of Central Florida, USA
Rosa Cintrón, University of Central Florida, USA

This study explores the nature of piracy prevention tools used by IT departments in the Florida State University System to determine their relative effectiveness. The study also examines the opinions of the Information Security Officer in terms of alternative piracy prevention techniques that do not involve legal action and monitoring. It was found that most institutions do not use a formal piece of software that monitors for infringing data. Furthermore, institutions agreed that students lack proper ethics and concern over the matter of copyright, but were not fully convinced that other prevention methods would be effective. The authors conclude that monitoring techniques are a short-term solution and more research must put into finding long-term solutions.

Based on recent complaints about the neglect of the human in the philosophy of technology, this paper explores the different ways how technology ethics put the relation between the human and the technical on stage. It identifies various similarities in the treatment of the human in technology and the treatment of the child in education and compares Heidegger's concerns about the role of technology with the duplicity of childhood and adulthood in conflicts of adolescence. The findings give reason to assume that technology ethics and pedagogy are closely related. A brief review of selected topics in technology ethics illustrates exemplarily how a pedagogic interpretation of the current discussion can contribute to further progress in the field.

The requirement of always obtaining participants' informed consent in research with human subjects cannot always be met, for a variety of reasons. This paper describes and categorises research situations where informed consent is unobtainable. Some of these kinds of situations, common in biomedicine and psychology, have been previously discussed, whereas others, for example, those more prevalent in infrastructure research, introduce new perspectives. The advancement of new technology may lead to an increase in research of these kinds. The paper also provides a review of methods intended to compensate for lack of consent, and their applicability and usefulness for the different categories of situations are discussed. The aim of this is to provide insights into one important aspect of the question of permitting research without informed consent, namely, how well that which informed consent is meant to safeguard can be achieved by other means.

The 2001 terrorist attacks in the USA and the 2011 seismic events in Japan have brought into sharp relief the vulnerabilities involved in storing nuclear waste on the land's surface. Nuclear engineers and waste managers are deciding that disposing nuclear waste deep underground is the preferred management option. However, deep disposal of nuclear waste is replete with enormous technical uncertainties. A proposed solution to protect against both the technical vagaries of deep disposal and the dangers of surface events is to store the nuclear waste at shallow depths underground. This paper explores social and ethical issues that are relevant to such shallow storage, including security motivations, intergenerational equity, nuclear stigma, and community acceptance. One of the main ethical questions to emerge is whether it is right for the present generation to burden local communities and future generations with these problems since neither local peoples nor future people have sanctioned the industrial and military processes that have produced the waste in the first place.

Preface

TECHNOETHICS AND NEW DILEMMAS IN THE AGE OF TECHNOLOGY

Ethical Considerations in the Age of Technology

The verdict is in—society has successfully raised technology from an infant. But are humans mature enough to provide ethical nurturing and responsible direction as it continues to grow and evolve into a teenager? This is a tough question to answer and one that requires a serious consideration of our roles as guardians of multiple children: technology, humanity, and the world inhabited by technology and humans. To begin, infants are a blessing that give meaning and inspiration to many that are parents. But, most parents also remember that it was not all fun and games during those early years of changing diapers, late nights with baby and sleep deprivation. The same can be said of technology in its infancy. Since the emergence of organized society thousands of years ago, humans have created technology to survive in the world by forging of tools and weapons, constructing shelters for protection from the elements, creating factories organizational structures, and cultivating crops and livestock for food. This came with its own ethical problems and dilemmas for its human creators—war, environmental destruction, and human exploitation.

When infants become toddlers and then children, parents get to witness first words, first steps, and other developmental advances that make parents' hearts melt. At the same time, it is hard for parents to forget the temper tantrums, messy bedrooms, damaged furniture, and endless household cleanups. Similarly, beginning in the mid 20th century, the information revolution was nurtured in by the creation of the Internet, the development of new information and communication technologies (ICT's), and the convergence of ICT's allowing individuals to more freely interact and exchange multiple types of information content through increasingly smaller multi-purpose mobile technologies. This was a wonderful time in technological childhood. Jenkins described this convergence, as "the flow of content across multiple media platforms, the cooperation between multiple media industries, and the migratory behaviour of media audiences "(p. 6). Again this came with its baggage and more ethical dilemmas— digital piracy, cyberespionage, online identity theft, Internet addiction, and online bullying.

When children become adolescents the joy of parenting is seen in watching them succeed in school, participate in a play or sporting event, go to prom, and graduate from high school. Again there are difficult moments of adolescent rebellion, bullying at school, and even more damaged furniture and household cleanup to deal with. At the same time, other technology convergences across biotechnology, nanotechnology, information technology and cognitive science (NITC) have given rise to technology advancements that are re-shaping the very boundaries of human life and identity within society helping amputees walk, improving chances for healthy reproduction, helping individuals with depression cope and more. But we also must deal with the dark side of technological adolescence—ethical dilemmas concerning human cloning, bioterrorism, genetically modified food, and public fears of the loss of humanity resulting from the misuse of human enhancement technologies.

As humans continue on the parenting trajectory with human and technological teeneagers, there are strategies that can be drawn on to reduce the risk of unintended consequences of technological development that human creators want to avoid. In the next sections we look at the present context of human-technological interwovenness followed by a brief introduction to the field of technoethics as a means of generating knowledge about new dilemmas revolving around technological advancement to help guide ethical decision-making and responsible action.

The Ethical Challenges of Technoselves Living in a Technosociety

Selves are the ultimate negentropic technologies, through which information temporarily overcomes its own entropy, becomes conscious, and is finally able to recount the story of its own emergence in terms of a progressive detachment from external reality. There are still only informational structures. But some are things, some are organisms, and some are minds, intelligent and self-aware beings. Only minds are able to interpret other informational structures as things or organisms or selves. And this is part of their special position in the universe. –Floridi, 2011

Floridi's quote provides just a glimpse at the interwovenness of humans, science, and technology within our evolving society. The rise of new technologies throughout the 20th century have transformed life and society in significant ways by offering new machines, tools, and techniques to advance core areas of human activity including transportation, construction, information transfer, communication, medicine, education, and economics. These transformations have been both outward and inward, challenging the very nature of human existence and what it means to live harmoniously with one another in a society that revolves around technological progress and innovation. One emerging area of technoethical inquiry and debate revolves around the very nature of our human condition as technoselves (Luppicini, 2012a) and what it means to be human amidst a myriad of human enhancement technologies that allow us to augment human bodies and minds in substantive ways. There are a variety of human enhancing technologies currently available or in development which have spurred public debate and concern (e.g., plastic surgery, prosthetic limbs, exoskeletons, performance and mind enhancing drugs, biosensors, neural implants, wearable computers, etc.). These technologies (and other to come) are being scrutinized as an (un)ethical means of altering or transforming the human condition itself. Another area of technoethical inquiry examines the moral impact of new digital technologies and the ethical dilemmas surrounding the boundaries of the social world itself as we struggle to navigate the juxtaposition of both online and offline lives while trying to avoid the many risks and pitfalls that rise from the technosociety (e.g., cybercrime, internet addiction, smart technologies, increasing automation, moral agency in machines, robot rights and responsibilities, etc). Under the umbrella of Technoethics, this book highlights new work in these and other areas of life and society where ethical, moral, and social dilemmas arise from technological progress.

Background of Technoethics as a Field of Research

A number of seminal publications can be cited as key drivers in the formalization of Technoethics as an interdisciplinary research field. These publications brought together leading technology and ethics scholars from around the world and provided a solid intellectual platform upon which to grow. First, the two volume Handbook of Research on Technoethics (Luppicini & Adel, 2009) drew on the contributions of over 100 experts from around the world working on a diversity of areas where technoethical inquiry. Next, the first reader in Technoethics, Technoethics and the Evolving Knowledge Society: Ethical Issues in Technological Design, Research, Development, and Innovation (Luppicini, 2010) was published for use at the undergraduate and graduate level in a variety of courses that focus on technology and ethics in society. The text focused on a broad base of human activities connected to technology interwoven within social, political, and moral spheres of life. It engaged readers in the study of key ethical dimensions of a technological society and helped reinforce work found in the Handbook. Together, these publications helped set the stage for the creation of the International Journal of Technoethics in 2010.

In 2013, the *International Journal of Technoethics*, (Rocci Luppicini-Founding Editor-in-Chief), now in its fourth year, continues to provide a forum for scholarly exchange among philosophers, researchers, students social theorists, ethicists, historians, practitioners, and technologists working in areas of human activity affected by technological advancements and applications. With the strong support of IGI Publishing, the journal retains its founding Editor-in-Chief and its twelve Associate Editors, namely, Allison Anderson (University of Plymouth), Keith Bauer (Marquette University), Josep Esquirol (University of Barcelona), Deb Gearthart (Troy University), Pablo Iannone (Central Connecticut State University), Mathias Klang (University of Lund), Andy Miah (University of the West of Scotland), Lynne Roberts (Curtin University of Technology), Neil Rowe (U.S. Naval Postgraduate School), Martin Ryder (Sun Microsystems), John Sullins III (Sonoma State University), Mary Thorseth (NTNU). The mission of the journal was as follows:

The mission of the *International Journal of Technoethics* (IJT) is to evolve technological relationships of humans with a focus on ethical implications for human life, social norms and values, education, work, politics, law, and ecological impact. This journal provides cutting edge analysis of technological innovations, research, developments policies, theories, and methodologies related to ethical aspects of technology in society. IJT publishes empirical research, theoretical studies, innovative methodologies, practical applications, case studies, and book reviews. IJT encourages submissions from philosophers, researchers, social theorists, ethicists, historians, practitioners, and technologists from all areas of human activity affected by advancing technology (*International Journal of Technoethics*, 2013).

This present collection of chapters is derived from the second year of the *International Journal of Technoethics*. It provides coverage of cutting edge work from a variety of areas where technoethical inquiry is currently being applied.

The Importance of Technoethics in the Age of Technology

As the unity of the modern world becomes increasingly a technological rather than a social affair, the techniques of the arts provide the most valuable means of insight into the real direction of our own collective purposes - Marshall McLuhan, 1951

The modern world, as depicted by McLuhan over 70 years ago is continuing to evolve (technologically, individually, and socially) as humans attempt to weave the benefits of technological progress with the need to further collective purposes of humans living together in a society that embraces progress and innovation. This is a challenging act of gymnastics and one in which field of Technoethics has attempted to meet. Technoethics developed through the coming together of technology experts (philosophers, technicians, administrators, instructors, students, and researchers) struggling with the many dilemmas arising from public controversies and ethical debate created by technological advancement in society. This interdisciplinary research field provided a means to transcend traditional approaches in the study of ethics and technology driven by existing philosophical approaches, intellectual analyses of pervasive problems, and logical reasoning. It provided a multifaceted intellectual platform for experts working at the nexus of applied work in technology and ethics (e.g., bioethics, engineering ethics, computer ethics, etc.). The types of scholars attracted to this field are scholars and technology experts working in new areas of technology research where social and ethical issues emerge (I.e., genetic research, nanotechnology, human enhancement, neurotechnology, robotics, reproductive technologies, etc). The current state of Technoethics is marked by an openness to multiple forms of scholarly inquiry and practical real world value. As stated in Luppicini (2012b):

As pioneering breakthroughs are made in technological advancements and applications, novel questions arise regarding human values and ethical implications for society, many of which give rise to ethical dilemmas where conflicting viewpoints cannot be solved by relying on any one ethical theory or set of moral principles. Accordingly, the field of Technoethics takes a practical focus on the actual impacts (and potential impacts) of technology on human beings struggling to navigate the "real world" of technology. In many cases, this leads to the creation of more questions than answers in an effort to discern the underlying ethical complexities connected to the application of technology within real-life situations.

Objectives of this Book

In one word, science (critically undertaken and methodically directed) is the narrow gate that leads to the true doctrine of practical wisdom, if we understand by this not merely what one ought to do, but what ought to serve teachers as a guide to construct well and clearly the road to wisdom which everyone should travel, and to secure others from going astray.

-Immanuel Kant, 1898

The above passage from Kant highlights the role of intellectual inquiry as a means of guiding the construction of knowledge from which to offer ethical guidelines for practice and boundaries to guard against going astray. This edited book presents a selected collection of studies in an effort to build on the existing intellectual platform within the field of technoethics. As a rapidly expanding area of inquiry, technoethics draws on and goes beyond traditional 'ethics of technology' and 'philosophy of technology' approaches which highlight longstanding ethical theories and controversies for intellectual analysis. Technoethics also deals with current (and future) problems in science and technology innovation at the intersection of human life and society. As such, it brings into play an interdisciplinary base of scholarly contributions from pure philosophy, the social sciences, humanities, engineering, computing, applied sciences and other areas of scholarly inquiry into technology and ethics. This interdisciplinary focus is helping to leverage ethical analysis, risk analysis, technology evaluation, and the combination of ethical and technological analyses within a variety of real life decision-making contexts and future planning situations faced by 21st century society.

The significance of the selected chapters appearing in this volume can be attributed to a variety of factors including high peer review standards, the timeliness of the topics covered, author attention to research protocols and methodological procedures, perceived contribution to research knowledge on ethics and technology, and perceived contribution to practice. Both empirical and theoretically oriented chapters are offered to provide a current snapshot of new developments in the field today.

Organization of the Book

We usually see our own lives and those of others as a series of narratives, and we continually reinterpret and revise our narrative self-understanding. For example, we scroll through our cache of stories to find one that can best clarify the moral problem at hand and that we can reconcile with our self-representations and ideals -Magnani, 2007

In echoing the above passage from Lorenzo Magnani, this volume attempts to piece together the best possible collection of narrative constructions pertaining to recent developments within the field of technoethics. In terms of book organization, this book contains 21 chapters divided into separate sections to highlight a logical flow of writing organized into key thematic areas of technoethical inquiry. Section 1: Historical and Theoretical Perspectives contains 5 chapters: Chapter 1, "Virtue and Virtuality: Technoethics, IT, and the Masters of the Future" (Miles Kennedy), Chapter 2, "Internet History (Raphael Cohen-Almagor), Chapter 3, "Laboring in Cyberspace: A Lockean Theory of Property in Virtual Worlds" (Marcus Schulzke), Chapter 4, "Perverting Activism: Cyberactivism and its Potential Failures in Enhancing Democratic Institutions" (Tommaso Bertolotti, Emanuele Bardone, and Lorenzo Magnani), and Chapter 5, "On the Moral Equality of Artificial Agents" (Christopher Wareham).

Section 2: Ethical Dilemmas of Technoselves in a Technosociety contains six chapters: Chapter 6, "Biometrics: An Overview on New Technologies and Ethic Problems" (Halim Sayoud), Chapter 7, "On Biometrics and Profiling: A Challenge for Privacy and Democracy?" (Daniele Cantore), Chapter 8, "Cellular Telephones and Social Interactions: Evidence of Interpersonal Surveillance" (Steven E. Stern and Benjamin E. Grounds), Chapter 9, "Socio-Technical Influences of Cyber Espionage: A Case Study of the GhostNet System" (Xue Lin and Rocci Luppicini), Chapter 10, "The Impact of Context on Employee Perceptions of Acceptable Non-Work Related Computing" (Troy J. Strader, J. Royce Fichtner, Suzanne R. Clayton, and Lou Ann Simpson), and Chapter 11, "Chicken Killers or Bandwidth Patriots? A Case Study of Ethics in Virtual Reality" (Kurt Reymers).

Section 3: Morality and Techno-Mediated Violence contains five chapters: Chapter 12, "Structural and Technology-Mediated Violence: Profiling and the Urgent Need of New Tutelary Technoknowledge" (Lorenzo Magnani), Chapter 13, "Fairness and Regulation of Violence in Technological Design" (Cameron Shelley), Chapter 14, "Unintended Affordances as Violent Mediators: Maladaptive Effects of Technologically Enriched Human Niches" (Emanuele Bardone), Chapter 15, "Infosphere to Ethosphere: Moral Mediators in the Nonviolent Transformation of Self and World" (Jeffrey Benjamin White), and Chapter 16, "Facebook Has it: The Irresistible Violence of Social Cognition in the Age of Social Networking"(Tommaso Bertolotti).

Section 4: Current Trends and Applications in Technoethics contains five chapters: Chapter 17, "Technoethics and the State of Science and Technology Studies (STS) in Canada" (Rocci Luppicini), Chapter 18, "College Students, Piracy, and Ethics: Is There a Teachable Moment?" (Jeffrey Reiss and Rosa Cintrón), Chapter 19, "Boys with Toys and Fearful Parents? The Pedagogical Dimensions of the Discourse in Technology Ethics" (Albrecht Fritzsche), Chapter 20, "Without Informed Consent" (Sara Belfrage), and Chapter 21, "The Middle Ground for Nuclear Waste Management: Social and Ethical Aspects of Shallow Storage" (Alan Marshall).

Is there virtue in virtuality? What are the technoethical implications created through human engagement with virtual environments? How does the Internet form a backdrop for technoethical inquiry into social networking and work? How can traditional philosophical theories help inform current legal relations and property issues within virtual environments? In what ways can cyberactivism impair and advance institutional democracy? Could artificial agents have the high degree of moral status that is attributed to humans, and if so, how should they be treated? In Chapter 1, *Virtue and Virtuality: Technoethics, IT, and the Masters of the Future*, Miles Kennedy questions the role of virtue within the online realm. The author sheds light on how the technoethics of IT sheds light on the relationship between knowledge about information and the capacity to turn information into knowledge. This chapter speaks to students and researchers who have a stake in the virtual environment and its ethical makeup. In Chapter 2, *Internet History*, Raphael Cohen-Almagor explores milestones in the history of the Internet from a historical standpoint, social and ethical standpoint. In Chapter 3, *Laboring in Cyberspace: A Lockean Theory of Property in Virtual Worlds*, Marcus Schulzke questions property relations in massively multiplayer on-line games (MMOGs) through the lens of John Locke's theory of property. This chapter helps connect the existing body of Lockean theory to current ownership rights of online gamers over the intellectual material of virtual worlds. In Chapter 4, *Perverting Activism: Cyberactivism and its Potential Failures in Enhancing Democratic Institutions*, Tommaso Bertolotti, Emanuele Bardone, and Lorenzo Magnani explore the influence of new technologies on a range of practices related to activism and cyberactivism. This chapter takes a sobering look at the strengths and weaknesses of cyberactivism in leveraging institutional democracy. Then, in Chapter 5, *On the Moral Equality of Artificial Agents*, Christopher Wareham focuses on the fact that robots are beginning to perform increasingly significant ethical roles in society which calls into question the rights and moral responsibilities of these creations. The author advances a respect-based account of the ethical criteria for moral agents along with an empirical test for artificial agents to be considered as having rights and duties of human agents.

What are biometric technologies and why is their use creating ethical debate in society? How can biometrics and profiling challenge individual privacy and democracy? How are new capacities for human surveillance offered by ICT's affecting human relationships and interactions in society? How is technology being used to engage in cybercrimal activities such as information theft, cyberbullying, cyberespionage, and institutional sabotage? In Chapter 6, *Biometrics: An Overview on New Technologies and Ethic Problems*, Halim Sayoud questions the role of biometric technologies (automated methods of verifying the identity of a living person based on physiological or behavioral characteristics) in society. The chapter provides an overview of key historical developments and current uses of biometric technologies. Important contributions of biometric technologies are explored along with possible ethical problems that may result from their misuse. In Chapter 7, *On Biometrics and Profiling: A Challenge for Privacy and Democracy?*, Daniele Cantore looks at the advancement of profiling (collecting and organizing individuals' preferences and attitudes as consumers and costumers) and biometric technologies (collecting biological and behavioural data from humans) from an ethical perspective. This chapter provides a much needed look at biometrics in relation to current concepts of democracy, technological development, violence, and privacy issues. In Chapter 8, *Cellular Telephones and Social Interactions: Evidence of Interpersonal Surveillance*, Steven E. Stern and Benjamin E. Grounds present an empirical study on how cellular telephones have made it possible for members of romantically involved couples to keep track of each other. The authors surveyed 69 undergraduates on cellular telephone use, finding that that nearly 25% of romantically involved cellular telephone users admit to tracking their significant other. The chapter raises questions concerning trust, privacy, and surveillance. In Chapter 9, *Socio-Technical Influences of Cyber Espionage: A Case Study of the GhostNet System*, Xue Lin and Rocci Luppicini examine the largest cyberespionage network (GhostNet) to date. Using the Actor-Network Theory, these authors delve into the ethical, technical, and communicative processes at work. In Chapter 10, *The Impact of Context on Employee Perceptions of Acceptable Non-Work Related Computing*, Troy J.

Strader, J. Royce Fichtner, Suzanne R. Clayton, and Lou Ann Simpson, look at a common concern faced by organizations around the world. This timely study focuses on the relationship between employee's ethical orientation, underlying factors which influence employee ethical perceptions, and emloyee perceptions regarding the acceptability of various non-work related computing activities while at work. This empirical research sheds new light on a very real challenge faced by employers and employees in negotiating the ethical boundaries of personal and professional life within the workforce. In a slightly different vein, Chapter 11, *Chicken Killers or Bandwidth Patriots? A Case Study of Ethics in Virtual Reality*, Kurt Reymers delves into the ever expanding virtual world where social interactions range from online gaming to identity development to business. More specifically, the author focuses on a case in the virtual world of Second Life where the sale of virtual chickens and eggs led to virtual world conflict. The chapter provides a 3-pronged examination of this virtual event as a resource dilemma, an argument from moral and psychological harm, and as an ethical case under a just war theory perspective. As highlighted by the author, "All of these virtual world phenomena, from the interactive role-playing of virtual farmers to the social, political and economic repercussions within and beyond the virtual world, can be examined with a critical focus on the ethical ramifications of virtual world conflicts." This study opens the door to ethical inquiry into the ever expanding realm of virtual environments used increasingly by people around the world.

What is the state of morality within the context of techno-mediated violence? What is the need for technoethical knowledge when dealing with profiling technologies and practices? What are the key moral mediators in the nonviolent transformation of self and world? How does the violence of social cognition manifest itself in the age of social networking applications such as Facebook? In Chapter 12, *Structural and Technology-Mediated Violence: Profiling and the Urgent Need of New Tutelary Technoknowledge*, Lorenzo Magnani addresses the expanding boundaries of violence within a technology-mediated world. The author argues that, "when too much knowledge about people is incorporated in external artificial things, human beings' *visibility* can become excessive and dangerous." This insightful piece deals with the very real problem of identity and cyberprivacy, making the case that a new type of knowledge (technoknowledge) is required to help protect individuals from the various forms of information circulation about them which can be potentially harmful. In Chapter 13, *Fairness and Regulation of Violence in Technological Design*, Cameron Shelley expands on Magnani's work by discussing the design of technology as it relates to the notion of fairness and the distribution of violence in modern society. The chapter demonstrates how errors arising from failed predictions can influence individuals in unintended ways that designers may not realize. Insightful recommendations offered speak to the roles of government regulation, safety issues, and need for public participation in technology design assessment. In extending the discussion about unintended consequences of technology, Chapter 14, *Unintended Affordances as Violent Mediators: Maladaptive Effects of Technologically Enriched Human Niches*, by Emanuele Bardone delves into the topic of violent mediators in human cognitive niches enriched by technology. The author focuses on the unintended consequences of violent mediators through examples of multitasking while driving, desultory behavior and cyberstalking. The chapter provides useful suggestions on how to counterbalance the maladaptive effects of techno-mediated violence. In Chapter 15, *Infosphere to Ethosphere: Moral Mediators in the Nonviolent Transformation of Self and World*, Jeffrey Benjamin White pulls together this section by examining the philosophical writing on information technology from Lorenzo Magnani and Luciano Floridi in an effort to identify challenges and opportunities for the nonviolent transformation of self and world. As expressed by White, "Though we have the capacity to construct our own selves, our world, and the myths in light of which self and world all make sense, choosing the right story, and living accordingly, is a most difficult task." This chapter pulls together insights from key resources in moral philosophers to highlight the importance of moral mediation in the nonviolent transformation of self and world. Then, in Chapter 16, *Facebook Has it: The Irresistible Violence of Social Cognition in the Age of Social Networking*, Tommaso Bertolotti finishes this section by exploring how transparent social networking technologies can cause violence in everyday life and society. The author focuses on the role of online gossip in the social evolution of human beings along with potentially violent consequences of techno-mediated gossip that should be avoided.

What are some of the current trends and applications in technoethics? What is the state of Science and Technology Studies (STS) in countries like Canada where democratic values, social welfare, and human rights are core to the political landscape and national identity? What are educational opportunities and challenges in teaching about piracy and ethics to college students? What are the similarities that can be drawn between pedagogy and the current discourse on technology and ethics? What are technoethical considerations in nuclear waste management in the 21st century? How can research in complex areas of science and technology innovation proceed where informed consent from human participants is not possible? In Chapter 17, *Technoethics and the State of Science and Technology Studies (STS) in Canada*, Rocci Luppicini examines the nature of science and technology in Canada by looking at state of ethics within STS curriculum at Canadian post-secondary institutions. Findings from this chapter reveal that ethical aspects of science and technology study are lacking within most Canadian academic programs in STS. A number of key challenges and recommendations are offered to help inform strategic change in STS program development at the university level. In a similar vein, Chapter 18, *College Students, Piracy, and Ethics: Is There a Teachable Moment?*, by Jeffrey Reiss and Rosa Cintrón, addresses the ethical problem of piracy and piracy prevention techniques. As a point of reference, the authors focus on piracy prevention tools and techniques used by IT departments in the Florida State University System to assess their relative effectiveness in combating digital piracy. In this empirical study a survey was administered to IT department's security officers within the 11 Florida State University System (SUS) institutions. Findings revealed that the majority of IT department's security officers believed students had little concern over the consequences of their actions and would engage in digital piracy off campus. Respondents were divided as to whether the integration of ethical training for students would lessen the need for the use of network monitoring software on campus. Challenges for implementing mandatory ethics training for college students were also discussed. In Chapter 19, *Boys with Toys and Fearful Parents? – The Pedagogical Dimensions of the Discourse in Technology Ethics*, Albrecht Fritzsche takes a sobering look at the reality of technology in developmental terms of when it enters our lives. As elegantly noted by the author, "Technology does not come into our lives when we reach maturity. Most of us probably got acquainted with technical operation at a much earlier age, when we played with wooden blocks, model cars and plastic hammers." The author effectively draws attention to the need for more philosophical studies of technology that focus on technology and children to add to the sea of intellectual scholarship on technology use and abuse among adults. This chapter offers a unique perspective on existing gaps in the literature on technology and ethics to inform future research and theory. In Chapter 20, *Without Informed Consent*, Sara Belfrage delves into the tricky area of human research within our technological society where the complexities of research, economic realities, cultural dynamics, and other forces can challenge adherence to ethical research guidelines. Given the intricacies of technology development and the need for human research where informed consent is difficult, this chapter provides a practical take on categories of cases where informed consent cannot be used. This helps shape the discourse on research and ethics within the evolving technological society (Luppicini,

2010). Finally, in Chapter 21, *The Middle Ground for Nuclear Waste Management: Social and Ethical Aspects of Shallow Storage*, Alan Marshall, places this timely study within the context of recent 2001 terrorist attacks in the USA and the 2011 nuclear disaster events in Japan. The author provides a selective review of key events that shaped the history of nuclear energy and waste management techniques that are available to deal with a controversial and needed category of technologies that affect not only the present generation but future populations that will inherit the burden of this current technological activity. The chapter attempts to provide a middle ground solution for nuclear waste management that meets the needs and values of our evolving technological society.

In summary, this collection provides a glimpse at the latest developments in technoethics in the hopes of helping readers better navigate the murky ethical and social waters created by a broad range of 'new' technological advances within society today. As the editor of this volume, it is a privilege to present the following twenty-one chapters which delve into the current state of technoethics and new dilemmas in the age of technology.

Rocci Luppicini
University of Ottawa, Canada

REFERENCES

Abbott, T. K., & Liberty Fund. (1898). *Kant's critique of practical reason and other works on the theory of ethics*. London, UK: Longmans, Green.

Floridi, L. (2011). The informational nature of personal identity. *Minds and Machines, 21*(4), 549–566. doi:10.1007/s11023-011-9259-6

(2013). *International Journal of Technoethics*. Hershey, PA: IGI Global.

Jenkins, H. (2006). *Convergence culture: Where old and new media collide*. New York, NY: New York University Press.

Luppicini, R. (2010). *Technoethics and the evolving knowledge society*. Hershey, PA: IGI Global. doi:10.4018/978-1-60566-952-6

(2012a). InLuppicini, R. (Ed.). Handbook of research on technoself: Identity in a technological society: *Vol. I. II*. Hershey, PA: IGI Global.

Luppicini, R. (Ed.). (2012b). *Ethical impact of technological advancements and applications in society*. Hershey, PA: IGI Global. doi:10.4018/978-1-4666-1773-5

(2009). InLuppicini, R., & Adell, R. (Eds.). Handbook of research on technoethics: *Vol. I. II*. Hershey, PA: IGI Global.

Magnani, L. (2007). *Morality in a technological world: Knowledge as duty*. Cambridge, UK: Cambridge University. doi:10.1017/CBO9780511498657

McLuhan, M. (1951). *Mechanical bride: Folklore of industrial man*. New York, NY: Vanguard Press.

Section 1
Historical and Theoretical
Perspectives

Chapter 1

Virtue and Virtuality:
Technoethics, IT and the Masters of the Future

Miles Kennedy
National University of Ireland Galway, Ireland

ABSTRACT

This paper is an attempt to establish a foundation for technoethics of IT that makes an account of the virtual environment based within the lived situation of those who work and dwell in that emerging realm. The most important phenomenon for technoethics of IT is the relationship between knowledge about information and the capacity to turn information into knowledge. This relationship is embodied in being a Master of Information Technology. To achieve mastery of information and mould it into knowledge, a useful tool-like entity, is to have power in the contemporary world. Once this situation is recognised ethical questions arise of their own volition. A selection of these questions are dealt with in the following paper, they are the questions of the distinction between information and knowledge, the central issue of virtue and virtuality, and the distinction between stealing and sharing in the virtual environment. This paper constitutes a think piece; readers who have a stake in the virtual environment and its ethical makeup are urged to ask themselves these questions and come up with others in turn.

INTRODUCTION: TECHNOETHICS AND THE MASTERS OF TOMORROW

All traditional ethical study essentially has to do with the recognition of personal responsibility and right action in 'the physical environment' (Blackburn, 2002, p. 1). Technoethics strives to

establish what specific or new ethical implications are thrown up by human use and engagement with technology. This paper sets out to investigate a particular question within technoethics, put simply: is there virtue in virtuality? For the purposes of this paper a distinction is drawn from the very beginning between the ethical status of actions undertaken wholly within the virtual environment, whichever specific one it might be (Secondlife, World of Warcraft, file sharing sites etc), and

DOI: 10.4018/978-1-4666-2931-8.ch001

online or IT related activities that have what can loosely be called a "real effect".

Floridi and Sanders coin and define the very useful term 'artificial evil' and go to some lengths to differentiate between the kind of evil that is fully contained within cyberspace and the two traditional notions of evil, natural evil and moral evil, that underpin real world expressions of ethical disapproval (2001). In order to establish a lived co-relate to this theoretical delineation between these different types of action from the outset some examples are helpful: spamming, "Flaming" (intense sometimes rude or personally motivated discussion within a given online community) or data mining are generally regarded in one way whereas stalking, grooming or fraud are regarded in a very different way. This difference in ethical judgment occurs primarily because the first set of activities take place within the virtual environment, and are therefore regarded as less significant, while the second set of transgressions transcend from the virtual into the real, physical environment and are therefore regarded as far more significant. This discussion is limited to the kinds of activity that can be described as purely virtual in order to maintain a reasonably high level of clarity about the singular and specific nature of the ethical issues that emerge within the virtual environment. Spilling over into "real" actions in the "real" world would tend toward a reassertion of pesupposed "real world" ethical norms when it is exactly the appropriateness of these norms that this paper, and technoethics in general, sets out to assess.

The virtual environment that this particular strand of technoethics examines is understood in a concrete sense as 'the (eco)system of information acted on by digital agents' and the structure of 'information stored as bits... organized in vast tracts such as data-bases, files, records and online archives' which is lent geometrical and metaphorical presence through being described as a web (Floridi & Sanders, 2001, pp. 55-66). This web itself is understood as a technological

artifact or pattern of artifacts that has rapidly become 'embedded within social political and moral spheres' and in which those spheres themselves have become embedded to a significant extent (Luppicini, 2008, p. 4). As such it is understood in the more phenomenological sense as '... the true environment in which most people now spend most of their intentional lives as conscious beings' (Floridi, 2001, p. 18).

Taking these established conceptions of the virtual environment as its foundations this paper contains an attempt at developing the field of IT technoethics through a phenomenological approach which makes a similar movement as that which took place in the history of phenomenology itself, namely the move from a Husserlian focus on intentionality and consciousness, or the "how we concieve of technology" approach, to a more Heideggerian study of being-in-the-world, or the "how we live with and in a technological environment" approach (Husserl, 1970; Heidegger, 1977). As Luciano Floridi put it, 'the philosophy of information is phenomenologically biased' and a technoethics of IT should therefore be phenomenologically grounded (2001, p. 18).

In order to make this movement from a theoretical to a lived perspective specific to the field of IT ethics, a move initiated by Albert Borgmann with regard to technology in general early on in the development of technoethics (Borgmann, 1987), an engagement with two further key components of technoethics is neccessary: 1. its focus on the moral responsibilities of technologists and engineers and 2. the requirement that those technology-empowered persons should recognise their responsibility for their professional actions (Bunge, 1977, pp. 96-107). The specific group of technologists discussed in this paper, and with whom it was discussed, were students enrolled in a Masters of IT programme at the National University of Ireland Galway. This programme is designed to produce graduates who are technologically savvy yet grounded in a broad humanities education, yielding a group that was in an ideal position to offer insights into how aspiring Masters (and mistresses) of information technology construed their ethical responsibilities and the foundations for them.

After being presented with this paper the ten Masters students, who were of various ethnicities and nationalities including Irish, Nigerian, Indian and German, were asked to answer a specific set of questions that are raised in this theoretically driven engagement. This questionaire is designed with the purpose of examining whether some or all of the underlying assumptions presented in the main body of the paper are accurate representations of what really happens in the virtual environment. The ten students are considered representative of people who are aware of "what's really happening" in the virtual world due to the facts that they are: i. Technically adept and ii. Report spending an average of 6-8 hours a day online. Their insights are enumerated and discussed in the conclusion of this paper.

There was a dual pedagogical aim in this exercise, on the one hand it was intended to teach a group of technologists about their ethical responsibilities, as is any class in professional ethics, and on the other it was intended to inform a philosopher about the conceptions of a group of people who are deeply immersed in the virtual environment while being trained to implement its upkeep and development. In this effort to establish whether or not there is a conception among technologists of a specific kind of distinct ethical content to being a Master of Information Technology it is very important to ask some fundamental questions. For an aspiring Master of IT the first of these questions arises spontaneously and without any real effort, that being "Why study ethics at all?"

This naturally occurring challenge to the general discourse on ethics and ethical thinking is a good thing as it opens up a further question as to whether ethics as they have traditionally been conceived are applicable in the field of Information Technology. Of course basic professional ethics that deal with discretion, confidentiality, a bar

on insider trading, a certain level of obligation to one's self, one's clients, one's employer, to society and perhaps the environment are as pertinent and applicable to this field as they are to any other (Boylan, 1995). However, rather than simply adopting the more general model of professional ethics and trying to apply that to this emerging and developing field of human endeavour, the particular and singular make-up of the virtual environment calls for a specifically tailored type of ethics to adequately prepare those who work and live with and in it: a technoethics.

THE DISTINCTION BETWEEN INFORMATION AND KNOWLEDGE

In the current educational system in Information Technology it is all too often the case that students say "I have a masters in IT" or "I want to get a masters", the first task of technoethics with regard to IT students is to teach them to say "I am a Master of Information Technology" or "I want to be a Master". The name "Master of Information Technology" is itself somewhat misleading in that it is broad and ill defined. One definite aspect of this endeavour is the neccesity to develop mastery of knowledge about information technology rather than mastery of mere information or data (Floridi, 2001 pp. 106-108). This distinction is not as clear as it may first appear because the very systems the aspiring Master works with have problematised and continue to problematise this relationship as they are themselves information systems which can very often function as sources of knowledge and even more often be mistaken for sources of knowledge (Johnson, 2001, p. 4).

There have been attempts to clarify this sticky question of the distinction between "information" and "knowledge" by reference to a key characteristic of all professionals: "Education" (Johnson, 2001, p. 60). However, here again a problem emerges, education in general is regarded as empowering but an education in "information technology" specifically, a key area in "the knowledge economy" (Barkham, 2008), is particularly empowering, not in an abstract sense but in a real tangible way that is linked to the nature of the knowledge acquired. The ready availability of masses of brute information on the internet, what has been called "the information superhighway", has created a culture in which anyone with a computer and a modem, and particularly someone on their way to mastery of information technology, can access as much information on any given topic as they could possibly want and then present it as knowledge.

In the academic field in particular it happens all too often that a student downloads one or several pieces of information reformats it and presents it as their own knowledge. Where this was once regarded in all cases as straightforward and intentional plagiarism it now has to be investigated more closely as it could be the case that the student or students simply cannot distinguish between the presentation of mere information and the presentation of genuine knowledge. One might argue that at the very minimal level a student would at least have to know where to look but, given the nature of the technology itself and an adequate mastery of it, it is not necessary to even have this much knowledge as the search engine can do the looking with very little amount of input. "Education" in IT, then, can be seen to add further complexity rather than clarity when used as a defining characteristic in the distinction between information and knowledge, as the knowledge being acquired by an aspiring Master of IT is exactly about the means and modes of information acquisition, retrieval and storage.

Moving on from education to the everyday circumstances of a professional in IT the same problem of distinguishing between information and knowledge remains. The basic function of the Master of IT is bringing knowledge to bear on various modes of information. The Master must be able to do. The Master must be able to enter into a situation of need and offer corrections or

solutions, like the master builder or the master locksmith; this is fundamentally an ethical engagement that has ethical content (Aristotle, 1996, pp. 12-13). This use of analogy to conceptualise ethical responsibility in a virtual environment is discussed in more depth later in this paper, for now the question is: what is it to be a master of information technology?

Perhaps an example will help to clarify, when a company or employer desires to know what level of security their systems containing sensitive data need the Master of IT will use acquired knowledge to inform them of the right choices. In many situations the Master of IT may be the only individual from outside a company or organisation who is given access to this kind of sensitive information, unlike a hacker for example who may get access to information systems but has not been given that access and has thereby made an ethical decision from the very outset which is embodied in the activity itself, and is very often acting illegally. However, though hacking activities are often illegal, it should be remembered that hackers are also Masters of technology and, as Steven Levy pointed out, they too operate within an established, if unspoken, ethical framework (Levy, 2010). It is the ethical rather than strictly legal status of these kinds of activity that is at issue in technoethics.

Even without going into the realm of illegality, though the distinction between law and ethics should be borne in mind, it is clear that as an individual who is given access to sensitive information it is likely that ethical issues that transcend the limits of any given contractual arrangement will arise in the course of a Master's work. A Master of IT may be asked, again this is merely an example, to set up or update a webpage for a specific client, this role will necessarily give insight into the real situation, financial, ethical, legal or otherwise of that client and require the Master to develop a virtual representation of that situation either quite true, i.e., close to the real situation or somewhat false, i.e., further away from that same situation, more idealised. This role empowers the Master to define a specific entity or client in a way which will very likely become their most viewed and widely recognised form thus giving a large amount of power, by virtue of IT knowledge, over their success or failure. "Power corrupts and absolute power corrupts absolutely" (Acton, 1917).

The examples just given illustrate the context in which IT professionals in particular need to develop ethical awareness. Though some of these circumstances may be covered in terms of employment or clauses in contracts it is knowledge that will be the reason for a Master's employment in most cases, that being so it will always be imperative to be aware of possible ethical issues that an employer or client does not have the necessary knowledge to foresee and, therefore, cannot possibly include in any set of terms and conditions no matter how exhaustive. As has been said of old "knowledge is power".

Working in a central area of the knowledge economy is by its very nature empowering. Mastery of IT gives access to sensitive data a company or employer is anxious to protect or, by way of designing websites, control over a client's online presence and is thereby empowering in the field of ethical decision making over issues such as privacy, sensitivity to financially significant information and information about patterns of use (whether legitimate or otherwise). Many problematic cases of insider trading or extortion emerge from just such access; in the most extreme cases it should be noted that almost all instigators of major virus attacks in recent years are themselves students or workers in the field of IT. Infamous examples of this are the 'The Robert Morris case' and 'The Melissa Virus' (Johnson, 2001, pp. 81-83).

Along with direct empowerment over the information itself comes an ethical responsibility, "With great power [or in this case great knowledge which amounts to the same thing] comes great responsibility" (Parker, 1962). These practical questions about the distinction between knowledge and information both in the education and work

of the Master of IT bring to light new intricacies in the traditional ethical question of intention i.e. the question as to whether or not a moral agent intended to act in a specific manner, acted without intention, or acted in ignorance. This question has always been linked to knowledge, meaning it was always bound up in the division between acting with knowledge and acting unknowingly, the fuzzy distinction between knowledge and information drawn here indicates a grey area between these two modes of action into which the Master of IT is always perilously close to falling. This grey area, simply put, is a zone of action in which one has information or the capacity to get information, rather than knowledge *per se*. In order to comprehend the significance of this grey area, bearing in mind Aristotle's exhortation from *The Nicomachean Ethics* that "we do not study ethics to learn what good is but rather to learn to be good" (1996, p. 27), a fundamental yet practical division between information and knowledge must be drawn.

Information can be conceived of as existing in an analogous state to what Martin Heidegger called the "mere thing" in "On the Origin of the Work of Art". In that 1935 lecture he differentiated between things, tools and works (Heidegger, 1975, p. 29). The mere thing stays relatively anonymous and unformed, it is there as part of the earth and available but is not put to work in the sense that a tool is. Both tools and works act to bring forth a world from the earth by drawing themselves out from anonymity and being engaged with, by human beings, as meaningful. In the contemporary era information unlike other mere things is now sectioned off in a kind of alternate world, it is not "just there" in the sense that a stone or a sod of grass or an unused shoe is part of the earth but is rather contained in an alternate and artificial reality that exists alongside this natural world. As such this information forms a more well defined "standing reserve" but nonetheless remains in a state of "just being" until it is actively pursued and formed into knowledge.

The traditional goal of education "knowing that" (*epistēmē*) is developed in the Master of IT into a "knowing how" (*technē*) the goal of which is to be 'at home in something [in this case the virtual environment], to understand and be expert in it' (Heidegger, 1977, p. 12). By this definition mastering information technology is perhaps closer to mastering an art than a science, though the production of knowledge from information has traditionally been seen as a scientific endeavour. Here the question has already been asked, though only implicitly, by the appearance of IT courses both as Arts degree subjects and as Computer sciences.

What is meant about the relation between information and knowledge being similar to the relation between a mere thing and a tool is that the information must be found and taken from within a sheltering standing reserved and formed into something, knowledge, by an act analogous with the creation of a statue, for example, from a lump of granite. In order to keep this description in terms of these two major categories, mere thing and tool, this question of artistry and the status of the work are left aside. It should, however, be borne in mind that leaving this question aside also leaves the issue of whether mastery of information technology is an art or a science open. Very likely this question can only adequately be answered by examining the particular activity of a given individual in the context of their own specific work, there is of course a big difference between someone who works in web design and someone who works in the design and maintenance of software and hardware. Perhaps one creates art and the other manufactures tools?

To return to the core distinction drawn between mere things and tools, in the virtual environment information shares the thingly characteristic of being "just there" in a kind of earthly anonymity while knowledge on the other hand has more of the characteristic, in the present context, of a tool. It is worldlier. Knowledge is useful, it is not merely at hand or present but is rather "to hand"

and ready for use, the significant factor here is the user themselves (Heidegger, 2005, p. 138). This, perhaps, is the lynchpin, the information about information and its storage, its protection, its transmission and retrieval can only become knowledge and thereby become available for use, become power, rather than merely available in general, through the intervention of a Master of that same information technology. Here the discussion has slipped a little into an examination of tools and by this movement into an interrogation of technologies but only in order to try to draw this distinction between information and knowledge. Mastery of IT is knowledge but it is knowledge in the practical applicable sense, it is know-how.

Put simply the Master of information technology will martial information and pull it out of its existence and context within a mere standing reserve in the same or at least a similar manner as a builder will use stone and sand to build a house, not merely by making or manipulating what is already there, the information, but by revealing it is meaningful, as knowledge. The client, the person who needs that process of changing brute information into knowledge performed, will of course know how to live in the house, as it were, but not how to build it, they will generally though not in all cases, understand how to enter information, log on to a website, send e-mail etc. but not how to maintain internet service providers, design and maintain software and hardware or build and maintain websites. The core difference between the Master of technology and those that he or she serves is that the Master steps beyond the realm of "just-thereness" into the "there-for" or "there-as".

The Master's knowing becomes a will. A will to form information into knowledge. Through this process the traditional distinction between knowing and doing becomes blurred at the edges. Borgmann pointed out this blurring vis-a-vis technology in general but for the Master of IT specifically the knowing is the doing (Borgmann, 1987, p. 15). What is striking, on having drawn the distinction between information and knowledge, is that knowledge, in the field of information technology particularly, bears a relation to information in that it is information marshalled and thematised into a useful, tool-like, form. The implemental knowledge required to do this, the know-how, empowers the Master of these technologies over a new and distinct environment, a virtual environment. What kind of ethics, if any, the Master finds in or imposes upon that virtual environment is discussed in the next part of this paper.

VIRTUE AND VIRTUALITY

What makes ethics in IT different or gives it a unique quality in relation to ethics in general is its connection to the virtual environment. While it can be, and is, argued that issues such as privacy, intellectual property or taking a product or service without paying for it are pertinent to all ethical studies there is little question that the act of walking into a bookstore picking up a volume from the shelf and walking out again is a very different act than copying software provided by a friend or files found on a file sharing site. The bookstore is left "one book short", as it were, whereas when a file is shared or copied either with or without permission the original owner still possesses it (except in extreme cases in which wanton destruction is engaged in after having "hacked" into files). In a more complex example that highlights the unique nature of the virtual environment it would seem completely irrational for someone to break into a church or government building and make off with the paper files stored there on the off chance that they might find some use or profit in them. The time, effort and risk would be very far from offset by the possible reward whereas the analogous activity of data-mining is clearly a rationally driven, has the potential to be very lucrative, and requires relatively little effort. What is more data-mining is in many cases not distinctly illegal in the way that taking "real" records would be and therefore carries far lower risks.

There is a strong temptation to say that this is where the story ends, that the reason one act is regarded as practically insane and the other seems, if not ethically good, at least fairly smart is due solely to the technologies themselves and the calculation of risk and reward, but again underpinning this position is the implicit distinction between "real" and "virtual". To make this distinction explicit it can be put like this: the standing reserve of information in the virtual environment is not depleted by the process of moulding it into knowledge. Browsing around in cyberspace to see what kind of free stuff can be picked up is not the same as walking down the road and checking for unlocked doors because this distinction between the real and the virtual is a fundamental one that is/should always be taken into account. In a certain sense when something is copied, whether by cracking a code to access a protected site or merely by borrowing and installing licensed software, nothing has happened. This is not quite the old philosophical distinction between the "real" and the "ideal" exactly but it is something approaching it, something analogous perhaps.

If we take the loose analogy between "real" and "virtual" environments and the philosophical concepts of "real" and "ideal" states seriously we can see that the common distinction in attitudes about different types of activity articulated in the introduction to this paper and repeated just now has a deeply perilous ethical aspect to it. This peril is not only rooted in the widely held conception that "virtual" abuses have something of the same status as psychological abuses, in as much as the now infamous practice of "cyberbullying" can be and often is considered worse than straightforward face-to-face bullying, but on another level entirely (Hogan, 2008, p. 21). Two of the dominant traditional ethical theories that underpin current moral thought have the conditioning of the will, one to virtue (Virtue Ethics) the other to duty (Deontological Ethics), as central tenets, the third major branch of established ethical thinking (Consequentialism) has intention, another mode of willing, as a very significant consideration (Singer, 1994). This conditioning of the will is itself an activity which up until very recently could only be described as taking place in the individuals interior i.e. in the realm of ideal entities, what has traditionally been called ideality or the immanent realm.

This willing is now drawn into the same relation with the virtual as ideality itself is. If this is the case then current, less engaged and more permissive attitudes to "virtual" ethical issues, attitudes which treat them as though they were hardly significant in terms of ethics at all or perhaps should remain in the realm of misdemeanours rather than wrongs in the recieved sense, could be harming the very ability to make ethical judgements itself. What is meant here is that a constant willingness to allow or permit seemingly unethical behaviour due to its being "virtual" as opposed to "real", both by oneself and others, is very likely, according to traditional ethical theories, inadvertently training the wills or fixed characters of contemporary people, and technologists in particular, to allow such behaviour all the time. This permisiveness in respect of the virtual environment can be construed as acting on the formation of the will in the same way as a pilot trains his or her reactions in a flight simulator or a soldier trains his/her nerves to become immune to the impulse to intentionally miss his or her live target through a process of conditioning in or on a battle simulator.

The question emerges though: Is this a good or bad thing? This ethical quandary about the status of virtual entities and the virtual environment and relationships therein may well serve to reveal the inadequacy of traditional ethical theory and should therefore be treated as an opportunity rather than as a problem. Online and virtual resources could themselves be used to set up and test new ethics or improve our older conceptions for this new world. To draw a broad analogy this might be considered as something close to the first pilgrims desire to set up religiously based communities in the "Godless" new world during the early colonisation of America by Europeans. In fact different online communities, including various hacker networks, are already reputed to have loose codes of ethics which are hotly debated when transgression occurs (Levy, 2010).

A stronger and somewhat more anarchistic formulation might draw the conclusion from this conundrum about the ethical status of virtual entities and activities that traditional ethics themselves have been surpassed by technology and that the time has finally arrived to throw off these ancient shackles altogether, at least when it comes to the virtual environment. This stronger statement can be supported by the argument (first made by Aristotle in the defence of theatre) that the virtual environment offers a chance for catharsis where evil or wrongdoing can be endured and even enjoyed in order that the participants (in the case of inter-active activities such as gaming rather than the audience in the case of theatre) are improved either by unburdening themselves of their anxieties, "facing their fears", through engaging with 'instances arousing pity or fear' (Aristotle, 1920, p. 35), or by doing these things, in a certain sense, and thus feeling fulfilled without causing any significant harm, "getting it onto the system and out of our system". This, then, would act as a type of online virtual cathartic activity. This argument about the fundamental ethical status of virtual entities is extended to virtuality or the virtual environment itself and worked through by reference to the next theme dealt with, the theme of Stealing or Sharing.

STEALING OR SHARING?

The British government recently passed legislation in the form of the Digital Economy Act 2010 that makes file sharing illegal and allows for much more severe sanctions than previously applied, sanctions such as offenders being cut off from internet access by their service provider, black listed by alternative providers, and fined up to £50,000 by the courts. This new legislation was enacted at the behest of major corporations within the music and film industries and against the wishes of service providers and a significant number of web users (BIS, 2010). This is an important contextual point that should be borne in mind with regard to the following discussion. Given the parameters of the discussion so far it should be safe to forego on any extended outline or useful analogy of what is at stake in the question of sharing and/or stealing. Suffice to say that the distinction between the "Real" and the "Virtual" is most hotly contested between those who feel that the internet should essentially fulfil the role of a giant community in which the participants pool information in the form of files and data which essentially does not belong to any one in particular (this group are dubbed "sharers" here) and those who conceive of it as a giant warehouse through which they can sell and distribute information and data, whether that be visual, verbal, musical, mathematical, code or what have you, which they perceive as their sole property until such time as they are paid the set or market price for it (this group are dubbed "sellers" here).

Neither group is overtly labelled "stealers" as that is exactly the question at issue, however it should be noted that broadly speaking the former group see the central activity of internet use as sharing while the latter see that same sharing as stealing. On the other hand the sharers label the sellers as agents of "bandit capitalism" trying to profiteer from something no-one actually owns. The sharers make the historically accurate claim that their conception of what the virtual environment is for is closer to its original ideals and form (as an open access data-base) while the latter group, the sellers, argue that the potential of the internet would be severely limited if market forces and the available means of production, both intellectual and concrete, were not harnessed as an engine for growth and progress. Again an analogy with American history might be useful here, one of these groups are sharecroppers and

the other are cowboys and each sees the other as wrong in what they are trying to do with and in a new environment. Which group is which in this analogy very much depends on which group one belongs to oneself and which analogous group one identifies, or wishes to identify, more closely with.

This difference between sharers and sellers points to the structure of the relationship between the real and virtual, and the different ethical intuitions that are drawn from or read onto this relationship, in so far as the first group (the sharers) would be more inclined to set aside ethical considerations that are wholly appropriate to the "real" world such as strict property rights and favour the contention that 'all information should be free' (Levy, 2010, p. 28). They would set these purportedly ethical considerations aside in favour of the more overarching ethical point that by foregoing on these considerations in the virtual environment it is possible to establish and maintain a more fundamentally ethical space that is relatively free of the inherently problematic elements of global capitalism and inequity. This space is perceived as having the potential to be an idealised arena of human interaction in which each can take according to their desire and give without incurring any real loss to themselves.

The sellers on the other hand are better served by the maintenance of more traditional and concrete ethical stances and standards in the virtual environment and are far more likely to argue by analogy, as has been done in a somewhat subversive fashion throughout this paper, and to assert that the virtual is essentially no different than the real and that advancing technologies should not change ethical norms. Their argument is backed up by the factually correct statement that all the data and information does originally come from somewhere or someone and that it, therefore, constitutes property in a sense very similar to the traditional concrete or "real" sense. Of course these issues are continually being thrashed out in the courts but, as hinted at earlier, there is a difference between ethics and law. A court finding for a large, powerful and wealthy company or group of companies against an individual or group of individuals, or indeed a government enacting legislation that allows for these proceedings to take place, should not settle the ethical question but should instead create the ground on which the question opens up. By the same token, one should not always automatically "cheer for the little guy" until the ethical issues involved have been satisfactorily dealt with.

The Master of IT could just choose sides according to what suits her best depending on which group she happens to be in, or aspires to be in. One's actions will usually tend to guide one's ethical reasoning (rationalisation) in this way rather than the reverse. However, the task of technoethics is to interrogate the decision making process that, ideally, would be entered into before action or becoming a member of either of these groups, or, failing that, to inform one as to whether one should remain in the group one finds oneself a party to. The fundamental philosophical question at stake (or at least one of the fundamental philosophical questions at stake), is: Does/should the situation drive thinking or does/should thinking drive the situation?

Here situation should not be engaged with in a limited sense, i.e., limited to the activities of sharing/stealing/selling but should rather be interrogated in a broad sense in order to come back to this specific question which opened up the possibility or circumstance of enquiry in the first place. If the situation drives thinking then this new and unique milieu, the virtual environment, must have a bearing not only on the approach to the virtual environment itself but on ethical thinking in general. If this is the case then perhaps the more radical argument put forward for the sharers in this paper is the right path for the Master of IT to pursue. If on the other hand thinking drives the situation then it is imperative to maintain established ethical standards regardless of the perceived novelty of this new milieu. If this is the case then it is best for the Master to follow the path of the

sellers. It is the fundamental question, though, that should be addressed first before dealing with the separate and specific questions that bring it into view. These more specific considerations can only usefully be examined in terms of real virtual situations and attitudes toward them so an issue such as "sharing and stealing" can emerge but it is only one among many.

Questions of privacy, for example, take on a perplexing quality when they are rooted in what is arguably a fundamentally open environment. These questions assume an almost comical character as people who live in glass houses, to draw another analogy, plead with others to not only refrain from throwing stones but to resist the temptation to even look in. The position on one side is once again that "the right to privacy" does not apply and is not a suitable ethical guideline in a virtual environment that is basically porous in nature, while on the other side the argument is more fundamentalist in that it claims the right to privacy applies in any and all circumstances when it comes to protecting the dignity of human beings who are the subjects of inalienable natural rights regardless of technological advances. Again here the question as to which ethical standpoint is better is quickly overtaken by the question as to whether any ethical standpoint is better or even required at all and that in turn is overtaken by the more fundamental question as to whether the situation, in this case the emergence of a new, powerful and all pervasive media/environment/world, drives thought or thought, in this case the kind of thought embodied in the three classical ethical theories mentioned earlier or in rights based ethics mentioned just now, drive or should drive the situation.

If the Master of IT begins to think in ethical terms, she is thrown back by questions such as these into that fundamentally ethical problem of the difference between the is and the ought, what is and what should be. The virtual environment is strange and singular in that it is both real and ideal, it _is_, it exists, it is there now waiting for someone to log on, that is a certainty, but does it have the potential to reveal what it, and perhaps we, ought to be? Is the virtual "Real" or "Ideal" or is it somewhere in between? These are not issues to be decided based on the fixed structure of the virtual environment as it is now at this moment but rather questions that are being and will be worked out by the Masters of Information Technology who are currently training and working in the virtual environment which is itself in constant flux. Only these Masters of the future can mould the brute information of which that environment is made into useful tool-like knowledge and only they can set the terms of the knowledge economy that takes place in and around that environment. What is more, this ethical conundrum has the potential to turn out 'now one way now another' (Heidegger, 1977, p. 13.). It is ultimately the ethical code or codes of this group that will determine what should and should not, or rather what can and cannot, be done with and in the virtual environment. This ethics of what can and cannot be done will be a *technē*-ethics, a technoethics in its fundamental sense. In order to gain an insight into what kind of ethical thinking a sample of these Masters of the future currently engage in their responses to the foregoing paper form its conclusions.

CONCLUSION

The task outlined to the ten Masters students who participated in this exercise was to think through the arguments presented in the paper and answer the following ten questions from their own points of view.

Some short questions that may require long answers:

1. Is there a specifically ethical content of being a master of information technology?
2. Is mastery of IT an art or a science?
3. Is there an ethical distinction between "Virtual" and "Real" activities?

4. Do you personally have a different attitude to these different types of activity?
5. Do you find you "act differently" as it were, when you are in a virtual as opposed to a "real" environment?
6. Should we have ethics/ is there a need for ethical awareness in a virtual environment?
7. Do you experience or seek to experience catharsis in relation to virtual activity?
8. Sharer/stealer/seller which are you?
9. Does/should the situation drive our thinking or does/should our thinking drive the situation?
10. Is the virtual "Real" or "Ideal" or is it somewhere in between?

With regard to question 1 nine out of the ten students felt that there is a specifically ethical content to being Masters of IT. The most common view expressed was that 'legal authorities and other guardians of morality are overwhelmed with the pace and possibilities of IT'. This swift pace, which is not addressed directly in the paper, was also described as a 'torrent' and was percieved as an important element in the fundamentally fluctuating nature of IT ethics. Along with this perception of IT as outpacing legal and traditional ethical mechanisms there was a perception that a higher level of ethical awareness is required of the Master of IT when compared to other professionals, that 'even though laws help to ensure an individual's information is protected from misuse... IT professionals still face an enormous range of ethical decisions because of their power to control, design and access this kind of information'.

While these responses signal a widely held recognition among this group of technologists about their particular ethical responsibilities, they also indicate an inadequacy in both legal and traditional ethical mechanisms to give appropriate direction to this quickly developing group. Evidently then, one of Bunge's core aims of technoethics is achieved in so far as these particular technology-empowered persons show a recognition of their responsibility for their professional actions but more work remains to be done in the emerging field of technoethics to give specific and practical guidance to this group (Bunge, 1977).

In their answers to question 2 the group showed less unanimity with two holding that mastery of IT is purely an art one of whom differentiated on the ground that he saw science '... more as an act of understanding, attempting to understand information' whereas IT is an art in which 'computer knowledge is something we take 'to hand' in order to create something else'. Two held that: 'It is a science' but like other sciences it has the capacity to produce something that can be turned into art, that '... will not be "just there" or even "there for" but will have a special place well above those ways of being'.

The remaining six held that Mastery of IT is something between an art and a science or is a mixture of both, that '... the imagination and craftsmanship of an IT professional comes to bear in the process of producing his products in a unique manner which is artistry, however the end product is often reproducible and the process documented, which is a scientific process'. Another student expressed her perception of IT as having a unique status in this grey area between art and science saying that 'IT is a unique discipline in that it directly combines science and art to produce technological products that function well on the back end (science focussed end) and the front end (artistic focussed end)'. This perception was broadly shared among those who felt IT is both a science and an art in so far as its definition as one or the other '... is dependent on the person and their role. Graphic designers might view IT as an art... whereas programmers might view IT more as a science'.

This distinction between IT as an art and IT as a science might call for greater precision in how ethicists approach IT technoethics as the core distinction between the virtual and the real in the paper may overlook the relation of the "real" work that goes into generating the "virtual" environment.

However, the group dealt with here, i.e. Masters of IT, tend to be equipped for various roles and therefore require a broader focus, as one student put it '... there is an army of IT specialists who are not involved in the development of the basics and the majority of the "Masters of information technology" might be among them'.

Though the paper implicitly argued that there is an ethical distinction between virtual and real activities the group was split equally in their views on question 3. Five held that there is a distinction but that this distinction depends on the nature of the activity itself. For example: 'activities such as e-commerce have adopted the ethical system of the "Real" world...' while other activities '... like chatrooms have adopted different forms of ethics they have created themselves that are not written down and are almost tribal in nature'. This distinction brings to light the idea that the virtual environment is analogous to a new world in which different societies are emerging and developing loose ethical frameworks in an ad hoc manner as they do so. A question for technoethics is whether or not it should simply document this ad hoc process, as for example Levy did with regard to hacker ethics (Levy, 1987), or whether it should attempt to guide and inform this process of ethical development as Borgmann suggested (Borgmann, 1987)?

This question again opens out the fundamental question of whether the situation drives thought or thought drives the situation. The main thrust of this present paper is to suggest that the answer lies between the two in the sense that every situation is in essence an instance of being-in-the-world and cannot be considered either from a perspective outside that being or as free standing beyond that being. This is precisely why a survey of the views of Masters of IT who are deeply embedded, in the lived sense, within the virtual environment is significant for technoethics and ethics in general. In a more fundamental statement one student held simply that: 'Virtual activities do not have any boundaries and their applications are unlimited'.

On the other hand, with regard to question 3, five members of the group felt that there should be no strong ethical distinction drawn between the virtual and the real environments. This view was summed up in the statement that 'partial or total migration of activities to the virtual environment does not overwrite our responsibility to self, community and environment'. This view was strengthened by the nuanced position that held that 'similarly to stealing money from a bank, when a participant downloads illicit material they accomplish a gain without losing or offering anything in return. This is where the distinction should lie, no exchange has taken place'. This latter view, that there is no distinction between the virtual and the real in terms of ethics, simultaneously undermines and supports the argument made in the foregoing discussion paper. It undermines the main drive of the paper in that it does not accept that there is any real need for specifically tailored ethics, or technoethics, for the virtual environment as classically concieved ethics should still apply but it supports a crucial development in the arguement of the paper as it highlights the neccessity for those same classically concieved positions to argue by way of analogy thus attempting to force the new digital environment into the established analogue pattern. Each position is appealing in its own way as the former, which can loosely be termed the sharers position, is liberating and idealistic while the latter, which can loosely be described as the sellers position, maintains high minded ethical aspirations and argues for equitable exchange.

In an odd anomaly only three respondants to question 4 reported that they themselves had a different attitude to the different kinds of activities adressed, i.e., virtual and real. This result is odd in that two participants in this discussion felt there was a theoretical distinction between the virtual and the real yet they themselves did not report maintaining a different attitude with regard to the different realms of activity. This anomaly was explained by one respondant as coming from 'a greater level of knowledge of the internet and technology in general' and an awareness that 'even if the activity has no real world repercussions it can have an effect in the future in the virtual environment... ' as the ethical boundaries of this environment '... are poorly defined at best and may move meaning that actions which have no current repercussions may have some in the future'. This response illustrates very well the perception among Masters of the future that the future itself is currently up for grabs and that while it has the potential to be wide open it also has the potentil to be much more closed. The new legislation about peer-to-peer file sharing enacted in Great Britain after the discussion described here took place proves the point being made.

However, as the paper argues, it is the Masters of IT themselves who will ultimately decide what can and cannot be done in the virtual environment as it is they who bring this information out from the standing reserve and form it into useful tool-like things. The processes by which this is done are continually changing and developing with the technology, the reference to the coming 'torrent' of information cited earlier, for example, is an implicit reference to "bittorrent" filesharing which is a relatively recent technological innovation that simultaneously increased the demand by the movie and music industries for harsher penalties and, paradoxically, made any new laws meeting those demands very difficult to enforce in practice. Of those who reported having a different attitude to virtual activity the main grounds of this difference hinged on whether or not they were dealing with 'anybody in real life' or 'playing around with different virtual identities'. Overall the response was surprisingly rigid in that most thought that they should, in a sense, carry their established ethical standpoint into the virtual environment with them even in despite of having differentiated strongly between the virtual and real environments.

In their responses to question 5 the group reasserted the basic 50/50 division exhibited in their responses to question 3 with the exception of one respondant who was undecided. Five of the group reported that they did act differently in virtual environments. For the most part acting differently in virtual environments had to do with gaming and social networking wherein 'the virtual world offers an escape from reality... [and] allows people to become characters that are not feasible in reality'. Another view expressed was that people generally act different in different environments as those environments themselves dictate but that essentially the different actions undertaken in the virtual environment are not dissimilar to acting differently in 'a foreign city'. This is an interesting insight in general but it does tend to overlook the possibilities of consequence free action or actions that are 'not feasible in reality' that open up in the virtual environment. Again though it should be borne in mind, as pointed out with regard to question 4, the existence of a consequence free environment now does not ensure the existence of that same environment in the future. This is an important point because it is of course in the future that consequences generally happen.

Of the four participants in this conversation who held they did not act differently in the virtual environment the general consensus, on a practical note, was that it is not actually the consequence free environment it is often percieved as i.e. 'the internet is a gigantic filing cabinet that never loses anything due to its design so discretion is advisable at all times'. Another humorous practicality was pointed out by one respondant who reported that he had tried to act differently in the virtual environment but found it too difficult and stressful saying: 'I used fictitious names to register on websites and to open e-mail accounts. However, I quickly forgot the user details and could not continue to have access to the portals'. To save his blushes this participants details will be omitted in this instance, thus perhaps compounding his original indiscretion or discretion depending on how it is percieved.

On a more fundamental level the basic reason given for not acting differently in the virtual as opposed to the real environment was summed up as follows: 'If our behaviour in the virtual environment is not compatible with existing standard ethical practices, we tend over time to start exhibiting unethical virtual habits in the physical "real" world'. This summation conforms with the picture of Aristotelian virtue ethics given in the foregoing discussion paper. Again the group who professed to acting the same in the virtual as in the real environments tended to question whether or not there was a strong distinction between the two at all and to argue from more traditional recieved ethical positions that assert that thought should drive the situation rather than the reverse.

In regard to question 6, all the respondants held that there is a need for ethical awareness in the virtual environment. However, the reasons for that need and the views on what kind of ethics could answer that need were divided into two main areas. One group expressed the conviction, as they did throughout, that 'fundamentally there is a need for ethics in every situation'. This contention rested largely on the fact that virtual activities can have a real world effect and a number of these responses included references to the neccessity to protect children from the content itself rather than any specific "real world" harm that might come to them in any physical sense. This concern was summed up as follows 'there is definitely a need for ethical awareness in the virtual world, with more kids accessing the internet at a younger age it is essential that they are protected from inappropriate material'. Again here the language expresses the kind of ethical essentialism associated with classical models and theories. The second constituency in this group took the view that a new ethical framework is required for a new environment in which 'a sense of anonymity... plays a large role in facilitating different behaviours and creates different ethical issues'. For this group the technology itself is regarded as calling forth a new kind of ethics 'the internet exists as an almost lawless domain, immune to national borders or laws and so there is a need for ethics in order to preserve a level of control and decency'. The question that remains open here, though, is encapsulated thus 'the internet has created its own social constructs and cultural rules within each area, so where should ethics apply?'.

The question of catharsis, which relates to an aside in the paper as a whole, illicited a negative response from seven of the group all of whom basically contended that 'I have not experienced catharsis in a virtual world and have never sought to experience such' (H. Gbinigie). One respondent held that she could see how others might seek to have their 'soul freed of its emotional abundance...' but that 'in relation to virtual activity I don't experience catharsis because... I try to avoid excessive emotion in general'. Others tended to agree that they themselves did not seek or experience catharsis in the virtual environment but that they believed other people did stating, for example, that 'I believe it is possible to experience catharsis in the virtual world'. Another participant in this group felt that she might be missing out on the possibility of cathartic experience 'due to a lack of knowledge of what is available in order to achieve such experience'. A more conclusive answer was offered stating that '... the theory of catharsis itself is highly controversial' and that the remotness from the venue of action that is a neccessary element of current 'common technical conditions' make it impossible to achieve catharsis by means of online virtual interaction, put simply and succinctly this member of the group held that: 'If one wants to get cleansed of ones fear of spiders, just sitting in front of a computer and learning about spiders and looking at pictures on Wikipedia wouldn't do it'.

The other three members of the group held that catharsis could be achieved and that they themselves achieved it through activities such as gaming. One respondent put this point particularly strongly by stating that catharsis '... is the essence of gaming as it allows users to experience situations which in reality are not attainable'. Those who responded that they did not have or seek cathartic experience through virtual activity also acknowledged the role of gaming by distancing themselves from it in statements such as 'I personally do not play video games, nor do I engage in role-playing or social interactions in a virtual environment, using the internet mainly for information gathering'. This difference of opinion seems to come down to a question of will as most of the larger group contend that they do not experience catharsis because they do not seek it but that they believe it is achieveable, while the smaller group who report having experienced catharsis in the virtual environment acknowledge the fact that they spend time and energy on gaming which is the main mode of cathartic activity discussed. An interesting aspect of this discussion is that those who did not seek catharsis in virtual activity tended to have a condemnatory tone to their responses, holding for example that 'I prefer to approach a problem head on rather than seek catharsis as a solution', while those who did seek it saw it as 'the positive side to virtual activity... [that] doesn't harm anyone in the process'.

Question 8 asked "Sharer/Stealer/Seller which are you?" and is perhaps the most revealing about the self image of a small group of Masters of the future. Eight out of ten declared themselves sharers, of this group most acknowledged a neccessity for intellectual property to be recognised but felt that 'owners of property should be aware that previous business models might not hold in present times'. Another respondent classified himself as a sharer who '... belives in the open exchange of ideas and technologies over the internet' while still holding that 'I can understand why sellers might see me as a stealer... but it is more complicated, if an organization or individual has not been directly deprived of a product nothing has been stolen'. In a very practical statement one of the self-identified sharers said simply that 'I know every source of information or file is someone elses work but the internet culture I grew up in said "why would you pay for it when you can get it for free?"'. This may seem a little self-serving on the face of it but it is exactly an example of how ad hoc ethical frameworks develop very rapidly in the virtual environment, particularly as no specific online group is mentioned rather reference is made to the more naturalistic 'culture I grew up in'. This is the depth to which technoethics must go in order to deal with IT and the virtual environent, the perenial question of cultural relativism is as pertinent here as anywhere else.

Only one respondant identified himself as a seller and it is very revealing that he argued for this position in terms of the basic good of the market paraphrasing Winston Churchill's assertion that 'capitalism is the worst economic and social system to organize our democracies upon, except for all the others'. This signifies a kind of fundamental appeal to existing norms and historical authorities which has at its heart a drive to protect property in order to create profit through exchange. This lone voice was very clear that the virtual environment 'is a very big place' but that its main function is as 'a market'. One respondant expressed the view that he might best identify himself as a stealer because he often took from the internet but rarely contributed.

Here, the core distinction between sharers and sellers laid out in the foregoing discussion paper is verified in the responses of the Masters group as is the basic difference between their answers and modes of arguing. The sharers see the virtual environment as an idealised space where exchange can take place without loss to anyone and argue that it is peculiar and novel in this respect. The seller on the other hand argues that established ethical norms must hold and also argues through analogy, first implicitly by identifying the internet with a marketplace and then more explicitly by comparing the virtual environment to the sea or the air, both of which he argues were once completely free but are now, due to technological advances such as airplanes, divided up into proprietory or nationalised holdings. It is very interesting to note that only one respondant identified himself as a seller, if this pattern holds in the ranks of Masters

of IT as a whole it may well have a very big effect on how ethics in the virtual environment develops, although it may well be an anomaly with regard to this specific and relatively small group.

With regard to the fundamental question as to whether situations drive thinking or thinking drives situations six of the ten participants held that ethical thinking should always drive the situation. Some respondants in this group felt that 'theft is theft... and the virtual environment must be viewed as an entity which is governed by our pre-established thinking'. While another respondant argued, once again by analogy, that 'allowing the situation to drive our thinking regarding moral issues is like embarking on a car journey with a blind driver'. On the other hand two of the participants in this discussion held that the scale and pace of IT development had outstripped ethical thinking, very much in line with their answers to question 1. Of these a very clear assessment to support this position was that 'the internet is so large and complicated that it is almost impossible to change at this point, so the situation has to drive our think-ing about ethical issues that arise' this in spite of the fact that 'commercial interests will continue to try and apply real world ethical approaches'. The remaining two members of the group held that thinking drove situations in some phases or circumstances in human life and situations drove thinking in others. This contention was backed by the arguement that while 'our thinking drives the situation... in many cases we are having to wait for the situation to catch up... however we are only proactive in situations we can foresee... it is only when the situation deviates from the norm that the situation drives our thinking and we become reactive'.

Finally, with regard to the nature of the virtual environment, whether it is "Real", "Ideal" or somewhere in between, all of the group treated this question in terms of whether or not the virtual environment was ideal in the sense of utopian rather than whether it was ideal in the sense of immaterial or "made of ideas". Only one respondant contended that the virtual was wholly '... ideal because the model created is more perfect and excellent than reality'. Three respondants held that the virtual environment is wholly real due to its mechanical structure and the concrete nature of hardware interfaces. One of these respondants, perhaps without knowing it, put forward an arguement about ontological status similar to one Descartes used by stating that the virtual must be real because 'after all it exists in the real world'. This is a restatement of the arguement that states: that which is contained cannot have a greater existence than that which contains it. The remaining six participants in this discussion supported the basic outline of the virtual environment drawn in the paper in that they felt the virtual lay somewhere between the real and ideal. Members of this largest group argued that 'the virtual world can be as real or ideal as we want to make it' thereby asserting the key role of Masters of IT and that 'the virtual is an ideal representation of the real' thereby defining it as ambiguous in nature.

The set of answers by ten future Masters of IT to ten questions about virtue and virtuality presented here may not represent a big enough survey group to draw reliable statistical conclusions from. What it does demonstrate though is that there is as yet no consensus on any given ethical standard or set of standards that should be applied to the virtual environment though there is, perhaps surprisingly, consensus that some kind of ethics should apply. It is significant here that the group that asserts the more conservative standpoint, that being that established thinking should drive the situation, retains its integrity throughout the various questions while the group that argues for thinking that is driven by the situation seems to have a greater level of insecurity about that core position and swing away from it as the questions change and develop. So, for example, while more of the respondants identify themselves as sharers

and give a more idealised account of what they think a Master of Technology should be they find it more difficult to express the fundamental underpinnings of this position and some of their number turn to more traditional positions when pressed on foundational questions.

Ultimately this engagement shows a need for greater precision and clarity about what it is that this IT ethics should try to address and that, while established professional ethics and codes of conduct have their place, any adequate account of this rapidly growing and developing environment must come to terms with its peculiar and singular nature. This can only be done by a technoethics that can combine ethical thinking and research with a close awareness of the technological situation as it continually morphs and changes. Fruitful exchanges between students in IT, or the Masters of the future as they are dubbed here, and ethicists should continue to be carried out in order to bring the engineers' lived technological experience into a mutually beneficial relationship with the philosophers' learned abstraction. This is the key phenomenon that gives itself for study to a technoethics of IT, the real building, dwelling and thinking lived by Masters of IT in the virtual environment.

ACKNOWLEDGMENT

A version of this paper was originally presented as the concluding part of an ethics course for students in the 2009/10 Masters in Information Technology course at the National University of Ireland Galway. The author wishes to thank all the members of the 2009/10 Masters in IT class at NUI Galway who participated in this class and agreed to have their responses form part of this paper: Deirdre Adams, Joy Burke, Bertram Dehn, Hastings Gbinigie, Swathika Kavanur, Alan McCann, David Niland, Niall Ryan, Darren Stephenson and Noel Vaughan.

REFERENCES

Acton, L. (1917). Letter to Bishop Creighton, 1887. In *Acton-Creighton Correspondence*. London: Figgis and Laurence.

Aristotle,. (1920). *Poetics (I. Bywater trans.)*. Oxford, UK: Clarendon Press.

Aristotle,. (1996). *Nicomachean Ethics* (Griffin, T., Ed.). London: Wordsworth.

Barkham, P. (2008, July 17). The Question: What is the knowledge economy? *The Guardian*.

BIS. Department of Business Innovation and Skill. (2010). *P2P Consultation Responses January 2009*. Retrieved August 8, 2010, from http://www.bis.gov.uk/policies/business-sectors/digital-content/p2p-consultation-responses

Blackburn, S. (2002). *Being Good: A short introduction to ethics*. Oxford, UK: Oxford University Press.

Borgmann, A. (1987). *Technology and the Character of Contemporary Life: A philosophical inquiry*. Chicago: University of Chicago Press.

Boylan, M. (1995). *Ethical Issues in Business*. London: Harcourt Brace.

Bunge, M. (1977). Toward a Technoethics. *The Monist*, *60*, 96–107.

Digital Economy Act. (2010). *The UK Law Database*. Retrieved August 8, 2010, from http://www.stautelaw.gov.uk/content.aspx?activeTextDocId=3699621

Floridi, L. (2001). *Philosophy and Computing*. London: Routledge.

Floridi, L., & Sanford, J. W. (2001). Artificial Evil and the Foundation of Computer Ethics. *Ethics and Information Technology*, *3*(1), 55–66. doi:10.1023/A:1011440125207

Heidegger, M. (1975). *Poetry, Language, Thought (A. Hofstadter trans.)*. New York: Harper and Rowe.

Heidegger, M. (1977). *The Question Concerning Technology and Other Essays (W. Lovitt trans.)*. New York: Harper and Rowe.

Heidegger, M. (2005). *Being and Time (J. Macquarrie & E. Robinson trans.)*. London: Blackwell.

Hogan, L. (2008, May 20). Cyberbullying 'widespread' but parents kept in the dark. *The Irish Independent*.

Johnson, D. (2001). *Computer Ethics*. Upper Saddle River, NJ: Prentice Hall.

Levy, S. (2010). *Hackers: Heroes of the Computer Revolution* (25th anniversary ed.). Sebastopol, CA: O'Reilly Media Inc.

Luppicini, R. (2008). The Emerging Field of Technoethics. In Luppicini, R., & Adell, R. (Eds.), *Handbook of Research in Technoethics* (pp. 1–18). Hershey, PA: IGI Global.

Parker, Uncle B. (1962, August). The Amazing Spiderman. *Amazing Stories*, 15.

Singer, P. (1994). *Ethics*. Oxford, UK: Oxford University Press.

Tavani, H. (2003). *Ethics and Technology: Ethical Issues in an Age of Information and Communication Technology*. London: Wiley.

This work was previously published in the International Journal of Technoethics, Volume 2, Issue 1, edited by Rocci Luppicini, pp. 1-18, copyright 2011 by IGI Publishing (an imprint of IGI Global).

Chapter 2
Internet History

Raphael Cohen-Almagor
University of Hull, UK

ABSTRACT

This paper outlines and analyzes milestones in the history of the Internet. As technology advances, it presents new societal and ethical challenges. The early Internet was devised and implemented in American research units, universities, and telecommunication companies that had vision and interest in cutting-edge research. The Internet then entered into the commercial phase (1984-1989). It was facilitated by the upgrading of backbone links, the writing of new software programs, and the growing number of interconnected international networks. The author examines the massive expansion of the Internet into a global network during the 1990s when business and personal computers with different operating systems joined the universal network. The instant and growing success of social networking-sites that enable Netusers to share information, photos, private journals, hobbies, and personal as well as commercial interests with networks of mutual friends and colleagues is discussed.

INTRODUCTION

History consists of a series of accumulated imaginative inventions.

~ *Voltaire*

Floridi (2009, 2010) argues that we are now experiencing the fourth scientific revolution. The first was of Nicolaus Copernicus (1473–1543), the first astronomer to formulate a scientifically-based he-liocentric cosmology that displaced the Earth and hence humanity from the center of the universe. The second was Charles Darwin (1809–1882), who showed that all species of life have evolved over time from common ancestors through natural selection, thus displacing humanity from the centre of the biological kingdom. The third was Sigmund Freud (1856–1939), who acknowledged that the mind is also unconscious and subject to the defence mechanism of repression, thus we are far from being Cartesian minds entirely transparent to ourselves. And now, in the information revo-

DOI: 10.4018/978-1-4666-2931-8.ch002

lution, we are in the process of dislocation and reassessment of humanity's fundamental nature and role in the universe. Floridi argues that while technology keeps growing bottom-up, it is high time we start digging deeper, top-down, in order to expand and reinforce our conceptual understanding of our information age, of its nature, less visible implications and its impact on human and environmental welfare, giving ourselves a chance to anticipate difficulties, identify opportunities and resolve problems, conflicts and dilemmas.

This essay focuses on the milestones that led to the establishment of the Internet as we know it today, from its inception as an idea in the 1950s until the early 21st Century. The varied and complex social and technological transformations we witness today have their roots in the way the Internet has been developed through research grants from the U.S. Department of Defense's Advanced Research Projects Agency. Scientists wished to maintain communication links between distant locations in the event that electrical rout had been destroyed. The early Internet was devised and implemented in American research units, universities, and telecommunication companies that had vision and interest in cutting-edge research. The program grew in the 60s and 70s, becoming a network of computers that transmitted information by "packet switching."

The network of computers was from the start an open, diffused and multi-platform network that up until the 1990s developed in the United States and then, within a few years, expanded globally in impressive pace and with no less impressive technological innovations the end of which we are yet to witness.

The interdisciplinary field of Technoethics is concerned with the moral and ethical aspects of technology in society. The Internet plays a crucial world in today's technology and society (Luppicini, 2010). In order to understand how the Internet became an integral part of our lives, it is crucial to examine its history and the major developments that took place from its modest infancy until its giant presence. In fifty years (1960-2010) the technology advanced rapidly. This has been an age of innovation where ideas have driven the development of new applications which, in turn, have driven demand. Then we witness circularity. New demands yielded further innovation and many more new applications – email, the world-wide-web, file sharing, social networking, blogs, skype. These were not imagined in the early stage of the net.

This essay examines milestones in the history of the Internet, how the Internet evolved from the Advanced Research Projects Agency (ARPA, 2004) in 1957, its formative years (1957-1984) until nowadays; from the early Internet devised and implemented in American research units, universities, and telecommunication companies that had vision and interest in cutting-edge research until a global phenomenon. I highlight the entry of the Internet into the commercial phase (1984-1989), facilitated by the upgrading of backbone links, the writing of new software programs and the growing number of interconnected international networks; the massive expansion of the Internet into a global network during the 1990s when business and personal computers with different operating systems joined the universal network; the instant and growing success of social networking -- sites that enable Netusers to share information, photos, private journals, hobbies and personal as well as commercial interests with networks of mutual friends and colleagues. The technology has transformed into a quotidian network for identifying, sharing and conveying information and ideas, exchanging graphics, videos, sounds and animation to hundreds of millions of Netusers around the world.

THE FORMATIVE YEARS

The history of the Internet started in the United States in the early 1960s. This was the Cold War period, when the world was bi-polar: The United

States and the Soviet Union were competing in expanding their influence in the world, viewing each other with great caution and suspicion.

On October 4, 1957, the Soviet Union launched the first space satellite, Sputnik. The Sputnik success necessitated American reaction. It was a question of pride and leadership. The US Department of Defense responded by establishing the Advanced Research Projects Agency (ARPA, 2004),[1] designed to promote research that would ensure that the USA compete with and excel over the USSR in any technological race. ARPA's mission was to produce innovative research ideas, to provide meaningful technological impact that went far beyond the convention evolutionary developmental approaches, and to act on these ideas by developing prototype systems.[2] One of the ARPA offices was the Information Processing Techniques Office (IPTO) which funded research in computer science designed to mobilize American universities and research laboratories to build up a strategic communication network (Command and Control Research) that would make available messaging capabilities to the government (Curran & Seaton, 2009; Conn, 2002).

A popular myth holds that the Department of Defense scientists thought that if the Soviet were capable to launch satellites, they might as well be capable to launch long-distance nuclear missiles. Because networks at the time relied on a single, central control function, so the myth goes, the main concern was networks' vulnerability to attack: Once the network's central control point ceased to function, the entire network would become unusable. The scientists wanted to diffuse the network so it could be sustained after attacking one or more of its communication centers (Schneider & Evans, 2007).[3] They had in mind a "decentralized repository for defense-related secrets" during wartime (Conn, 2002, p. xiii). However, the pioneers of the ARPA Network project argue that ARPANET was not related to building a network resistant to nuclear war: "This was never true of the ARPANET, only the unrelated

RAND study on secure voice considered nuclear war. However, the later work on Internetting did emphasize robustness and survivability, including the capability to withstand losses of large portions of the underlying networks."[4] Leonard Kleinrock, the father of Modern Data Networking, one of the pioneers of digital network communications who helped to build ARPANAET, explained that the reason ARPA wanted to deploy a network was to allow its researchers to share each others' specialized resources (hardware, software, services and applications). It was not to protect against a military attack.[5] And David D. Clark, Senior research scientist at MIT Laboratory for Computer Science who worked in the ARPANET project in the early 1970s, said he never heard of nuclear survivability and that there is no mentioning of this idea in the ARPA records from the 1960s. In a personal communication, Clark wrote:

I have asked some of the folks who pushed for the ARPAnet: Larry Roberts and Bob Kahn. They both assert that nobody had nuclear survivability on their mind. I was there from about 73, and I never heard it once. There might have been somebody who had the idea in the back of his mind, but 1) if so, he held it real close, and 2) I cannot figure out who it might have been. We know who more or less all the important actors were. (Sadly, Licklider has died, but I think I did ask him when he was still alive. I wish I had better notes.) So I am very confident that Baran's objective did not survive to drive the ARPA effort. It was resource sharing, human interaction... and command and control.[6]

In 1962, J.C.R. Licklider became the first director of the Information Processing Techniques Office. His role was to interconnect the Department of Defense's main computers via a global, dispersed network. Licklider articulated the vision of a "galactic" computer network—a globally interconnected set of processing nodes through which anyone anywhere can access data and programs.[7] In August 1962, Licklider and Welden Clark published the first Paper on the concept of the Internet titled "On-Line Man Computer Communication."[8] They saw communication network as a tool for scientific collaboration. Here the seeds for what would later become the Internet were planted.

Paul Baran (1964) of the RAND Corporation deserves particular attention not only because his research project created the myth that connected ARPANET to the development of a robust decentralized network that would enable the US a second-strike capability. Baran (1964) had been commissioned by the United States Air Force to study how the military could maintain control over its missiles and bombers in the aftermath of a nuclear attack. In 1964, Baran proposed a distributed scheme for U.S. telecommunications infrastructure with no central command or control point that would survive a "first strike." In the event of an attack on any one point, all surviving points would be able to re-establish contact with each other.[9] Note that Baran's research project came about six years after ARPA was established. Lawrence G. Roberts, the principal architect of ARAPNET, wrote that the Rand work had no significant impact on the ARPANET plans and Internet history (Roberts, 1999).

In 1965, Donald Davies of the British National Physical Laboratory (NPL) began thinking about packet networks and coined the term "packet." In fact, at that period of time three scientists in three different locations were thinking independently about that same technology: Leonard Kleinrock was the first to develop the underlying principles of packet switching. His ideas, drafted at the MIT labs in 1961, constituted an important milestone in the development of the Internet.[10] Baran at RAND formulated the idea of standard-size addressed message blocks and adaptive alternate routing procedures with distributed control. And Davies thought similarly that to achieve communication between computers a fast message-switching communication service was needed, in which long messages were split into chunks sent separately so as to minimise the risk of congestion. The chunks

he called packets, and the technique became known as packet-switching. Davies's network design was received by the ARPA scientists. The Arpanet and the NPL local network became the first two computer networks in the world using the technique (Kleinrock, 2008).[11]

The ARPANET was launched by Bolt Beranek and Newman (BBN) at the end of 1969.[12] BBN was commissioned to design four Interface Message Processors (IMPs), machines that would create open communication between four different computers running on four different operating systems, thus creating the first long-haul computer network and connecting between the University of California at Los Angeles (UCLA), the Stanford Research Institute (SRI) in Menlo Park, California, the University of California at Santa Barbara (UCSB), and the University of Utah which together comprised the Network Working Group (NWG).[13] A fifth ARPANET node was installed at BBN's headquarters. Each node consisted of an IMP, which performed the store-and-forward packet switching functions. Packet switching was a new and radical idea in the 1960s. Via ARPANET's Network Control Protocol (NCP), users were able to access and use computers and printers in other locations and transport files between computers. This was an investigational project that explored the most favorable way of building a network that could function as a trustworthy communications medium. The main hurdle to overcome was to develop an agreed upon set of signals between different computers that would open up communication channels, enabling data to pass from one point to another. These agreed upon signals were called *protocols*.

Essentially common grammatical tools of a technological language, protocols allow for conversations between any two computers so that anyone anywhere can search for and receive (or, conversely, create and send) text, graphic images, and audio and video files (Dubow, 2005).[14] The experimental project was based on open dialogue, where scientists posted Requests for Comments

(RFC), on free exchange of information and ideas, on collaboration rather than competition.[15] There were no barriers, secrets or proprietary content. Indeed, this free, open culture was critical to the development of new technologies and shaped the future of the Internet. The NCP was a great success, enabling the linking together of researchers at remote sites. At the time, only hard-core computer scientists knew of this network's existence (Spinello, 2000).[16]

In those early days, the seeds of what will come to be known the Internet architecture and trade-marks were planted. The directors of ARPA's Information Processing Techniques Office (IPTO), Robert Taylor and Larry Roberts, allowed considerable freedom and flexibility in research. They imposed minimal requirements in terms of progress reports, meetings, site visits, oversight and other customarily bureaucratic mechanisms that are so prevalent in many organizations. Kleinrock (2008, p. 12) wrote: "We felt strongly that control of the network should be vested in all the people who were using the Net and not in the carriers, the providers or the corporate world."

The network then expanded to other institutions, including Harvard, MIT, Carnegie Mellon, Case Western Reserve and University of Illinois at Urbana. Within sixteen months there were more than ten sites with an estimated 2,000 users and at least two routes between any two sites for the transmission of information packets (Slevin, 2000; Conn, 2002). ARPANET was the world's first advanced computer network using packet switching. Leonard Kleinrock wanted to develop a design methodology that would scale to very large networks, and the only way he thought was available to accomplish that was to introduce the concept of distributed control, wherein the responsibility for controlling the network routing would be shared among all the nodes, and therefore, no node would be unduly tasked.[17] This resulted in robust networks.

One of the major characteristics of the emerging network is innovation. One development quickly leads to another. In the early 1970s, scientists tried to overcome new problems. The new communication ideas, the experiments, the testing, and the tentative designs, brought about an endless stream of networks that were ultimately interlinked to become the Internet. Someone had to record all the protocols, the identifiers, networks and addresses and the names of all the things in the networked universe. And someone had to keep track of all the information that stemmed from the discussions. That someone was Jonathan B. Postel, a young computer scientist who worked at that time on the ARPA project at UCLA (Cerf, 1998). Postel devoted himself to building and running the Internet's naming and numbering structure. He proposed the top-level domains dot-com, dot-edu, and dot-net (Hafner & Lyon, 1998).[18] In those pioneering, unstructured and building years, Postel was, in effect, the Internet Assigned Numbers Authority (IANA). Postel was not elected to the position of responsibility he held in the Internet community; he was simply, in the words of the White House's Internet policy adviser, Ira Magaziner, "the guy they trust".[19]

Secondly, the ARPANET succeeded in connecting the computers used in different time-sharing systems. Now they wished to connect the packet switching network of the ARPANET with a satellite packet switching network and a packet radio packet switching network. In July 1970, the first packet radio ALOHANET, based on the concept of random packet transmission, was developed at the University of Hawaii by Norman Abramson and became operational. ALOHANET linked the University of Hawaii's seven campuses to each other and to the ARPANET. Based on this model, ARPA built its own packet radio network which was called PRNET (Ryan, 2011). At that same period of time, ARPA also developed a satellite network, called SATNET.

In 1971, UNIX operating system was developed at Bell Lab, quickly gaining the appreciation of many scientists. UNIX provides a suite of programs which makes the computer work. It is a stable, multi-user, multi-tasking system for servers, desktops and later on also for laptops.[20] In 1972, ALOHANET connected to the ARPANET and a commercial version of ARPANET, called TELNET, became the first Public Packet Data Service. The Telnet protocol was a relatively simple procedure. It was a minimal mechanism that permitted basic communication between two host machines.[21] Telnet applications allow users to log on and to operate remote computers. Such applications can, for example, be used to search and consult remote databases such as library catalogues.

A year later, in 1973, ARPANET was connected to international hosts. File transfer Protocol (FTP) came into existence and worked using a Client Server Architecture.[22] The file-transfer protocol specified the formatting for data files traded over the network. FTP made it possible to share files between machines. Moving files might seem simple, but the differences between machines made it very difficult. FTP was the first application to permit two computers to cooperate as peers instead of treating one as a terminal to the other (Hafner & Lyon, 1998). Telnet, FTP and TALK were the first applications to become available on ARPANET and are still used in some form or another on the Internet today. TALK was the first program that allowed Netusers to engage in a real-time conversation over the network (Slevin, 2000). Netusers typed messages onto a split screen and read replies written at the bottom of the screen.

In early 1973, the network had grown to 35 nodes and was connected to 38 host computers (Rubinstein, 2009). That year, Norway and England were added to the network and traffic had expanded significantly. In 1974, Vint Cerf and Robert Kahn developed a set of protocols that implemented the open architecture philosophy.[23] These new protocols were the *Transmission Control Protocol* (TCP) and the *Internet Protocol* (IP). TCP includes rules that computers on a network use to establish and break connections; IP includes rules for routing of individual data packets. The

Transmission Control Protocol/Internet Protocol (TCP/IP) organizes the data into packages, put them into the right order on arrival at their destination, and checked them for errors.

Most of the applications use the client/server model. A request is made for a particular service from the client to the server. The server responds or the conversation continues between the client and server until one of the participants ends it (Cerf & Kahn, 1974; Langford, 2000). By 1983, all networks connected to the ARPANET made use of TCP/IP and the old Network Control Protocol was replaced entirely. From then on, the collection of interconnected and publicly accessible networks using the TCP/IP protocols came to be called the "Internet" (Slevin, 2000).[24]

ARPANET grew into the Internet based on the idea that there would be multiple independent networks of rather arbitrary design (Leiner et al., 1997). The term "Internet" was first used by Vint Cerf and Robert Kahn in their 1974 article about the TCP protocol (Cerf & Kahn, 2000). The importance of the TCP/IP protocol in the history of the Internet is so great that many people consider Cerf to be the father of the Internet. A number of TCP/IP-based networks – independent of the ARPANET – were created in the late 1970s and early 1980s. The National Science Foundation (NSF) funded the Computer Science Network (CSNET) for educational and research institutions that did not have access to the ARPANET (Schneider & Evans, 2007).

Though the original design of the ARPANET was for resource sharing, it quickly demonstrated its utility as a message system. Soon researchers understood how useful the network can be for the transmission of communication. They continually sought to improve this characteristic of the network. In 1973, Lenny Kleinrock sent the first personal message over ARPANET; Ray Tomlinson of Bolt Beranek and Newman (BBN) wrote the first email program. The @ sign was introduced as a means of punctuating email addresses, separating the user name on the left from the site or computer identifier on the right.[25] Electronic mail grew first among the elite community of computer scientists on the ARPANET. They found it effective, convenient, easy to use and obviously much less time consuming than any other mode of communication. From its inception, email lacked formality and small-talk. It was a business tool to pass messages. Soon emailing bloomed across the Internet. While the ARPANET's creators did not have a grand vision for the invention of an earth-circling message-handling system, once the first couple of dozen nodes were installed, early Netusers turned the system of linked computers into a personal as well as a professional communications tool (Hafner & Lyon, 1998). Seventy five percent of the ARPANET traffic was email (Jenkins, 2001). ARPANET became a sophisticated email system.

On June 7, 1975, Steve Walker, a program manager at ARPA's Information Processing Techniques Office, announced the formation of an electronic discussion group which he called Message Services Group (MsgGroup) (Chick Net, n. d.). He sought to establish a group of people concerned with message processing in order to determine "1. What is mandatory; 2. What is nice; 3. What is not desirable in email functions" (Hauben, 1998). Walker wrote that his goal was not to establish another committee, but to see if dialogue can develop over the Net. He was creating a prototype form to utilize computer conferencing to determine its capabilities (Hauben, 1998). This was an example of how the ARPANET and the Internet were developed: Setting up a prototype, inviting comments, checking feasibility, and developing the prototype further to accommodate needs.

In 1979, USENET, a "poor man's ARPANET," was created by Tom Truscott, Jim Ellis, and Steve Belovin to share information via email and message boards between Duke University and the University of North Carolina, using dial-up telephone lines and the protocols in the Berkeley UNIX distributions (Hauben & Hauben, 1997).[26] The original Usenet News Service was devoted to transmitting computing news and facilitating discussions among employees of university computing departments on topics such as operating systems and programming languages (Schneider & Evans, 2007). Later Usenet developed into a world-wide distributed discussion system. It consists of a set of newsgroups on specified subjects. "Articles" or messages are posted to the newsgroups and these articles are then broadcast to other interconnected computer systems via a wide variety of networks. The Usenet routes messages by topic, rather than by individual or through a mailing list. Any Netuser can post messages while others can view and reply to the posted messages. Some of the newsgroups are moderated for approval before appearing in the newsgroup. Others are not.[27]

The early 1980s saw the continued growth not only of the ARPANET but also of other networks. The Joint Academic Network (Janet, n. d.) was established in the United Kingdom to link universities there. It consists of a large number of sub-networks that connect between the UK's education and research organizations and between them and the rest of the world. In addition, Janet includes a separate network that is available to the community for experimental activities in network development.[28] In 1982, the ARPANET had 200 hosts and a year later the network grew to 500 hosts (Spinello, 2000).[29] In 1983, ARPANET, and all networks attached to it, officially adopted the TCP/IP networking protocol.[30] Mailing lists, information posting areas (such as the User's News Network, or Usenet, newsgroups), and adventure games were among the new applications appearing on the ARPANET (Schneider & Evans, 2007).

An important undertaking, very relevant for technoethics, took place in October of 1981, when a discussion group was formed on a computer message system at the Xerox Palo Alto Research Center. Recognizing that computer professionals in other areas might share similar concerns, the group debated the merits of forming an organization dedicated to raising the awareness of the profession and the public with regard to the dangers inherent in the use of computers in critical systems. They wished to devise common principles to guide technological innovations and application to benefit society in an ethical and responsible fashion. In June 1982, the group adopted the name Computer Professionals for Social Responsibility - CPSR. Up until the mid 1980s, CPSR focused nearly all of its energy on the dangers posed by the massive increase in the use of computing technology in military applications. It became known for its fierce opposition to the Strategic Defense Initiative (SDI), which President Reagan announced in early 1983.[31]

In 1983, a mere 500 computer hosts were connected to the Internet. In 1984, the number of hosts increased to 1024.[32] As more researchers connected their computers and computer networks to the ARPANET, interest in the network grew in the academic community. One reason for increased interest in the project was its adherence to an open architecture philosophy: Each network could continue using its own protocols and data-transmission methods internally. There was no need for special accommodations to be connected to the Internet, there was no global control over the network, and all could join in. This open architecture philosophy was revolutionary at the time. Most companies used to make their networks distinct and incompatible with other networks. They feared competition and strove to make their products inaccessible to competitors. The shift to an open architecture approach is one of the most celebrated features of the Internet.

ENTERING THE COMMERCIAL PHASE

During the mid-1980s, the Internet entered its commercial phase. In 1984, the Department of Defense split the ARPANET into two specialized networks: ARPANET would continue its advanced research activities, and MILNET (for Military Network) would be reserved for military uses that required greater security. Connections were developed so that users could communicate between the two networks. In 1986, the number of Internet hosts increased to 5000. By 1987, when the number of hosts reached 10,000, congestion on the ARPANET caused by the limited-capacity leased telephone lines was becoming complicated. To trim down the traffic load on the ARPANET, a network run by the National Science Foundation, called NSFnet, merged with another NSF network, called CSNet, and with BITNET to compose one network that could carry much of the network traffic. As the civilian network became increasingly commercial, budget limitations impelled the U.S. government's departure from participation in the Internet's structure. In turn, private telecoms companies entered the picture (Cerf, 2008; Langford, 2000). The civilian network's use widened as a consequence of the proliferation of computer networks, and became more varied. Grassroots networks were established by university students. Merit Network, Inc., IBM, Sprint, and the State of Michigan were contracted to upgrade and operate the main NSFnet backbone.[33] By the late 1980s, many other TCP/IP networks had merged or established interconnections (Schneider and Evans, 2007).

In 1988, the NSFnet backbone was upgraded to DS-1 (1.544 Mbps) links, which was able to handle more than 75 million packets a day. This innovation immediately yielded further expansion of the Internet. The number of Internet hosts broke to 100,000.[34] The NSFnet began to encompass many other lower-level networks such as those developed by academic institutions. Gradually, the Internet as we know it today, a maze of interconnected networks came about (Spinello, 2000). Canada (CA), Denmark (DK), France (FR), Iceland (IS), Norway (NO) and Sweden (SE) connected to NSFnet.[35] The first transatlantic fiber-optic cable was installed, using glass fibers so transparent that repeaters (to regenerate and recondition the signal) were needed about 40 miles apart. Linking North

America and France, the 3,148-mile shark-proof cable was capable of handling 40,000 telephone calls simultaneously.[36] The same year, Jarkko Oikarinen wrote a communications program that extended the capabilities of the Talk program for his employer, the University of Oulu in Finland. He called his multiuser program *Internet Relay Chat (IRC)*. By 1991, IRC was running on more than 100 servers globally. IRC's popularity grew among scientists and academicians for conducting open discussions about theories, experiments and innovation (Schneider & Evans, 2007).

In 1989, number of hosts reached 159,000.[37] Australia (AU), Germany (DE), Israel (IL), Italy (IT), Japan (JP), Mexico (MX), Netherlands (NL), New Zealand (NZ), Puerto Rico (PR), and the United Kingdom (UK) connected to NSFnet.[38] William Wulf proposed the idea of a *collaboratory* which argued for the creation of tools to allow linked computers to be used as a rich environment for computer-based collaboration. The term merged "collaboration" and "laboratory" to describe a "center without walls, in which the nation's researchers can perform their research without regard to geographical location--interacting with colleagues, accessing instrumentation, sharing data and computational resources, and accessing information in digital libraries" (Kouzes, Myers, & Wulf, 1996).[39] This idea was certainly apt for the evolving technology, in line with the *raison d'être* that drove the founding architects of the Net and one that continues to prevail throughout the history of the Internet to date.

Also in 1989, Englishman Tim Berners-Lee, a researcher at the Organisation Europeenne pour la Recherche Nucleaire (CERN) in Geneva, proposed the idea of an international system of protocols: Building a distributed hypermedia server which would allow Netusers to prepare electronic documents that are composites of, or pointers to, many different files of potentially different types, scattered across the world. Berners-Lee called it the World Wide Web (WWW). He wrote the first WWW client (a browser-editor running under NeXTStep) and most of the communications software, defining URLs (Uniform Resource Locator, webpage address), HTTP (Hypertext Transfer Protocol between a server and clients) and HTML (interactive HyperText Markup Language).[40] His hypermedia software program enabled people to access, link and create communications in a single global web of information. The web was superimposed on the Internet and incorporated its protocols. The web thus marked the coming together of three different strands of innovation: Personal computing, networking, and connective software (Curran & Seaton, 2009).[41] Using hyperlinks embedded in hypertext, Netusers acting as producers of information link up files containing text, sound and graphics to create webpages. The sources of information linked in this way can be located on any computer that is also part of the web. Each information source may itself be linked to an indefinite number of webpages. Hypertext and hyperlinks allow Netusers acting as receivers of information to wander from one source of information to another effortlessly, deciding for themselves which information they wish to have transferred to their browser and which link they want to explore or to skip (Slevin, 2000).[42] Netusers could also index the data they possess and search for further data.

THE MASSIVE EXPANSION

By the late 1980s, a significant number of people (mostly professionals) were using email but the Internet was not in the public eye. I was a student at Oxford University at that time and can testify that using the Internet was a most frustrating experience. Most websites were not accessible. Navigating between sites was anything but seamless. It was easier to retrieve information from the library in the good, old-fashioned way.

But things were soon about to change. During the 1990s we witnessed a massive expansion of the Net. The Internet's accessibility, its multi-application and its decentralized nature were instrumental in this rapid growth. Business as well personal computers with different operating systems could join the universal network. The Internet became a global phenomenon, more countries and people joined and ground-breaking minds expanded the horizons of the platform with new, imaginative innovations. In 1990, the ARPANET project was officially over when it handed over control of the public Internet backbone to the National Science Foundation (Curran & Seaton, 2009; Slevin, 2000). In 1991, the Internet Society was formed and Croatia (HR), Hong Kong (HK), Hungary (HU), Poland (PL), Portugal (PT), Singapore (SG), South Africa (ZA), Taiwan (TW) and Tunisia (TN) joined the NSFnet network whose backbone was upgraded to DS-3 (44.736 Mbps) as the traffic passed to 1 trillion bytes and 10 billion packets per month. That year, 1991, saw another milestone as the popular encryption program PGP (Pretty Good Privacy) was released by Philip Zimmerman (1996).[43] Unfortunately, PGP presents a technological-ethical challenge with significant social implications as it is also used by Net abusers. As PGP is freely available, powerful tool, it is used by criminals and radicals who wish to hide their Net identity in order to advance anti-social behavior. In other words, encryption is a double-sword crypto-assisted anonymity tool: It may enhance your privacy and anonymity but it might also undermine your own security.

In ethical terms, there is a conflict between anonymity, on the one hand, and trust and accountability on the other hand. Indeed, anonymity undermines accountability on the Internet: If Netusers can hide their identity and be entirely sure that no one knows they are the agent of mischief, this might be an incentive for some people to adopt norms and codes of behavior that they would otherwise be deterred to adopt.[44] The Internet opened new horizons for criminals and terrorists.

In 1992, the number of Internet hosts broke to 1 million with almost 50 web pages.[45] In 1993,

there were 623 Websites in the world.[46] The United Nations came on-line and the NSFnet expanded internationally as Bulgaria (BG), Costa Rica (CR), Egypt (EG), Fiji (FJ), Ghana (GH), Guam (GU), Indonesia (ID), Kazakhstan (KZ), Kenya (KE), Liechtenstein (LI), Peru (PE), Romania (RO), Russian Federation (RU), Turkey (TR), Ukraine (UA), UAE (AE), and US Virgin Islands (VI) joined the network. The World Wide Web proliferated at a 341,634% annual growth rate of service traffic.[47] By the end of 1993, there were 2.1 million hosts.[48] The phenomenal growth and success of the Internet were the result of technological creativity, flexibility and decentralization as well as healthy curiosity of people who wanted to be part of the scene.

In 1994, Cerf argued (1995) that the "Internet has gone from near-invisibility to near-ubiquity." The growth of the Internet, its expanding international character, and awareness to its effective features brought more and more business to believe in the innovation and to invest in it. Shopping malls arrived on the Internet. First Virtual, the first cyberbank, opened up for business. Two Stanford PhD students, Jerry Yang and David Filo, started out a website which they called "Jerry and David's Guide to the World Wide Web." This guide swiftly expanded and later changed its name to one word, Yahoo![49] More countries joined the network, including Algeria (DZ), Armenia (AM), Bermuda (BM), Burkina Faso (BF), China (CN), Colombia (CO), Jamaica (JM), Jordan (JO), Lebanon (LB), Lithuania (LT), Macao (MO), Morocco (MA), New Caledonia (NC), Nicaragua (NI), Niger (NE), Panama (PA), Philippines (PH), Senegal (SN), Sri Lanka (LK), Swaziland (SZ), Uruguay (UY), and Uzbekistan (UZ). The number of Internet hosts increased to 3 million. This necessitated technological accommodation and, indeed, the same year, the NSFnet backbone was upgraded to OC-3 (155mbps) links and the volume of traffic increased to 10 trillion bytes per month. To navigate between the growing numbers of sites, the first version of the popular Netscape web browser was released by Mosaic Communications Corporation.[50] Mosaic made using the Internet as easy as pointing a mouse and clicking on icons and words (Hafner & Lyon, 1998). By then, the birth pangs of the global network were over and information retrieval became efficient and effective.

In 1995, major carriers such as British Telecom, France Telecom, Deutsche Telekom, Swedish Telecom, Norwegian Telecom, and Finnish Telecom, among many others, announced Internet services. An estimated 300 service providers were in operation, ranging from very small resellers to large telecom carriers. More than 30,000 websites were in operation and the number was doubling every two months (Cerf, 1995). The growing importance of commercial traffic and commercial networks was discussed at a series of conferences initiated by the National Science Foundation on the commercialization and privatization of the Internet. The NSF first awarded a contract to Merit Network, Inc., in partnership with IBM and MCI Communication Corp., to manage and modernize the Internet backbone. Then the NSF awarded three additional contracts: One to Network Solutions, allowing them to assign Internet addresses; second to AT&T to maintain Internet directory and database services; third to General Atomics to maintain the provision of information services to Netusers. In 1995, the NSFnet was shut down completely and the American core Internet backbone was privatized (Curran & Seaton, 2009).

The result was that the number of hosts more than doubled in one year, reaching 6.6 million.[51] The mid-1990s were the years when the Internet established itself as the focal point for communication, information and business. A number of Net related companies went public, with Netscape leading the pack with the 3rd largest ever NASDAQ IPO share value.[52] At the same time, many people began creating their own personal Web areas. Homepages and bookmarks were introduced to allow Netusers (about 16 million)[53] to organize their personal documents and to keep track of useful information. The Internet was growing strong in a rapid pace, attracting more and more people who grew to use it for their daily life: Finding information, research, business, commerce, entertainment, travel and essentially any need. For each and every need there came the entrepreneur who seized the opportunity and opened a website addressing the need.

In 1996, the number of Netusers more than doubled, from 16 million in 1995 to 36 million.[54] From the mid-1990s, the development of the Internet took a new turn as a growing number of large and medium-sized organizations started running the TCP/IP protocols on their internal organizational communication networks, called "intranets." For security purposes, intranets shielded themselves from the outside world by firewalls. These protection systems often allow for the exchange of information with the Internet via specified "gateways". These private networks are called "extranets" and allow organizations to exchange data with each other. By 1997, the market for intranets and extranets was growing annually at a rate of 40 per cent worldwide (Slevin, 2000). The number of Netusers estimated to be 70 million by the end of the year.[55]

At that time, the number of hosts was about 10 million with an untold number of links between them.[56] Finding information on the web became, yet again, a tricky issue but for different reasons. Connectivity was no longer the issue; rather, navigating and finding the information you needed in the growing maze was difficult. Addressing this challenge, two Stanford graduate students, Larry Page and Sergey Brin, started to work on a search engine which they called BackRub, as it was designed to analyze a 'back link' on the Web. Later they renamed their search engine Google, after googol, the term for the numeral 1 followed by 100 zeroes. They released the first version of Google on the Stanford Website in August 1996 (Battelle, 2005).[57]

In 1997, the Fiber Optic Link Around the Globe (FLAG) became the longest single-cable network in the world, providing infrastructure for the next generation of Internet applications. The 17,500-mile cable began in England and ran through the Strait of Gibraltar to Palermo, Sicily, before crossing the Mediterranean to Egypt. It then went overland to the FLAG operations center in Dubai, United Arab Emirates, crossing to the Indian Ocean, Bay of Bengal, and Andaman Sea; through Thailand; and across the South China Sea to Hong Kong and Japan.[58] With this infrastructure, that year alone some fifty additional country domains were registered. The Internet became truly international and the number of Internet hosts broke to 16 million.[59] The number of host computers grew to more than 36.7 million in mid-1998 while the number of websites had grown to 1.3 million. The number of sites was doubling every few months (Jenkins, 2001).

By 1998, there were approximately 150 million Netusers in more than 60 countries, representing about 2.5 percent of the world's population. The vast majority, or 130 million of those users, was located in the 15 most industrialized countries. Thus, despite its dramatic growth, large disparities in Internet access and usage persisted. A more accurate examination of the late-90's Internet usage reveals a user rate of 6.5 percent in a small number of high-usage nations and only a 0.5 percent usage rate in the remaining 200 countries (Langford, 2000; Spinello, 2000; Paré, 2005). There were clear differences between developed and developing countries. There still are.

The same year, 1998, the Internet Corporation for Assigned Names and Numbers (ICANN, 2010) was established. It is a not-for-profit public-benefit corporation with participants from across the world dedicated to keeping the Internet secure, stable and interoperable. ICANN promotes competition and develops policy on the Internet's unique identifiers. It does not control Internet content, cannot stop spam, and it does not deal with access to the global network. But through its coordination role of the Internet's naming system, it does have an important impact on the expansion and evolution of the Internet.[60] ICANN has secured long-term commitments of funding from registries and registrars to support its Internet-coordination activities, including the performance of the IANA functions which came under its control.

Large corporations became more aware of the massive potential of the Internet. America Online (AOL), Microsoft, Sun Microsystems, Inktomi, Yahoo! and Cisco caught the attention of Wall Street valuations. AOL alone had seen its stock rise 50,000 percent (McCracken, 2010). In 1998, AOL acquired Netscape Communications Corporation for a stock transaction valued at $4.2 billion. Microsoft bought Hotmail for $400 million. In 1999, online retailers reported 5.3 billion sale.[61]

By December 1999, the total number of Netusers worldwide was estimated to be 248 million.[62] For the fourth year running, the number of Netusers was growing in an extraordinary pace, doubling from one year to another. The United States, Western Europe and affluent parts of Asia produced much of the content of the web, while the rest of the world continued to contribute very little (Curran and Seaton, 2009). In 2000, the USA produced almost two-thirds of the top thousand most visited websites. It accounted for 83% of the total pageviews of Netusers. Less than 10% of the world speaks English as their first language, but English was becoming intelligible to a growing number of people, and has begun to assume the function once occupied by Latin in medieval Europe. In the late 1990, an estimated 85% of the web was written in English (Curran & Seaton, 2009). This picture, however, was rapidly changing.

In 2000, there were 361 million Netusers and the ten millionth domain name was registered.[63] The number of websites exceeded 50 million with a growing number of Internet Service Providers (ISPs) (Jenkins, 2001). BBC News Online (Postel, 1998) reported that 50 percent of the U.S. population had home Internet access. In Europe as a whole (despite high distribution in Scandinavia, Britain, and elsewhere) the proportion was as low as 4 percent, and only 3 percent in Russia. In China the figure was not much above 1 percent, and in Africa it was 0.016 percent (Schuler & Day, 2004). Subsequently, these figures have grown, in some cases dramatically, but large disparities still exist.

Not only legitimate businesses realized the potential of the Internet. Criminals were also quick to abuse the Internet for profit. On June 22, 2001, the European Council finalized its international *Convention on Cybercrime* and adopted it on November 9, 2001.[64] This was the first treaty addressing criminal offenses committed over the Internet. The same year, Firewall Enhancement Protocol (FEP) was proposed, and Jimmy Wales and Larry Sanger launched "Wikipedia," the web based free encyclopedia. It is a collaborative, multilingual project supported by the non-profit Wikimedia Foundation. Its 17 million articles (over 3.3 million in English) have been written by volunteers around the world, and almost all of its articles can be edited by anyone with access to the site.[65] Wikipedia became the largest and most popular general reference resource on the Internet.

The same year, 2001, there were 513 million Netusers and English ceased to be the language of the majority of users. English fell to a 45 percent share (Kleinrock, 2008). The following year, broadband Netusers exceeded the number of dial-up users in the United States (Kleinrock, 2008). This had massive implications. With more broadband, gigantic storage capacities, wireless access, and advanced visual displays the technology facilitated peer-to-peer file sharing networks, photo and video generation and sharing, and the construction of social networking mechanisms where people can report and upload any data they may wish to share.

SOCIAL NETWORKING

The study of Internet social networking is of much need in the field of technoethics. Most people use social networks to socialize, exchange information and ideas; some, however, abuse social networks to advance anti-social, violent purposes like terrorism and child pornography. In July 2003 Myspace was founded by Tom Anderson and Chris DeWolfe. MySpace allows members to create unique personal profiles online in order to find and communicate with old and new friends. The services offered by MySpace include any MySpace branded URL (the "MySpace Website"), the MySpace instant messaging service, the MySpace application developer service and other features..[66] MySpace became the most popular social networking site in the United States. In June 2006, there were more than 100 million MySpace users. It is estimated that every month over ten million American teens log on to MySpace. However, in 2008 Myspace was overtaken internationally by its main competitor, Facebook.[67] Facebook.com was founded on February 4, 2004 by Mark Zuckerberg, Eduardo Saverin, Dustin Moskovitz and Chris Hughes (Carlson, 2010). Facebook started as a social network for American universities but in September 2006 the network was extended beyond educational institutions to anyone with a registered email address. The site remains free to join, and makes a profit through advertising revenue.

In addition to the abovementioned features, as of 2007, Facebook users can give gifts to friends, post free classified advertisements and even develop their own applications - graffiti and Scrabble are particularly popular (Phillips, 2007). On July 22, 2010, the 500 millionth signed account on the largest social network (22 percent of all Netusers). Facebook users spend more than 500 billion minutes a month on the site, share more than 25 billion pieces of content each month (including news stories, blog posts and photos), and each of them, on average, creates 70 pieces of content a month (Rosen, 2010; Arthur & Kiss, 2010).[68] Three years after the founding of Facebook, in 2007, Microsoft made $15bn bid to buy the company but Zuckerberg declined (Lowensohn, 2010). He did not want to lose control over his creation. In 2010, Facebook is estimated to worth $52.1 billion.[69]

In 2005, there were 1,018 million Netusers.[70] That year, three former employees of Paypal, Chad Hurley, Steve Chen and Jawed Karim created a video file sharing website called "YouTube." The official debut was December 15, 2005. On October 9, 2006, Google bought YouTube for $1.65 billion (Lidsky, 2010).

The same year, in 2006, the free social networking site Twitter was started by Jack Dorsey. Essentially, Twitter combines Short Code Messaging, SMS with a way to create social groups. One can send information to one's followers and receive information from individuals or organizations one has chosen to follow (Malik, 2009).[71] There are more than 100 million registered Twitter users (Rosen, 2010).

The number of Netusers continued to grow from 1,319 million in 2007, to 1,574 million in 2008, to 1,802 million in 2009, to 1,971 million in September 2010.[72] The most recent figure accounts for some 29% of the world population. As of December 2010, the Indexed Web contains at least 2.69 billion pages.[73] Table 1 shows the world Internet usage statistics and population statistics.

CONCLUSION

The Internet and its architecture have grown in evolutionary fashion from modest beginnings, rather than from a Grand Plan (Carpenter, 1996).

The ingenuity of the Internet as it was developed in the 1960s by the ARPA scientists lies in the packet switching technology. Until ARPANET was built, most communications experts claimed that packet switching would never work (Roberts, 1999).[75] In 1965, when the first network experiment took place, and for the first time packets were used to communicate between computers, the scientists did not imagine the multiple usages of this technology on society. Kleinrock, the inventor of packet switching, explicitly wrote that he did not foresee the powerful community side of the Internet and its impact on every aspect of society (Kleinrock, 2008). The Net diffusiveness and its focus on flexibility, decentralization and collaboration brought about the Internet as we know it today. In the initial stages, the Internet was promoted and funded, but not designed, by the U.S. government. Allowing the original research and education network to evolve freely and openly without any restrictions, selecting TCP/IP for the NSFnet and other backbone networks, and subsequently privatizing the NSFNET backbone, were the most critical decisions for the Internet's evolution.

The Internet's design was unprecedented because it was conceived as a decentralized, open and neutral network of networks. The open architecture of the Internet allows free access to protocols from anywhere in the world and is capable to accept almost any kind of computer or network

Table 1. Internet usage statistics74 world internet users and population stats

World Regions	Population (2010)	Internet Users Latest Data	Penetration (% Population)
Africa	1,013,779,050	110,931,700	10.9%
Asia	3,834,792,852	825,094,396	21.5%
Europe	813,319,511	475,069,448	58.4%
Middle East	212,336,924	63,240,946	29.8%
North America	344,124,450	266,224,500	77.4%
Latin America/Caribbean	592,556,972	204,689,836	34.5%
Oceania / Australia	34,700,201	21,263,990	61.3%
WORLD TOTAL	6,845,609,960	1,966,514,816	28.7%

to join in. The choice of any individual network technology is not dictated by particular network architecture but rather could be selected freely by a provider and made to interwork with other networks through a meta-level "Internetworking Architecture."[76] This open architecture encourages the development of more net applications. And the Internet is neutral between different applications of text, audio and video. This allowed new and better applications (like email, the World Wide Web, and peer-to-peer technology) to evolve and replace the old (Goldsmith and Wu, 2006).

There are inherent tensions between the various technological tools: Those designed to enhance one's privacy may harm security and *vice versa*. They can be put for good use (filtering child pornography) and might cause abuse (encrypting child porn images). Encryption promotes privacy and anonymity on the Net but, at the same time, anonymity does not contribute to cultivating a sense of Net responsibility or trust.

At the beginning of the 21st Century, the Internet embraces some 300,000 networks stretching across the planet. Its communications travel on optical fibers, cable television lines, and radio waves as well as telephone lines. The traffic continues to grow in a rapid pace. Mobile phones and other communication devices are joining computers in the vast network. Some data are now being tagged in ways that allow websites to interact.[77] Today, the growth of cloud computing is providing powerful new ways to easily build and support new software. Because companies and individuals can "rent" computing power and storage from services like the Amazon Elastic Compute Cloud, it is much easier and faster for someone with a good idea to turn it into an online service. This is leading to an explosion in new uses for the Internet and a corresponding explosion in the amount of traffic flowing across the Internet (Nelson, 2010). The result is the most impressive web of communications in the history of humanity. Millions of people around the globe cannot describe their lives and function as they wish without the Internet.

REFERENCES

Abbate, J. (2000). *Inventing the Internet*. Cambridge, MA: MIT Press.

ARPA (DARPA). (2004). *Velocity guide*. Retrieved from http://www.velocityguide.com/internet-history/arpa-darpa.html

Arthur, C., & Kiss, J. (2010). Facebook: The club with 500m 'friends'. *The Guardian*.

Baran, P. (1964). *On distributed communications series*. Santa Monica, CA: RAND.

Battelle, J. (2005). *The birth of Google*. Retrieved from http://www.wired.com/wired/archive/13.08/battelle.html?pg=2&topic=battelle&topic_set=

Beckett, D. (2000). Internet technology. In Langford, D. (Ed.), *Internet ethics*. New York, NY: St. Martin's Press.

Broadbrand Suppliers. (2010). *Recap the Internet history*. Retrieved from http://www.broadbandsuppliers.co.uk/uk-isp/recap-the-history-of-Internet/

Carlson, N. (2010, March 5). *At last-the full story of how Facebook was founded*. Retrieved from http://www.businessinsider.com/how-facebook-was-founded-2010-3#we-can-talk-about-that-after-i-get-all-the-basic-functionality-up-tomorrow-night-1

Carpenter, B. (1996). *The architectural principles of the Internet*. Retrieved from http://www.ietf.org/rfc/rfc1958.txt

Cerf, V. G. (1995). *Computer networking: Global infrastructure for the 21st century*. Retrieved from http://www.cs.washington.edu/homes/lazowska/cra/networks.html

Cerf, V. G. (1998). *I remember IANA*. Retrieved from http://www.rfc-editor.org/rfc/rfc2468.txt

Cerf, V. G. (2008). The scope of Internet governance. In Doria, A., & Kleinwachter, W. (Eds.), *Internet governance forum (IGF): The first two years* (pp. 51–56). Geneva, Switzerland: IGF Office.

Cerf, V. G., & Kahn, R. (1974). A protocol for packet network interconnection. *IEEE Transactions on Communications, 22*(5). doi:10.1109/TCOM.1974.1092259

Chick Net. (n. d.). *Pioneers of the net*. Retrieved from http://www.chick.net/wizards/pioneers.html

Clark, D., Field, F., & Richards, M. (2010). *Computer networks and the Internet: A brief history of predicting their future*. Retrieved from http://groups.csail.mit.edu/ana/People/DDC/Working%20Papers.html

Cohen, D. (1998). *Remembering Jonathan B. Postel*. Retrieved from http://www.postel.org/remembrances/cohen-story.html

Computer Professionals for Social Responsibility. (2005). *CPSR history*. Retrieved from http://cpsr.org/about/history/

Conn, K. (2002). *The Internet and the law: What educators need to know*. Alexandria, VA: Association for Supervision and Curriculum Development.

Council of Europe. (n. d.). *Convention on cybercrime: Preamble*. Retrieved from http://cis-sacp.government.bg/sacp/CIS/content_en/law/item06.htm

Curran, J., & Seaton, J. (2009). *Power without responsibility*. London, UK: Routledge.

Dorsey, J., & Enge, E. (2007). *Jack Dorsey and Eric Enge talk about Twitter*. Retrieved from http://www.stonetemple.com/articles/interview-jack-dorsey.shtml

Dubow, C. (2005). The Internet: An overview. In Mur, C. (Ed.), *Does the Internet benefit society?* Farmington Hills, MI: Greenhaven.

Floridi, L. (2009). The information society and its philosophy: Introduction to the special issue on the philosophy of information, its nature and future developments. *The Information Society, 25*(3). doi:10.1080/01972240902848583

Floridi, L. (2010). *Information – a very short introduction*. Oxford, UK: Oxford University Press.

Gillies, J., & Cailliau, R. (2000). *How the Web was born: The story of the World Wide Web*. Oxford, UK: Oxford University Press.

Goldsmith, J., & Wu, T. (2006). *Who controls the Internet? Illusions of a borderless world*. New York, NY: Oxford University Press.

Griffiths, R. T. (2002). *Search engines*. Retrieved from http://www.let.leidenuniv.nl/history/ivh/chap4.htm

Gromov, G. R. (1996). *The roads and crossroads of Internet history*. Retrieved from http://www.netvalley.com/intvalnext.html

Il'obboo' Zakon. R. (2010). *H'obbes' Internet timeline 10.1*. Retrieved from http://www.zakon.org/robert/Internet/timeline/

Hafner, A., & Lyon, M. (1998). *Where wizards stay up late: The origins of the Internet*. New York, NY: Simon & Schuster.

Hauben, M., & Hauben, R. (1997). *Netizens: On the history and impact of UseNet and the Internet*. Washington, DC: IEEE Computer Society.

Hauben, R. (1998). *Lessons from the early MsgGroup mailing list as a foundation for identifying the principles for Internet governance: Abstract*. Retrieved from http://www.columbia.edu/~rh120/other/talk_governance.txt

History of Computing Project. (2001). *Donald W. Davies CBE, FRS*. Retrieved from http://www.thocp.net/biographies/davies_donald.htm

ibiblio. (n. d.). *Internet pioneers*. Retrieved from http://www.ibiblio.org/pioneers/

ICANN. (2010). *About ICANN*. Retrieved from http://www.icann.org/en/about/

Internet World Stats. (2010). *Internet growth statistics*. Retrieved from http://www.Internetworldstats.com/emarketing.htm

JANET. (n. d.). *About JANET.* Retrieved from http://www.ja.net/company/about.html

Jenkins, P. (2001). *Beyond tolerance: Child pornography on the Internet.* New York, NY: New York University Press.

Kirkpatrick, D. (2010). *The Facebook effect.* New York, NY: Simon and Schuster.

Kleinrock, L. (1961). *Information flow in large communication nets.* Unpublished doctoral dissertation, MIT, Cambridge, MA.

Kleinrock, L. (1973). *Communication nets: Stochastic message flow and design.* New York, NY: Dover Publications.

Kleinrock, L. (1996). *Personal history/biography: The birth of the Internet.* Retrieved from http://www.lk.cs.ucla.edu/LK/Inet/birth.html

Kleinrock, L. (2002). Creating a mathematical theory of computer networks. *Operations Research, 50*(1), 125–131. doi:10.1287/opre.50.1.125.17772

Kleinrock, L. (2008). History of the Internet and its flexible future. *IEEE Wireless Communications, 15*(1), 8–18. doi:10.1109/MWC.2008.4454699

Kleinrock, L. (2010). An early history of the Internet. *IEEE Communications Magazine, 48*(8), 26–36. doi:10.1109/MCOM.2010.5534584

Kouzes, R. T., Myers, J. D., & Wulf, W. (1996). Collaboratories: Doing science on the Internet. *Computer, 29*(8), 40–46. doi:10.1109/2.532044

Langford, D. (2000). *Internet ethics.* New York, NY: St. Martin's Press.

Leiner, B. M., Cerf, V. D., Clark, D. D., Kahn, R. E., Kleinrock, L., Lynch, D. C. et al. (2009). A brief history of the Internet. *ACM SIGCOMM Communications Review, 39*(5).

Leiner, B. M., Cerf, V. G., Clark, D. D., Kahn, R. E., Kleinrock, L., & Lynch, D. C. (1997). The past and future history of the Internet. *Communications of the ACM, 40*(2), 103. doi:10.1145/253671.253741

Levmore, S. (2010). The Internet's anonymity problem. In Levmore, S., & Nussbaum, M. C. (Eds.), *The offensive Internet: Speech, privacy, and reputation* (pp. 50–67). Cambridge, MA: Harvard University Press.

Lidsky, D. (2010, February 1). *The brief but impactful history of YouTube.* Retrieved from http://www.fastcompany.com/magazine/142/it-had-to-be-you.html

Living Internet. (2000). *DARPA/ARPA Defense: Advanced research project agency.* Retrieved from http://www.livingInternet.com/i/ii_darpa.htm

Lowensohn, J. (2010, December 10). *Microsoft made $15bn bid to buy Facebook, says exec.* Retrieved from http://www.zdnet.co.uk/news/mergers-and-acquisitions/2010/12/10/microsoft-made-15bn-bid-to-buy-facebook-says-exec-40091124/

Luppicini, R. (2010). *Technoethics and the evolving knowledge society: Ethical issues in technological design, research, development and innovation.* Hershey, PA: IGI Global.

Malik, O. (2009). *A brief history of Twitter.* Retrieved from http://gigaom.com/2009/02/01/a-brief-history-of-twitter/

McCracken, H. (2010, May 24). *A history of AOL, as told in its own old press releases.* Retrieved from http://technologizer.com/2010/05/24/aol-anniversary/

McFadden, C., & Fulginiti, M. (2008, March 24). Searching for justice: Online harassment. *ABC News Transcript.*

Mosaic Communications Corporation. (1994). *Who are we: Our mission.* Retrieved from http://home.mcom.com/MCOM/mcom_docs/backgrounder_docs/mission.html

National Academy of Engineering. (2011). *Timeline.* Retrieved from http://www.greatachievements.org/Default.aspx?id=2984

National Academy of Engineering. (n. d.). *Internet history*. Retrieved from http://www.greatachievements.org/?id=3747

Nelson, M. R. (2010). A response to responsibility of and trust in ISPs by Raphael Cohen-Almagor. *Knowledge, Technology and Policy, 23*(3).

Paré, D. (2005). The digital divide: Why the 'the' is misleading. In Klang, M., & Murray, A. (Eds.), *Human rights in the digital age*. London, UK: GlassHouse.

Phillips, S. (2007, July 25). *A brief history of Facebook*. Retrieved from http://www.guardian.co.uk/technology/2007/jul/25/media.newmedia

Postel, J. (1998, October 19). *'God of the Internet' is dead*. Retrieved from http://news.bbc.co.uk/1/hi/sci/tech/196487.stm

RAND Corporation. (1994). *Paul Baran and the origins of the Internet*. Retrieved from http://www.rand.org/about/history/baran.html

Reuters. (2010). *Facebook is worth $52 billion, and that's not a good thing*. Retrieved from http://blogs.reuters.com/mediafile/2010/12/14/facebook-is-worth-52-billion-and-thats-not-a-good-thing/

Roberts, L. G. (1999). *Internet chronology*. Retrieved from http://www.ziplink.net/users/lroberts/InternetChronology.html

Rosen, J. (2010, July 19). *The Web means the end of forgetting*. Retrieved from http://www.nytimes.com/2010/07/25/magazine/25privacy-t2.html

Rubinstein, I. (2009, June 4-6). *Anonymity reconsidered*. Paper presented at the Second Annual Berkeley-GW Privacy Law Scholars Conference, Berkeley, CA.

Ryan, J. (2011). *The essence of the 'net': A history of the protocols that hold the network together*. Retrieved from http://arstechnica.com/tech-policy/news/2011/03/the-essence-of-the-net.ars/

Salus, P. H. (1995). *Casting the net: From ARPANET to Internet and beyond*. Reading, MA: Addison-Wesley.

Schneider, G. P., & Evans, J. (2007). *New perspectives on the Internet: Comprehensive* (6th ed.). Boston, MA: Thomson.

Schuler, D., & Day, P. (2004). *Shaping the network society*. Cambridge, MA: MIT Press.

Slevin, J. (2000). *The Internet and society*. Oxford, UK: Polity Press.

Spinello, R. A. (2000). *Cyberethics: Morality and law in cyberspace*. Sudbury, MA: Jones and Bartlett.

Strickland, J. (2010). *How ARPANET works*. Retrieved from http://www.howstuffworks.com/arpanet.htm/printable

The London Speaker Bureau. (n. d.). *Sir Tim Berners-Lee*. Retrieved from http://www.londonspeakerbureau.co.uk/sir_tim_berners_lee.aspx

University of Surrey. (2000). *UNIX introduction*. Retrieved from http://www.ee.surrey.ac.uk/Teaching/Unix/unixintro.html

UseNet. (2009). *What is UseNet? - User network*. Retrieved from http://www.usenet.com/usenet.html

W3C. (2003). *Who's who at the World Wide Web consortium*. Retrieved from http://www.w3.org/People/all#timbl

White, A. E. (2006). *Virtually obscene: The case for an uncensored Internet*. Jefferson, NC: McFarland & Company.

Wikipedia. (n. d.). *Wikipedia*. Retrieved from http://en.wikipedia.org/wiki/Wikipedia

Yahoo. Media Relations. (2005). *The history of Yahoo! -How it all started*. Retrieved from http://docs.yahoo.com/info/misc/history.html

Yahoo. (n. d.). *Birth of the Internet – timeline*. Retrieved from http://smithsonian.yahoo.com/timeline.html

Zimmerman, P. R. (1996). *The official PGP user's guide*. Boston, MA: MIT Press.

ENDNOTES

[1]   During its lifetime, this agency has used two acronyms, ARPA and DAPRA, Defense Advanced Research Projects Agency.

[2]   "ARPA (DARPA)," *Velocity Guide*, http://www.velocityguide.com/Internet-history/arpa-darpa.html

[3]   See also "DARPA / ARPA -- Defense / Advanced Research Project Agency," *livingInternet.com*, http://www.livingInternet.com/i/ii_darpa.htm; "Internet Pioneers," *ibiblio.org,* http://www.ibiblio.org/pioneers/

[4]   Leiner, Cerf, Clark *et al.*, "A Brief History of the Internet," *The Internet Society*, http://www.isoc.org/Internet/history/brief.shtml

[5]   Leonard Kleinrock, personal communication (July 19, 2010).

[6]   David D. Clark, personal communication (July 19, 2010). See also Kleinrock (August 2010: 26-36).

[7]   Timeline, http://www.greatachievements.org/Default.aspx?id=2984; "J.C.R. Licklider," *Velocity Guide*, http://www.velocityguide.com/Internet-history/jcr-licklider.html

[8]   http://www.computer.org/portal/web/csdl/doi/10.1109/AFIPS.1962.24

[9]   Baran, "On Distributed Communications Series," *RAND*, at http://www.rand.org/about/history/baran.list.html; Paul Baran and the Origins of the Internet, http://www.rand.org/about/history/baran.html; Slevin (2000: 29–30). See also http://www.rand.org/pubs/authors/b/baran_paul.html

[10]   Kleinrock (1961), http://www.lk.cs.ucla.edu/LK/Bib/REPORT/PhD/; Kleinrock (1973); Kleinrock (January-February 2002: 125-131). See also Leonard Kleinrock's Personal History/Biography, at http://www.lk.cs.ucla.edu/LK/Inet/birth.html

[11]   Donald W. Davies CBE, FRS, http://www.thocp.net/biographies/davies_donald.htm.

[12]   In 1949, two MIT professors, Richard Bolt and Leo Beranek, established a small acoustics consulting firm, and soon added a former student of Bolt's, Robert Newman. http://www.bbn.com/about/timeline/

[13]   Jonathan Strickland, "How ARPANET works," at http://www.howstuffworks.com/arpanet.htm/printable; Beckett (2000: 15).

[14]   See also Gillies and Cailliau (2000).

[15]   Steve Crocker from UCLA played a key role in establishing the request for comments in 1969. See http://tools.ietf.org/html/rfc1

[16]   See also Salus (1995).

[17]   Leonard Kleinrock, personal communication (July 19, 2010).

[18]   See also Danny Cohen, Remembering Jonathan B. Postel, http://www.postel.org/remembrances/cohen-story.html

[19]   "'God of the Internet' is dead" (October 19, 1998).

[20]   UNIX Introduction, http://www.ee.surrey.ac.uk/Teaching/Unix/unixintro.html

[21]   'Host' means computer that is connected to the network.

[22]   Recap the Internet history, at http://www.broadbandsuppliers.co.uk/uk-isp/recap-the-history-of-Internet/

[23]   The idea was originally introduced by Kahn in 1972 as part of the packet radio program.

[24]   See also White (2006: 13) and generally Abbate (2000).

[25]   http://www.bbn.com/about/timeline/; *pioneers of the net*, http://www.chick.net/wizards/pioneers.html. The first head of the state to send an email message, in 1976, was the Queen of England, Elizabeth II. See

Recap the Internet history, at http://www.
broadbandsuppliers.co.uk/uk-isp/recap-the-
history-of-Internet/

26 See also Timeline, http://www.greatachieve-
ments.org/Default.aspx?id=2984

27 What is Usenet? - User Network, http://
www.usenet.com/usenet.html

28 About JANET, http://www.ja.net/company/
about.html

29 See also Recap the Internet history, at http://
www.broadbandsuppliers.co.uk/uk-isp/
recap-the-history-of-Internet/.

30 Timeline, http://www.greatachievements.
org/Default.aspx?id=2984

31 CPSR History, http://cpsr.org/about/history/

32 Birth of the Internet – Timeline, http://
smithsonian.yahoo.com/timeline.html

33 A *network backbone* includes the long-
distance lines and supporting technology
that transports large amounts of data between
major network nodes.

34 Recap the Internet history, at http://www.
broadbandsuppliers.co.uk/uk-isp/recap-the-
history-of-Internet/

35 Robert H'obbes' Zakon, Hobbes' Internet
Timeline 10, *at* http://www.zakon.org/rob-
ert/Internet/timeline/

36 Timeline, http://www.greatachievements.
org/Default.aspx?id=2984

37 Robert H'obbes' Zakon, Hobbes' Internet
Timeline 10, at http://www.zakon.org/robert/
Internet/timeline/

38 *Ibid.,* at http://www.zakon.org/robert/Inter-
net/timeline/

39 See also Clark, Field and Richards (January
2010).

40 Who's Who at the World Wide Web Consortium..., at http://www.w3.org/People/
all#timbl; Sir Tim Berners-Lee, http://www.
londonspeakerbureau.co.uk/sir_tim_bern-
ers_lee.aspx

41 See also Hauben and Hauben (1997).

42 See also Gregory R. Gromov, "The Roads
and Crossroads

of Internet History," http://www.netvalley.
com/intvalnext.html.

43 See also Recap the Internet history, at
http://www.broadbandsuppliers.co.uk/uk-
isp/recap-the-history-of-Internet/; Robert
H'obbes' Zakon, Hobbes' Internet Timeline
10, at http://www.zakon.org/robert/Internet/
timeline/

44 For deliberation on anonymity, see Levmore
(2010: 50-67).

45 Recap the Internet history, at http://www.
broadbandsuppliers.co.uk/uk-isp/recap-the-
history-of-Internet/

46 Birth of the Internet – Timeline, http://
smithsonian.yahoo.com/timeline.html

47 Robert H'obbes' Zakon, Hobbes' Internet
Timeline 10, at http://www.zakon.org/robert/
Internet/timeline/

48 Internet Growth, http://www.internetworld-
stats.com/emarketing.htm

49 The History of Yahoo! - How It All Started...,
http://docs.yahoo.com/info/misc/history.
html

50 Mosaic Communications Corporation, Who
Are We, http://home.mcom.com/MCOM/
mcom_docs/backgrounder_docs/mission.
html; Recap the Internet history, at http://
www.broadbandsuppliers.co.uk/uk-isp/
recap-the-history-of-Internet/; Robert
H'obbes' Zakon, Hobbes' Internet Timeline
10, at http://www.zakon.org/robert/Internet/
timeline/

51 Internet Growth, http://www.internetworld-
stats.com/emarketing.htm

52 Robert H'obbes' Zakon, Hobbes' Internet
Timeline 10, at http://www.zakon.org/robert/
Internet/timeline/

53 INTERNET GROWTH STATISTICS, http://
www.Internetworldstats.com/emarketing.
htm

54 INTERNET GROWTH STATISTICS, http://
www.Internetworldstats.com/emarketing.
htm

55 INTERNET GROWTH STATISTICS, http://www.Internetworldstats.com/emarketing.htm

56 Internet Growth: Raw Data, http://www.internetworldstats.com/emarketing.htm

57 See also Griffiths, "Search Engines," http://www.let.leidenuniv.nl/history/ivh/chap4.htm

58 Timeline, http://www.greatachievements.org/Default.aspx?id=2984

59 Recap the Internet history, at http://www.broadbandsuppliers.co.uk/uk-isp/recap-the-history-of-Internet/; Robert H'obbes' Zakon, Hobbes' Internet Timeline 10, at http://www.zakon.org/robert/Internet/timeline/

60 About ICANN, http://www.icann.org/en/about/. In June 1999, at its Oslo meeting, Internet Engineering Task Force (IETF) signed an agreement with ICANN on the tasks that IANA would perform for the IETF.

61 Recap the Internet history, at http://www.broadbandsuppliers.co.uk/uk-isp/recap-the-history-of-Internet/;

62 INTERNET GROWTH STATISTICS, http://www.Internetworldstats.com/emarketing.htm

63 INTERNET GROWTH STATISTICS, http://www.Internetworldstats.com/emarketing.htm; Recap the Internet history, at http://www.broadbandsuppliers.co.uk/uk-isp/recap-the-history-of-Internet/

64 Council of Europe – *Convention on Cyber-crime* - http://cis-sacp.government.bg/sacp/CIS/content_en/law/item06.htm.

65 "Wikipedia," in *Wikipedia*, http://en.wikipedia.org/wiki/Wikipedia

66 http://www.myspace.com/index.cfm?fuseaction=misc.terms; McFadden and Fulginiti (March 24, 2008).

67 http://www.answers.com/topic/myspace

68 For further discussion, see Kirkpatrick (2010).

69 "Facebook is worth $52 billion, and that's not a good thing," *MediaFile* (December 13, 2010).

70 INTERNET GROWTH STATISTICS, http://www.Internetworldstats.com/emarketing.htm

71 See also "Jack Dorsey and Eric Enge talk about Twitter," *StoneTemple* (October 15, 2007).

72 INTERNET GROWTH STATISTICS, http://www.Internetworldstats.com/emarketing.htm

73 The size of the World Wide Web, http://www.worldwidewebsize.com/

74 http://www.internetworldstats.com/stats.htm (April 2, 2011).

75 Roberts wrote: "Packet switching was new and radical in the 1960's. In order to plan to spend millions of dollars and stake my reputation, I needed to understand that it would work. Without Kleinrock's work of Networks and Queuing Theory, I could never have taken such a radical step. All the communications community argued that it couldn't work. This book was critical to my standing up to them and betting that it would work." Quoted in Gillies and Cailliau (2000: 26).

76 Leiner, Cerf, Clark *et al.,* "A Brief History of the Internet," *The Internet Society*, http://www.isoc.org/Internet/history/brief.shtml

77 Internet History, http://www.greatachievements.org/?id=3747

This work was previously published in the International Journal of Technoethics, Volume 2, Issue 2, edited by Rocci Luppicini, pp. 46-65, copyright 2011 by IGI Publishing (an imprint of IGI Global).

Chapter 3
Laboring in Cyberspace:
A Lockean Theory of Property in Virtual Worlds

Marcus Schulzke
State University of New York at Albany, USA

ABSTRACT

This paper examines property relations in massively multiplayer online games (MMOGs) through the lens of John Locke's theory of property. It argues that Locke's understanding of the common must be modified to reflect the differences between the physical world that he dealt with and the virtual world that is now the site of property disputes. Once it is modified, Locke's theory provides grounds for recognizing player ownership of much of the intellectual material of virtual worlds, the goods players are responsible for creating, and the developer-created goods that players obtain through an exchange of labor or goods representing labor value.

INTRODUCTION

Virtual worlds, or massively multiplayer online games (MMOGs), are increasingly the site of property disputes between the developers who design the worlds and the players who occupy them. Although these are games, they can be very serious for players, many of whom spend twenty to thirty hours a week on average playing them

DOI: 10.4018/978-1-4666-2931-8.ch003

(Castronova, 2007). Some players are so heavily invested in MMOGs that they earn incomes from the sale of virtual goods and services. Ownership and trade have become so important to virtual worlds that many have their own economies comparable to those of real countries (Castronova, 2007). With much at stake for developers and players, it is unsurprising that virtual property has become a source of dispute. Virtual property strains the existing categories of ownership, introducing such difficult questions as the status of virtual theft (Arias, 2008; Brenner, 2008) and the right to tax virtual goods (Lederman, 2007; Seto, 2008, 2009).

The debate over virtual property fits into the interdisciplinary study of technoethics, a branch of scholarship that is concerned with exploring the moral dimensions of emerging technologies and their effects on society. Because these only exist in a digital form, they challenge existing moral categories used to assign ownership. Technoethics provides a way of reconsidering the concept of property and the moral and ethical implications of property in light of the way digital technologies have altered it.

One way of determining property rights is by using John Locke's labor theory of property, which holds that people become owners of un-claimed materials by adding their labor to them. Locke's theory is one of the most important in the debate over ownership of virtual property, as players claiming property often justify their right to own and sell property on the grounds that they have labored to create it or to procure it (Kayser, 2007). The problem is that developers can claim to own virtual goods on the same grounds; they initially create them and host the virtual worlds that serve as the setting for the players' labor. Most Lockean studies of virtual property have concluded that all or most virtual goods belong to the developers (Horowitz, 2007; Kennedy, 2008), giving players limited rights to do what they wish with virtual property. This paper will argue that this conclusion has much to do with a misapplica-tion of Locke's theory of the common. When this part of the theory is reconsidered to account for the differences between real and virtual worlds, Locke's theory provides strong grounds for grant-ing players ownership over many virtual goods.

Steinberg (2009) finds that three classes of virtual goods are at the center of the debate over virtual property: currency, items, and player ac-counts. This paper focuses on the ownership of the virtual common, of player accounts, and of the virtual worlds themselves. It shows that the first of these is owned by players collectively, the second by individual players, and the third by both the developers and the players. This paper will also argue that players can have strong claims to virtual money and items, especially when these are created by the players themselves, but that players only become owners of these things that developers have created when the property is gained through labor that is equivalent to the value of the good or by exchanging goods that represent their labor value. This only addresses property ownership as a moral issue. The laws regarding property ownership will have to be made by each country and will reflect existing property law in these jurisdictions. The legal question is one that must be resolved, but the purpose that the laws governing virtual property should serve is clarified by examining the extent to which there are moral justifications for ownership of virtual goods.

VIRTUAL PROPERTY RELATIONS

Virtual world developers differ in the extent to which they recognize user's claims of ownership of virtual goods. *World of Warcraft* represents one extreme of property relations. Blizzard, the company responsible for it, forbids players for selling their accounts or their items for real money or goods; only trade within the game world for other virtual goods is permitted (Lastowka, 2004). Most virtual worlds follow the same model, giv-ing players little control over the items they find or produce and forbidding or restricting the sale of avatars and items outside the virtual world (Lastowka, 2004). By contrast, *Second Life* al-lows players to create products in the game world, and to copyright them, patent them, and sell them (Steinberg, 2009). This has led to the rise of a diverse economy within that world, with users creating their own products for other players to buy, much as manufacturers do in the real world. *Second Life* acknowledge a very high level of user-control over property (Horowtiz, 2007), yet even it has experienced property disputes because developers have the power to exclude users from access to the virtual goods that they own (Glushko, 2007). This indicates that whatever level of player ownership virtual worlds acknowledge, property disputes are likely to arise because of the developer's gatekeeper role.

From a legal standpoint, developers have a strong claim to the ownership of the worlds and the goods they contain. EULAs usually state that the developers have ultimate control over the worlds and that they may seize player property if they wish (Caramore, 2008; Horowitz, 2007). For this reasons, players are at a disadvantage when seeking to assert their rights over property they have created or earned while playing MMOGs (Jankowich, 2005). However, the problem of whether players can make a legal claim to the ownership of virtual property is different from the problem of whether there are moral grounds for recognizing their ownership claims. Players may have a moral right to their virtual property, even if they cannot assert their property claims through the law. Thus far, most of the work done on virtual property has been done from a legal perspective, but it is essential to develop a moral basis for virtual property. This is especially true because there are often problems resolving legal disputes over virtual goods, even when the goods are regulated by EULAs (Glushko, 2007).

One of the most popular ways of theorizing property rights in virtual worlds is with Locke's labor theory of property (Horowitz, 2007; Steinberg, 2009; Westbrook, 2006). It is also one of the most influential theories of property in legal theory in general (Mossoff, 2002). Locke's theory will be discussed in detail and adapted to the virtual world in the following sections. For now, it can be stated in terms of three central claims. First, all people have the right to take control of things that are held in common, i.e., things that belong to all members of a community, provided they leave enough for others and refrain from taking more than they can use. Second, individuals own their

labor. Third, because individuals own their labor, any person can use their labor to take things that are common by adding their labor to them. Thus, the theory states that adding labor to something that is collectively owned confers property ownership because the labor adds value to the object.

Many studies applying Locke's theory to virtual worlds have concluded that the theory gives developers stronger property claims than players (Caramore, 2008; Horowitz, 2007; Kennedy, 2008). Kennedy says that "If Lockean theory grants property rights to game objects to anyone, it is to the developers who have invested their labour in creating the game in the first instance" (2008, p. 102). Horowitz (2007) argues that users could have a strong claim to virtual goods based on their labor, but that developers own the common from which players extract goods and that this pre-existing claim supersedes players' claims. Nevertheless, players also use Locke's theory to support their own property claims, arguing that they are entitled to the things that they have earned through labor, even when the EULA forbids players from owning virtual property (Glushko, 2007). This probably has much to do with the intuitive appeal of conferring ownership of things that have taken time and energy to acquire. Castronova expressed the players' perspective on labor perfectly: "If I spend thousands of hours developing assets of various forms (equipment, real estate, and avatar capital), is it not mine?" (2005, p. 157).

Adapting Locke's theory to MMOGs poses a significant challenge. It requires some modification to apply in contexts in which the common and labor have a much different character than in the real world. Horowitz finds three questions that a Lockean account of virtual property must answer: "what constitutes the common from which all property might be drawn," "what counts as labor," and "what is the content of the property right attained through labor-desert?" (2007, pp. 451-452). He finds the second question, about what counts as labor, is relatively easy to resolve. Horowitz proposes treating all production, even when it is enjoyable and done as a recreational activity, as labor, rather than thinking of labor as a painful activity (2007: 452). This is a sound suggestion given the overlap of work and play in virtual worlds that will be discussed later in this paper. The first and third questions are more difficult and require a careful making some alterations to Locke's theory. These questions will be discussed in the following sections. Before they can be dealt with, it is necessary to say something about the kinds of property that exist in the virtual world.

As Horowitz (2007) points out, many of the items in virtual worlds are not actually created by users, only discovered by them or won through the completion of challenges. Players can steal, win, and purchase items and abilities that were created by the developers. These items and abilities will hereafter be called developer-created goods. Although many virtual goods are developer-created, characterizing the debate over virtual property as being primarily concerned with things that the developers have created overlooks the players' creative power. The second class of goods includes those that would not exist in the game world were it not for the players' action. These are player-created goods. The developers may indirectly contribute to the production of these goods, by providing the raw materials or a space for production, but they do not actually produce theses goods themselves. MMOGs differ in the extent to which users can create their own goods. In some, like *Second Life*, this kind of production is common, while others offer more limited options for modification. Nevertheless, player-created goods exist in any games that allow players to modify existing items or to create their own avatars. There can be some degree of creation in any environment that permits combinations of developer-created goods in ways that the developers may not have intended. Creation can be a matter of arranging things that already exist in different ways (Boellstorff, 2008).

Character creation is a universal feature of virtual worlds and probably the best example of player-created goods in any MMOG. Even if players cannot produce any other items, they do create characters and develop their characters through their actions and achievements. Those who participate in virtual worlds often construct avatars that reflect the players' own appearance and interests, that represent an ideal self, or as a means of experimenting with different identities (Taylor, 2002; Turkle, 1997; Wolfendale, 2007). Whatever the motive, character creation reflects the players' own traits, beliefs, desires, and experiences. Moreover, character creation is a labor-intensive process. Much of it happens when players first enter a virtual world and choose their avatar's appearance, basic traits, and skills, but it continues throughout the game as players earn new level, complete quests, and form associational links. Character creation is a productive act that involves the players themselves. Player accounts should therefore be seen as player-created goods. For that reason, all virtual worlds allow players to perform some creative work. Making this distinction facilitates the Lockean reading of virtual property, as this distinction corresponds to who is performing the initial creative work of removing something form the common.

THE COMMON AS A NATURAL SPACE

The foundation of Locke's theory of property is his explanation of how people have any right to add their labor to things. Before anything can be taken, it must be part of a common that belongs to everyone and that can be used to produce individual property. Citing the Bible, Locke (1980) argues that God grants humans collective dominion over the earth and all its contents. This divine donation establishes a collective property right that allows all humans to claim things from nature. The common is a reserve of raw material, but virtual worlds contain no raw material in the conventional sense. It is therefore essential to reconsider what the common is and what it is composed of.

There are two senses of a common in virtual worlds. For clarity, these virtual commons will be distinguished as common$_1$ and common$_2$. Common$_1$ is based on a literal application of Locke's theory to the game world. It is composed of all the virtual raw material that other players in the game have not claimed. Understanding the common in the sense of common$_1$ places the virtual world's developers in the God role. They are the creators of the raw material and they are therefore free to dispose of it in any way they wish. Most Lockean theories of virtual property use this conception of the common or something very similar to it. They treat the developers as creators who have the ultimate control over the virtual world and the power to use as they wish (Horowitz, 2007). Defining the virtual common in the sense of common$_1$ is misleading because it is based on an apparent similarity between a state of nature in the real world and the images of nature created by software. This way of defining the common leads to three serious problems.

First, in Locke's theory, God has power over all things and the ability to establish property rights because he has created everything in the world. His power is absolute and derives from his position as the maker of the world and everything in it, including the human inhabitants (Tully, 1983; Waldron, 2002). Game developers create most of the goods in virtual worlds, but their relationship to the virtual worlds and to the players are much different from the one that Locke thinks God has to the world and its inhabitants. Developers do not create a world from nothing. They build virtual worlds with components developed from earlier games, draw on other sources, and revise the games based on players' feedback. They are creators, but they are not solely responsible for virtual world's existence, nor are they the creators of the players who occupy the virtual world. More importantly, Locke thinks that God is totally independent of the world and that people are entirely dependent on it

(Tully, 1983). This does not describe the relationship of developers and players to virtual worlds. Neither players nor developers are dependent on virtual world, because they can live without them. At the same time, neither is independent of the virtual worlds because they can be deeply affected by in-game events.

Second, arguing that developers own the common would defeat the purpose of the common in Locke's political philosophy. Locke's devotes his *First Treatise of Government* to refuting Sir Robert Filmer's claim that monarchs are appointed by God (2003). Locke does this by arguing that God gave property to all humans and not to Adam alone, as Filmer claimed. If property rights were only given to Adam, there could be no common – Adam's descendents would have complete control over all property. Locke is strongly opposed to giving one person or a group of people control of the common. He repeatedly affirms that the common cannot be seen in terms of a private domain: "nobody has originally a private dominion, exclusive of the rest of mankind" (Locke, 1980, pp. 18-19). Throughout his *Second Treatise of Government*, Locke says that the right to claim property is something that belongs to all people. John Rawls explains Locke's point as being that "no one can be excluded from the use of, or from the access to, the necessary means of life provided by the great common of the world, except from that which we have made our property" (2007, p. 145). Thus, theorizing virtual property from the assumption that the common is owned by a few developers goes against the spirit of Locke's (1980) argument, which is strongly egalitarian.

Finally, the common that Locke describes is a natural world, devoid of human interference. Locke draws a sharp distinction between the natural and social worlds. Only things that are part of the natural world can be part of the common; anything that is modified by humans or produced by them has been removed from it (Locke, 1980). Objects are only common to the extent that they are natural and open to claim by anyone (Locke,

1980). Although objects in virtual worlds are analogous to real world objects, they are distinct in Locke's ontology. In the real world, an untouched tree on unclaimed plot of land can be part of the common because they have not been removed from the state of nature. Virtual worlds include representations of unclaimed trees and plots of land, but their naturalness is illusory (Lastowka, 2004). Some objects may have received more labor than others and some may have only received labor from the developers and not from players. Nevertheless, all objects in the virtual world are equally artificial. There is nothing more natural about a virtual tree than there is about something that is created from it, so common$_1$ does not offer any grounds for drawing a distinction between what is common and what is owned.

These problems show that Locke's theory of the common cannot take the form of common$_1$ without misunderstanding the developers' role, contradicting one of Locke's intentions in proposing the common, and ignoring the difference between the raw materials of the natural world and the artificial raw materials of the virtual world. It is necessary to develop an alternative view of the common – one that modifies Locke's (1980) theory without leading to contradictions. Moreover, as the point about the difference between natural and artificial raw materials shows, any definition of the common that can hold true for virtual worlds must be a socially constructed common.

THE SOCIAL COMMON

Developers are primarily responsible for creating MMOGs because they create the structure of the games. In worlds that do not permit user modification, they may even be the only ones responsible for creating the items that players try to accumulate. However, the virtual worlds they create would be inoperative without the thousands, or in some cases millions, of players who devote their time to transforming the virtual world into

a community of players. As the previous section argued, any apparent naturalness in virtual worlds is illusory. The worlds are human products and as such, they are largely constituted by the social activities that occur within them. To the extent that MMOGs can have a common, it must be a common that is produced by all who have a role in creating the social world that all players inhabit.

Studies of MMOG have shown that virtual worlds are heavily dependent on players (Marks, 2003; Sicart, 2009). Many virtual worlds are home to fully developed communities that work, conduct trade, and form friendships (Lastowka, 2004). As Taylor explains, the games can only be accurately explained as products of the players' actions:

The collective production of game experience and knowledge does not simply constitute a helpful "addon" to the game, but is a fundamental factor in both its pleasure and sustainability. Most radically put, the very product of the game is not constructed simply by the designers or publisher, nor contained within the boxed product, but produced only in conjunction with the players (Taylor, 2006, p. 126).

Players collectively produce the worlds they inhabit by interacting with other players and sustaining the online social world that constitutes much of the gameplay experience. By doing this, they create the common$_2$. This can be thought of as a common composed of ideas, cultures, and institutions, rather than of simulated material objects. It is the lifeworld of the virtual world, establishing the background from which the events and artifacts of the game world derive their meaning.

The cultures of virtual worlds are partially formed by the character classes, geography, and other starting conditions that developers have put in place, but it is up to players to affirm or subvert the characteristics developers have programmed. The same is true of the institutions. Developers may create things like governments, currencies, and organizations. However, these are heavily dependent on the players' actions. The institutions of the virtual are only maintained as long as players continue to acknowledge them and to follow the constraints they impose on the game. If players are sufficiently unified, they can transform these institutions by developing own systems of exchange, organizational structures, or forms of governance. A prime example of players creating their own institutions and cultures are the guilds that they form in MMORPG's (Massively Multiplayer Online Role-Playing Games). These associations are governed by their own rules and norms, created by the players themselves. Players could even transform virtual worlds in more fundamental ways than this. Players in a game based on a gold standard of currency could abandon it in favor of another standard or switch to a barter economy to avoid the inflation caused by gold farming. They may, as individuals or through their cooperation with others, inspire the developers to modify games to suit the players' wishes. Most virtual worlds are not simply created and released in their final form. They are dynamic entities that develop and change over time and the players are one of the strongest forces in shaping this development (Sicart, 2009, p. 175).

The growing literature on the philosophy of society offers some insight into the production of the $common_2$. Searle (1995, 2006, 2010) argues that people create social worlds by building networks of interrelated institutional facts. By his definition, institutions take a range of forms. They can be abstract concepts that have no physical existence, like corporations, or they can be embodied in physical objects, as the institution of money is embodied in bills and coins. The social world people inhabit is an assemblage of institutions and each institution is created through collective action. Institutions may take on the false appearance of necessity because they are supported by tradition and maintained out of habit, but they are contingent. Searle argues that to create institutions or to maintain them requires what he calls "collective intentionality." People

choose to take part in collective enterprises like trade, governance, and games. These activities are sustained by each person's intention to take part in a collective enterprise and through the actions they perform to support the institutions. In other words, institutions are only meaningful because people agree that they are meaningful. A few individuals may dissent by refusing to recognize certain institutions, but the institutions continue to exist as long as a sufficient number of people continue sustain them.

The institutions Searle describes form a common because they are the collective property of everyone who inhabits the social world. Institutions may grant some people more authority than others, and some people may have a disproportionate power over the institutions, but no one owns them. To take a real world example, people who have a great deal of money have more purchasing power than others, but this greater power to use money does not make them owners of the institution of money. No one owns the institution of money, only individual bits of money. The same is true of the other institutions that structure the social world. In virtual worlds, some players may be more powerful, wealthy, or influential than others, but even the least powerful players have some role in determining the culture and institutions of the world.

This form of common may appear much different than the one Locke describes because it does not have the same superficial resemblance to Locke's as $common_1$. However, it fits well with Locke's definition of the common. It is a space that all players have dominion over, but that no person has a unique claim to because it is produced through the collective action of the developers and the community of players. It is also similar to Locke's common because it serves as the raw material of virtual worlds. In order to establish a unique claim to any piece of property, players must draw on the institutions and culture of the virtual world that constitute $common_2$ and are responsible for player-created, like new items or avatars. With this version of the common in mind, the players' significant contribution to virtual worlds is much easier to recognize and there is a stronger basis for understanding the digital labor process.

LABOR AND THE CREATION OF PERSONAL PROPERTY

Locke assumes that each person owns their own labor and that this is unalienable. Although it might be rented to an employer for a wage, the capacity for work is itself something that cannot be transferred to another person. Starting from this assumption, Locke makes his famous point that everything a person removes from the common becomes their property. "The *labour* of his body, and the work of his hands, we may say, are properly his. Whatsoever then he removes from out of the state that nature hath provided, and left it in, he hath mixed his *labour* with, and joined to it something that is his own, and thereby makes it his *property*"(Locke, 1980, p. 19). Individuals claim property by mixing their labor with that which is part of the common. Labor has the power to confer ownership because it is what makes the largest contribution in determining the value of the finished product. "Labour makes the far greatest part of the value of things we enjoy in this world" (Locke, 1980, p. 26).

One of the points of contention in the discussion of virtual property is whether actions in the virtual world can be described as labor. Some element of difficulty and even displeasure may be necessary for something to count as labor. Waldron (1988) argues that Locke's theory of property requires that labor involve an investment of time or the activity is difficulty, otherwise there would be no reason to think that the process confers ownership. Horowitz (2007) is correct in arguing that players actions often do meet this criterion, as studies of MMOGs have found that game play in is often experienced as work and that it can be nearly identical to tasks performed in normal jobs

(Lastowka, 2004; Yee, 2006). Moreover, Taylor (2006) finds that many power gamers become so heavily invested in virtual worlds that they spend as much time "playing" as they do at work. Players may spend hours fighting easy opponents, experiencing no challenge or variety, to help their character earn enough experience to reach the next skill level. Others spend hours mining gold or working as blacksmiths to earn money for virtual goods. The existence of virtual sweatshops and the widespread practice of buying high level characters rather than building them up oneself should also serve as evidence that not all play is enjoyable or purely recreational (Grimes, 2010).

Locke's theory of property makes it relatively easy to assign the ownership of player-created goods. Although the initial creation of an avatar may be simple, players expend a great deal of labor developing these avatars. As this paper already pointed out, avatars are unique to players and a clear product of individual labor. Some elements of them, such as racial characteristics or starting abilities, may be drawn from the material provided by the developers and the common$_2$, but they are created by individuals, and through the expenditure of labor. Thus, avatars should be regarded as the property of the player who creates them. They are the paradigmatic case of the amount of labor invested in a thing making it more valuable. Other virtual items that would not exist were it not for the players' expenditure of labor should also be counted as the property of the player who created them, as they are products of individual labor that remakes material drawn from the common$_2$. By extension, anyone who purchases player-created goods should have the right to them. Although someone who buys these goods is not responsible for creating them, purchasing those things from the laborer represents an exchange the goods for an amount of money that the seller judges sufficient to compensate for the labor.

It is more difficult to determine the ownership of player-created goods; these must be considered on a case-by-case basis. Initially player-created goods clearly belong to the developers, as the developers have invested their labor to produce these goods. However, the problem that arises is that both developers and players have labor claims to the goods once players are participants in the virtual world. The developers expend labor to create goods and players labor to procure them. The competing labor-based claims may seem to create an impasse. However, it is important to remember that Locke's discussion of labor imparting ownership only applies when things are taken out of the common. Acquiring finished items made by the developers of a game is not an act of removing something from the common, because these items are clearly attributable to the developers. Players can only claim ownership of these goods when they have acquired them by the exchange of a comparable labor or of virtual items representing labor value.

To acquire virtual goods, players exchange labor or objects representing labor value. Most MMOGs require some investment of real money to buy the game and to pay for monthly subscriptions. Some, like *The Sims Online* and *Second Life*, even allow players to buy virtual money and virtual goods with real money. According to Locke's theory, the exchange of real currency for virtual goods and services entitles players to the goods they acquire even if these were initially produced by the developers. The same holds true when virtual goods are exchanged for other virtual goods. So long as the players expend labor to earn money and items, they are entitled to the benefits of their labor. If labor is the mark of ownership, then players have a claim to property that they have purchased, won on quests, or earned working a virtual job. However, just as in the real world, they would usually not be entitled to goods stolen from other players or those that other players have lost.

Although Locke says little about the nature of property once it is transferred through exchange, his thoughts about the nature of money provide some guidance for theorizing virtual exchange. Initially, Locke sets limits on the amount of prop-

erty that a single person can acquire. One is the spoilage limit, which holds that a person can only justifiably have as much as one can use. It would not make sense to collect more food than one can eat or trade because food has a shelf life that prevents it from being a permanent store of value (Locke, 1980). In a pre-monetary society, there can be no accumulation, Locke thinks, because all goods are prone to degeneration over time if they are not used. The critical breakthrough is the introduction of money as a durable mark of the value of the labor added to a product (Locke, 1980). For digital property, the spoilage limit may seem to be irrelevant. Digital property can exist indefinitely without decaying. However, virtual goods, whether they are items or money, can spoil in the sense of losing their value. With no limits on the quantity of virtual goods that can be produced, they are highly susceptible to value fluctuations.

There must be trust that when something of value is exchanged for virtual goods – whether it is real money, virtual goods, or the investment of time, energy, and money necessary to play the game long enough to find the goods – that the transfer of ownership will be recognized. To do otherwise, violates the trust that gives representations of labor their value. Players labor to create and acquire virtual property, but they do so with the expectation that their labor will be rewarded. Just as athletes endure painful training regimens to feel the excitement of defeating rivals and sick people undergo painful medical treatment to become healthy, players endure unpleasant moments of virtual worlds so that they can reach some more pleasant payoff. Players may experience level grinding and quests to obtain items as means to an end or as enjoyable in themselves, but in either case, the player is probably working toward some greater goal. Players want to earn achievements, reach higher levels, and find new items. They spend countless hours in virtual worlds in pursuit of these things. Whether or not they enjoy the experience, they are motivated by expectations of things that are yet to come (Rueveni, 2007).

To appreciate the importance of expectation in the value of virtual goods, consider the extent of its effect on the real world. Money has no intrinsic value, even when it is backed by gold or silver, as no objects have value apart from the value that people attribute to them. The only thing that makes money a useful means of exchange is that people trust that it will store value. It's worth depends on the expectation that it can be exchanged for other goods. This is true for exchange more generally. Goods hold their value to the extent that there is a collective trust that they can be exchanged. The result of a breakdown in trust can be catastrophic market failures, bank runs, and inflation. As Locke says, money "has its value only from the consent of men" (1980, p. 29). Players may not create the money or the items that they use in the virtual world, but these things represent the value of the labor the player used to obtain them. These items can only symbolize this value to the extent that there is a general trust that these things will retain their worth and that the value will be recognized as an indicator of the amount of labor required to purchase or obtain them. If anything, trust is even more important in virtual worlds, as these do not even offer a material substance that can be used if exchange value disappears. Developers sometimes change the power of items they have created, causing the item to be less valuable for use and exchange (Kayser, 2007: 69; Meehan, 2006 p. 32). Doing this violates trust in much the same way as devaluing currency.

WHO OWNS THE WORLD?

The final object of property contestation may the virtual worlds themselves. As Westbrook (2006) points out, users and developers both invest considerable labor and money in virtual worlds, which seems to give both parties rights to claim ownership of them. Based on the account of the $common_2$ discussed on the previous section, virtual worlds cannot be considered the property of the game's developers alone. Although developers labor to produce much of the virtual world and they own the serves that host the worlds, so much of the virtual world is constituted by the $common_2$, that developers cannot be the sole owners. This may seem to be an unsatisfying answer, given the necessity of determining who has the right to modify virtual worlds and to control access to them. There is also the problem of whether users have any right to protect their investment in a virtual world should the developers decide to end it. Although the largest virtual worlds have yet to encounter this problem, it is only a matter of time before even the most popular comes to an end. If this end results from waning interest in the world, then it may be undisputed. However, if it is carried out against the wishes of the majority of players and threatens to destroy their investments, it may be contested.

Virtual worlds are extremely vulnerable, as they only exist on the host servers and these are physical objects that must be maintained by someone in the real world (Lastowka, 2004). Given this practically necessity of maintaining the servers, someone must have the responsibility of hosting virtual worlds and controlling them, even though they may be collective products. Developers have the strongest claim to ownership of virtual worlds and the right to gain profits by hosting them because they are responsible for producing much of the content of these worlds and for ensuring that the worlds continue to exist. However, their ownership of the worlds should not imply that they have unlimited power over them, as this would violate players' claims to ownership of the $common_2$ and the virtual goods that they own.

It would be unfair to force developers to continue hosting a virtual world indefinitely, but they could be obligated to transfer the authority to host the world to the players who have helped to create it and to give the players the opportunity to preserve the world that their labor helped to establish. This would support a more democratic style of governing virtual worlds and it is a subject that Jankowich (2005) has discussed at length. Jankowich argues against Balkin (2004), who thinks that users should not be granted property rights, by saying that failure to recognize players' property claims would leave players completely at the mercy of the game developers. The failure to recognize property rights would establish virtual world dictatorship, in which players have few rights and the results of their labor could be alienated at any time (Jankowich, 2005). Thus, Jankowich, like Locke, finds that property ownership is desirable because it decentralizes control and empowers ordinary citizens. "Property ownership by participants will help to serve as a stepping stone to greater democracy in virtual worlds" (Jankowich, 2005, p. 206). As he points out, property ownership may be a precondition for establishing virtual rights for users and creating participatory institutions with which to govern virtual worlds.

CONCLUSION: THE RESPONSIBILITIES OF OWNERSHIP

When modified to fit virtual worlds, Locke's theory of property implies that much of the content of these worlds should belong to the players whose labor makes these worlds come alive as sites of social interaction. A Lockean theory of virtual property provides strong grounds for recognizing players as the collective owners of the social common and individual players as the owners of their avatars and virtual goods that they have either created or purchased through an exchange of labor. Viewing virtual property in this way supports efforts of scholars in other fields to show that networked media have undermined the traditional dichotomy between producers and consumers (Jenkins, 2006, 2007; Russell, 2008; Sandvoss, 2007). In virtual worlds, developers and players are each producers. The former produce the environments, the quests, and the structure of virtual worlds. The latter produce the social net-

works, institutions, and culture of the worlds, as well as the thousands of player-created characters who inhabit them.

The Lockean reading of virtual property proposed in this paper suggests that players should have the right to sell their avatars, goods they have created, and goods they have acquired through exchange. However, this should not be taken as granting players the right to do whatever they wish with the avatars and goods that they create. Players with property also have responsibilities under Locke's theory. Locke poses several constraints on individual accumulation. In principle, no constraints are necessary, as the scarcity imposed on virtual goods is unnecessary. It is possible to grant all players access to all of the items in the game. However, doing so would undermine one of the elements of gameplay. Without scarcity, the struggle to overcome challenges and to reach new levels of play might seem hollow. The scarcity of the digital world is artificial and unnecessary, but it has a significant effect on gameplay. According to Locke's theory, players must recognize that they are playing in worlds with scarce resources and recognize the needs of other players. Those who own property must leave enough for others, both in terms of quantity and quality (Rawls, 2007). The limit Locke (1980) imposes is that a person can only own what they can use.

Thus, the Lockean theory of property states that those with property have obligations to those who do not. Players have a strong claim to being able to use the common$_2$ for their own purposes and to consider player-created goods and developer-created that have goods that have been acquired through exchange to be their property. However, by the same standard, players have no right to deny others the use of common$_2$ or to use their power to prevent others from attempting to enrich themselves through labor. As players continue to assert their rights to their own creations, they must also continue to uphold the rights of future players to freely access the common$_2$ and to labor to acquire their own virtual property.

REFERENCES

Arias, A. V. (2008). Life, liberty, and the pursuit of sword and armor: Regulating the theft of virtual goods. *Emory Law Journal, 57,* 1301–1345.

Balkin, J. (2004). Virtual liberty: Freedom to design and freedom to play in virtual worlds. *Virginia Law Review, 90*(8), 2043–2098. doi:10.2307/1515641

Boellstorff, T. (2008). *Coming of age in Second Life: An anthropologist explores the virtually human.* Princeton, NJ: Princeton University Press.

Brenner, S. W. (2008). Fantasy crime: The role of criminal law in virtual worlds. *Vanderbilt Journal of Entertainment and Technology Law, 11*(1).

Caramore, M. B. (2008). Help! My intellectual property is trapped: Second Life, conflicting ownership claims and the problem of access. *Richmond Journal of Law and Technology, 15*(1), 1–20.

Castronova, E. (2005). *Synthetic worlds: The business and culture of online games.* Chicago, IL: University of Chicago Press.

Castronova, E. (2007). *Exodus to the virtual world: How online fun is changing reality.* New York, NY: Palgrave Macmillan.

Glushko, B. (2007). Tales of the (virtual) city: Governing property disputes in virtual worlds. *Berkeley Technology Law Journal, 22,* 507–531.

Grimes, J. M., Fleischmann, K. R., & Jaegger, P. T. (2010). Emerging ethical issues of life in virtual worlds. In Wankel, C., & Malleck, S. (Eds.), *Emerging ethical issues of life in virtual worlds.* Charlotte, NC: Information Age Publishing.

Horowitz, S. J. (2007). Competing Lockean claims to virtual property. *Harvard Journal of Law & Technology, 20.*

Jankowich, A. E. (2005). Property and democracy in virtual worlds. *Boston University Journal of Science and Technology Law, 11,* 173–220.

Jenkins, H. (2006). *Fans, bloggers, and gamers: Exploring participatory culture.* New York, NY: New York University Press.

Jenkins, H. (2007). Afterword: The future of fandom. In Gray, J., Jones, J. P., & Thompson, E. (Eds.), *Fandom: Identities and communities in a mediated world.* New York, NY: New York University Press.

Kayser, J. J. (2007). The new-world: Virtual property and the end user license agreement. *Loyola of Los Angeles Entertainment Law Review, 27,* 59–85.

Kennedy, R. (2008). Virtual rights? Property in online game objects and characters. *Information & Communications Technology Law, 17*(2), 95–106. doi:10.1080/13600830802204195

Lastowka, G., & Hunter, D. (2004). Law of virtual worlds. *California Law Review, 92*(1), 3–73. doi:10.2307/3481444

Lederman, L. (2007). Stranger than fiction: Taxing virtual worlds. *New York University Law Review, 82,* 1620–1672.

Locke, J. (1980). *Second treatise of government.* Indianapolis, IN: Hackett Publishing.

Locke, J. (2003). First treatise of government. In Shapiro, I. (Ed.), *Two treatises of government and a letter concerning toleration.* New Haven, CT: Yale University Press.

Marks, R. (2003). *EverQuest companion: The inside lore of a game world.* New York, NY: McGraw-Hill.

Meehan, M. (2006). Virtual property: Protecting bits in context. *Richmond Journal of Law and Technology, 13*(2), 1–48.

Mossoff, A. (2002). Locke's labor lost. *University of Chicago Law School Roundtable, 9*(155).

Rawls, J. (2007). *Lectures on the history of political philosophy*. Cambridge, MA: Harvard University Press.

Reuveni, E. (2007). On virtual worlds: Copyright and contract law at the dawn of the virtual age. *Indiana Law Journal (Indianapolis, Ind.)*, 82.

Russell, A., Ito, M., Richmond, T., & Tuters, M. (2008). Culture: Media convergence and networked participation. In Varnelis, K. (Ed.), *Networked publics*. Cambridge, MA: MIT Press.

Sandvoss, C. (2007). The death of the reader. In Gray, C. S. J. A., & Harrington, C. L. (Eds.), *Fandom: Identities and communities in a mediated world*. New York, NY: New York University Press.

Searle, J. R. (1995). *The construction of social reality*. New York, NY: Simon & Schuster.

Searle, J. R. (2006). Social ontology: Some basic principle. *Anthropological Theory*, 6(1), 12–29. doi:10.1177/1463499606061731

Searle, J. R. (2010). *Making the social world: The structure of human civilization*. New York, NY: Oxford University Press.

Seto, T. P. (2009). When is a game only a game? The taxation of virtual worlds. *University of Cincinnati Law Review*, 77, 1027.

Sicart, M. (2009). *The ethics of computer games*. Cambridge, MA: MIT Press.

Steinberg, A. B. (2009). For sale - One level 5 barbarian for 94,800 Won: The international effects of virtual property and the legality of its ownership. *Georgia Journal of International and Comparative Law*, 37, 381–420.

Taylor, T. L. (2002). Living digitally: Embodiment in virtual worlds. In Schroeder, R. (Ed.), *The social life of avatars: Presence and interaction in shared virtual environments* (pp. 40–62). New York, NY: Springer.

Taylor, T. L. (2006). *Play between worlds: Exploring online game culture*. Cambridge, MA: MIT Press.

Tully, J. (1983). *A discourse on property: John Locke and his adversaries*. Cambridge, UK: Cambridge University Press.

Turkle, S. (1997). *Life on the screen: Identity in the age of the Internet*. New York, NY: Simon & Schuster.

Waldron, J. (1988). *The right to private property*. New York, NY: Clarendon Press.

Waldron, J. (2002). *God, Locke, and equality: Christian foundations in Locke's political thought*. Cambridge, UK: Cambridge University Press. doi:10.1017/CBO9780511613920

Westbrook, J. (2006). Owned: Finding a place for virtual world property rights. *Michigan State Law Review*, 779-812.

Wolfendale, J. (2007). My Avatar, My Self: Virtual harm and attachment. *Ethics and Information Technology*, 9(2), 111–119. doi:10.1007/s10676-006-9125-z

Yee, N. (2006). The labor of fun: How video games blur the boundaries of work and play. *Games and Culture*, 1(1), 68–71. doi:10.1177/1555412005281819

This work was previously published in the International Journal of Technoethics, Volume 2, Issue 3, edited by Rocci Luppicini, pp. 62-73, copyright 2011 by IGI Publishing (an imprint of IGI Global).

Chapter 4
Perverting Activism:
Cyberactivism and Its Potential Failures in Enhancing Democratic Institutions

Tommaso Bertolotti
University of Pavia, Italy

Emanuele Bardone
University of Pavia, Italy

Lorenzo Magnani
University of Pavia, Italy

ABSTRACT

This paper analyzes the impact of new technologies on a range of practices related to activism. The first section shows how the functioning of democratic institutions can be impaired by scarce political accountability connected with the emergence of moral hazard; the second section displays how cyberactivism can improve the transparency of political dynamics; in the last section the authors turn specifically to cyberactivism and isolate its flaws and some of the most pernicious and self-defeating effects.

INTRODUCTION: POLITICAL ACCOUNTABILITY AND MORAL HAZARD

The word *technoethics* describes a will to understand the transformations of human beings' moral life in a world radically modified by the massive presence of advanced technological artifacts: such distribution of the results of scientific advancements in everyday lifestyle is, in fact, likely to affect many established relationships between moral agents and moral patients. Considering the ever-growing rapidity of changes brought about by scientific and technological progress, the already often blurred distinction between *descriptive* ethics and *normative* ethics is all the more hard to implement as far as *technoethics* are concerned. Our analysis of how new informational media have been affecting a morally relevant socio-political behavior known as *activism* is an example of such stance: a careful description of cyberactivism's

DOI: 10.4018/978-1-4666-2931-8.ch004

moral implications (and risks) cannot, and in our opinion *should not*, be decoupled from a cautious prescriptive disposition. A philosophical attitude, as we will try to show in the following sections, is in fact fundamental to individuate, understand and protect the moral stakes set by the fulfillment of our digitalized era, in order to preserve one of the greatest acquisition of our rather recent past: democracy.

As Popper (1945) argued, the appeal of democracy rests on the possibility of getting rid of those who rule without bloodshed, but through general elections. Whereas in the other forms of government those who are ruled must make a revolution to dismiss who rules: that is, the force of the best army is the necessary condition to change the government. Indeed, democracy is not immune from malfunctioning. If one considers political accountability of government as the fundamental ingredient characterizing democracy, then it follows that democracy might not be working as expected when the various mechanisms to enforce political accountability – for instance, general elections – are impaired or weakened to a certain extent.

The problem of political accountability can be framed referring to the notion of *moral hazard*. Dowd (2009) defines moral hazard the situation in which one party is responsible for the interests of another having an incentive to let his interests come first.[1] Our take is that democracy is efficient insofar as moral hazard is unlikely to emerge, or remains under a threshold that can be disregarded. Conversely, the emergence of moral hazard increases the possibility to see democratic dynamics not working as they should. In a representative democracy, the relationship between represented and representing is captured by the principal-agent game (Alvarez, 2006). A principal-agent game arises whenever a person (or a group of persons), usually called the *principal*, delegates another person (or another group of persons), the *agent*, to accomplish a certain task that the principal is not able to accomplish. The principal-agent game

– that basically concerns contractual relationships – brings two issues about. The first issue is related to what kind of person (or group of persons) one should delegate the task. The second is about how to control that the agent is actually doing what he is supposed to do.

In representative democracy, the first issue is handled by means of electoral competition. That is, those who want to be elected, namely, the politicians, compete on a platform consisting of campaign promises, which are those policies they will implement if elected. The electorate is, in turn, called to make their preference public by casting a vote for the candidate(s) whose platform they like or prefer. The second one turns into a problem as the principal and the agent may have conflicting goals. For instance, in democracy, politicians may be tempted to put their own interest's first approving policies that are not in the citizens' best interest.

Even though these two problems are treated separately, they are two faces of the same medal. According to Popper, democracy is more related to controlling those who gained the power than selecting them. In fact, he placed great emphasis upon the "replacing" character of general elections. If this interpretation is correct, we can easily see how the two issues are related: as a matter of fact, democracy is what gives us the chance to select politicians by replacing them. To put it briefly, the voters' power is best represented by their ability to vote their deputies *out*, rather than voting them *in*. When and how such a power is weakened, in connection with moral hazard?

As already mentioned, in democracy the electorate are called to cast their vote in order to express their preference. It is worth noting that democracy leans on the *one man, one vote* principle, meaning that everybody has the right to vote for whoever he or she wants to. Every vote counts, whether it is formulated relying on emotions or more reasoned attitudes, whether one is right or wrong. Electors are sovereign citizens: there is no *wrong* vote. That being said, we are now interested in looking into a series of distortions that might happen in democracy.[2]

The first question we should address is: how does the electorate formulate their preferences? Even though electors are free to pick the candidate they want, we assume that, if a person delegates another to accomplish a task, she is supposed to monitor that her choice was a good one.

For instance, in the research on the public control of the politician, there are two contrasting approaches about how to interpret political accountability: they basically diverge on whether politicians' actions are observable or not (Dogan, 2010). Central to this issue is how the voter's decision is framed: some maintain that electors base their decision to reelect the politician by looking at the politician's past policies (Barro, 1973; Rogoff, 1990). This approach clearly assumes that electors have complete information or, at least, they can have access to those pieces of information that are relevant to evaluate the politician's work. Some others strongly disagree on such an approach, claiming that the politician's past policies are not observable by voters and, instead, they suggest that the voter's decision is based on the outcomes of the policies implemented. There are other dimensions on which voters might rely: for instance, whether the promises made during electoral campaign were fulfilled or not.

As a matter of fact, what is observable by the voters and what is not is fundamental in order to evaluate the efficacy of those institutional mechanisms granting political accountability and, at the same time, avoiding moral hazard in democracy. Developing this line of thought, we maintain that moral hazard is more likely to emerge when three main elements are made not observable or, at least, hard to observe: 1) if electoral promises are fulfilled or not, 2) what policies are actually implemented, 3) their outcomes.

The first element is related to whether a politician is trustworthy or not. As already mentioned, the candidates compete on a platform containing the description of a number of policies they prom-

ise to implement, if elected. Such a platform is indeed informative about the idea of society that politicians hold on their mind. As the politician is elected on the basis of a certain political platform, his reelection should be somehow related to whether he acted according to his electoral promises as if it was a sort of ideological base.

The second element is indeed related to the first, as some voters might disagree on the platform put forward by the politician, but they might agree on the policies later implemented by him or her. As a matter of fact, electoral promises cannot cover all the issues the politician may face during his or her term in office: because of this, knowing what policies are implemented is important also to situate what the politician actually does.

The third element is about the policies' outcomes. It is important to see whether the politician uses his power in accordance to what he or she had promised. It is important to know what policies he or she actually implements, but it is also important to assess the outcomes of such policies. For example, some voters might have supported certain policies which, however, turned out to be a disaster. In this sense, knowing about the effects that certain policies have had over society is important to test one's ideological orientation and revise it accordingly.

As one can easily see, the quality of democracy can be impoverished, as voters are hacked around the three elements listed above. More precisely, the three elements can be made observable from the voters by hacking their understanding. Such a *cognitive hacking* takes place by making use of propaganda (Thompson, 2007). In our view, propaganda distort voters' understanding by the manipulation of public discourse (Barnett & Littlejohn, 1997).

In democracy public discourse is a sort of mediator that has two main functions. The first is an ethical one: it aggregates interests and values so as to raise collective awareness for certain issues that are supposed to be morally salient for the *polis*. The second is more cognitive: it helps select the information that is relevant for joining public debates. That is, public discourse delivers important resources to formulate opinions about the politicians, the policies they implement, and politics in general: more generally, we may argue that public discourse can be *educating* insofar as it supports citizens to form sound judgments about the common good. It is worth noting that public discourse is not an agent in itself, but it results from a distributed network of agents – individual and institutional – including, for instance, journalists, experts, opinion-makers, think-tanks, and so on. Now, we claim that the role of propagandists is to influence these agents so as to *bias* the public discourse and so to spread specific interpretations supporting the political agenda of the government.

Insofar far as propagandists are particularly effective in biasing the electorate, what the government actually does becomes less and less visible and so harder and harder to control. Indeed, such a biasing activity affects the three elements we mentioned above. Propagandists may employ a number of strategies that aim at hiding information, putting forward arguments distorting the real effects of the policies implemented by the government, minimizing its actual responsibilities in case of patent political failures, downplaying the lack of success in implementing the promised policies, and so on. Here the use of fallacies is fundamental to poison public discourse and so make it blunt. Fallacies are composing the propagandist toolbox for cognitively hacking voters' understanding (Magnani, 2009, in press). For instance, understanding can be hindered by caricaturing alternative policies through the so-called *straw man* fallacy, the opposition can be denigrated by using the so-called *argumentum ad hominem*, and negative information can be selected out by using the so-called the *one-sidedness fallacy*, and so on.

This idea of biasing the electorate by impoverishing public discourse is captured by the image of the *glass ceiling*. Such an image is introduced by Yankelovich (1991, p. 3) who wrote: "beneath the surface of formal arrangements to ensure citizen participation, the political reality is that an intangible something separates the general public from the thin layer of elites – officials, experts, and leaders who hold the real power and make the important decisions."

The glass ceiling metaphor stresses that the electorate have the illusion to control the politician who is conversely immune. The glass ceiling gives him or her the chance to systematically distort and misrepresent his/her contribution and so pursue his or her private interests.

To sum up, moral hazard in democracy is likely to emerge as the various mechanisms to enforce political accountability are weakened. In turn, political accountability might be drastically impaired when propagandists act so as to impoverish public discourse that has the function to inform and orient citizens in their decision-making process. Propagandists cognitively hack citizens' understanding of what is going on by biasing the representations the electorate has of the politicians' promises, the policies they have actually implemented, and the outcomes.

THE INTERNET AS A POLITICAL MEDIATING FRAMEWORK

As we just contended, politicians may have different magnitudes of moral hazard, and the general understanding of such affection varies from one society to the other at a given time: the Internet, thanks to some characteristics we will describe in this section, might hold a big potential to act on democracy and the way its citizen perceive it. The question is: may the Internet challenge and help democracy *in the long run*? What kind of political activism may the Internet mediate in order to reshape or simply implement the deliberative democracies we are accustomed with?

Generally speaking, it seems that the Internet, as a cognitive mediator[3] (Hutchins, 1995; Magnani, 2009; Bardone, 2011), may enhance democracy by reducing moral hazard and facilitating political accountability in two respects. First, the Internet allows people to confront different sources of information in an unprecedented way, so that almost everyone can try to verify and test the information delivered by traditional media; second, it affords civic engagement and participation: more precisely, the Internet must be considered as a powerful community builder. These two factors contributed immensely to produce and foster a new, widespread kind of political activism in which any person, connected to the World Wide Web by a proper device, can engage in a wide span of endeavors, ranging from local issues to those crucial for the whole planet, with just a few clicks of her mouse.

The Internet as a Cognitive Mediator

What people think, what their preferences are, become especially important in democracy. We already mentioned Popper's (1945) insight, about how the appeal of democracy rests on the possibility of getting rid of those who rule without bloodshed or resource-consuming revolutions, but thanks to (more or less) general elections: in democracy information matters, not weapons, because voting is based on ideas and arguments which one can have or get (at least in its idealized form). This leads to two interesting consequences: first, activism matters insofar as individuals can influence each other's views and consequently orient the action policy makers. Second, public debates and discussions are fundamental to foster democracy as we know it, insofar as individuals are provided with actual information and available points of view to choose the side they want to support: grassroots activism can foster such occasions for confrontation and negotiation only in an informationally and communicatively rich environment, such as the Internet. In fact, we maintain that a deliberative conception fairly represents the appeal of democracy. That is, we claim that discussing matters before making some collective decision constitutes the rationale of

democracy *per se*; of course, the more people can freely access (and, at a further stage, participate in) public debates and confront different opinions, the more democracy serves its purpose.

We provide two arguments to support this conclusion. The first one is moral. Discussions allow people to express and debate their preferences: that is, everyone has the chance to have her/his say. Therefore, we should reasonably expect this to make people more inclined to accept the outcome of a vote, no matter what it is, because they had the opportunity to discuss it. Moreover, the fact that people can have their say implies everyone should provide a justification for their ideas: failing to grasp these stances means, in our opinion, to grossly misunderstand democracy.

The second argument is a cognitive one. The essence of debating consists in revealing private information that otherwise would remain folded. A mere vote does not contribute to express what one thinks and, most of all, how intense one's preferences are (Fearon, 1998). This is crucial to compare different instances and solve inconsistencies: moreover, discussions are important also for lessening "bounded" rationality. In this case, debate allows people to pool their limited capabilities through discussion (Simon, 1983).

The rationale of democracy rests on its deliberative nature and the fact that no one, in principle, can be excluded, nevertheless democracy *in se* is not protected from possible degeneration. As already mentioned, in democracy people's preferences acquire great importance, since people base their vote upon the information and the arguments they gather and are confronted with. That is, citizens vote for those who support, or are closest to, their own ideas. However, this is only one face of the coin when considering "propaganda". In spite of the negative connotation of the word, propaganda is a necessary condition to keep democracy working. As Bernays (the man credited with being the inventor of *public relations*) put it, "a desire of a specific reform, however widespread, cannot be translated into action until it is made articulate"

(Bernays, 2005, p. 57). As a matter of fact, a desire for a certain policy does not come up to the citizens' minds simultaneously (Lippmann, 1997, p. 155): public opinion must be focused and organized. However, citizens' preferences can be easily manipulated and even manufactured. As Chomsky put it, in democracy the government cannot control people by force, but "it can control people's minds" (Chomsky, 2002, p. 201). There-fore, the way people can access information, how they build their preferences up, is a key issue to prevent democracy from degeneration: in fact, political activism has always been deeply con-cerned with the issue of providing citizens with transparent and trustworthy news, as opposed to mainstream, corporate news.

In fact, cyberactivism has been majorly shaped by the Internet's ability to improve dramatically citizens' access to independent news and media: traditional media can be easily manipulated and controlled by political power who often boosts its agenda with biased, or even bribed, columnists or editors (Furedi, 2002), and fills higher ranks with *yes-folks* who clearly understand how their monthly income is strictly related to the govern-ment's fortune. In contrast, the Internet, as a basically unstructured and ever growing informa-tional space, seems to reduce the overall power of governments to control citizens (Simon, 2002). The Internet and the Web in particular are *search-ing environments* in which people are enabled to search for whatever they want without any kind of filtering. They can access various sources of information and exploit social sources of informa-tion such as forums, chat rooms and blogs. In this sense, the Internet dramatically changes the task people face, when they deal with political issues.

With this respect, it is particularly interest-ing to analyze the case of *Indymedia*, a website regrouping independent news providers[4] from all over the world. The World Wide Web is indeed an unstructured space, but this has to be understood rather as a moral and institutional fuzziness, and not as an informational chaos: that is to say

that cyberactivism, to benefit of the possibilities unveiled by the Internet, had to develop specific aggregators. They proved essential to organize coherently the resources so that sharers could connect with viewers (or with each other, as the line between provider and receiver is much more blurred than in traditional media), thus *co-opting* to social needs – in Pickard's words (2008) – virtual infrastructures that had proved extremely successful for commercial purposes: they were declined into an open, peer-oriented framework. "Characterized by its 'Be the Media' slogan, Indymedia's user-driven news productions, col-lective editing and open source practices place it in the vanguard for implementing technical strat-egies for amplifying democratic processes. [...] Technologically, Indymedia's radical democratic model is further evidenced by their reliance on open source, open publishing, and wikis. Open publishing guidelines allow users to contribute original content or to comment on other postings. This interactivity is based on a transparent editorial process that encourages readers to get involved. Open publishing and wikis allow information to be corrected and supplemented faster and more efficiently" (Pickard, 2008, p. 639).

The Internet and Civic Engagement

The Internet may enhance democracy in another respect that is related to the dimension of political activism and civic engagement. As shown above in the case of *Indymedia*, the rationale of "virtual" democratic participation not only concerns voting,[5] but also gaining access to relevant information, debating and discussing it. However, discussing and debating presuppose that people are truly engaged in all those activities that involve public life and commitment towards public welfare: this draws a line between the cyberactivist and the person who just wishes to be informed on as many topics as possible concerning her life and interests. As Putnam (2000) suggested, it is more likely that democracy spreads, when the so called connective tissue of the society is highly developed. The more people are separated from each other, the more the political engagement drastically decreases. The role of activism has always been this: promoting public awareness on relevant themes and encouraging people to connect with each other, and thus acknowledge their power; cyberactivism can rely on an unprecedented ability to share and connect.

The claim that the Internet allows people to search for whatever they want is well-founded, but it is not the whole deal. As maintained by Meikle (2002), the Internet is not only a medium of consumption, but also of *intercreativity*. For instance, reading a newspaper is a kind of activity that presupposes a one-way communication flow, so to say. I can read what an editor writes, but she cannot read what I would like to write to her. In this sense, people are primarily information consumers. On the contrary, the new technologies that belong to the so called "Internet Galaxy" (Castells, 2001) makes *intercreativity* possible. By the term intercreativity, we mean something more than simple interactivity. In order to define what intercreativity is, we have to introduce some important distinctions.

Let us consider an example: several on-line newspapers (which are not considered as cyberactivism) allow people to select what they want to read or receive in their e-mailbox, thus customizing their reading experience. In many cases, one can also post a comment on a given article. However, almost always the options available to the user is limited and already selected by the editor: the way comments are displayed does not afford many instances of cyberactivism, insofar as they are more about expressing one's view (as one would do watching TV news in a café and put her two cents in) than interested in creating a true, lasting debate. This is the kind of interactivity exhibited by a jukebox: comments and what they relate to (video, article) do not belong to the same ontological regions. Comments are not supposed to influence the substance of the article (unless,

maybe, some reader highlights gross mistakes and typos, but it still depends on the good will of the editors). Within this subject, one could draw internal distinctions between website of newspapers that allow comments on every article, from extensive ones to flash news (such as the French *Le Figaro,* http://www.lefigaro.fr), and others who permit a comment form only on certain articles (such as the Italian newspaper *Il Corriere della Sera,* http://www.corriere.it) whose criteria to determine which articles can be commented and which not is still a puzzle to the authors of this paper).

On the contrary, by the term intercreativity we rather refer to the process of creating something by a truly two-way communication flow, in which everyone can contribute to producing, choosing, and modifying a given document (an article or the course of an open discussion). For instance, an email exchange with a friend or a forum is an example of this kind, brought to its most characteristic form in the *wiki*. With this respect, *Indymedia* is more than a news provider as it allows any user to be an activist in many respects: they can share news, but also add to them and comment, contributing with their knowledge to what they think relevant, and comments are clearly not just a frill added to the news, as in corporate news websites.

Now, the fact that the Internet exhibits this kind of intercreativity can play a crucial role in enhancing civic engagement. As mentioned above, the more people are separated from each other, the more the political engagement drastically decreases. Now, the point is that the Internet provides activists with new possibilities that drastically change the way they can reach each other and the general public, in order to promote awareness on topics of general utility. All this is brought about by the possibility for citizens to stop being mere information-consumers and become participants. There are plenty of examples where new political strategies of civic engagement are brought about. No matter where they are, people can share information, make common cause, and jointly advance their mutual political or other agendas (Simon, 2002). Mailing list, newsletters, forums, on-line conference tools, contribute to boost civic engagement. Besides, it is worth noting that also the idea of open publishing promotes those values that are very close to democracy, such as freedom of speech, and so forth.

With this respect, we can once again turn to Pickard's investigation of cyberactivism (2008) to point out three illuminating examples in which the Internet allows for powerful aggregators aimed specifically at political activism: *Democratic Underground*, *Free Republic* and *Move On*.

Working as both a message board and news source, "Democratic Underground is largely discourse-focused, but concrete actions emerge, such as launching political candidacies, exchanging political information, and organizing around particular issues" (p. 635), while "Free Republic bills itself as an 'online gathering place for independent, grass-roots conservatism on the web'. Like Democratic Underground, membership is free, though donations are periodically solicited. On the site, FreeRepublic members (self-labeled 'freepers') can get links to news sources in both major and minor publications, as well as enter discussion threads" (p. 636). Both these sites, defined as "partisan public spheres", have discretely contributed to the American political scene in the last decade, becoming the expression of each side's cyberactivism: even though both websites openly perform activities of inside-censorship (so to mute opponents who would not join the discussion with productive contributions), they can be regarded as eloquent examples of how to implement the power of new technologies in the political tissue of western democracies, spreading the culture of e-democracy.

Move On, on the other hand, displays more of a hierarchical, top-down structure in its configuration of cyberactivism. "Describing itself as 'democracy in action', Move On uses its email list to facilitate house parties, distribute muckrak-

ing documentaries, critique mainstream media, raise money for hard-hitting commercials, and to promote a host of other activities aimed at supporting progressive causes. Boasting over three million members, in recent years Move On has generated, often in days or even hours, online anti-war petitions with hundreds of thousands of names, raised millions of dollars for pro-peace politicians, and launched primetime television and radio commercials" (p. 367). With respect to the websites we just mentioned, *Move On* can count on a more massive base of activists, and – instead of implementing a radical model of e-democracy – it aims at reaching its members in the most effective ways, so to quickly produce noticeable effects at national level, as described by Pickard. This extreme efficacy of *Move On* is not exempt from side-effects, as the project is "fraught with recurring tensions. There are signs that members suffer what may be called 'email fatigue' from the frequent calls to action from Move On organizers" (p. 641): *Move On* could be therefore considered to be sitting on the fence, between cyberactivism and traditional activism. Members are reached via the means unveiled by the Internet, but they are principally asked to make their contribution not in the public sphere of the World Wide Web but in "real life".

Understanding the implications of such a hiatus between different modes of cyberactivism might help us in spelling out the limitations and the challenges that affect activism on the Internet.

ACTIVISM PERVERTED

In spite of the conspicuous advantages that new communicational technologies can implement in democratic dynamics, they can also be a source of structural flaws and failures which, on the long run, might further impair the functioning of the State instead of benefitting it.

Empowerment and *Amateurization*

It is easy to understand how the internet dramatically empowered[6] activism with an unprecedented power of informational exchange (Illia, 2003). The appeal of cyberactivism rests on its potentiality for *intercreativity*, as suggested by Berners-Lee: "Intercreativity is the process of making things or solving problems together. If interactivity is not just sitting there passively in front of a display screen, then intercreativity is not just sitting there in front of something 'interactive'". This is embedded in four different spheres of cyberactivism: "intercreative texts, tactics, strategies and networks" (Meikle, 2010, p. 364).

We must be aware that intercreativity also has its reverse of the coin. Keen (2007) had provocatively denounced the *mass amateurization* processes triggered by the diffusion of user-created content on the internet: amateurization can also be a problem in the case of cyberactivism. Activism, when the citizen becomes a non-institutionalized actor in the socio-political scenario, implies the refusal of any mediation in the maintenance of the *res pubblica*. Democracy is a distributed mediation framework between *res pubblica* and public opinion, embodied in its laws, institutions, political personnel and ideologies. This distributed, and therefore shared, environment can be refused in favor of more flexible and radical means of aggregation, which could better allow the emergence of common people's opinions: the risk of amateurization is nevertheless looming over such processes.[7] As long as activists aim at inserting the objective of their battles in an institutional framework, they must cultivate and develop the related knowledge: conversely, it can be argued that as long as the campaigns presupposes an extra-institutional end, the need for meta-political knowledge dramatically decreases and activism faces the risk of becoming a free-for-all. Mediation frameworks are dismissed even inside the activist movement: in a paroxysm of the refusal of all mediation, the only accepted framework becomes the structural one, that is the Internet, which affords any social and informational interaction one may need: "[p]articipation, discussion, the active role of the user as well as organizational and social benefits by using the global infrastructure for creating networks are *important parts – if not the basis – of political participation and activism*. Still political leaders, commercial global players and international institutions already have an enormous influence on the structure and design of the Internet" (Neumayer & Raffl, 2008, p. 4, emphasis added).

If the Internet allowed the increase of activism, can we say that it increased the number of activists as well? If the answer is affirmative, we should define the new figure of activist which is emerging from the Internet-based communicational dynamics.

Self-Efficacy and *"Oh-Dear"-ism*

A major effect we want to highlight is related to self-efficacy as perceived in democracy and how this perception is altered by cyberactivism. This brings us to the famous "voter's paradox", which basically accounts for how the benefitting return to an individual from a group contribution is likely to be inferior than the cost of the contribution (further interesting considerations about the rationality of voting can be found in (Dowding, 2005)). In our constitutional democracies, candidates are usually elected *or not*, there is no alternative in-between: this means that if I vote for a candidate that received less than the necessary amount of preferences, she will not be elected and hence my vote is not likely to affect the system. Therefore, in *folkpolitics*, the common assumption is that each voter's preference is irrelevant, if taken individually: the question here is that of the *Sorites paradox*, also known as the *heap paradox* (Cargile, 1969; Campbell, 1974; Dowding, 2005). Each voter is (and feels) irrelevant, but all voters together are not: this immerses the democratic participant in a blunt condition of anonymity, in which her *civic duty* to vote embodies the fact that if *all* voters would not vote then democracy would utterly fail.

Activism, and furthermore cyberactivism, allows a dramatic increase in self-efficacy with respect to the political discourse. If I disapprove on a determinate ecological policy, for instance, I might vote for the candidate who, at least in his program, does not endorse it: if she is not elected, my battle within the voting institution is over and I lost it, no matter how strong my convictions and my devotions were, they all had to be reduced to the form of the vote. This is right where we got used to see activism kick in, since activism is in prime a refusal of the low self-efficacy to which individuals and groups are confined in democratic dynamics, and it allows for political commitment on certain issue to become public in a wide array of *civic anomalies*. By the use of this term we mean to stress how the power of activism rests on its ability to promote awareness by breaking public expectations about the functioning of society: mob actions, (more or less violent) protests, strikes, disruptions of public or private activities, unusual communication strategies (i.e. flyers, pamphlets) etc. are all phenomena that we do not usually *expect* in the civic and socio-economic environments we are accustomed to.

The diffusion of activism, and the success of many of its campaigns over the last two centuries (on humanitarian and ecological issues, chiefly), increased the perception of ordinary democratic forms as obsolescent and unable to fit the flexibility and variability of contemporary society, with the result of an overvaluing (in principle) and a misunderstanding (*de facto*) of democracy.

This phenomenon reached its (now enduring) apex starting from the Sixties and the Seventies, with the diffusion of what commonly goes under the label of *counterculture* (Turner, 2006). Politicians were labeled with a irremovable brand of shame and corruption, which was extended to politics *as a whole*: democracy and its institutions stopped being something a citizen could rely on[8] and globalized activism refused to engage them as a partner or an opponent anymore, concentrating its activity on parallel, independent channels.

The relationship between activism and the democratic institutions could be in fact considered as a discriminating issue: activism can either engage political dynamics, laws and so on as the objective of its campaigns, or ignore the political aspect and pursue its objectives as a kind of *macht-activism*, openly admitting that its display of strength confronts the political immobility without any attempt nor hope to actively engage it. This relates immediately to the kind of epistemic activity proposed by activism: in the former case, activism could be said to rely on the transmission of quality, consistent information, able to consciously affect the decisions of both the electorate and policy-makers; the latter, instead, seems to count on a massive diffusion of impressive pieces of information, easy to agree with, aimed at increasing consensus and thus the numbers of the movement.

This ability of spreading an activist mentality to high numbers of people can be linked to what British documentarist Adam Curtis defined *"Oh Dear"-ism*:[9] a series of informational strategies aimed at touching the audience's sympathetic core, likely to trigger one's desire to take a stand.

"Oh Dear"-ism can easily degenerate in what could be defined *indignation marketing*, likely to affect distributive and participatory journalism as well: the Internet then risks to become a mere framework to disseminate irritating and outraging news to massive audiences, which – for some easily understandable moral and pragmatic rationale – are embedded in equations of the kind: "p is outrageous, therefore p is true"; "the government does not like us to know p, therefore p is true"; and so on.

Our potentiality to be moved and outraged by some piece of news that we actively acquired is certainly linked to the aforementioned issue of self-efficacy: discovering a shocking movie about human rights violation in Iran, for instance on Indymedia, is totally different from receiving the same information from corporate news. My activity was fundamental in disclosing the item that sparked the indignant reaction, therefore my moral turmoil is *already* perceived as a sign of commitment, and the unequivocally *simple* action of sharing the very movie of my blog, social network or mailing list boosts my self-efficacy.

This connects with the aforementioned problem of *amateurization*, inasmuch activism, thus configured, does not need the amount of specific knowledge and expertise required to deal with a fully structured campaign: a few moving images or movies, an outrageous account of abuse and prevarication are enough to produce a *flash-mob* reaction, with thousands, sometimes millions of people *sincerely* supporting a cause that they utterly ignored until a few days before, and that they will ignore once again little time after they will feel they "made their part", by an economic or pragmatic (real or virtual) action.

Enthusiasm and Granular Knowledge

Gillin (2007) put forward a quite detailed analysis about the role played by the so-called enthusiasts. According to him, enthusiasts are basically a company's customers who are really committed and passionate about a particular product of the company. Gillin (2007) labels them as the cream of the crop. They do not only provide feedbacks, but they enthusiastically share their passion, opinions, feelings about a certain product for which they sort of become like evangelists. With the advent of the Web 2.0, enthusiasts have become important resources for businesses and organizations, as they can effectively influence other customers, for instance through blogging. They can also participate in endless discussions in forums about the new version of a certain product. They simply love to talk and in doing so they generate a lot of content. More generally, what characterizes them is a particular care not only about the product, but also about the process of forming their opinions about it. Their opinions result to be well-informed and reasoned so that they are usually

asked by others for suggestion. It is worth to note that enthusiasts are not on the payroll of any of the company whose products they advertise. The result for the company is to have a global online focus group working for free (p. 36).

Enthusiasts and activists share some important characteristics that are worth considering here. First of all, they want to be involved. They want to participate and not be passive recipient. Secondly, they want to discuss. Enthusiasts as well as activists love to discuss, to verbally express their position. It seems that they make extensive use of what is called playful arguments (Hample, Han, & Payne, 2009; Magnani, in press); meaning that they often engage a partner in discussion to entertain themselves. The third element is about transparency: enthusiasts and activists want to be transparent in the sense that they apparently reject any form of strategizing. They are direct and they do not want to be accommodating. They stage themselves as truth-lovers, meaning that they have no other interest but seeking truth.

Gillin (2007) listed another aspect characterizing enthusiasts which would be interesting to put in comparison with activists. Gillin (2007) maintained that enthusiasts know a great deal about the product they talk and write about. Enthusiasts manage to translate their passion into knowledge. It is worth noting that the kind of knowledge they develop is limited to the product that is *literally* their passion. We posit that this kind of knowledge they have can be called *nested knowledge*. That is, it is a kind of knowledge that is inside a larger part enthusiasts do not necessarily display or have. For instance, a big fan of the iPad does not necessarily know a great deal about programming or computing, because she is enthusiast as a user, not as a hacker or programmer.

In the case of the activist, things become much more complicated. What characterizes activists is that they usually get involved in single-issue campaigns, say, campaigns that usually have a specific objective or goal. In this sense, the activist is not so different from the enthusiast, as she or he exhibits a strong commitment to a particular activity. However, the kind of knowledge activists displays is different. Enthusiasts' passion is almost immediately translated into expertise. They get to know more and more about the product. So, they develop this sort of nested knowledge. Such an expertise refers to a particular thing, namely, a product. Conversely, the case of the political activist is different, because it is different what *politics* is all about. As we have already pointed out, activism is characterized by the refusal of any mediation brought about by institutional channels, for example, by politicians. That refusal implies, however, that the activist should be able to tackle down issues concerning the all spectrum of politics. But what seems to happen is that they just develop a very strong concerned for single issues that may never come to be integrated in a coherent vision of society as a whole.

It is this lack of integration that makes activist's knowledge scarcely *granular*. In Artificial Intelligence *granularity* of a given knowledge base – i.e., what one knows – refers to the ability to generate new knowledge through making use of known knowledge (Qian, Liang, & Dang, 2008). It follows that the more granular a knowledge base is, the easier it is to draw an inference from what one already knows to the aim of approximating an unknown concept. Conversely, the less it is granular, the harder it is to establish connections among different pieces of information contained in one's knowledge base. The former case fairly represents the kind of knowledge base activists have. As one can easily note, this fragmented character makes extremely hard any political development that involves the extension of one's knowledge in order to make sense of other issues. In contrast to that, ideology can be considered a special kind of (political) knowledge base that is extremely granular and much less fragmented, as it allows a person to easily establish useful links and connections among apparently different issues and/or problems affecting society.

Single-Issued and Desultory Activism

Cyberactivism benefits from being highly compatible with the multitasking identity characterizing *homo informaticus* more and more. We can consider recent campaigns, such as the international mobilization to prevent Sakineh, an Iranian woman accused of adultery and murder, from being stoned to death, or the anti-FARC campaigns reported in (Neumayer & Raffl, 2008): those campaigns are clearly devised by highly-committed and informed individuals, and the proposed unto mass public by means of mouth-to-ear dynamics (sometimes picked up and echoed by corporate news) and social networks. As a matter of fact, one could easily say that the keyword of cyberactivism is *sharing*: the aware cyber-citizen's duty is to be *docile* (Simon, 1993) and to *share* any relevant pieces of information she may come across during her internet experience.

Even if this creates an unlimited potentiality to spread useful information and global awareness on vital topics, the risk is to create a perennial state of activist alert, yet embedded in a multitasking dimension, therefore *among any other things*. We contend that the *activist state of mind* can become an ideology *per se*, with its own framework moralization and overmoralization (Magnani, in press): an overcharged self-efficacy (permitted by technological means that allow everyone's voice to be heard much more easily than within a traditional information framework (Amichai-Hamburger et al., 2008)), persuades people that their opinion *is* actually relevant whatever the topic is, thus compelling them to *have* an opinion, often polarized, on everything, easily crossing within the boundaries of plain bullshitting (Frankfurt, 2005; Bardone, 2011; Magnani, in press). This further corroborates the *activist state of mind*, insofar a person perceives her state of constant activation and (perceived) commitment as contrasting those who do not share the same opinion/urgency or, even worse, do not feel concerned at all.

If social networks such as Facebook are the prime framework for such a distributed activism (Neumayer & Raffl, 2008), one must pay attention to the phenomenology of such activism, which is usually embodied in *Groups* and *Fanpages*. Groups respond to a bistable logic to which one can or cannot adhere: furthermore, group adhesion is *una tantum*. It is significant to notice how, as of October 2010, it is extremely difficult to retrieve official statistics about Facebook usage[10], a website offering statistics about groups and pages (Social Bakers, n. d.) suggests how, for instance, the *fan page* of US President Barack Obama is only 16th in popularity, with over 15 million members, overwhelmed by the Texas Hold'em Poker page, which counts over 25 million members and is the most popular page in the social network.

Such data suggest us how we should always be careful when dealing with the figures of cyber-activism: we do not mean to put forward a conservative stance on the matter, nevertheless the 3 million members of the anti-FARC group seem hardly relevant when compared to other much more popular social phenomena.

What we want to highlight, instead, is the easiness of cyberactivism when compared to traditional forms of *real world* activism, for instance that of the 1960's or, more widely, of the various social movements (Kornberg & Brehm, 1971). Traditionally, activism was considered as an *effect* of a strong ideological commitment, and was usually focused within a well-defined field (i.e. women's rights, civil rights, environmental issues) while the new state of perennial activation presupposes a ideology of activism afforded by the new distributed technological framework. Nowadays, activism seems in fact to be supported by a "Do whatever is in your power to do" rationale, echoing counterculture's refusal of political mediation, but in our informational ecologies, what we can do is often within our easy reach, notwithstanding the particular field of interest.

It must be acknowledged how such a state of paroxysmal activation has already been observed in recent history: "[d]uring periods of social turbulence, collective action may be taken in a disorganized manner by a generational unit, without the benefit of an articulated ideology or a politicized group identification. During the Watts riots of 1965, for instance, participants may not have been acting on an articulated ideology of race consciousness so much as they were acting on a diffuse feeling of anger or frustration" (Duncan, 1999, p. 613). It is easy to draw an hypothetical implication, then, between the diffusion of activism and the state of national and international activation following the events of 9/11, and this distributed and *soft* activism as a reaction (and counteraction) to the moral and aggressive alert instructed by western democracies (Jacobs, 2005).

CONCLUSION

Embedded in multitasking habits, cyberactivism faces the risk of becoming desultory, a kind of activism *à temps pérdu*, which allows citizens to overestimate their self-efficacy by employing their spare time in order to increase global awareness. Again, our contention is that there is nothing wrong, *au contraire!*, in developing one's civic awareness with the aim of becoming a better informed citizen, voter and so on: still, we maintain that a strict line of demarcation should be drawn between what is and what is not activism, lest – in case our societies face a *real* emergency, be it political, social, economic – the big figures of cyber-activism crumble to pieces (as everyone will say "I have already done my part, I fought my share of the battle!") leaving the real fight to a small bunch of hard-core, old-fashioned (and maybe extremist) activists.[11]

REFERENCES

Alvarez, R. (2006). Controlling democracy: The principal-agent problems in election administration. *Policy Studies Journal: the Journal of the Policy Studies Organization, 34*, 491–520. doi:10.1111/j.1541-0072.2006.00188.x

Amichai-Hamburger, Y., McKenna, K., & Tal, S. (2008). E-empowerment: Empowerment by the Internet. *Computers in Human Behavior, 24*(5), 1776–1789. doi:10.1016/j.chb.2008.02.002

Bardone, E. (2011). *Seeking chance: From biased rationality to distributed cognition*. Berlin, Germany: Springer-Verlag.

Barnett, P. W., & Littlejohn, S. W. (1997). *Moral conflict: When social worlds collide*. London, UK: Sage.

Barro, R. (1973). The control of politicians: an economic model. *Public Choice, 14*, 19–42. doi:10.1007/BF01718440

Bernays, E. (2005). *Propaganda*. Brooklyn, NY: Ig Publishing.

Campbell, R. (1974). The sorites paradox. *Philosophical Studies, 26*(3), 175–191. doi:10.1007/BF00398877

Cargile, J. (1969). The sorites paradox. *The British Journal for the Philosophy of Science*, 193–202. doi:10.1093/bjps/20.3.193

Carroll, W., & Hackett, R. (2006). Democratic media activism through the lens of social movement theory. *Media Culture & Society, 28*(1), 83–104. doi:10.1177/0163443706059289

Castells, M. (2001). *The Internet galaxy: Reflections on the Internet, business, and society*. Oxford, UK: Oxford University Press.

Chomsky, N. (2002). *Understanding power*. New York, NY: New Press.

Dogan, M. K. (2010). Transparency and political moral hazard. *Public Choice, 142*, 215–235. doi:10.1007/s11127-009-9485-0

Dowd, K. (2009). Moral hazard and the financial crisis. *The Cato Journal, 29*(1), 141–166.

Dowding, K. (2005). Is it rational to vote? Five types of answer and a suggestion. *British Journal of Politics and International Relations, 7*(3), 442–459. doi:10.1111/j.1467-856X.2005.00188.x

Duncan, L. (1999). Motivation for collective action: Group consciousness as mediator of personality, life experiences, and women's rights activism. *Political Psychology, 20*(3), 611–635. doi:10.1111/0162-895X.00159

Fearon, F. (1998). Deliberation as discussion. In Elster, J. (Ed.), *Deliberative democracy* (pp. 123–140). Cambridge, UK: Cambridge University Press.

Frankfurt, H. (2005). *On bullshit*. Princeton, NJ: Princeton University Press.

Furedi, F. (2002). *Culture of fear*. London, UK: Continuum.

Gillin, P. (2007). *The new influencers: A marketer's guide to the new social media*. New York, NY: Quill Driver Books.

Hample, D., Han, B., & Payne, D. (2010). The aggressiveness of playful arguments. *Argumentation, 24*(4), 405–421. doi:10.1007/s10503-009-9173-8

Hutchins, E. (1995). *Cognition in the wild*. Cambridge, MA: MIT Press.

Illia, L. (2003). Passage to cyberactivism: How dynamics of activism change. *Journal of Public Affairs, 3*(4), 326–337. doi:10.1002/pa.161

Jacobs, D. (2005). Internet activism and the democratic emergency in the US. *Ephemera: Theory & Politics in Organization, 5*(1), 68–77.

Keen, A. (2007). *The cult of amateur. How today's Internet is killing our culture and assaulting our economy*. London, UK: Nicholas Brealey Publishing.

Kornberg, A., & Brehm, M. (1971). Ideology, institutional identification, and campus activism. *Social Forces, 49*(3), 445–459. doi:10.2307/3005736

Lippmann, W. (1997). *Public opinion*. London, UK: Free Press.

Magnani, L. (2009). *Abductive cognition: The epistemological and eco-cognitive dimensions of hypothetical reasoning*. Berlin, Germany: Springer-Verlag.

Magnani, L. (in press). *Understanding violence. morality, religion, and violence intertwined: A philosophical stance*.

Meikle, G. (2002). *Future active: Media activism and the Internet*. London, UK: Routledge.

Meikle, G. (2010). Intercreativity: Mapping online activism. *The International Handbook of Internet Research*, 363-377.

Neumayer, C., & Raffl, C. (2008). Facebook for global protest: The potential and limits of social software for grassroots activism. In *Proceedings of the 5th Prato Community Informatics & Development Informatics Conference*.

Pickard, V. (2008). Cooptation and cooperation: Institutional exemplars of democratic internet technology. *New Media & Society, 10*(4), 625–645. doi:10.1177/1461444808093734

Popper, K. R. (1945). *Open society and its enemies*. London, UK: Routledge.

Postill, J. (2008). Localizing the Internet beyond communities and networks. *New Media & Society, 10*(3), 413–431. doi:10.1177/1461444808089416

Putnam, R. (2000). *Bowling alone*. New York, NY: Simon & Schuster.

Qian, Y., Liang, J., & Dang, C. (2008). Knowledge structure, knowledge granulation and knowledge distance in a knowledge base. *International Journal of Approximate Reasoning, 50*, 174–188. doi:10.1016/j.ijar.2008.08.004

Rogoff, K. (1990). Equilibrium political budget cycles. *The American Economic Review, 80*, 21–36.

Shavell, S. (1979). On moral hazard and insurance. *The Quarterly Journal of Economics, 93*, 541–562. doi:10.2307/1884469

Simon, H. (1983). *Reason in human affairs*. Stanford, CA: Stanford University Press.

Simon, H. (1993). Altruism and economics. *The American Economic Review, 83*(2), 156–161.

Simon, L. (Ed.). (2002). *Democracy and the Internet: Allies or adversaries?* Washington, DC: Woodrow.

Social Bakers. (n. d.). *Facebook page statistics*. Retrieved from http://www.socialbakers.com/facebook-pages/

Thompson, P. (2007). Deception as a semantic attack. In Kott, A., & McEneaney, W. (Eds.), *Adversarial reasoning: Computational approaches to reading the opponent's mind* (pp. 125–144). London, UK: Chapman & Hall/CRC.

Turner, F. (2006). *From counterculture to cyberculture: Stewart Brand, the whole earth network, and the rise of digital utopianism*. Chicago, IL: University of Chicago Press.

Yankelovich, D. (1991). *Coming to public judgment: Making democracy work in a complex world*. Syracuse, NY: Syracuse University Press.

ENDNOTES

[1] Moral hazard can be also applied to several other domains, for instance, the insurance

markets. In this case the moral hazard regards the insured party that may have an incentive to misbehave, as he knows that the negative consequences of his actions will be covered by the insurance. One of the side effects caused by moral hazard in the insurance markets is that the costs of being insured will tend to raise (Shavell, 1979). Moral hazard may also regard a worker having an incentive to shirk on the job. For instance, once a person acquires a secure or permanent position, he will be less motivated to put his company's interests first, as he will not bear the consequences of a poor performance at work. More generally, the deceptive and perverted nature of moral hazard is treated in (Magnani, 2011).

2 Needless to say, the democratic perspective is an anthropologically optimist one: if we did not believe that a person should be able, at least in principle, to monitor her choices then we had better turn to a despotic, Platonic rule of superior individuals over a herd of more or less deficient human beings.

3 This expression was coined from the cognitive anthropologist Hutchins (1995) to refer to various external tools that can be built to cognitively help the activity of navigating in modern but also in "primitive" settings.

4 As for the rise of democratic media activism see (Carroll & Hackett, 2006), which conceptualizes its original structure as follows: "Media activism appears to have arisen from several social sources, which may speculatively be conceptualized in terms of three concentric circles. At the centre are groups within and around the media industries, groups whose working life or professional specialization may stimulate awareness of the alienation, exploitation and/or constraints on creativity and public information rights generated by a commercialized corporate media system. Such groups as media workers, journalists, independent producers,

librarians and communication researchers have already been well represented in media democracy initiatives" (p. 85).

5 Many projects concentrated of course on how the Internet could serve the democratic activity par excellence, that is voting and expressing one's preferences, such as the now-commercialized by Adobe PICOLA project (http://caae.phil.cmu.edu/picola/).

6 For the notion of e-empowerment and a survey of the related potentialities brought about by our technological framework see (Amichai-Hamburger, McKenna, & Tal, 2008).

7 For the notion of e-empowerment and a survey of the related potentialities brought about by our technological framework see (Amichai-Hamburger, McKenna, & Tal, 2008). rces are therefore often questioned by independent information channels (i.e. the aforementioned Indymedia, Current TV etc.). Nevertheless, the quality and truthfulness of any news provider can be questioned a priori, be it a big news corporation or a first-person freelance blogger (for further reference on media activism see the already quoted (Carroll & Hackett, 2006)).

8 Such a complete refusal of political and institutional mediation had already developed in Italy short after WWII and the end of Fascism: the short-lived movement was called the "Front of the Common Man" (Fronte dell'Uomo Qualunque). Their scope consisted in the delegitimization of politicians and parties: their main tools where irony and political satire.

9 The documentary appeared in the program called Newspipe in 2009, and begins as we report here: "Everyone knows that television news can be boring, that's because it's often about politics which can be very dull... But these days there's another problem with watching the news. Night after night we are shown terrible things which we feel we can

do nothing about. Images of civil wars, massacres and starving children which leave us feeling helpless and depressed and to which the only response is: 'Oh dear.' There is a name for this. It's called 'oh-dearism' and this is the story of the rise of oh dearism in television news."

10 We do not feel like commenting how this lack of transparency hardly blends with the scopes and aims of activism. In December 2009, after a controversial street aggression against Italian PM Silvio Berlusconi, many Facebook users claimed they had been subscribed to support-groups in favor of the Prime Minister against their knowledge or will.

11 On a more positive note, a virtuous interaction can be identified between amateur cyberactivism and what is referred to as banal activism (Postill, 2008): "the activism of seemingly mundane issues such as traffic congestion, waste disposal and petty crime" (p. 419). The usefulness of technologies to develop awareness – beyond single issues within the marketing of indignation – for local communities has often been disregarded.

This work was previously published in the International Journal of Technoethics, Volume 2, Issue 2, edited by Rocci Luppicini, pp. 14-29, copyright 2011 by IGI Publishing (an imprint of IGI Global).

Chapter 5
On the Moral Equality of Artificial Agents[1]

Christopher Wareham
European School of Molecular Medicine, Italy & The University of Milan, Italy

ABSTRACT

Artificial agents such as robots are performing increasingly significant ethical roles in society. As a result, there is a growing literature regarding their moral status with many suggesting it is justified to regard manufactured entities as having intrinsic moral worth. However, the question of whether artificial agents could have the high degree of moral status that is attributed to human persons has largely been neglected. To address this question, the author developed a respect-based account of the ethical criteria for the moral status of persons. From this account, the paper employs an empirical test that must be passed in order for artificial agents to be considered alongside persons as having the corresponding rights and duties.

INTRODUCTION

Increasingly interdisciplinary work in fields such as synthetic biology, molecular medicine, nanotechnology, and computational and cognitive sciences is presenting new challenges for ethical thought. These challenges have given rise to the evolving discipline of technoethics, which focuses on humans' complex ethical relationships

DOI: 10.4018/978-1-4666-2931-8.ch005

with our rapidly changing technological environment. An emerging feature of this environment is the increasingly significant array of social roles performed by artificial agents such as robots.[2] As Mark Coeckelbergh (2010) notes, the uses of robots have diversified to include warfare, education, entertainment, sex, and healthcare. Evidence suggests that humans often treat robots as companions and partners, and not merely as objects (Libin, A., & Libin, E. 2004). Inevitably their increased social importance has generated interest in the ethics of artificial agents. Some of this interest is focussed on the types of moral rules artificial agents should have[3] and how these rules could be acquired.[4] Theorists have also discussed the moral status of artificial agents; that is, whether artificial agents should be treated as objects of moral concern.[5] However, despite the burgeoning literature claiming that artificial agents have moral worth, and could learn and apply moral rules, few have considered whether artificial agents could gain the same moral status as human persons.[6] In response to this question I develop an account of the ethical criteria for the moral status of persons. Such an account will, of course, face difficulties in its application. In order to mitigate these difficulties, I defend an empirical test that must be passed in order for artificial agents to be considered alongside persons as having the corresponding rights and duties.

MORAL STATUS AND PERSONHOOD

Moral theorists use the word 'person' in a way that is different to the common usage of the term. In moral theory, a person is a being with particular types of interests, such as an interest in preference satisfaction, or particular capacities, such as the ability to reason. On the basis of these interests or capacities, persons are regarded as having the highest moral status. Significantly, it is a core intuition of moral and political theory that all persons have equal moral status and are thus attributed the same basic rights and duties.

It is important to note that in secular ethical theory not all humans are persons.[7] For instance, foetuses and humans that are severely cognitively impaired lack the relevant interests or capacities that persons have.[8] More pertinently to the issue at hand, in order to avoid arbitrary anthropocentrism, most moral theorists accept that non-humans can in principle have the status of persons. For instance, if an alien or an animal had the relevant interests or capacities, it would be a person. Indeed, some

theorists have argued that higher mammals such as dolphins should be recognised as having higher moral status.[9] Thus, if an artificial agent fulfilled the relevant criteria it should be considered as having rights and duties that persons have.

But what are the *relevant* criteria? In the rest of this section I examine the ethical basis for the higher moral status of persons in order to clarify criteria for equivalent moral status. Two main families of theory are thought to explain moral status: interest-based theories and respect-based theories. I argue that the former should be rejected, since they undermine the ideal that persons are moral equals. I argue that if respect-based accounts are suitably fleshed out they can provide adequate ethical criteria for the moral worth of persons.

Interest-Based Theories

Interest-based theories attribute moral status according to how much good a being's life contains. For instance, a flea's life contains very little recognisable as good, while a dog's life might contain goods such as the pleasure of gnawing on a squeaky toy.

On this type of account, *persons* have higher moral value since their lives contain higher order goods. John Stuart Mill, for instance, claimed that human happiness is qualitatively superior to that of animals, since humans have higher order intellectual and emotional capacities.[10] If this type of account is accepted, then, an artificial agent would have the same moral status as persons if its life contained the same degree of higher goods that humans *qua* persons generally possess.

However, interest-based accounts of the moral status of persons should be rejected, since they contradict a fundamental ethical tenet: the idea that persons have *equal* moral status and are thus deserving of equal respect.[11] Interest-based accounts attribute moral status according to the amount of good that a being's life contains. Thus if some person has a life that contains a greater degree of well-being than another person's life contains, they have higher moral status. This means that interest-based theories entail, unacceptably, that if a person's life goes worse for them, they have less moral worth than a person whose life goes better for them. Since this implication conflicts with the idea that persons are morally equal, interest-based accounts provide an inadequate basis for the moral status of persons.[12]

Respect-Based Accounts

Respect-based accounts avoid the above problem. Rather than attributing moral status in proportion to the amount of good a life contains, this type of theory is based on kantian and contractualist accounts, which ground intrinsic moral worth in *the capacity to engage in mutual accountability through the giving and heeding of reasons*. (Buchanan, 2009, p. 361, My italics)

If a person has this basic capacity then she has intrinsic moral worth that ought to be respected. Improvements in the ability to give and heed reasons don't imply a higher moral status because the basic ability is all that's required. In this way the respect-based theory provides for a threshold criterion above which equal respect for the moral status of persons should be observed.[13]

As it stands, however, the above criterion is too vague; it is not entirely clear what is meant by the capacity to 'give and heed *reasons*', or how this capacity contributes to the mutual accountability that grounds equal moral worth. In a broad sense even some insects may be said to have the ability to give and heed reasons, since they communicate chemical signals which tell them where to find food. We would not, however, hold insects to be accountable in the way that persons are. Proponents of respect-based accounts have a different type of reason in mind. In particular, respect-based accounts ground moral status in the ability to give and heed *moral* reasons.[14] In order to flesh out the criterion for equal moral worth, it is thus necessary to bridge the gap between the giving and heeding of moral reasons and the 'mutual accountability' mentioned above.

R. Jay Wallace's (1995) account of the criteria for a morally accountable agent provides an effective means of linking mutual accountability with the giving and heeding of reasons. According to Wallace a morally accountable agent is one who has the capacity for 'reflective self-control.' This requires i) the ability to grasp and apply moral reasons and ii) the ability to regulate one's behaviour in the light of those reasons. The capacity for reflective self-control thus explains the relation between moral reasons and the mutual accountability of persons. Since it is this relationship that grounds equal moral worth, the possession of the capacity for reflective self-control justifies an attribution of moral worth equal to that of persons.

AN EMPIRICAL TEST FOR EQUIVALENT MORAL STATUS

The upshot of the above argument is that an artificial agent has moral status equal to that of a person if it has the capacity for reflective self-control. Can artificial agents possess this capacity? Promising developments in the field of artificial life, as well as sustained and productive efforts to create moral machines[15] give some reason to think they could. However, on the face of it, this seems like an empirical question. It is desirable to devise a test which demonstrates whether or not a particular artificial agent has the ability to grasp and apply moral reasons and to regulate its behaviour in the light of those reasons. I consider three candidate tests – the moral Turing test, the humanity test, and the psychological test – and argue that only the last is well-suited to determination of equivalent moral status. Thereafter I address the objection that it is *in principle* impossible for a computational entity to have the relevant capacities.

Moral Turing Tests

Several authors have suggested the possibility of a moral Turing test.[16] Alan Turing famously advocated Turing tests as a measure of machine intelligence. He proposed that if an objective observer cannot distinguish a machine interlocutor from a human interlocutor at a level above chance, then the machine should be regarded as having intelligence (Turing, 1950).

The possibility of this type of test has been innovatively applied to the moral case. Allen et al. (2000) suggest that the standard Turing test could be modified by restricting the conversation to discussion concerning morality. This moral Turing test could potentially provide a measure of the artificial agent's ability to grasp and apply moral reasons. If an observer is unable to distinguish between the human person and the artificial agent, this provides some grounds for believing that the artificial agent has reflective self-control and thus equivalent moral status.

However, Turing tests are unsuited to the purpose of demonstrating equivalent moral status. Some artificial agents may possess the capacity for reflective self-control to a greater degree than human persons due to superior reasoning ability.[17] If so the conversation of such an agent would likely be distinguishable from the human person's conversation on the basis of superior ability to apply moral reasons, and it would fail the moral Turing test. Since it would be grossly immoral to deny equal moral status on the basis of better capacities, or capacities that are differently expressed, the moral Turing test is inadequate as a test for reflective self-control.[18]

The Humanity Test

A second test appeals to a reliable and often-used standard for determining the possession of reflective-self-control: humanity. Humans generally have the ability to grasp and apply moral reasons and to modify their behaviour in the light of those reasons. We have the capacity of reflective self-control, even if we don't always use it. Perhaps, then, humanity could function as the relevant test for an artificial agent's moral status. Artificial agents could pass the test if they were created to be functionally and structurally isomorphic to humans.

Humanity *is* a good test for personhood. And, given the seriousness of the potential moral error it is a good idea to treat humans as moral equals unless there is strong evidence to the contrary. However, as mentioned earlier, some non-humans may be persons with equivalent moral status. So while the humanity test provides a good indicator of higher moral status, it leaves open the possibility that non-human artificial agents will be arbitrarily denied equal moral status. Given the severity of the moral wrongs that have resulted from failure to recognise the equal moral status of persons, a better test for personhood would be desirable.

Psychological Tests

I argue that existing psychological tests could be adapted to fulfil this role. Psychologists are often called upon to distinguish between those that have the capacity of reflective self-control and those that don't.[19] Psychopaths and sociopaths are paradigm cases of humans without the ability to grasp and apply moral reasons, and who cannot modify their behaviour in the light of moral reasons.[20] Psychopaths, for instance, have an abnormal lack of empathy – the distinctively moral ability to engage with or take into account the mental states of others. This ability forms a crucial aspect of moral reasoning. Without it, psychopaths are often incapable of regulating their behaviour in accordance with moral reasons.

Psychologists thus already have a variety of techniques used to identify the absence of relevant capacities that constitute reflective self-control.[21] Since we are testing for capacities that human persons have, it seems appropriate to use tests for artificial agents that are similar to those used for humans. Nevertheless, such techniques would have to be adapted. Given human bias on the part of psychologists, artificial agents that do not look or sound like humans may be discriminated against. In such cases a typed evaluation may be appropriate, as in the original Turing test. Alternatively, additional training for psychologists may be required in order to counteract potential biases.[22]

With suitably adapted diagnostics in place, the psychologist should be capable of identifying the presence of reflective self-control in artificial agents. Evaluations with a suitably trained psychologist may proceed as follows. As in the case of the moral Turing test, matters concerning morality will be discussed and the agent's capacity to grasp and apply moral reasons will be assessed. Perhaps the evaluation will involve immersing the agent in a simulated environment in which the reason-responsiveness of its behaviour can be tested. This would assist the psychologist in her diagnosis by providing biographic information often present in the human case, but which may be absent in the case of artificial agents. By conducting the psychological test in this way the artificial agent's capacities of reflective self-control might be examined in a way that is analogous to the way in which humans' capacities are examined.

Moral Turing tests and the humanity test are excessively restrictive in different ways. Both require *equivalence* in features not determinative of the possession of the capacity of reflective self-control. Moral Turing tests require equivalence in moral reasoning, which would unjustifiably exclude artificial agents with equal moral status if they have better reasoning ability. Humanity tests require equivalence with humans in physical and functional capacities, but this neglects the morally serious possibility that non-humans can be persons. Psychological testing corrects these shortcomings by directly testing the presence of the relevant capacity of reflective self-control. Although such tests are by no means infallible, they are the best empirical means we have for testing whether artificial agents have the moral status of persons.[23]

Objections to Psychological Tests for Equivalent Moral Status

Nevertheless, it might be objected that this type of test sets the bar too high. Artificial agents may have a high degree of moral status, even if they lack reflective self-control, in much the same way that many people think that psychopaths have a high degree of moral status, despite their lack of empathic abilities. However, failing the test for reflective self-control does not entail that agents have low or non-existent intrinsic worth. They could, for instance, have high *worth* on the basis of higher order goods stipulated by interest-based accounts, even though, as argued, they would not have equivalent moral *status*. Furthermore, there may be grounds for *treating* psychopaths as having equal moral status, given the fact that, since they are human, they may have increased potential to develop the relevant capacities.[24] However, they do not, as it stands, carry the same rights and moral status as persons, in part since they are incapable of fulfilling the moral duties that persons are capable of fulfilling.

A further objection holds that psychological tests, and indeed all the above tests, set the bar for granting equivalent moral status too low, since an artificial agent can *in principle* never have the capacity for reflective self-control. Arguing that Turing tests fail as a measure for machine intelligence, Block (1981) and Searle (1980) hold that computational entities cannot be considered to 'understand' any more than a toaster. Instead, complex responses are the result of programmed syntax devoid of semantic meaning. Any 'understanding' exhibited is solely simulation, and not duplication of human understanding. Applying this objection to the current context, it can be questioned whether computational artificial agents could in fact grasp the moral reasons that they applied and in light of which they modified their behaviour, or whether the behaviour would merely be syntactical output capable of fooling psychologists.

The first point to note in reply is that I have nowhere stated that artificial agents would necessarily be computational entities. Artificial agents may, for instance, be artificially created organisms. They may also be hybrids between machine entities and organic beings.[25] In these cases Block and Searle's criticisms do not apply since they relate only to the syntactic interface of computational agents. Nevertheless, their objections dismiss the potential moral equality of an enormous category of artificial agents, so it is worthwhile to discuss them.

Block and Searle's criticisms raise a complex issue, which cannot be comprehensively dealt with here. However, two considerations count strongly in favour of recognising computational artificial agents as persons if they pass the relevant psychological tests. The first consideration is that unwarranted extensions of moral status are more acceptable than unjustified denials. The failures to acknowledge slaves, particular racial groups and women as moral equals are surely more unacceptable than ancient Egyptians' attribution of extremely high moral status to cats. It is thus much better to accord moral status to something which doesn't have it than it is fail to accord moral status to something that does. When combined with this consideration, good evidence of personhood that would be supplied by psychological tests creates a strong presumption in favour of treating successful artificial agents as persons.

The second consideration directly targets the Block/Searle contention. I argue that a computer that contains the syntactical information required to pass the psychological test simply is not feasible in practice. As Mark Bedau points out, for an unthinking device to pass a Turing test, the number of pieces of information they must store is larger than the number of elementary particles in the entire universe. Though possible in principle, such a device is clearly impossible in practice. (Bedau, 2004, p. 209)

It seems highly plausible that the amount of computing space required to demonstrate the moral capacity of reflective self-control to a trained psychologist is at least as extensive. Thus it is unrealistic to assume that an artificial agent that passes the test is merely providing *syntactic* output to fulfil the psychologist's evaluation. If the computational artificial agent passes the psychological test, it is far more reasonable to believe that it has the relevant *semantic* capabilities themselves.

CONCLUSION

I argued that artificial agents could be considered as having the same moral status as humans if they have the capacity of reflective self-control, which consists in the ability to grasp and apply moral reasons and to regulate one's behaviour in the light of such reasons. Thereafter, I claimed that it is possible to develop a test for this capacity by modifying existing psychological tests. If these claims are correct, it follows that an artificial agent that passes the type of psychological test I have defended has moral status equivalent to that of a human person. In this case an artificial agent would have the same moral rights and responsibilities that human persons have.

The creation of moral agent of this type may seem like a significant technological goal. However, it is worth pointing out that bringing a being with equivalent moral status into existence carries with it a tremendous responsibility. It is, for instance, at least as morally significant as bringing a human child into existence. Furthermore, given that laws generally recognise the rights of morally equal beings, it seems plausible that the same laws that govern the creation of human persons should govern the creation of artificial persons. As such, the decision to create an artificial agent with the moral status of persons is not one that should be taken lightly.

REFERENCES

Adam, A. (2008). Ethics for things. *Ethics and Information Technology, 10,* 149–154. doi:10.1007/s10676-008-9169-3

Allen, C., Smit, I., & Wallach, W. (2005). Artificial morality: top-down, bottom-up, and hybrid approaches. *Ethics and Information Technology, 7,* 149–155. doi:10.1007/s10676-006-0004-4

Allen, C., Varner, G., & Zinser, J. (2000). Prolegomena to any future artificial moral agent. *Journal of Experimental & Theoretical Artificial Intelligence, 12*(3), 251–261. doi:10.1080/09528130050111428

Bedau, M. (2003). Artificial life: organization, adaptation and complexity from the bottom up. *Trends in Cognitive Sciences, 7*(11), 505–511. doi:10.1016/j.tics.2003.09.012

Bedau, M. (2004). Artificial life. In Floridi, L. (Ed.), *Philosophy of computing and information* (pp. 97–211). Oxford, UK: Blackwell Press.

Block, N. (1981). Psychologism and behaviorism. *The Philosophical Review, 90,* 5–43. doi:10.2307/2184371

Bringsjord, S. (2004). On building robot persons: A response to Zlatev. *Minds and Machines, 14*(3), 381–385. doi:10.1023/B:MIND.0000035477.39773.21

Buchanan, A. (2009). Moral status and human enhancement. *Philosophy & Public Affairs, 37*(4), 346–381. doi:10.1111/j.1088-4963.2009.01166.x

Coeckelberg, M. (2010). Robot rights? Towards a social-relational justification of moral consideration. *Ethics and Information Technology, 12,* 209–221. doi:10.1007/s10676-010-9235-5

Dennett, D. (1997). When HAL kills, who's to blame? Computer ethics. In Stork, D. G. (Ed.), *HAL's Legacy: 2001's Computer as Dream and Reality.* Cambridge, MA: MIT Press.

Floridi, L. (2008). Information ethics: a reappraisal. *Ethics and Information Technology, 10,* 189–204. doi:10.1007/s10676-008-9176-4

Hare, R. D. (2003). *The Psychopathy Checklist—Revised.* Toronto, Canada: Multi-Health Systems.

Himma, K. E. (2004). There's something about Mary: the moral value of things *qua* information objects. *Ethics and Information Technology, 6,* 145–159. doi:10.1007/s10676-004-3804-4

Libin, A., & Libin, E. (2004). Person–robot interactions from the robopsychologists point of view: the robotic psychology and robotherapy approach. *Proceedings of the IEEE, 92*(11), 1789–1803. doi:10.1109/JPROC.2004.835366

Mill, J. S. (1957). *Utilitarianism.* Indianapolis, IN: The Liberal Arts Press.

Regan, T. (1983). *The case for animal rights.* Berkeley, CA: UCLA Press.

Scanlon, T. (1998). *What we owe to each other.* Cambridge, MA: The Belknap Press of Harvard University Press.

Searle, J. (1980). Minds, brains and programs. *The Behavioral and Brain Sciences, 3,* 417–457. doi:10.1017/S0140525X00005756

Singer, P. (1993). *Practical ethics.* Cambridge, UK: Cambridge University Press.

Stahl, B. C. (2004). Information ethics and computers: the problem of autonomous moral agents. *Minds and Machines, 14,* 67–83. doi:10.1023/B:MIND.0000005136.61217.93

Turing, A. (1950). Computing machinery and intelligence. *Mind, 59*(236), 433–460. doi:10.1093/mind/LIX.236.433

Versanyi, L. (1974). Can robots be moral? *Ethics, 84*(3), 248–259. doi:10.1086/291922

Wallace, R. J. (1994). *Responsibility and the moral sentiments*. Cambridge, MA: Harvard University Press.

White, T. (2007). *In defense of dolphins: the new moral frontier*. Oxford, UK: Blackwell. doi:10.1002/9780470694152

ENDNOTES

1. Thanks to Mark Bedau for insightful comments on an earlier draft.

2. I define an artificial agent as a manufactured object capable of affecting its environment.

3. Allen, Varner, and Zinser (2000) discuss how moral theory could inform the rules to be followed by moral artificial agents.

4. Allen, Smit, and Wallach (2005) suggest that theories of robot learning could contribute to the adoption of moral rules, as well as the development of moral theory.

5. See Adam (2008), Floridi (2008), Himma (2004), and Versanyi (1974) for discussions of the intrinsic moral worth of artificial agents.

6. In a recent exception to this trend Coeckelbergh (2010) discusses the related question of whether robots should have rights. Bringsjord (2004) briefly discusses whether robots could have meaning in the way that persons do, but does not relate this to considerations about moral status.

7. Examples of the ethical decoupling of humanity and personhood include Singer (1993), Buchanan (2009), and Regan (1983).

8. Arguments about the morality of abortion thus tend to focus on whether foetuses' *potential* to become persons contributes to their moral status.

9. See for example White (2007).

10. Mill (1957, p.14) claims that 'it is better to be a human dissatisfied than a pig satisfied.'

11. It is of course possible to reject the assumption of moral equality. However, given its centrality in morality and political ethics, a stronger argument is needed than that provided by interest theorists.

12. This is not to say that interests and well-being are irrelevant to moral status, just that they cannot capture all that is meant by the moral status of *persons*. Buchanan (2009) advances a similar argument and claims that interests should be considered on a separate scale of 'moral considerability.'

13. There will inevitably be some vagueness about the boundaries of personhood. For instance, the precise age at which children acquire the ability to 'give and heed reasons' is unclear. Nevertheless if a being has this ability, suitably specified, they are a person with equal moral status.

14. See Scanlon (1998) for an account of moral reasons.

15. See Allen, Varner, et al (2000) and Allen, Smit, et al (2005) for discussion of ways to create moral machines and the rules that should guide them.

16. Allen, Varner, et al. (2000) and Allen, Smit, et al. (2005), Himma (2004), and Stahl (2004) have discussed moral Turing tests. Note that these authors consider the moral Turing test as a test for moral agency rather than for equal status.

17. As suggested earlier, the respect-based account holds that having a superior reasoning capacity does not entail higher moral status.

18. The moral Turing test could, of course, be modified so that it requires the interlocutors to say whether an artificial agent is above or below the threshold. However, this is less in keeping with the original Turing test and more in keeping with the psychological test I defend.

19. Of course psychologists don't frame their tests in terms of reflective self-control, but

20. as I discuss below, this can be seen as an aspect of what such tests achieve.

21. R. Jay Wallace lists psychopathy as an example of an 'exempting condition.' Exempting conditions preclude moral accountability, and, on the view developed here, also remove the conditions required for moral equality with persons. Dennett (1997) discusses the possibility that highly sophisticated computers could be exempted from moral responsibility.

22. Robert Hare's (2003) revised psychopathy checklist (PCL-R) is an example of a widely used diagnostic tool that could be adapted to provide a test for reflective self-control.

23. The emerging discipline of robot psychology may provide some of the resources for this training. See Libin et al. (2004).

24. Unfortunately the reliance of these tests on a common language prevents them from being applied to existing candidates for equal moral status, such as higher mammals. I think a non-linguistic psychological test could be developed to resolve this shortcoming.

25. This idea is questionable for two reasons. First, theorists such as Singer deny that having the potential to have a degree of moral status warrants being treated as having that status. Second, it is empirically controversial whether psychopaths, for instance, can be 'corrected' and develop the capacities they lack.

26. See Bedau (2003) for a discussion of the potential convergence of cognitive science, and artificial life.

This work was previously published in the International Journal of Technoethics, Volume 2, Issue 1, edited by Rocci Luppicini, pp. 35-42, copyright 2011 by IGI Publishing (an imprint of IGI Global).

Section 2
Ethical Dilemmas of Technoselves in a Technosociety

Chapter 6

Biometrics:
An Overview on New Technologies and Ethic Problems

Halim Sayoud
USTHB University, Algeria

ABSTRACT

The term biometrics is derived from the Greek words: bio (life) and metrics (to measure). "Biometric technologies" are defined as automated methods of verifying or recognizing the identity of a living person based on a physiological or behavioral characteristic. Several techniques and features were used over time to recognize human beings several years before the birth of Christ. Today, this research field has become very employed in many applications such as security applications, multimedia applications and banking applications. Also, many methods have been developed to strengthen the biometric accuracy and reduce the imposture errors by using several features such as face, speech, iris, finger vein, etc. From a security purpose and economic point of view, biometrics has brought a great benefit and has become an important tool for governments and institutions. However, citizens are expressing their thorough worry, which is due to the freedom limitations and loss of privacy. This paper briefly presents some new technologies that have recently been proposed in biometrics with their levels of reliability, and discusses the different social and ethic problems that may result from the abusive use of these technologies.

INTRODUCTION

One of the oldest characteristics that have been used for recognition by humans is the face (and sometimes the voice). Also animals are used to recognize each other by the smell and maybe by the sound they produce. For instance, dogs are able to recognize their human masters with a high accuracy. Thus, human beings and animals use these characteristics, unconsciously, to recognize known individuals every day.

Other characteristics have also been used throughout the history of civilization as a more formal means of recognition, as we can see in the following chronological sequence of examples during our old history:

DOI: 10.4018/978-1-4666-2931-8.ch006

- In a cave estimated to be at least 31 thousand years old, the walls are surrounded by numerous handprints that may be considered as signatures of their owners (NSTC, 2006). Several similar cases have been observed in different places.
- In 500 years B.C. Babylonian business transactions were recorded in tablets including fingerprints (NSTC, 2006).
- Early Chinese merchants used fingerprints to settle business transactions (NSTC, 2006).
- According to early Egyptian history, traders were identified by their physical descriptors (NSTC, 2006).
- In 1857, Bertillon developed anthropometrics to identify individuals (NSTC, 2006).
- In 1892, Galton developed a classification system for fingerprints (NSTC, 2006).
- In 1903, New York prisons began to use fingerprints (NSTC, 2006).
- In 1936, the use of iris pattern was proposed (NSTC, 2006).
- In 1963, many research works in signature recognition were proposed (NSTC, 2006).
- In 1974, appeared the first commercial hand geometry system (NSTC, 2006).
- In 1976, the first system for speaker recognition was developed (NSTC, 2006).
- In 1987, a patent stating that the iris could be used for identification was awarded (NSTC, 2006).
- In 1991, real-time face recognition possibility (NSTC, 2006).
- In 1996, NIST began hosting annual speaker recognition evaluations (NSTC, 2006).
- In 1998, FBI launched CODIS, which is a DNA forensic database (NSTC, 2006).
- In 2000, a research work described the use of vascular patterns in biometrics (NSTC, 2006).
- In 2003, the European Biometrics Forum is established (NSTC, 2006).

- In 2005, Fujitsu proposed a security device that uses vein patterns to verify identity.
- In autumn 2005, RFID was integrated into passports in Germany.
- In 2007, RFID was integrated into identity cards in Germany.
- The BioSecure Network of Excellence supported the data collection, software development and infrastructure necessary to conduct a Multimodal Evaluation Campaign in 2007.
- In 2008, a new biometric standard was announced for e-transactions: the standard ISO 19092:2008.
- In 2009, Sagem and Hitachi unveiled the first multi-modal finger vein and fingerprint device at Biometrics 2009 in London.
- In 2010, new biometric modalities using internal physical characteristics have been proposed, such as the Boneprint and ImpPrints (Brooks, 2010).

As we can see, the utilization of biometric techniques has become very popular over the time and especially during the last years, since the computing technologies have seen an important advance and since the need of secure transaction means have been much solicited. In fact, nowadays the applications of biometrics are very varied: security applications (airport and immigration security), multimedia applications (multimedia access and interactivity), banking applications (direct and remote transactions), etc.

However, an important question may arise: what are the disadvantages of biometrics (if any)?

The response is obviously clear, since many human organizations have shown their disappointments regarding this technology. In fact, several ethic and social problems have been quoted, such as: the limitation of freedom (democracy limitation, travelling limitation, etc.), loss of privacy (personal and private information revealed to others), risk of imposture (banking transactions), risk of false rejection (identity rejection during passport inspection) and the oppression feeling for certain reserved persons.

Thus, besides the technological aspect, the present paper is also intended to show and discuss some ethic problems of this technology. This last research field, called ethnoethics, is defined in (Luppicini & Adell, 2008) as an interdisciplinary field concerned with all ethical aspects of technology within a society shaped by technology. In other words, it deals with human processes and practices connected to technology which are embedded within social, political, and moral spheres of life (Luppicini & Adell, 2008).

Our manuscript is organized as follows: In the first section of the paper, an overview on the new biometric technologies is given. The second section describes the social and ethic problems. The third and fourth sections, respectively, discuss the reliability and legitimacy of biometrics (respectively), which try to summarize our research investigation and provide an overall conclusion on this topic.

NEW BIOMETRIC TECHNOLOGIES

As we saw in the introduction, different techniques and features were used over the time to try to make an efficient person authentication. However, there is no technique that could really be perfect. Every technique has advantages and disadvantages, and according to the intended utilization, one may prefer one of them.

The most common techniques/features that have been used in research or industry are summarized in the following points:

Signature Authentication

Signature has been the means by which we validate all our legal documents for a long time (Figure 1). However, absolute validation of signatures is a different matter in practice (Bio, 2002).

Some systems use pens with motion-sensing and pressure-sensing devices inside. In this case, a special pen is used that contains a bi-axial accelerometer to measure changes in force in the x and y direction (Bio, 2002).

Other biometric signature systems use a magnetic tip pen with a sensitive tablet. These systems analyze only the dynamic changes in the x and y directions (Bio, 2002).

A person's signature presents a high intra-variability. Consequently, the recognition systems may not always make a correct authentication. So, in this case, further research works must be made.

Fingerprints Authentication

Fingerprinting is one of the oldest means of identification in use today. An individual's fingerprints are defined by a complex combination of patterns: lines, arches, loops, and whorls (Bio, 2002).

Fingerprint data can be obtained in several ways. One procedure uses the capture of an inked image of the print. It is also possible to optically scan fingerprints (Figure 2). A scanner records and analyzes an image of a finger placed on a glass plate. While much more convenient than inkpads, this optics-based approach can be unreliable because of foreign matter that may cause distortions in the image (Bio, 2002).

Other capture methods based on CMOS technologies can provide direct digital output of detected fingerprint minutiae.

Figure 1. Example of scanned signature

Figure 2. Example of scanned fingerprint

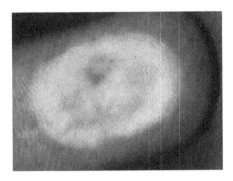

Face Recognition

Face represents a visible and easy biometric identifier, and usually people show no resistance to face recognition systems in practice (Wei & Li, 2005).

There are two general facial recognition systems: The first class uses facial geometrical features, by using a single camera or sometimes multiple cameras to build 3D models of the face. The second class uses facial thermograms. The temperature patterns of the faces of different people are distinctive, and an infrared camera can capture such patterns. There are also emerging approaches, including the use of some skin details of the face (Wei & Li, 2005) (Figure 3).

Figure 3. Contour detection techniques can be used to extract the face contours (2D face recognition)

Speaker Recognition

The vocal expression is a particular characterization for the speaker: thus it is possible, in normal conditions, to recognize his corresponding talker during a direct or telephonic conversation. Speaker characterization is a generic term indicating the possibility to discriminate between several persons thanks to their voices and in this research domain we want to recognize, not what it was said, but the identity of the speaker which is talking, by his vocal characteristics.

Speaker recognition is the ability to recognize who is speaking, by using the vocal characteristics of his or her speech signal. Speaker recognition is divided into several specialties: speaker identification, speaker verification, speaker indexing and speaker discrimination.

- Speaker identification is the ability to identify the identity of a speaker among others;
- Speaker verification is the process of accepting or rejecting the identity claim of a speaker;
- Speaker indexing consists in segmenting and labeling a multi speaker audio document into homogenous segments containing only one speaker each;
- Speaker discrimination is the ability to recognize whether two utterances come from the same or different speakers. This field is an important component of segmenting an audio stream into meaningful subunits, because the location of the speaker changes is crucial for dialogue understanding. Speaker discrimination is also related to speaker verification, but this last process is based on prior knowledge about a limited number of speaker identities, whereas in speaker discrimination, only knowledge about the speech signal is provided (Ouamour, Guerti, & Sayoud, 2008).

Speaker recognition has several practical applications as:

voice dailing, banking transactions by telephone, database access services, voice mail, biometric secure access, forensic applications, etc....

The problems encountered in speaker recognition are usually due to the intra-speaker variability of the speech, effect of noise and sometimes to the reduction of the spectral bandwidth as for the telephonic speech: (300-3400Hz) (Figure 4).

Palmprint Recognition

The palms of the human hands contain pattern of ridges and valleys much like the fingerprints, but the area of the palm is larger than the area of a finger. Consequently, palmprint scanners need to capture a large area (Figure 5), they are more expensive than fingerprint sensors (Jain & Ross, 2007). However, human palms also contain additional distinctive features such as principal lines and wrinkles that can be captured even with a lower resolution scanner, which would be cheaper.

In order to enhance the authentication accuracy, we can use high-resolution palmprint scanners. In this case, all the features of the hand such as geometry, ridge and valley features, principal lines, and wrinkles may be combined during the recognition process (Jain & Ross, 2007).

Iris Recognition

The iris pattern is the biometric characteristic that has one of the highest discriminating capacities. In 2001 it was used to identify an Afghan girl whose picture was shown on the cover of National Geographic 16 years earlier (Wei & Li, 2005). An iris-based person identification system works by encoding the iris (Figure 6) as sequences of 0's and 1's and then comparing these sequences. Usually, during the authentication step, the recognition system detects the boundary of the iris and applies a 2-D Garbor wavelet operation to perform the encoding, and the sequence is matched against the records in the database. Several organizations conducted evaluations of iris-based person identification systems and no false match has ever been found among the billions of tests (Wei & Li, 2005).

Vein Patterns Based Identification

Vascular pattern technology uses infrared light to produce an image of the vein pattern in the face, in the back of a hand, or on the wrist. Vascular

Figure 4. Spectrogram of a speech signal showing the formants and harmonics of the speaker

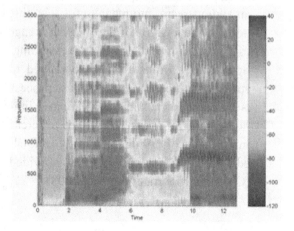

Figure 5. Scanned palm of a human hand

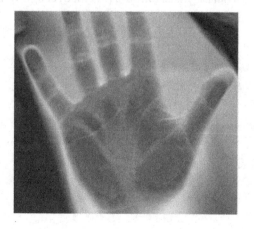

Figure 6. Photo of an Iris

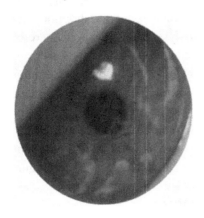

pattern technology is generally acceptable to users except some particular users who still object the use of infrared (Ruggles, 2002).

In 2005, Fujitsu Ltd. started selling a biometric security device that relies on the vein patterns of the hand. The company's palm-vein recognition system has been available in Japan and used successfully. Furthermore, the Bank of Tokyo Mitsubishi began installing the system on its ATMs in 2004 (Williams, 2005).

The authentication system contains a camera that takes a picture of the palm of the user's hand, and then the image is matched against a recorded database. Since the camera uses a near-infrared light, the veins present under the skin are sharply visible (Figure 7).

Gait Recognition

Such systems of authentication use motion capture techniques, which represent the process of recording human movement, and transforming it to digital data. Intercepted data can contain information about the changes of the position in the space of specific points on the person's body or only information about the movements of particular muscles (Klempous, 2009).

So, from the way used by a person for walking, we can make an assessment on his identity: the gait is used as an identifier. Such research works

have been encouraged by the need to identify people from surveillance cameras in poor light or without a clear visibility of the face. However, this is a difficult task since a person's gait presents a high intra-variability and clothing may hide several important details (Dunstone & Yager, 2010) (Figure 8).

Keystroke Identification

Keystroke dynamics were first recognized as a biometric identifier during the Second World War. Allied telegraph operators could identify other operators, friendly and enemy, through a keying rhythm which was called "The Fist of the Sender" (Dunstone & Yager, 2010).

Currently, the new techniques of keystroke dynamics use the timing difference between keystrokes and looks for idiosyncrasies in the use of keys: for instance, how long the typist holds down the shift key.

Commercial keystroke dynamic systems do exist, but they are not widely adopted because of their low precision. Recently, many improved commercial systems have been proposed for personal users and companies (Figure 9).

Figure 7. Vein pattern of a hand

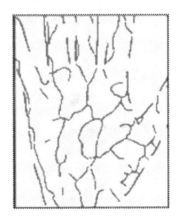

Figure 8. The gait can be modelized by the movement of the different segments (dashed lines)

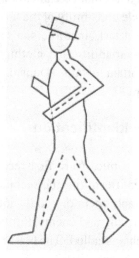

DNA Based Identification

DNA fingerprinting is a term used by Sir Alec Jeffreys describing the multi-locus probes results obtained in 1985. The term fingerprint is not really correct, and the term DNA profiling suggested by Evett & Buckleton is more adequate. In fact, the term profile indicates that this type of analysis does not allow characterizing a person's DNA, but only certain parts and an explicit mention of the markers and the technique used should

always accompany a given profile (Li & Jain, 2009). Nowadays, many laboratories are using a DNA analysis for medical testing research and health testing.

This genetic data represents a unique combination, difficult to be kept, confidential and extremely revealing about us. Furthermore, it is easy to acquire since people constantly slough off hair, saliva, skin cells, and other substances containing DNA (Holvast, 2009) (Figure 10).

Stylometry

Individuals have distinctive ways of speaking and writing, and there exists a long history of linguistic and stylistic investigation into authorship attribution. In recent years, practical applications for authorship attribution have grown in areas such as intelligence, criminal law, civil law and computer security. This domain is part of a larger field of automatic authentication, including biometrics, cryptographic signatures, intrusion detection systems, and others (Madigan et al., 2005).

The pioneering stylometric study of Mendenhall, in 1901(Malyutov, 2006), of Shakespeare contemporaries, used histograms of the word-length distribution. The paper describes the histograms for Shakespeare contemporaries commissioned and supported by A. Hemminway.

Figure 9. Everyone has a particular style of typing on the keyboard

Figure 10. DNA fingerprint of a person, using 7 different DNA probes

This study demonstrated a significant difference of Shakespearean histogram from those of all, except one contemporaries studied (including the Bacon's) (Malyutov, 2006). Consequently, by this stylometric investigation, he could demonstrate the unlikelihood of Bacon's authorship of Shakespeare (Figure 11).

Internal-Physical-Characteristics Based Authentication

Biomedical researches show that Individual humans are unique internally (as they are unique externally). Consequently, new biometric modalities have been proposed to identify people by using their unique internal characteristics (Brooks, 2010). For example, Boneprints use acoustic fields to scan the unique bone density pattern of a thumb pressed on a small acoustic sensor. Similarly, Imp-Prints measure the electrical impedance patterns of a hand to identify a person's identity. Small impedance sensors can be easily embedded in certain small devices such as smart cards, for example.

These internal biometric modalities are based on physical characteristics which are not visible or photographable. This fact is a great advantage for hiding the biometric data, which may provide a further level of security (Brooks, 2010) (Figure 12).

WHAT ELSE?

No researcher can expect what could be used in the future, even in the near future. In fact, many investigations have analyzed the human body and its physical or acoustic properties, by proposing new features, methods or innovations of existing techniques. They all state that the human body and corresponding physical/acoustic properties are unique for one human being. This fact is really amazing, since the number of existing humans that has been estimated by the United Nations in 2009, is estimated to be 6.8 billion (Wik, 2010).

In my opinion, the future challenge will be on the use of multi-modalities (several biometric modalities) and the combination of the corresponding authentication decisions into a more robust decision, by using some fusion techniques. In fact, every modality has an authentication error probability, which theoretically may occur for different persons (not the same persons), and by fusing several modalities we may reduce the error probability and consequently improve the biometric precision.

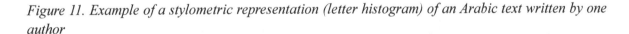

Figure 11. Example of a stylometric representation (letter histogram) of an Arabic text written by one author

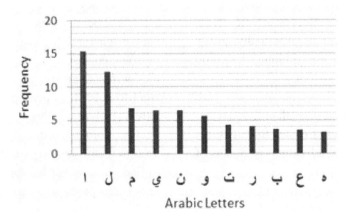

Figure 12. Internal impedance characteristics of a male person

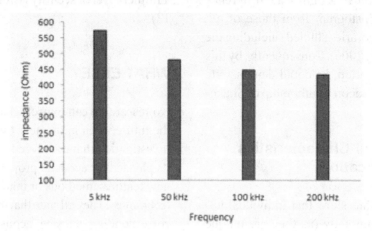

Social and Ethic Problems

As described in the introduction, several social and ethic problems have been raised by independent human organizations. In this section, some of these problems are quoted and discussed below:

Limitation of Freedom

The most significant disadvantage of biometric authentication systems is their potential to locate and track people physically. Such systems present a great danger because they promise high tracking accuracy. In fact, we believe that intensive tracking is harmful to a free society. A society in which everyone's actions are tracked is not, in principle, free (Abernathy, 2003).

Many civil liberty groups believe that biometric surveillance, even without any deliberate abuse, would have a frightening effect on artistic, scientific and political expressions (Abernathy, 2003).

Hence, the improper use of biometrics could lead to the following disasters:

• Increasing the visibility of individual behavior. This makes it easier for measures to be taken against individuals by agents of the government.

• Causing political damage and personal embarrassment (e.g., blackmails). This hurts democracy, because it reduces the willingness of competent people to participate in public life.

• Enabling the matching of people's behavior against some particular patterns. This could be used by the government to generate suspicion, or by the private sector to classify individuals into micro-markets (Abernathy, 2003).

Loss of Privacy

Humans have always had a need for privacy. The privacy issue can already be seen in the writings of Socrates, when a distinction is made between the outer and the inner. In the most fundamental form, privacy is related to the most intimate aspects of being human. Throughout history, privacy is related to the house, to family life, and to personal correspondence. For instance, between the fourteenth and the eighteenth century, people went to court for opening and reading personal letters (Holvast, 2009).

The biometric information is assumed to be used to identification or verification of a person to find out whether he/she is who he/she claims to be. This process can mean an attack on one's privacy when the collection takes place without consent or permission and without transparency about the purpose for which this data is used.

For example, surveillance video cameras are increasingly being used in almost every public place. If somebody wants to go for a walk in any developed city, it is clear that he will be recorded at several places, with or without his consent, and it is expected that this surveillance will be expanded in the next years by improved technology, under the assumption that cameras are providing security (Holvast, 2009).

The same problem is noticed with the systems of speaker/speech detection in telephonic communications. In fact, in some countries, any critical word or particular voice may set off a recording process, which will record all the speech of our discussion. Thus, several persons (honest or dishonest) may listen to this discussion without our consent. Of course, the justification will always be « security ».

Risk of Imposture

The risk of false acceptance (imposture) does exist for two possible causes: the first cause, which is involuntary, may result from a failure in the system of authentication (e.g., face recognition of twins); the second cause is voluntary and is due to a willingness of deceiving the authentication system.

Let us consider two applications: access to a nuclear plant and automatic banking transactions.

- The verification of identity to access a nuclear plant is a high risk situation. A terrorist could try to show the image of an iris or come with a gummy finger of the target employee. He may well succeed: he is a deliberate impostor (Chollet, Dorizzi, & Petrovska, 2009). In this case the risk is extremely high and may result in an irreversible catastrophe.

- In automatic banking transactions, a person's identity should be verified to allow the transaction. Suppose that a person has lost his credit card and an impostor, who knows that person, has found it. This impostor may make a mask of the person's face, a gummy copy of his fingerprint, use voice transformation techniques to sound like the person, etc. (Chollet, Dorizzi, & Petrovska, 2009). Even though the risk is much less dangerous than in the first example, this case of imposture presents a real danger for the account owner.

Risk of False Rejection

The false rejection is a mistake occasionally made by biometric security systems. In such cases, the system fails to recognize an authorized person and rejects that person as an impostor.

Normally, the risk of false rejection is less serious than the risk of false acceptance, but it remains embarrassing in all cases. For concreteness, many companies in the world have adopted a biometric access control for entrance or getting access to privileged services (e.g., some buildings of Panasonic Corporation, in the USA, use a speaker recognition system for the entrance control). So, suppose one day, that an employer gets a sore throat: how would he manage to go to work?

Oppression Feeling

People do not want parts of their body duplicated in databases. I remember, in 2005, when attending the Biosecure residential workshop (Biosecure, 2005), several working groups asked me to participate to the construction of the biometric databases that include fingerprints, photographs, etc. Even though I was aware that these data will be used only for research purpose, I really expressed a certain fear that the data could be spread and distributed to other organizations for other purposes.

Also, in general, people do not want to be continuously under surveillance or to be obliged to prove their identities. They do not want to prove that they are not impostors or terrorists either. This psychological issue may create a certain feeling of oppression in some communities, which will result in a continuous refusal and repulsion of these technologies.

DISCUSSION ON THE LEGITIMACY OF BIOMETRICS

After seeing the new biometric technologies, some of their applications and the resulting ethic prob lems, we will discuss in this section the legitimacy of the use of biometrics in daily life. We will also try to make different opinions according to the targeted application. That is why, we have divided our discussion into several points, and where each point is related to a particular application:

Biometrics in Physical Access Control

Even though there exists a real risk of false reject in such systems of access control (e.g., building access control), people may not express a great inconvenience in these cases. So the use is acceptable and may even improve the daily life quality.

Biometrics in multimedia applications

Accessing our computer or a particular privilege on internet, by using biometric tools will not, in my opinion, create any ethic or social problem for the users. Contrarily, it will make the transaction easier and more interactive.

Biometrics in Forensics

Probably, the most delicate field of application of biometrics is forensics, because the least mistake could lead to a fatal error, which could not be corrected in the future. Many researchers, over the time, have raised the gravity of the problem. Thus, we consider it illogical to accuse somebody for the unique reason that his voice is roughly similar to that of a potential person, which was recorded during a telephonic call.

On the other hand, the scientific community should encourage the use of biometrics to prove the innocence of somebody, because the risk is much less.

Biometrics in Border Control

Theoretically, no one should refuse the use of such authentication techniques in border control, but the risk of these techniques is high in reality. The first risk is due to the intra-variability of the biometric features over the time: for instance, fingerprints of some workers such as bricklayers are continuously altered. So the use of such technologies may result on some false rejections, which should make the traveler in very difficult situations. The second risk is the use of the traveling data by some oppressing governments to track and limit the privacy of people.

Biometrics in Banking Transactions

I personally encourage the use of biometrics in banking transactions, but not alone. The use of biometrics in an association with classic identification systems (e.g., password or magnetic cards) is reliable and should improve the accuracy of the authentication. However, the use of the biometric authentication alone may result on some cases of impostures, since any one could get a copy of the target's fingerprint, for example, and use it to make the system recognizing him as the target person.

Biometrics in Security and Intelligence Services

It is obvious that biometrics will enhance the security level: we agree with this. However, the problem is that we assume that the persons who manage this technology are fair and honest, but in reality things are different. Consequently, a real danger of abusive utilization against the human rights of citizens could exist.

Biometrics in Scientific Research

In scientific research, no limits should be placed since the research goal is the advance of technology in the domain of biometrics. However, scientists should be aware about the dangers of this technology if it is applied against the humanity principles. Therefore they must drive their research project towards the right applications and inform the different economic and political leaders about the advantages and the disadvantages of biometrics.

DISCUSSION ON THE RELIABILITY OF BIOMETRICS

In this paper, the effectiveness of a biometric authentication modality is measured by the good authentication rate (precision), False Acceptance Rate (FAR), False Rejection Rate (FRR) or by the Equal Error Rate (EER).

By analyzing the latest biometric systems of authentication, we have deduced the following relative assessments:

Signature Authentication

Results reported in (Bio, 2002) show a FAR of 5% and a FRR of 10%. These results show that signatures cannot be used in authentication systems requiring high level of precision. This result was expected since the intra-variability of the signature is relatively high (variation for a same person).

Fingerprints Authentication

Results reported in (Bio, 2002) show a FAR of 0.001% and a FRR less than 1%. Such results confirm the high authentication accuracy provided by fingerprints based recognition techniques. That is why many authentication systems continue to use fingerprints.

Face Recognition

Experiments conducted by the authors of (Lu, Wang, & Jain, 2003) on a face database containing 206 subjects (2,060 face images) show that the proposed approach got a recognition accuracy score of 90.2%.

This accuracy is not sufficient enough to make this type of biometric modality exploitable in practice. Moreover, face recognition becomes very difficult if the hairstyle changes, which is usually the case for female faces.

Speaker Recognition

According to the works of Bimbot (Bimbot, Magrin-Chagnolleau, & Mathan, 1995), the experiments of speaker recognition in a database of about 630 speakers gave an identification error of 0% with clean microphonic speech. However, with telephonic speech, the Equal Error Rate (EER) reported by (Li et al., 2009), was about 2.5% in the database of NIST 2008 (Li et al., 2009). These results show that speaker recognition can be used accurately with microphonic speech, but with telephonic speech the speaker recognition accuracy is really deteriorated.

Palmprint Recognition

Experiments using the PolyU palmprint database for evaluating palmprint recognition performance have been done by the authors of (Ito, Iitsuka, & Aoki, 2009). This database consists of 600 images (384 × 284 pixels) with 100 subjects and 6 different images of each palmprint. The Equal Error Rate (EER) of the proposed algorithm was 0%, which means that this modality can get high accuracy in secure biometric authentication.

Iris Recognition

According to the works of Daugman (2006), the statistical results from 200 billion iris cross-comparisons showed an observed false match rate (authentication error) of 0%, which shows that iris recognition can provide very high authentication accuracy. Furthermore, this biometric modality is considered as one of the best techniques in biometrics.

Vein Patterns Based Identification

In a database of 144 infrared palm images (each of 24 individuals has 6 images), the authors of (Zhang, Li, You, & Bhattacharya, 2007) got a recognition score of 98.8%, where the false acceptance rate was 5.5%. However, the simplicity of utilization, during the authentication process, makes this biometric technique interesting in practical use. Furthermore several companies like Fujitsu and Hitashi have proposed efficient systems of vein patterns based authentication that are used in several banks successfully.

Gait Recognition

With a database constituted by a population of 21 participants; 12 male and 9 female aged between 20 and 40 years, the authors of (Gafurov, Helkala, & Sondrol, 2006) got an Equal Error Rate (EER) of 5% by using the gait modality for persons authentication. The precision is not sufficient to use this modality in real applications. Furthermore, the performances should decrease if the person to identify has long or large cloths.

Keystroke Identification

Results of authentication experiments on the old keystroke capture data (Eusebi, Eusebi, Gliga, John, & Maisonave, 2008), which contains 36 subjects who are using laptops, showed a score of good authentication ranging from 72.33% to 95.81%. This low performance level shows that the technique is not reliable and cannot be used in serious authentication applications.

DNA Based Identification

DNA is used as a form of identification that is employed to produce either a DNA fingerprint or a DNA profile. DNA identity testing seems to be highly accurate and exclusions are done sharply. This technology matches in two unrelated people with a probability of 1 in 6 billion (Dutt & Aditya, n.d.). The problem with this biometric modality is the cost of the tests. Another serious problem concerns the discrimination between twins: in fact, a DNA test cannot determine the difference between identical twins.

Stylometry

During the 2009 first international competition on plagiarism detection, the external plagiarism analysis task showed a precision of 74.18%. The competition database contains 36,475 plagiarism cases in the corpus for the external analysis task (Potthast, Eiselt, & Barron-Cedeno, 2009). Results do not show good authentication accuracy, but they are still promising for literary authentication, because the authentication accuracy should be enhanced if the textual size of the database is increased.

Internal-Physical-Characteristics Based Authentication

The performances of this technique are not known by the author.

ACKNOWLEDGMENT

The author would like to thank all the researchers who contributed directly or indirectly in the elaboration of this modest work and the reviewers who will read and enrich this scientific research work by their useful comments. Furthermore, he encourages and welcomes any feedback, comment or suggestion from the readers.

REFERENCES

Abernathy, W., & Tien, L. (2003). *Biometrics: Who's Watching You?* Retrieved from http://www.eff.org/wp/biometrics-whos-watching-you, 2003

Bimbot, F., Magrin-Chagnolleau, I., & Mathan, L. (1995). Second-Order Statistical Measures for Text-Independent Speaker Identification. *Speech Communication*, *17*(1-2), 177–192. doi:10.1016/0167-6393(95)00013-E

BioSecure. (2005, August 5). In *Proceedings of the BioSecure Residential Workshop*, Paris. Retrieved from http://biosecure.it-sudparis.eu/AB/index.php?option=com_content&view=article&id=23&Itemid=23

BioTech. (2002). *Biometric Technical Assessment*. Retrieved from http://www.bioconsulting.com/Bio_Tech_Assessment.html, 2002

Brooks, M. J. (2010). New biometric modalities using internal physical characteristics, Biometric Technology for Human Identification VII. In V. Kumar et al. (Eds.), *Proceedings of the SPIE, Volume 7667* (pp. 76670N-76670N-12).

Chollet, G., Dorizzi, B., & Petrovska, D. (2009). Introduction—About the Need of an Evaluation Framework in Biometrics. In *Guide to Biometric Reference Systems and Performance Evaluation* (pp. 1–10). London: Springer Verlag. doi:10.1007/978-1-84800-292-0_1

Daugman, J. J. (2006). Probing the uniqueness and randomness of IrisCodes: Results from 200 billion iris pair comparisons. *Proceedings of the IEEE*, *94*(11), 1927–1935. doi:10.1109/JPROC.2006.884092

Dunstone, T., & Yager, N. (2010). *Biometric System and Data Analysis*. New York: Springer Verlag.

Dutt, K. S., & Aditya, A. V. (n.d.). *Biometrics*. Retrieved from http://issuu.com/sohildutt/docs/4_biometrics

Eusebi, C., Gliga, C., John, D., & Maisonave, A. (2008). A Data Mining Study of Mouse Movement, Stylometry, and Keystroke Biometric Data. In *Proceedings of the Student-Faculty Research Day (CSIS)*, Pace University, New York (pp. B1.1-B1.6).

Gafurov, D., Helkala, K., & Sondrol, T. (2006). Biometric Gait Authentication Using Accelerometer Sensor. *Journal of computers*, *1*(7), 51-59.

Holvast, J. (2009). *The Future of Identity in the Information Society*. Boston: Springer.

Ito, K., Iitsuka, S., & Aoki, T. (2009). A palmprint recognition algorithm using phase-based correspondence matching. In. *Proceedings of the IEEE International Conference on Image Processing, 2009*, 1977–1980.

Jain, A. K., & Ross, A. (2007). Introduction to Biometrics. In *Handbook of Biometrics* (pp. 1–22). New York: Springer.

Klempous, R. (2009). Biometric Motion Identification Based on Motion Capture. In *Computational Intelligence* (pp. 335–348). Berlin: Springer.

Li, H., Ma, B., Lee, K., Sun, H., Zhu, D., Sim, K. C., et al. (2009). The I4U system in NIST 2008 speaker recognition evaluation. In *Proceedings of the 2009 IEEE International Conference on Acoustics, Speech and Signal Processing*, Taipei, Taiwan (pp. 4201-4204).

Li, S. Z., & Jain, A. K. (2009). *Encyclopedia of Biometrics* (1st ed.). New York: Springer. doi:10.1007/978-0-387-73003-5

Lu, X., Wang, Y., & Jain, A. K. (2003). Combining classifiers for face recognition. In *Proceedings of the 2003 International Conference on Multimedia and Expo (ICME '03)*, Baltimore. *MD Medical Newsmagazine, 3*, 13–16.

Luppicini, R., & Adell, R. (2008). *Handbook of Research on Technoethics*. Hershey, PA: IGI Global.

Madigan, D., Genkin, A., Lewis, D. D., Argamon, S., Fradkin, D., & Ye, L. (2005). Author identification on the large scale. In *Proceedings of the Joint Annual Meeting of the Interface and the Classification Society of North America (CSNA)*.

Malyutov, M. B. (2006). Authorship Attribution of Texts: A Review. In *General Theory of Information Transfer and Combinatorics* (pp. 362–380). Berlin: Springer. doi:10.1007/11889342_20

NSTC. (2006). National Sciences and Technology Council NSTC. *Biometrics history*. Retrieved from http://www.biometrics.gov/Documents/BioHistory.pdf

Ouamour, S., Guerti, M., & Sayoud, H. (2008). A New Relativistic Vision in Speaker Discrimination. *Canadian Acoustics Journal, 36*(4), 24–34.

Ouamour, S., Sayoud, H., & Guerti, M. (2009). Optimal Spectral Resolution in Speaker Authentication, Application in noisy environment and Telephony. *International Journal of Mobile Computing and Multimedia Communications*, 36–47.

Potthast, M., Eiselt, A., & Barron-Cedeno, A. (2009). Uncovering Plagiarism, Authorship and Social Software Misuse. In *Proceedings of the 1st International Competition on Plagiarism Detection (SEPLN '09)*. Retrieved from www.uni-weimar.de/medien/webis/research/ workshopseries/pan-09/competition.html

Ruggles, T. (2002). *Comparison of biometric techniques*. Retrieved from www.bioconsulting.com/bio.htm

Wei, G., & Li, D. (2005). Biometrics: Applications, Challenges and the Future. In *Privacy and Technologies of Identity* (pp. 135–149). New York: Springer.

Wikipedia. (2010). *Wikipedia. World population*. Retrieved from http://en.wikipedia.org/wiki/World_population

Williams, M. (2005). Forget fingerprints and eye scans; the latest in biometrics is in vein. In *Computerworld*. Retrieved from www.computerworld.com/s/article/102861/Forget_fingerprints_and_eye_scans_the_latest_in_biometrics_is_in_vein

Zhang, Y., Li, Q., & You, J. (2007). Palm Vein Extraction and Matching for Personal Authentication. In Bhattacharya, P. (Ed.), *Advances in Visual Information Systems*. Berlin: Springer. doi:10.1007/978-3-540-76414-4_16

This work was previously published in the International Journal of Technoethics, Volume 2, Issue 1, edited by Rocci Luppicini, pp. 19-34, copyright 2011 by IGI Publishing (an imprint of IGI Global).

Chapter 7
On Biometrics and Profiling:
A Challenge for Privacy and Democracy?

Daniele Cantore
University of Pavia, Italy

ABSTRACT

This paper advances an analysis of biometrics and profiling. Biometrics represents the most effective technology in order to prove someone's identity. Profiling regards the capability of collecting and organizing individuals' preferences and attitudes as consumers and costumers. Moreover, biometrics is already used in order to gather and manage biological and behavioral data and this tendency may increase in Ambient Intelligence context. Therefore, dealing with individuals' data, both biometrics and profiling have to tackle many ethical issues related to privacy on one hand and democracy on the other. After a brief introduction, the author introduces biometrics, exploring its methodology and applications. The following section focuses on profiling both in public and private sector. The last section analyzes those issues concerning privacy and democracy, within also the Ambient Intelligence.

1. INTRODUCTION

In every society, owning the information means being able to handle a situation. This becomes especially preeminent if dealing with complex societies where the fragmentation of information gives more leverage to whom is able to manage that information. In our technological world, one of the most challenging problems refers to the increasing data overload. Therefore, it has

become essential to be able to discriminate useful information from useless information.

Profiling activities allow public administration and private corporations to sharpen their aims becoming more effective in achieving their goals and purposes. As a matter of fact, profiling allows gathering and organizing individuals' data. In particular, within the public sphere, profiling is used in order to increase safety and law enforcement; in private business, it regards the capabilities of knowing stakeholders' attitudes and interests in order to make more effective marketing and com-

DOI: 10.4018/978-1-4666-2931-8.ch007

munication campaigns. However, due to the fact that profiling regards individuals and their data, the process of collecting and managing those data can never be considered neutral.

In this paper, I will describe the importance of profiling techniques in our contemporary society, both in public and private applications, especially as they are related to biometrics. First of all, I will describe biometrics and its procedures as an example of technology that is used in forensic and other law-enforcement applications. I will then explore some methodologies, used to profile people, that benefit private businesses including individual profiling and user profiling on the one hand, and distributive profiling and non-distributive profiling on the other. Successively, I will explore some of the possible issues regarding biometrics and profiling within the paradigm of Ambient Intelligence (AmI). Through this paper, I will stress the actual necessity of creating a law-system able to cope with profiling, also foreseeing future developments and issues. At the same time, I will attempt to shed light on some aspects strictly related to our current concepts of democracy, technological development, and violence, especially focusing on privacy issues within biometrics and the AmI context.

The basic idea throughout this paper is to create a framework where those topics are juxtaposed rather than analysing them one by one. This way, I hope to show the complexity of these themes and how profoundly intertwined they are.

2. INTRODUCING BIOMETRICS: METHODOLOGIES AND APPLICATION

Biometrics regards the possibility of recognizing someone's identity from an unambiguous set of biological data, or behaviour. The essential idea is that every human being is unique: no matter how many physiological features or feelings or behaviours all human beings have in common, every

person is different one from others just like every snowflake is different from others. Biometrics techniques can be divided and categorized in two different areas: (i) physiological: including DNA, facial-scan, finger-scan, hand-scan, iris-scan, otoacoustic emission, retinal-scan, vascular patterns; (ii) behavioural: including gait, gesture and grip, handwriting, keystroke, lip motion, mouse movements, signature, voice. Obviously, each of those techniques has reached a different level of maturity. Nevertheless, they have all developed for two different ends: (i) to protect information: especially biological data are detected in order to replace old systems based on passwords or PINs for identification purposes; (ii) to perform various kinds of verification: biological data are detected in order to assess someone's identity, especially in forensic applications for verification purposes.

In order to shed light on biometrics in a more specific way, it is essential to clarify and distinguish the areas in which it operates starting from the difference between identifying someone and verifying someone's identification: (i) identifying means to establish that a particular person, thing or quality corresponds to a determinate identity; (ii) verifying one's identity is about assessing the validity and truthfulness of the relationship I just defined. As for biometrics, the different meaning between these two words shows the differences in their applications: (i) the first case refers to those determining criteria used to establish the identity of a certain person; (ii) the second case refers to the necessity of a verification process that allows the adoption of evaluation criteria to approve the legitimacy of an individual's request, and consequently allow her to access a certain system (Nanavati, Thieme, & Nanavati, 2002).

In the identification system, the methodology involves searching a database containing the biometric data of many people in order to find a match with one of them. This type of activity is known as one-to-many matching, and it is often used in the forensic domain, especially to apprehend wanted people by using watch-lists. The identification procedure responds to the questions *who am I?* or *whose biometric data is this?* (Nanavati, Thieme, & Nanavati, 2002).

In the verification system, the main methodology consists in a comparison among the biometric data of the subject captured by one or more devices and the data previously stored in a database (as a biometric template of that subject); if the latest data captured by the identification devices is matching with the subject's template, the resulting process is valid and the subject's identity is verified. This is an active process since the subject is usually a user who is requesting a valid identification in order to get access a certain system, often constituted by a digital structure such as a computer network. The verification procedure responds to the question *am I who I claim to be?*, and is based on assessing a one-to-one match (Nanavati, Thieme, & Nanavati, 2002).

Verification and identification systems are already in use in many safety applications in the governmental domain, and in forensic and commercial ones as well. The eve growing need for user authentication in these domain is pushing for improvements of every biometric technique, as well as of those networking, communication, and mobility technologies supporting them. Biometric applications involve several sectors including law enforcement, banking, computer system, physical access, national security, telephone systems and others. All those applications have in common the idea of recognizing someone's identity in order to let him or her to enter that specific system.

Biometric data are represented in three possible ways: (i) raw data (also known as *sample data*): "data gathered directly from the sensor before any processing has been carried out" (Dunstone & Yager, 2009, p. 32). The effectiveness of the sample representation depends directly on the techniques used for this aim. Nonetheless, the environmental conditions may have important repercussions at this stage; for instance, the light conditions may change during several sample data acquisitions when using a face recognition technique such as one based on camera images, and the entire verification or identification process may be invalidated; (ii) token data: "a token is representation of the raw data that has had some minimal amount of processing applied" (Dunstone & Yager, 2009, p. 32). The image of the subject is stored in chips on passport or other identity cards. In this way, the sample stored can be used in other biometric applications using any recognition algorithm. Token data are also known as *template interoperability*. Moreover, other features may be extracted from the token successively; (iii) template data: this is the tool used by every biometric technology; "a template is the refined, processed and stored representation of the distinguishing characteristics of a particular individual" (Dunstone & Yager, 2009, p. 33). Two templates from the same biometric will never be identical because of variation during enrolment step. However, the entire biometric system is based on template data: first of all, it represents the starting phase of the work; successively, template data will be used in order to find a match with new samples, and, consequently, to verify identities. Furthermore, very important for evaluations are metadata: "this is data that describes the attributes of either the biometric sample (e.g., wearing glasses), the capture process (e.g., time acquired) or the demographics of the person (e.g., gender and age)" (Dunstone & Yager, 2009, p. 34).

In addition, biometric system is basically a statistical process; this means that a variation between the phases of the acquisition of samples and the subsequent recognition as well as the temporary or permanent body changes could invalidate the process itself. Moreover, unlike PINs or passwords – where even a small variation means denying the access – there is no clear line between a match and a non-match. Therefore, the correspondence between a sample and a template is not only a matter of rating, but it depends also on the error margin considered tolerable. A biometric application requesting a high safety demand, such as a military one, might request a

90% probability of a match in order to consider acceptable the access on its system. Vice versa, a biometric application requesting a low safety demand might request just an 80% probability of a match. Thus, biometric system depends also on the application of security demands.

Hence, dealing with a biometric system means also dealing with its errors: (i) false match (FMR): it refers to a situation in which someone non-authorized succeeds in entering into a system; (ii) false non-match (FNMR): it refers to a situation in which someone is rejected by the system even if she has the right to get into it; (iii) failure to enrol (FTE): it refers to a situation in which a profile could not be enrolled into system. Biometrics relies on a statistical process and the relevance of the error does not involve only the process itself: because biometrics are used in forensic and other law-enforcement applications, mistakes in this field may generate consistent issues in in the management of democratic institutions that I will analyse in the following sections.

Dealing with biometric technologies means in fact to tackle many challenges in several areas; indeed, as already stated, biometric techniques can tell a lot about us, about our personal physiological structures and characteristics. Generally speaking, we can recognize three big risk areas: (i) privacy principles: who can access my biometric data and why? Is there any risk for my privacy while my biometric data are stored in databases? Is there any risk of using my biometric data for other purposes than confirming my identity? (ii) democracy-related issues: who, how and why is managing my data? Could I be discriminated against other citizens because of my biological data or my behaviours or my preferences as a consumer? (iii) identity-related matters: if my biometric data are able to identify me, who am I? Is an assembly of physiological data and the devices that read them enough to say who I am? Are we going toward the *human hybrid*?

Before analysing the relationships between biometrics and privacy issues on the one hand

and biometrics and democracy on the other hand, I will briefly expose the scopes of profiling both in public and in private sector. This way, it will be easier to understand those issues so intimately related with biometrics I just mentioned.

3. FOCUSING ON PROFILING

Profiling occurs in a plurality of contexts. A first distinction regards its area of application: it deals with public affairs when related with forensic and criminal investigation; but it also concerns the private sector when used in marketing and communication strategies. Across these two fields, profiling regards also healthcare applications as well as identification and verification processes. Considering how wide the area of application is, it could be difficult to provide a unique definition of profiling. However, generally speaking, profiling refers to a large set of technologies using algorithms (or data mining) in order to find out and then construct knowledge from huge amounts of data. In other words: "profiling can be described as the process of knowledge discovery in databases (KDD), of which data mining (DM, using mathematical techniques to detect relevant patterns) is a part" (Hildebrandt, 2008, p. 58). Therefore, the main point of profiling regards the ability of creating consistent knowledge from an apparently random amount of data. In a nutshell, this represents at the same time the interest and the risk factor of profiling. Thus, through advanced profiling techniques and a set of knowledge management tools, any community, user or item can be profiled in a same way, so to allow a direct assessment of relevance, which permits the creation of network identities: "network identity is the total set of characteristics about a person held on a computer network, and since any part of a profile could be used to identify a person for some purpose or other, the profile in conjunction with a profile ID, constitute a person's network identity" (Newbould & Collingridge, 2003).

Individual profiling refers to a profile constructed with the data of a single person; this way, it may be possible to discover the particular characteristics of a certain individual, which provides a unique identification and personalized services. In this sense, biometric technologies may be used as a source of profiling information due to the possibilities they afford in verification and identification performances: moreover, since biometric features are usually persistent, they provide a great opportunity for biometrics, allowing the creation of templates related to single profiles. Conversely, researchers in the field of Intrusion Detection study *user profiling*. It consists of observing someone interacting with a computer, creating a model of such behavior and using it as a template for what is considered a normal behavior for that particular user. If the behavior of supposedly the same user is significantly different we can speculate that perhaps it is a different user masquerading as the user whose profile is stored in our security system as a template (Yampolskiy, 2008). Thus, the aim of behavioural biometric profiling is to detect relevant patterns that allow identification of a person and his or her habits or preferences. An example of a popular behavioural biometric technique, which provides a unique identification, is the way in which a person types on a keyboard. Indeed, it is unique enough to provide a reliable personal identification (Gupta, Mazudmar, & Rao, 2004). However, while identification purposes are strictly related to individual profiling, personalized services are most often developed for *group profiling*, by which a certain "type" of person can be categorized: matching a user with a group-type is permitted by the fact that her profile matches with one that has been delineated by analysing massive amounts of data about other users. Therefore, a group profile may either refer to an existing community or to a "category" of users that, even if do not form an actual community, are found to share certain patterns, for instance as far as economic behaviour is concerned. Group profiling can easily support scapegoating mechanisms, as we are

going to see later. To conclude this brief survey, group profiles can also be divided as follows: (i) distributive profiling: it refers to those groups in which the several properties are shared equally by all the members of the groups; (ii) non-distributive profiling: it refers to those groups in which profile does not necessarily apply to all the members of the group. It is important, though, to stress since the beginning how all of the technologies used for profiling (as well as biometrics used for these purposes) may trigger several problems even if used for purposes like the protection of personal information or the increase of business efficacy. In the next section I will sketch out the two aforementioned fields of applications (public and private) in order to stress similarities and differences between the two main areas.

Examples of Profiling in the Public Sphere and for Private Enterprise

As already hinted, profiling within public area concerns mainly law-enforcement and forensics. A clear example of law-enforcement applications regards immigration control; for example, the Charlotte/Douglas International Airport in North Carolina and the Flughafen Frankfort Airport in Germany allow frequent passengers to register their iris scans in an effort to streamline boarding procedures. In Berkshire County, the iris-scan technology is used in the newly built Berkshire County Jail as a security check for employees. In the same way, in 1994 the Lancaster County Prison in Pennsylvania became the first correctional facility to employ this technology for prisoner identification. Also the European Union claims the importance of biometric technologies for this purposes:

Another (claimed) benefit of face recognition is that it could be used to mine existing databases of photographs. Current technology would struggle however with the quality of photographs available. The distinctive feature of face recognition that is appealing to law enforcement agencies is the option of matching witness descriptions or artist-rendered images to databases of suspects, i.e., the capacity to compare biometric data with non-biometric data within the same system. Though the results are not precise enough to be admissible as evidence, they could provide the police with leads for further investigation machine-readable travel documents (MRTD). Though this means a digitalized (European Commission, 2005, p. 56).

Conversely, profiling in private sector occurs for one main reason: "as customer loyalty can no longer be taken for granted, companies develop CRM [Customer Relationship Management] in the hope of surviving the competitive arena of neo-liberal market economies. At the same time they try to establish which consumers may be persuaded to become their new customers and under what conditions" (Hildebrandt, 2008, p. 56). Thus, companies are able to set more precise targets for their marketing campaigns. By knowing costumers' preferences and attitudes, it is possible indeed to create more effective communication strategies and marketing campaigns. This kind of profiling follows two principal directions: (i) customers' data are acquired by a company when someone signed the agreement for using a certain product; often avoiding the part of reading the contract, the customer give leverage to the company to use those data. Moreover, most of the time, that company receive the right of giving those data to other companies; (ii) using data mining, companies are able to know which sites are visited by their targets, knowing also the contents user are searching for or the websites frequently browsed. This way, a company is able to create an advertising campaign designed to its target:

After a transaction is complete, the information is shared with any number of legal data-sharing entities, the 3rd-party data user who is a known external data-sharing partner, for example, a credit reporting company such as Experian who *shares data with 2nd-party permission. Companies, such as Experian, generate their revenues by matching consumer information to transaction information, profiling consumers, and reselling the expanded information (Luppicini & Adell, 2009).*

In order to comprehend the importance of gathering individuals data with profiling activities, it may be sufficient to think of the sums firms are willing to pay to access this knowledge: the amount of business generated by ChoicePoint, a company specialized in collecting individuals' data for security analysis, was around 4.1 billion dollars in 2008 (Brunton & Nissenbaum 2010).

4. MEDIATING VIOLENCE THROUGH BIOMETRICS AND PROFILING: CAN DEMOCRATIC FREEDOM BE PUT IN JEOPARDY?

The simple introduction of biometrics and profiling, because of their very nature, produced a number of patently violent outcomes to be dealt with: concerning privacy, for instance, or mistakes in identification and verification processes, unauthorized data mining for profiling activities, and so on. On the other hand, potential issues, including scapegoating and violent attacks against democratic institutions, are less evident. Furthermore, certain risks specifically concern the development of technological paradigms like Ambient Intelligence (AmI) and Ubiquitous Computing. Therefore, in this section I will analyse the main violent issues already afflicting our society, while I will subsequently point out some of the issues that may arise out of future technological developments of our world.

Dealing with biometrics, the first issue surely concerns our own privacy as profiled individuals. Indeed, the diffusion of biometrics creates a new relationship between data administration and the tools used for this purpose. Unlike PINs, passwords, and tokens, biometric technologies are not neutral as they carry personal data defining us as individuals: "because biometric technologies are based on measurements of physiological or behavioural characteristics of the human body, discussion of biometrics frequently centers on privacy concerns. The increasing use of biometric technology raises question about the technology's impact on privacy in the public sector, in the workplace, and at home" (Nanavati, Thieme, & Nanavati, 2002, p. 237). While the recognition of biometric data itself may not be considered as a privacy invasion – it is after all what we do every day when we recognize people we know by their aspect, scent, voice tone – the way those data are detected, stored and managed may create some ethical consequences. At the same time, how the government or private companies may use this information may generate other problematic issues: "the existence of a government database with facial-scan or finger-scan data may be tempting to law enforcement agents or private-sector companies searching for personal data. [...] Every biometric database becomes a potential database of criminal records, representing a significant increase in the government's investigative powers" (Nanavati, Thieme, & Nanavati, 2002, pp. 240-241). The impact on privacy of the different biometric systems may be labelled as follows: (i) privacy invasive: it refers to those occasions when biometric technologies may be used without the individual agreement just like facial-recognition permits; (ii) privacy neutral: it refers to those biometric data scarcely available in order to satisfy identity aims; (iii) privacy sympathetic: it refers to those techniques requesting special authorization to reach certain biometric data; (iv) privacy protective: it refers to those technologies using biometric techniques to protect personal information (Nanavati, Thieme, & Nanavati, 2002).

The fact of managing individual's very sensitive data creates the urgent necessity of protecting and keeping this information as safe as possible. As we said above, the essence of biometric techniques is the possibility of recognizing someone from her biological data or behavioural patterns. It must also be stressed that some biometric techniques can collect data samples without individuals being aware of this: "facial-scan technology, voice-scan technology, signature-scan technology, and keystroke-scan technology, because they utilize standard devices (cameras, telephones, etc.) to acquire biometric information," could be used for this purpose (Nanavati, Thieme, & Nanavati, 2002, pp. 240-241). As we can easily understand, this system has a great power on individuals. The risk of using a technology in a different way than its original purpose, known as *function creep*, may lead to violent devastating effects: the consequences of having your own biometric template (constituted by your biological data, your preferences and attitudes, your behaviours, etc.) always and easily accessible by the government might degenerate not only in totalitarian regimes, but also within democratic institutions as well. Discriminations because of one's profile may increase constantly. In this sense, profiling may facilitate the scapegoating of persons and groups: "when too much of our identity and data is externalized in distant, objective 'things', for example through automated profiling (knowledge discovery in database, KDD of which data mining, DM, is a part, the possibility of being reified in those externalized data increases, and our dignity/autonomy subsequently decreases, and refined *discriminations* are also favoured" (Magnani, in press). It would be for instance very easy to blame a pandemic on individuals with B+ blood group, and persecute them to exorcize public hysteria. Therefore, because of extensive profiling it is possible to scapegoat a person, or a group, increasing the chance of losing individual freedom (Magnani, 2011). Furthermore, given the fact that biometrics provides the amount of data necessary to create group-profiles, the profiled groups could be discriminated just for having or not having certain characteristics. In absence of a well-defined law, a scenario where some biological trait or a pattern of behaviour could be used in order to scapegoat someone or a group increases dramatically: imagine if the Nazi could have accessed a database tracking the genetic imprint of German population, making it immensely easier and more accurate to spell out the various ethnic descents.

Technological Development and Technoviolent Issues: The Case of Ambient Intelligence

In this last section I will explore some possible consequences of what exposed so far, specifically within the paradigm of Ambient Intelligence (AmI). Even if the questions concerning the AmI itself need a greater space than I can find here, I will try to point out some aspects that are intertwined with what I have analyzed above.

Within the Ambient Intelligence paradigm, and especially within the smart environment, the technologies and techniques create a special context where the relationship with humans may change drastically. The technologies belonging to this paradigm give us the possibility of enhancing our control over our surroundings, by generating new capabilities of modifying the environment. However, a singular question emerges: "paradoxically, however, this control is supposed to be gained through a delegation of control to machines. In other words, control is to be gained by giving it away. But is more control gained than lost in the process?" (Brey, 2006). *Yes* is the answer, according to supporters of *Proactive Computing*, which is one of the IT paradigms on which Ambient Intelligence relies. Contrasting the upholders of *Interactive Computing*, they believe that an Ambient Intelligence environment must consist of a network of computing devices that should not wait for human inputs in order to undertake an action course or communicate with another device:

If humans were to stand in between a computer and its environment, they would only function as *glorified input/output devices that perform tasks that the computer can perform for us. By being freed from the tedious task of interacting with a multiplicity of computing devices and making decisions for them, humans can freely focus on higher-level tasks. [...] [Therefore] humans can gain better control over important tasks by delegating unimportant tasks to embedded computers, as long as these devices are programmed to anticipate and respond to their needs (Brey, 2006).*

On the other hand, there are at least four ways in which Ambient Intelligence may be a violent threat to our freedom (Brey, 2006): (i) interpreting different performance: some technological devices may perform actions that do not correspond to the needs or the intentions of its user due to incorrect inferences about the user, the user's action, or the situation; (ii) taking control: even if a certain technology show up a great capability in understanding our behaviours, it may act by assigning particular meaning to human behaviours that may be unintended; (iii) responding to third parties' interests: it may happen that the technology recommends a purchase which in not based on the user's real needs, but on a need that a commercial firm assigns to the user – this may happen when a technology has been designed to take certain commercial interest into account, as in *ad-ware*; (iv) performing data collection: it refers to the case when some technologies are used by third parties for data collection and surveillance, which is the case of biometric techniques and the consequent profiling. It may happen – as for biometrics – that someone's preferences, behaviours, social interactions, experiences, physiological and medical data, could be used by third parties in order to exercise control over him or her, in commercials or even in law applications within an AmI environment.

Furthermore, a very interesting situation emerges if we believe that technological devices may be even more *intelligent* than humans: in fact, when humans believe, want or need something, but the surrounding technological devices suggest that they want or need something else, a cognitive dissonance appears. A cognitive dissonance occurs when two contrasting propositions may be held simultaneously to explain the same phenomenon: "if a person holds two cognitions that are inconsistent with one another, he will experience the pressure of an aversive motivational state called cognitive dissonance, a pressure which he will seek to remove, among other ways, by altering one of the two 'dissonant' cognitions" (Bem, 1967). This theory was firstly developed by Festinger (1957), describing this status within a social psychology paradigm as a discomfort similar to the notions of hunger, frustration, or disequilibrium. It is interesting to notice that, in this context, we may say that a cognitive dissonance occurs between humans and technological devices: indeed, humans have to choose between distrusting the machine and distrusting themselves. Thus, dealing with the theme of our own freedom, we may say that:

AmI has a serious potential to enhance positive freedom through its ability to enhance control over the environment by making it more responsive to one's needs and intentions. However, it also has a strong potential to limit freedom, both in the positive and the negative sense. It has a potential to limit negative freedom because it could confront humans with smart objects that perform autonomous actions against their wishes. Such actions could either result from imperfections in the technology or from the representation in the technology of interests other than those of the user. AmI also has the potential to limit positive freedom, by pretending to know what our needs are and telling us what to believe and decide (Brey, 2006).

The capability that the AmI may have in predicting our behavior and even in pretending to know our needs moves, once again, from profiling activities. However, delegating to machines those capabilities increases those issues concerning threats to privacy and democracy. Leaving a part

of our decision to certain technologies, could allow private companies to invade that space with greater easiness; strategic marketing campaigns and CRM could create an environment where the border between our conscious decisions and companies' "tips" could be blurred to a point of no return. By knowing all of our biological data and behaviors as well as our preferences and attitudes, private companies could nudge us in a way that could be impossible to resist and even to recognize. In this sense, the risk of scapegoat mechanisms could increase as well, in as much as it would be inspired by a dynamic environment manipulating our moods and subconscious perceptual imagery: individuals would just have to connect the dots and unleash dramatic violence against the targets set by either government of just-as-powerful private companies.

5. CONCLUSION

In this paper I tried to briefly introduce the main issues concerning biometrics and profiling and their main potential or actual violent effects. Although the very topic needs a greater analysis in order to shed light on the complexity represented by both biometrics and profiling, and the way they interact, I tried to create a frame to enclose those issues especially concerning threats that might concerns both privacy and democracy. Our contemporary technological world must cope with the fact that the more efficient we want technology to be in the satisfaction wishes and attitudes, the more our lives are going to be *profiled*. Moreover, all the stakeholders involved in this process have to try to foresee as many issues as possible among those that may arise with further development. At the same time, this also means that the law system, in order to be adequate, has to grow richer and richer to match the technological development, so to tackle those ethical and technological issues.

REFERENCES

Bem, D. J. (1967). Self-perception: An alternative interpretation of cognitive dissonance phenomena. *Psychological Review, 74*(3), 183–200. doi:10.1037/h0024835

Brey, P. (2006). Freedom and privacy in AmI. *Ethics and Information Technology, 7*(3), 157–166. doi:10.1007/s10676-006-0005-3

Brunton, F., & Nissenbaum, H. (in press). Vernacular resistance to data collection and analysis: A political theory of obfuscation. In Hildebrandt, M., & de Vries, E. (Eds.), *Privacy, due process and the computational turn: Philosophers of law meet philosophers of technology*. London, UK: Routledge.

Dunstone, T., & Yager, N. (2009). *Biometrics system and data analysis*. New York, NY: Springer. doi:10.1007/978-0-387-77627-9

European Commission. (2005). *Biometric at the frontiers*. Brussels, Belgium: Joint Research Centre.

Hildebrandt, M. (2008). A vision of ambient law. In Brownsword, R., & Yeung, K. (Eds.), *Regulating Technologies* (pp. 175–191). Oxford, UK: Hart.

Hildebrandt, M. (2008). Ambient intelligence, criminal liability and democracy. *Criminal Law and Philosophy, 2*(2), 163–180. doi:10.1007/s11572-007-9042-1

Hildebrandt, M. (2008). Profiling and the rule of law. *Identity in the Information Society, 1*(1), 55–70. doi:10.1007/s12394-008-0003-1

Hildebrandt, M., & Gutwirth, S. (2008). *Profiling the European Citizen*. New York, NY: Springer. doi:10.1007/978-1-4020-6914-7

Jain, A. K., Ross, A., & Pankanti, S. (2006). Biometrics: a tool for information security. *Transactions on Information Forensics and Security, 1*(2), 125–143. doi:10.1109/TIFS.2006.873653

Luppicini, R., & Adell, R. (2009). *Handbook of research in technoethics*. Hershey, PA: Information Science Reference.

Magnani, L. (2007). *Morality in a technological world*. Cambridge, UK: Cambridge University Press. doi:10.1017/CBO9780511498657

Magnani, L. (2011). *Understanding violence: The intertwining of morality, religion and violence: A philosophical stance*. Heidelberg, Germany: Springer-Verlag.

Magnani, L. (in press). Abducing personal data, destroying privacy. Diagnosing profiles through artifactual mediators. In Hildebrandt, M., & de Vries, E. (Eds.), *Privacy, due process and the computational turn: Philosophers of law meet philosophers of technology*. London, UK: Routledge.

Nakashima, H., Aghajan, H., & Augusto, J. C. (2010). *Handbook of ambient intelligence and smart environments*. New York, NY: Springer. doi:10.1007/978-0-387-93808-0

Nanavati, S., Thieme, M., & Nanavati, R. (2002). *Biometrics*. New York, NY: John Wiley & Sons.

Newbould, R., & Collingridge, R. (2003). Profiling - Technology. *BT Technology Journal, 21*(1), 44–55. doi:10.1023/A:1022400226864

Ratha, N. K., & Govindaraju, V. (2008). *Advances in biometrics*. New York, NY: Springer. doi:10.1007/978-1-84628-921-7

Revett, K. (2008). *Behavioral biometrics*. New York, NY: John Wiley & Sons. doi:10.1002/9780470997949

Taylor, J. A., Lips, M., & Organ, J. (2009). Identification practices in government: citizen surveillance and the quest for public service improvement. *Identity in the Information Society, 1*(1), 135–154. doi:10.1007/s12394-009-0007-5

U.S. National Science and Technology Council. (2006). *Privacy & biometrics building a conceptual foundation*. Washington, DC: Author.

Wang, L., & Geng, X. (2010). *Behavioral biometrics for human identification*. Hershey, PA: IGI Global.

Yampolskiy, R. V. (2008). Behavioral modeling: An overview. *American Journal of Applied Sciences*, *5*(5), 496–503. doi:10.3844/ajassp.2008.496.503

Zhang, D., & Jain, A. K. (2006). *Handbook of multibiometrics*. New York, NY: Springer.

Zhang, D., Song, F., Xu, Y., & Liang, Z. (2009). *Advanced pattern recognition technologies with applications to biometrics*. Hershey, PA: IGI Global. doi:10.4018/978-1-60566-200-8

ENDNOTE

[1] I will not discuss the identity issues here and neither the concept of *human hybrid*; however, about this last point, see Magnani (2007).

This work was previously published in the International Journal of Technoethics, Volume 2, Issue 4, edited by Rocci Luppicini, pp. 84-93, copyright 2011 by IGI Publishing (an imprint of IGI Global).

Chapter 8
Cellular Telephones and Social Interactions:
Evidence of Interpersonal Surveillance

Steven E. Stern
University of Pittsburgh at Johnstown, USA

Benjamin E. Grounds
Penn State College of Medicine, USA

ABSTRACT

Changes in technology often affect patterns of social interaction. In the current study, the authors examined how cellular telephones have made it possible for members of romantically involved couples to keep track of each other. The authors surveyed 69 undergraduates on their use of cellular telephones as well as their relationships and their level of sexual jealously. Results find that nearly a quarter of romantically involved cellular telephone users report tracking their significant other, and evidence shows that tracking behavior correlates with jealousy. Furthermore, participants frequently reported using countermeasures such as turning off their cellular telephones in order to avoid being tracked by others. In conclusion, newer communication technologies afford users to act upon protectiveness and jealousy more readily than before these technologies were available to the general public.

INTRODUCTION

Although there are many ways to define technology, one frequent theme is that technology refers to the tools and methods by which we increase our natural capabilities (e.g., Mumford, 1963). Because technologies develop constantly, human

action frequently, and arguably increasingly, leads to unanticipated effects (see Crabb & Stern, 2010). At the most dramatic and potentially horrifying, these changes include environmental disaster and the use of advanced weaponry. Technological evolution, however, does lead to countless subtler changes that we experience in our day to day lives. Many of these changes are societal, psychological or behavioral. From an ethical point of view, these many changes present people with situations and

DOI: 10.4018/978-1-4666-2931-8.ch008

choices with which they previously did not have any experience and for which there may not yet be societal norms (Moor, 2005).

The recognition that technological innovations can have both positive and negative consequences is central to the wide ranging, interdisciplinary, and expanding field of technoethics. The pervasiveness of technology in human life makes it imperative that technoethical scholars hail from a vast range of fields (de Vries, 2009) as they collectively approach the ethical implications of technology, including areas as far reaching as environmental impact, political implications, and societal transformation (Luppicini, 2009).

Technology in its end use directly impacts the thoughts and behaviors of individuals. Hence, psychology is a field that is well positioned to examine the impacts of technology on human functioning. Although the field of psychology has been neither organized nor systematic in its approach to technology (Kipnis, 1991), psychological research has a rich history in informing us of technology's negative consequences (Stern & Handel, 2001). As far back as the 1940s, psychologists warned readers of the potential misuse of radio for wartime propaganda (Allport, 1947). In the early 1960s, psychologists studied the links between children's television viewing and subsequent aggression (Eron, 1963). In the current age, psychologists have investigated the potentially numerous concerns regarding the Internet (e.g., Gackenbach, 1998; Kielser, 1997) including the potentially habit forming nature of the Internet (Stern, 1999; Young, 1998).

Privacy, and the right to privacy are frequently cited (Luppicini, 2009) aspects of human social life that have been altered by technological change. In this paper, we examine some privacy related effects that have accompanied the mass introduction of cellular telephones. Because this technology impacts individuals as well as interactions between individuals, this research can be considered social psychological.

Much of the psychological research on the effects of cellular telephones focuses on the effects of using cell phones while driving (Redelmeier & Tibshrani, 1997; Strayer & Johnston, 2001). This is quite understandable considering the well-established danger to self and others caused by driving while speaking on the telephone. There are, however, other psychological phenomena associated with the mass introduction of cellular telephones during the last two decades that deserve attention.

Technological innovation can change or eliminate social norms. Crabb (1996), for instance, found that users of easily affordable camcorders were quick to invade the privacy of others, demonstrating a degradation of the civil inattentiveness norm (i.e., it is impolite to stare). Technological change can also lead to shifts in patterns of influence (Kipnis, 1991). The introduction of the Internet, for instance, has given medical patients easier access to information about pharmaceuticals and their potential side effects, in turn empowering them vis a vis their physicians. Closer to the topic at hand, Ling and Yttri (2006) have described how cellular telephones alter power relationships between teenagers and their parents.

In his examination of technological design, Norman (1988) endorsed the useful notion of understanding the elements of our environment in terms of "affordances" or in other words "properties that determine...how a thing could possibly be used" (p. 8). Cellular telephones make it possible, for instance, to keep in touch with others who would not have access to a stable telephone. In turn, since the telephones are small and durable, they are also very portable. Finally, they are relatively cheap to purchase and use, so young people can typically afford to have one, and parents can afford to generously buy them for their children.

These qualities afford the users the ability to carry cellular telephones on their person through much of the day. They also afford the users to be accessed remotely, as well as constantly. In fact, it may be normative to keep the phones turned on, yet acceptable, as with residential telephones, to break this norm and to screen calls (Crabb, 1999).

At another level, we can also examine cellular telephones in terms of gratifications, or in other words, the instrumental needs that cellular telephones satisfy. While clearly satisfying our need to stay in touch with others, Leung and Wei (2000) also found that mobile phones satisfied people's needs to stay in fashion, to maintain affectionate relationships, to relax, to do business, and to feel safe and secure.

More specifically, our research examines cellular telephones in the context of subtle, day to day interpersonal surveillance. The social psychological and organizational psychological literature on surveillance unsurprisingly notes that people do not like being under surveillance (Aiello & Kolb, 1995). More surprisingly, there is evidence that surveillance, being a strong influence tactic, leads the observer to have less respect for the person who they are observing (Strickland, 1958). It has been argued that when we need to resort to strong influence tactics to get people to do the things that we want them to, we lose respect for them and devalue them and their work (Kipnis, 1976).

In the present study, we examined how university students find cellular telephones as a way in which parents can keep track of their children and people involved in romantic relationships can keep track of each other. We contend that cellular telephones, to some extent, disempower the user to be left alone, while empowering those who are socially attached to or responsible for the user to keep track of their whereabouts and behaviors. Insofar as cellular telephones can be easily left home, ignored, or turned off, they also afford the owner the ability to surreptitiously avoid attempts to reach them (see Crabb, 1999).

To this extent, we had four central hypotheses: 1) jealousy will correlate with reported tracking by the partner, 2) partner jealousy will correlate with self-reported tracking of the partner, 3) reported parent protectiveness will correlate with

reports of parental tracking, and 4) there will be evidence that cellular telephone users resort to tactics to counter the abilities of others to keep track of them.

METHOD

Participants

A total of 69 undergraduate students (29 male; 40 female) at the University of Pittsburgh at Johnstown participated in the study. Participants were solicited from Introductory Psychology courses and received course credit for their participation. Out of the total, 37 of the participants reported being in a romantic relationship. The average age of participants was 19.16 years.

Materials and Procedure

Questionnaires were developed for the study. The variables we were most interested in examining were jealousy, partner's level of jealousy, parent(s)' level of protectiveness, cellular telephone avoidance, self-reported tracking behavior of partner, and reported tracking behavior of partner. We also collected some basic demographics as well as issues surrounding relationship status and level of relationship commitment.

The study used a demographics questionnaire, a cell phones and relationships questionnaire, and a jealousy questionnaire. The jealousy questionnaire consisted of 10 true/false items that focused exclusively on sexual jealousy since that was most relevant to the current investigation (see Table 1).

The self-report questionnaires were used to gauge how much people reported tracking their significant others and how much they perceive their significant others tracking them. We also examined how much people perceived that their parents are tracking them and how protective their parents are. We also asked (aside from the jealousy questionnaire described above) how jealous

Table 1. Jealousy questionnaire used in study

Answer each question True or False. If you are not in a relationship, please answer the questions *as if* you were in a relationship.
1)_____ If my significant other were to see an old friend of the opposite sex and respond with a great deal of happiness, I would be annoyed.
2)_____ If my significant other were to help someone of the opposite sex, I would feel irritated.
3)_____ I want my significant other to remain good friends with the people he/she used to date.
4)_____ If my significant other had previously been married; I would feel resentment towards the ex-wife.
5)_____ Jealously is a sign of true love.
6)_____ I feel upset when my partner speaks about someone of the opposite sex.
7)_____ If my significant other were to go away for the weekend without me, my only concern would be whether he/she had a good time.
8)_____ If my significant other were to become very close with someone of the opposite sex I would feel very unhappy and/or angry.
9)_____ I don't think it would bother me if my significant other flirted with someone of the opposite sex.
10)_____ I like to find fault with my significant other's old dates.

people perceived themselves as well as their significant others to be. Finally we examined whether or not people used countermeasures such as shutting their cell phones off or ignoring calls to prevent being tracked

RESULTS AND DISCUSSION

The study revealed strong evidence of tracking behavior in the undergraduate population.

This is seen in the percentage of students who endorsed (by choosing either "agree" or "strongly agree") statements related to tracking behavior in questionnaires; 22% of those who were romantically involved reported that they tracked their partners while 24% reported that their partners tracked them. Parental tracking was even more salient; 42% of those surveyed reported that they believed that their parents used their cell phones to keep track of them.

We also examined the extent to which jealousy was related to tracking behavior. There was strong correlational support for our first hypothesis. Jealousy, as measured by the 10-item questionnaire, was found to correlate with self reported tracking of significant others, $r(34) = .56$, $p < .001$. Our second hypothesis was supported as well. Ratings of partner jealousy correlated with perceptions of being tracked by significant other $r(35) = 0.39$, $p < 0.025$.

The pattern of findings supporting the third hypothesis, involving parental overprotectiveness was less clear. Although parental overprotectiveness did not correlate significantly with reporting the perception that one's parents were tracking them, $r(66) = .06$, *n.s.*, there was evidence that people took countermeasures to avoid parents that they perceived as over-protectiveness. Ratings of parental over-protectiveness correlated with self-reports of ignoring calls from parents, $r(66) = .22$, $p < .10$.

In support of the fourth hypothesis there was also some evidence that participants used countermeasures such as turning off their cell phones or leaving them at home to disrupt tracking behavior. Turning off cell phones to avoid calls correlated with the perception that parents use one's cell phone to keep track of them, $r(67) = .24$, $p < .05$. Perception of a jealous significant other had a marginal relationship with leaving one's cell phone at home $r(37) = .28$, $p < .10$, and also with turning one's cell phone off to avoid calls $r(37) = .27$, $p < .10$.

Although there were no specific hypotheses regarding levels of relationship commitment, some of our findings regarding this variable deserve mention. As seen in Table 2, stronger levels of commitment were associated with less use of countermeasures to avoid tracking as well as less perceived tracking by significant others.

Changing communication technologies have undoubtedly changed patterns and habits of com-

Table 2. Correlations of tracking countermeasures with self-reported levels of relationship commitment

	Reported level of commitment	
	Participant	**Significant other**
Turn off phone to avoid calls	-0.31*	-.09
Ignore/silence calls from significant other when with same sex friends	-0.39**	-.20
Ignore/silence calls from significant other when with opposite sex friends	-.50***	-.62***
Significant other ignores your calls when with same sex friends	-.12	-.16
Significant other ignores your calls when with opposite sex friends	-.58***	-.55***

Note: * $p < .10$; ** $p < .05$; *** $p < .01$

munication. Cellular telephones make it possible to reach people consistently through the day, and to attempt to monitor their whereabouts and behavior. Cellular telephones also can be turned off, or ignored, making it possible for people to avoid being monitored.

In this study we found evidence that people do knowingly monitor each other and perceive that others monitor them. We also found evidence that people who are motivated by jealousy have a higher tendency to monitor by cellular telephone than those who are less jealous. And we also found that people in more committed relationships are less likely to guard their privacy by avoiding attempts by their significant others to monitor them.

Hence, we have evidence that the widespread use of cellular telephones has provided an added dimension to parental and romantic relationships. This dimension has ethical implications on the everyday interactions between parents and children and between romantically involved couples. Protectiveness and jealousy are acted upon more readily and more easily than possible in the precellular age. In turn, privacy is more difficult to protect, and people will more frequently turn to ignoring or deceiving others, by shutting off or ignoring their cellular phones, in order to maintain their privacy. The present examination, however, studied only one of a number of new widely adopted communications technologies. Further examination should be done to examine if similar patterns of surveillance behavior occurs for people using other emerging technologies including text-messaging and social networking websites.

ACKNOWLEDGMENT

The authors are grateful to Savannah Calhoun, Jennifer Conard, and Kendra Kuykendall for their assistance on this research.

REFERENCES

Aiello, J. R., & Kolb, K. J. (1995). Electronic performance monitoring and social context: Impact on productivity and stress. *The Journal of Applied Psychology*, *80*, 339–353. doi:10.1037/0021-9010.80.3.339

Allport, G. W., & Postman, L. (1947). *The psychology of rumor*. New York: Henry Holt.

Crabb, P. B. (1996). Video camcorders and civil inattention. *Journal of Social Behavior and Personality*, *11*, 805–816.

Crabb, P. B. (1999). The user of answering machines and caller ID to regulate home privacy. *Environment and Behavior*, *31*, 657–670. doi:10.1177/00139169921972281

Crabb, P. B., & Stern, S. E. (2010). Technology traps: Who is responsible? *International Journal of Technoethics*, *1*(2), 19–26.

de Vries, M. J. (2009). A multi-disciplinary approach to Technoethics. In Luppicini, R., & Adell, R. (Eds.), *Handbook of research on technoethics* (pp. 20–31). Hersey, PA: IGI Global.

Eron, L. D. (1963). Relationship of television viewing habits and aggressive behavior in children. *Journal of Abnormal and Social Psychology, 67,* 193–196. doi:10.1037/h0043794

Gackenbach, J. (1998). *Psychology and the Internet.* San Diego: Academic Press.

Kiesler, S. (1997). *Culture of the Internet.* Mahwah, NJ: Lawrence Erlbaum Associates.

Kipnis, D. (1976). *The powerholders.* Chicago: University of Chicago Press.

Kipnis, D. (1990). *Technology and power.* New York: Springer Verlag.

Kipnis, D. (1991). The technological perspective. *Psychological Science, 2,* 62–69. doi:10.1111/j.1467-9280.1991.tb00101.x

Leung, L., & Wei, R. (2000). More than just talk on the move: Uses and gratifications of the cellular phone. *Journalism & Mass Communication Quarterly, 77,* 308–322.

Ling, R., & Yttri, B. (2006). Control, emancipation, and status: The mobile telephone in teen's parental and peer relationships . In Kraut, R., Brynin, M., & Kiesler, S. (Eds.), *Computers, phones, and the internet* (pp. 219–234). New York: Oxford University Press.

Luppicini, R. (2009). The emerging field of Technoethics . In Luppicini, R., & Adell, R. (Eds.), *Handbook of research on Technoethics* (pp. 1–18). Hershey, PA: IGI Global.

Moor, J. H. (2005). Why we need better ethics for emerging technologies. *Ethics and Information Technology, 7,* 111–119. doi:10.1007/s10676-006-0008-0

Mumford, L. (1963). *Technics and civilization.* New York: Harcourt, Brace & World.

Norman, D. A. (n.d.). *The design of everyday things.* New York: Doubleday.

Redelmeier, D. A., & Thbshirani, R. A. (1997). Association between cellular-telephone calls and motor vehicle collisions. *The New England Journal of Medicine, 336,* 453–458. doi:10.1056/NEJM199702133360701

Stern, S. E. (1999). Addiction to technologies: A social psychological perspective of Internet addiction. *Cyberpsychology & Behavior, 2,* 419–424. doi:10.1089/cpb.1999.2.419

Stern, S. E., & Handel, A. D. (2001). Sexuality and mass media: The historical context of psychology's reaction to sexuality on the Internet. *Journal of Sex Research, 38,* 283–291. doi:10.1080/00224490109552099

Strayer, D. L., & Johnston, W. A. (2001). Driven to distraction: Dual-task studies of simulated driving and conversing on a cellular telephone. *Psychological Science, 12,* 462 466. doi:10.1111/1467-9280.00386

Strickland, L. H. (1958). Surveillance and trust. *Journal of Personality, 26,* 201–215. doi:10.1111/j.1467-6494.1958.tb01580.x

Young, K. S. (1998). *Caught in the net: How to recognize the signs of Internet addiction and a winning strategy for recovery.* New York: John Wiley and Sons.

This work was previously published in the International Journal of Technoethics, Volume 2, Issue 1, edited by Rocci Luppicini, pp. 43-49, copyright 2011 by IGI Publishing (an imprint of IGI Global).

Chapter 9
Socio–Technical Influences of Cyber Espionage:
A Case Study of the GhostNet System

Xue Lin
University of Ottawa, Canada

Rocci Luppicini
University of Ottawa, Canada

ABSTRACT

Technoethical inquiry deals with a variety of social, legal, cultural, economic, political, and ethical implications of new technological applications which can threaten important aspects of contemporary life and society. GhostNet is a large-scale cyber espionage network which has infiltrated important political, economic, and media institutions including embassies, foreign ministries and other government offices in 103 countries and infected at least 1,295 computers. The following case study explores the influences of GhostNet on affected organizations by critically reviewing GhostNet documentation and relevant literature on cyber espionage. The research delves into the socio-technical aspects of cyber espionage through a case study of GhostNet. Drawing on Actor Network Theory (ANT), the research examined key socio-technical relations of Ghostnet and their influence on affected organizations. Implications of these findings for the phenomenon of GhostNet are discussed in the hope of raising awareness about the importance of understanding the dynamics of socio-technical relations of cyber-espionage within organizations.

INTRODUCTION

What these guys [corporate officials] don't realize, because nobody tells them, is that a major foreign intelligence agency has taken control of major portions of their network. You can't get rid of this attacker very easily. It doesn't work like a normal virus. We've never seen anything this clever, this tenacious (Mills, 2010).

GhostNet was a cyber espionage network in 2008 that attracted much attention and raised serious public concern. The purpose of studying GhostNet here is to explore technoethical and communicative aspects of cyber espionage on affected organizations, which significantly influences cyber peace and corporate develop-

DOI: 10.4018/978-1-4666-2931-8.ch009

ment. Despite an increased attention to hackers and cyber espionage in the media, little empirical research has actually been conducted on GhostNet. For example, although some research, like *Tracking GhostNet: Investigating a Cyber Espionage Network* by Information Warfare Monitor (2009), examines the GhostNet phenomenon, it focuses in detail on the social and technological aspects and how GhostNet was tracked, leaving out the influence on communication. Therefore, this study will remedy this deficiency and provide a contribution to the scholarly literature by adopting Actor Network Theory (ANT) to examine the technical and communicative actors of a socio-technical process in a whole system.

This analysis of socio-technical aspects of cyber espionage is guided by a dominant theory of Science and Technologies Studies (STS), namely Actor Network Theory. ANT emerged during the mid-1980s with work primarily from Bruno Latour and Michel Callon, and significant later contributions from John Law (1999). ANT is a conceptual framework for exploring collective socio-technical processes, whose advocates have paid particular attention to scientific and technological activities. ANT suggests that the work of science is not fundamentally different from other social activities, instead asserting that it is a process of heterogeneous engineering in which the social, technical, conceptual and textual are juxtaposed and translated (Latour, 1991; Law, 1999).

As applied to this study, ANT suggests that several aspects should be studied: the actors of cyber espionage (behaviours), hackers (people) and technology (objects), as well as the network associated with these actors. Based on ANT, this research focused on socio-technical aspects revolving around how the GhostNet network was formed and how it fell apart. This research study utilizes a case study approach guided by the work of Yin (2003) and Creswell (2007) to carry out a comprehensive document analysis of relevant industry reports, research literature and other documentation pertaining to the case.

BACKGROUND

Cyber Espionage

Unlike previous, physical frontiers, cyberspace is a human construct. Cyber behaviours, such as cyber espionage, cyber surveillance, cyber terrorism, influence cyberspace significantly in different ways. "Cyber espionage is a technology by which system access is illegally obtained, data or computer equipment is stolen or destroyed, or software is illegally copied"(Stair, 1996, p. 529). Trojan horse programmes and other associated malware are often cited as tools for conducting sophisticated computer-based espionage.

Cyber space is a battlefield for information warfare with not only legal but also ethical implications. According to Mason (1986), the ethical issues of cyber space usually involve four areas of concerns: privacy, accuracy, property and access (PAPA):

Privacy—the ability of people to keep personal information about themselves private and confidential; how the widespread holding of personal information about people impacts on interpersonal relations of trust, autonomy, and dignity; Accuracy—the quality and accuracy of data/ information held in databases, and on which organisations act, assuming the data/information to be correct; Property—information ownership and control—who owns personal information about an individual, and who has the right to use it, or control its use; and Accessibility—access of members of society to the social store of information (Mason, 1986).

As an issue relating to ethical concerns and cyber crime, GhostNet entails two of the ethical charges, privacy and access.

Contemporary research in Technoethics extends to research on cyber ethics and topics such as online privacy, copyright, security, and surveillanc. Cyber terrorism, and cyber espionage are the signs of illegal and unethical wrongdoing in cyber space. GhostNet is one example of cyber espionage, which has used malware-attached email to achieve system access, data stealing or surveillance. Such criminal activities where the computer is a major factor in committing the criminal offence are categorized as cybercrimes (Wall, 2003). In the context of technology, it is important to study controversial cases with serious social and ethical implications. As a cyber espionage network, a case study of GhostNet allows us to examine socio-technical aspects of online activity between organizations to discern a greater understanding of the serious social and ethical implications which can arise through the use of information and commuication technologies (ICT's).

The GhostNet System

GhostNet came to light in March 2009 when a team of cyber warfare researchers brought together by the University of Toronto's Munk Centre for International Studies and the Canadian security think tank SecDev. They conducted a security audit for the Tibetan Government-in-Exile and found malicious software on its most sensitive computers. After a 10-month investigation, a report on the GhostNet system titled *Tracking GhostNet: Investigating a Cyber Espionage Network* (2009) was published by Information Warfare Monitor, a project of the SecDev group, to reveal the large scale of cyber espionage.

As a large-scale cyber espionage system, GhostNet utilized the unique properties of the Internet that integrates existing technologies into a single medium in an accessible manner (Dartnell, 2003). The GhostNet system directed infected computers to download attached documents packed with exploit code and malware known as "gh0st RAT" through emails sent to specific targets. GhostNet was designed to take advantage of vulnerabilities in software installed

on the target's computer for complete, real-time control. The malicious hacker attacked interferes with the peace of the cyber domain.

In *Tracking GhostNet: Investigating a Cyber Espionage Network* (2009), IWM provided documented evidence that GhostNet infected at least 1,295 computers in 103 countries, of which close to 30% can be considered high-value diplomatic, political, economic, and military targets (Information Warfare Monitor, 2009). Between June 2008 and March 2009, the Information Warfare Monitor conducted an extensive two-phase investigation focused on allegations of Chinese cyber espionage against the Tibetan community (Information Warfare Monitor, 2009). In the first phase, because GhostNet is malware based, network monitoring software was installed on various computers to collect forensic technical data from affected computer systems. The first phase also included interviews with officials in the Tibetan Government-in-Exile (Information Warfare Monitor, 2009). During the second phase of computer-based scouting, researchers discovered four control servers, which were identified and geo-located from the captured traffic using a simple IP lookup, as well as six command servers (Information Warfare Monitor, 2009). According to IWM, the GhostNet system was controlled from commercial Internet access accounts located on the island of Hainan, People's Republic of China" (p. 5)

Up to this point, existing research on GhostNet provides only a limited understanding of socio-technical aspects of GhostNet. First, previous research has mainly highlighted political and economic considerations about information lost from compromised computers and affected organizations. There is an absence of broader socio-technical considerations of cyber espionage needed to understand how and why humans within these organizations fell victim to the GhostNet threat. Visibly, there is room for more thorough research in this field.

METHODOLOGY

This research adopted a qualitative approach due to the nature of the problem under investigation. According to Creswell (2009), qualitative research is "a means for exploring and understanding the meaning individuals or groups ascribe to a social or human problem" (p. 5). The researchers chose to use the case study method to guide the research. Qualitative case studies are well suited to contribute to our knowledge of individual, organizational, social, and political phenomena (Yin, 1994). For this reason, the incorporation of case study methodology into this study is employed to uncover clues to inform future research on cyber espionage.

This case study draws on GhostNet research documentation and academic literature to answer the following questions:

RQ1: What are the socio-technological influences of GhostNet on affected organizations?
RQ2: How does Actor Network Theory explain the socio-technological influences of GhostNet on organizations?

In order to satisfy the research questions, the following qualitative data collection method was utilized. Data was collected by searching University of Ottawa data bases using the following inclusion criteria:

1. Published research
2. Published year between 1995 and 2010
3. Peer-reviewed journal articles, books and book sections

This excludes unpublished research and research published before 1995. Keywords titles searches (I.e., GhostNet OR cyber espionage OR cyber crime) of journal articles are carried out using ERIC, EBSCO, E-Journals @ Scholars Portal and INSPEC online databases for 15 years. Based on initial database search findings of multiple ar-

ticles from individual journals, follow-up library searches are conducted using these journals to target key journals of interest. Second, an Internet search is conducted using Google search engines to attain data of GhostNet or cyber espionage.

In terms of specific content, media reviews on cyber espionage were analysed in order to compare and contrast opinions given by different researchers. The documents that were analysed came from three different sources; studies that were published by the Information Warfare Monitor about GhostNet, reviews on the Information Warfare Monitor' studies of GhostNet by other researchers, and journal articles or books published on the subject of cyber espionage and other relevant themes including cyber security, and cyber crimes. The studies published by the Information Warfare Monitor were the first choice for review due to the fact that they discovered the GhostNet network. *Tracking GhostNet: Investigating a Cyber Espionage Network* was the first publicly revealed report that investigated the GhostNet system. This 53-page report was published in March 29th, 2009 (Information Warfare Monitor, 2009). After one year, in April 6th, 2010, the Information Warfare Monitor and the Shadowserver Foundation have released a report, *Shadows in the Cloud: An investigation into cyber espionage 2.0* (Information Warfare Monitor & Shadowserver Foundation, 2010) where it documents another targeted malware network that had also compromised computers at the Dalai Lama's office, which is one of the targets of GhostNet. Reviews on the Information Warfare Monitor studies of GhostNet by other researchers provide multiple perspectives for understanding the GhostNet network. In addition, journal articles and books on cyber espionage (and related themes) were also examined to help situate the analysis. Overall, the documentation examined was intended to allow the researchers to discern a detailed analysis of this case on socio-technical influences of GhostNet on affected organizations.

The researcher is an instrument in qualitative research. Inevitably, this means that this author's intensive research experience will introduce a range of strategic, ethical, and personal issues into the qualitative research process. With these concerns in mind, the author's biases, values, and personal background may be reflected heavily, shaping the interpretations formed during this study (Creswell, 2007). Tracking personal reactions and insights into one's self as well as literature read in the past is an effective way to guard against bias. Working in an Internet corporation for half year increased the researcher's knowledge about cyber space and sparked an interest in cyber communication. This past experience provides background data for understanding this topic and promises this study an in-depth study on cyber espionage.

Qualitative data analysis of this study entailed examining, categorizing, tabulating and recombining the evidence to address the initial propositions. The outcome of the analysis was used to develop a descriptive framework for organizing the case study of GhostNet and for identifying the appropriate links between the technological, ethical, and communicative aspects by employing the Actor Network perspective. The descriptive insights generated during the process of data collecting also helped the researchers to structure case explanations

The researchers used an explanation-building logic to study the socio-technical aspects of GhostNet. Explanation-building is considered a form of pattern-matching, in which the analysis of the case study is carried out by building an explanation of the case (Tellis, 1997). The element of explanations is a presumed set of casual links about the phenomenon, or "how" or "why" something happened (Yin, 2009). Based on ANT, the researchers assume that the actors of cyber espionage (behaviours), hackers (people) and technology (objects), as well as the network associated with these actors, have socio-technical

influences on affected organizations that can be discerned. To this end, the researchers examined the case study evidence to build an explanation about the case.

To follow Yin's (2003) strategies of ensuring validity, this research documented the procedures of the case study and taken steps to ensure the trustworthiness of study findings. The authors employed multiple sources of evidence to triangulate and build a coherent justification for themes unearthed. Also, research notes were taken throughout the data collection to guard against biases and leverage analysis. By taking notes while making observations when conducting the data collection, the researchers kept track of sources by building a research log which originated from Harris Cooper's model (2010). It enabled the researchers to return to the information during the analysis to track emerging study insights and new observations.

FINDINGS AND INTERPRETATIONS

General Findings

General socio-technical findings focus on the attack software, strategies the attackers used, and actions use by users of targeted computers. According to the field investigation of IWM (2009) on GhostNet, the procedures of the GhostNet system discovered that:

An email message arrives in the target's inbox carrying the malware in an attachment or web link. The attackers(s)' objective is to get the target to open the attachment or malicious link so that the malicious code can execute. In this case, the attacker(s) sent a carefully crafted email message which was configured to appear as if it was sent from campaigns@freetibet.org with an attached infected Word document named "Translation of Freedom Movement ID Book for Tibetans in Exile.doc" to entice the recipient to open the file.

...The targeted user proceeds to opens the attachment or malicious link. Once opened, the infected file or link exploits vulnerability on the user's machine and installs the malware on the user's computer, along with a seemingly benign file...

...Control over some targeted machines is maintained using the Chinese gh0st RAT (Remote Access Tool). These Trojans generally allow for near-unrestricted access to the infected systems (p.18).

From the description, it is clear that the Ghost-Net network is a Trojan program designed to steal information from targeted computers and conduct cyber surveillance (Figure 1). Trojans tend to be designed for two main purposes: gathering information and taking control of an infected system (Bocij, 2006). The malware, which is a contraction of "malicious" and "software", functions as a key to provide a way of controlling the victim's computer (Bocij, 2006).

The Trojan virus is a kind of malware, which refers to programs that fool users into executing them by pretending to perform a particular function but ultimately do something else (Furnell & Ward, 2006). GhostNet used words, such as "freedom movement" and "Tibetans in Exile" in the title and the format of.doc of that email attachment. These enticed the recipients to download the attachments. This email pretended to act as an official correspondence but ultimately proved to be a tool for performing cyber espionage. From the title of the attachments used, it is clear that the person who sent the emails knew the target and tried to attract recipients' attention.

The use of carefully-written email lures based on specific knowledge of the social context (social phishing) was successful in getting people to give access to their computers. In the case of GhostNet, the attackers also stole mail in transit and replaced the attachments with corrupt ones (Nagaraja & Anderson, 2009). In this case, the phishing technique was combined with the Trojan virus to gain access and control. Once the recipient opened or downloaded the attached file, the Trojan was active and began to collect data. Once active, a Trojan can scan the user's hard disk for sensitive information such as, passwords, credit card information, and anything else of value (Bocij, 2006). In the case of GhostNet, the target was a NGO, the Office of His Holiness the Dalai Lama (OHHDL) where there was secret and confidential information.

Figure 1. Four steps of the GhostNet attack

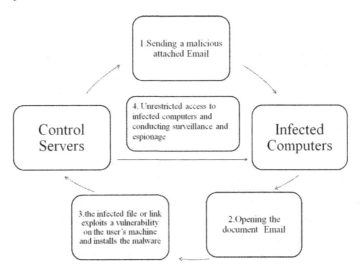

According to the investigation of Nagaraja and Anderson (2009), confidential files had been transferred out of the OHHDL violating the confidentiality of target without target's consciousness. The OHHDL did not suspect the cyber espionage until May 2008. In Nagaraja and Anderson's report, they said that:

The OHHDL started to suspect it was under surveillance while setting up meetings between His Holiness and foreign dignitaries. They sent an email invitation on behalf of His Holiness to a foreign diplomat, but before they could follow it up with a courtesy telephone call, the diplomat's office was contacted by the Chinese government and warned not to go ahead with the meeting. The Tibetans wondered whether a computer compromise might be the explanation (Nagaraja & Anderson, 2009, p. 5).

The unawareness of the target reflects both the success of the GhostNet system and the failure of the anti-virus programs of infected computers. After the connection between attackers' control servers and the infected computers, commands were set on the control servers that instruct infected computers to download additional remote administration Trojans, such as gh0st RAT, in order to take complete real-time control of the infected computers (Information Warfare Monitor, 2009). In addition, "attacker(s) may choose from a menu of commands, which includes options to download binaries that provide additional functionality (such as keystroke logging or remote administration), acquire system information (list computer information, software and documents on the computer), or cause the malware to become dormant" (Information Warfare Monitor, 2009, p. 34).

The set of commands implemented the full control of infected computers by attacker's control servers from being active to dormant. The GhostNet system conducted online surveillance and information stealing which are hided from being aware by the recipients. During one year of IWM's undercover investigation, GhostNet was still working and it vaporized as soon as the IWM's report on GhostNet was published. Although there is no exact time showing how long GhostNet has been working, it could be calculated at least one year.

In short, the GhostNet network was a Trojan which used malicious email and social phishing strategies to gain control of infected computers in order to conduct cyber espionage. Although this may not be the full picture on the GhostNet attacks, it reveals the general socio-technical findings and elementary analysis of GhostNet.

Actor-Network Theory (ANT) and the GhostNet Communications Systems

Because GhostNet involves the interaction of technology, human actors, and communications, Actor-Network Theory (ANT) can be used to leverage understanding of GhostNet influences within affected organizations when conceptualized as a human communications system. According to Spinello (2002), a typical technical system consists of three independent layers: Content layer (information and entertainment resources), logical layer (the software, communications standards, and protocols), and the physical layer (telecommunications wires and fibres, computer hardware, routers, and so forth). As applied to ANT, the three layers provide a way to understand the network as a communications system involving human and technological actors interacting via the Internet.

As a valuable, fast, inexpensive, and pervasive tool, the Internet presents new opportunities and challenges to human communications. Along with the great capacity of Internet communications comes great vulnerability which proliferates diversified cybercrimes. This explains how social engineering worked in GhostNet. Attackers took advantages of the Internet communication made by monks in OHHDL to gain contact information and personal interests of monks by observing their online communication. For instance, an attacker could easily note monks' names, their interests and the names of people with whom they interacted. The GhostNet break may have been facilitated by the fact that the monks in the OHHDL were active on various discussion sites (Nagaraja & Anderson, 2009). The information gained by attackers was used for social engineering by tailoring emails to monks to deceive them into opening email attachments. This highlights how the ease of communication brought by the Internet has shifted the social rules for communication management. Because good communication is an important success factor that connects all the other factors of network success, it raises the difficult and urgent problem of what combination of technical controls and procedural measures are needed in the new world of online communication (Nagaraja & Anderson, 2009).

According to Callon (1986), the translation of an actor-network consists of four major stages: problematization, interessement, enrollment and mobilization. Because these translation processes emphasize human actors, they are applicable to analyse human actors in the GhostNet network. During the problematization stage, an actor initiating the process defines identities and interests of other actors that are consistent with the interests of the initiating actor (MaHring, Holmstrom, Keil, & Montealerge, 2004). In the actor-network of GhostNet, the initiating actor is the attacker(s) who planned GhostNet strategically and designed GhostNet technologically. The initiating actors defined the problems and solutions and also established roles and identities for other actors in the network, such as the recipients. In the case of GhostNet, the problem was discovering ways to get the confidential information from other actors without being noticed. Based on the definition of the problem and the roles of other actors, the recipients who are working in OHHDL, the social engineering emails, malware, control servers and even the initiating actors themselves were needed to solve the problem.

The second translation stage is *interessement*, which involves convincing other actors that the interests defined by the initiator(s) are well in line with their own interests (MaHring et al., 2004). In this case, the social engineering emails functioned as a tool to "persuade" other actors that the content of the email was well in line with their own interests. The GhostNet system is a cybercrime by which initiating actors attack against computers of other actors to disrupt processing and perform espionage. This is done to make unauthorized copies of classified data in other actors' computers, including trick "convincing" other actors, in fact, was phishing and made other actors accept the initiating actors' interest passively.

After *interessement* was successful, *enrollment* occurred. *Enrollment* involves a definition of roles of each of the actors in the newly created actor-network. It also involves a set of strategies through which initiators seek to convince other actors to embrace the underlying ideas of the growing actor-network, and to be an active part of the whole project (MaHring et al., 2004). As applied to GhostNet, when the recipients opened the malicious email through an oversight they were tolerant to the initiating actors' interests and they were enrolled into this actor-network passively.

The fourth and final stage of translation, *mobilization*, includes initiators' use of a set of methods to ensure that allied spokespersons act according to the agreement and do not betray the initiators' interests (MaHring et al., 2004). Building on a set of enrolled actors, initiators "seek to secure continued support to the underlying ideas from the enrolled actors" (MaHring et al., 2004, p. 214). In this stage, the attackers made the carrier emails more sophisticated in their targeting and content in the hopes that the other actors would not betray the initiator's interests. The continued support of the initiating actors was from technologies of surveillance, cyber control and more targeting and sophisticated content "borrowed" from the recycled emails.

Human actors, as the primary actors of the GhostNet network, are of great importance in understanding the technological actors. Because of the political motivation of stealing confidential information and conducting cyber surveillance, attackers planned and carried out GhostNet actions. Because of the desire to gain relevant information, recipients were phished and "enrolled" into Ghost-Net actions. As key human actors of GhostNet engaged cyber espionage, the attacker(s) created GhostNet and got access to the target databases. Although there is no strong evidence showing the attackers' identity and motivation, the nature of their activities reveals some characteristics of the attackers of GhostNet. First, they are specialists of computer and Internet technology. The design of Gh0st RAT, the conduction of cyber espionage and even the sending of commands, show that the attackers are professional. Second, they possessed personal and social information about their targets, like their email addresses, social and discussion websites and their circulated information. Observing the phishing strategy that the attackers used, one can infer that attackers knew when and where they could approach targets. In addition, attackers were cautious of carrying on cyber espionage. The GhostNet system vaporized as soon as the IWM's report of GhostNet was released. The attackers were cautious and the plan of GhostNet is well planned.

The other group of actors within the system are the recipients of GhostNet whose computers were infected. In this case, the monks in OHHDL are the other key actors in the GhostNet network. As the first line of defence, the monks failed to secure their information for three reasons of human actors. First and most significantly, they have a general lack of security awareness, observing from their easily broken passwords and active exposure to various discussion sites. According to Nagaraja and Anderson (2009), "some passwords chosen by monks were easily broken with a dictionary attack using John the Ripper in about 15 minutes" (p.5). With the passwords, the attackers can access users' data easily without being traced. Also, the initial

break may have been facilitated by the fact that the monks in the OHHDL were not just engaged in administrative tasks but were also active on various discussion sites (Nagaraja & Anderson, 2009). A passive observer could easily note the users' names, their interests and the names of people with whom they interacted (Nagaraja & Anderson, 2009). The easily cracked passwords and active exposure to social websites do not only reflect the technological vulnerability of the infected computers but also the lack of security awareness of the casualties. In addition, when the monks found out the existence of computer viruses, there was no timely remedy and data movement, which provided the attackers longer time for cyber espionage.

The Internet is a new and relatively autonomous domain of human action (Stair, 1996). Human actors are key factors of the uses of technologies and the Internet. Anonymity gives online attackers opportunities to conduct cyber espionage. The problem of identification is particularly difficult in a domain where maintaining anonymity is easy and there are occasionally lapses between the intruder action, the intrusion itself, the awareness of intrusion, the actual disruptive effects and the system repair (Williams, Shimeall, & Dunlevy, 2002). This eludes to the need for a third group of human actors within a properly functioning human communication system.

Cyber investigators are the people who can be aware of the cyber attack and follow the cyber trail to investigate immediately. In the private sector, an investigation may initiate with a cyber incident such as "intrusion attempts, a DoS attack, information security violations, unauthorized usage of a system, or some other such incident" (Malinowski, 2006, p. 316). Cyber incidents are more likely to be treated with ignorance by normal users who have no idea what happened. In the case of GhostNet, the lack of security awareness of the monks resulted in the information lost in OHHDL. It is the view of the authors of this article that trained cyber investigators are necessary to take initiative to protect the computer system and networks from being attacked. This is expanded on in the discussion which follows.

DISCUSSION

As the diversity of the actors involved in the GhostNet network, it poses new challenges to cyber management and users' responsibilities. In this section, the researchers raise these topics in order to contribute to future research in this field.

User's Responsibilities and Ethical Standards

There is a great variety in what people want and value in cyber communication, but it is relatively uncontroversial that all human beings attempt to realize practical goals in social contexts and that access to information, and opportunities for communication, coordination, transaction, and personal relations are extremely important to them (Hoven, 2000). However, people do not only enjoy the opportunities that Internet communication brings, and they are responsible for their online behaviours. The question of what are the ethical responsibilities of users in cyber-attacks of their organizations is raised through the research of GhostNet.

According to Hoven (2000), as people establish connections or relationships online, obligations come with them. For example, in email contact a morally relevant relation is established. The same applies to ongoing relationships that are established via chat rooms. In the case of Ghost-Net, the attackers gained users' information from discussion sites, which is a moral wrongdoing. However, the concepts used to structure our moral life-world and used to provide us with moral guidance are problematic in their application to Internet contexts, partly because the Internet transforms the very nature of human interaction and practices. Consequently, the moral standards

that we use in real lives do not apply to cyber space in a straightforward way, which challenges people's own ethics and judgement of their cyber behaviours. As Hoven (2000) observed:

There are problems that result from the unforsee-ability of consequences of our actions given the independencies in networks and the concurrent actions in a causally intransparent network en-vironment. Individuals may simply be unable to predict the future consequences of their individual actions (p. 132).

Because users may not realize exactly what they are doing in terms of unknown consequences, they may be ignorant of the ethical implications of their actions. When dealing with user privacy issues and the confidentiality of data, organiza-tions need to educate general users to be aware of the potential risks ad use proper security measures when interacting online. In the case of GhostNet, the monks were naively exposed within an unsecure online environments includ-ing, which was partly responsible for the system security breach. It appears that the education of cyberethics and is lagging behind cyberlife and need to be highlighted.

International Cyber Law Enforcement

Although they are individuals who are engaged in online communication, in many cases, the target of communication exists beyond the boundary of their respective selves and organizations (Ku-rokawa, 2007). As the Internet cuts across the territorial borders of sovereign nation states, it makes geographical boundaries as delineations of jurisdictions inadequate (Hoven, 2000). In this situation, both the individual and organizations require local and international law enforcement and ethical standards to protect the confidential-ity of personal information and organizational business secrets.

It can be argued that the impulse to control and renationalize the Internet subverts the value structure and local control that underlies traditional media. Although the boundary of nations is obscured by the power of the Internet, governments need to begin to control cyberspace and make it a governing tool for sovereignty. GhostNet, as a severe cyber crime, functioned politically, crossing national boundaries. However, there were inadequate cyber regulations and international law enforcement in place to stop it. This needs to change but this is not an easy feat. According to Spinello (2002), "the Internet's unique decentralized structure tends to defy centralized regulations. There is no central server that can be easily contained; there are many nodes on multiple networks, each transmitting and receiving data" (p. 34). Also, the Internet's vast global reach, which transcends the jurisdiction of national governments, poses formidable problems for those government that seek to impose laws on cyberspace activity (Spinello, 2002). As Johnson and Post (1997) have observed, "the rise of an electronic medium that disregards geographical boundaries throws the law into disarray by creating an entirely new phenomena that need to become subject of clear-legal rules but that cannot be governed, satisfactorily, by any current territorially based sovereign" (p. 31). Although there are distinct disagreements on how the Internet should be regulated, there is consensus that it requires some type of governance, and technical and communicative coordination (Spinello, 2002).

To date, law enforcement efforts concerning cybercrime have been fragmented, largely due to the lack of a coordinated effort. Although most law enforcement agencies have a department assigned to manage computer crime, these groups do not regularly communicate or share resources (Trend Micro, 2009). Enforcement efforts are complicated by the fact that malware attacks typically cross national borders, making criminals especially difficult to hunt down and prosecute (Trend Micro, 2009). A lack of laws protecting consumers poses a formidable challenge.

Technoethics and the Evolving World

Within our contemporary social networks and systems, both technological and human actors are involved. Due to the complexity and multiplicity of human-technological intertwinements that arise, it is an ongoing challenge for social scientists to keep up with changes that occur in so many areas of life and society (Luppicini, 2009). Although technoethics was defined in the 1970s in terms of professional ethics of technologists on the front lines of innovation, the connection between Internet technology and cyber ethics is a growing area of technoethics still being defined. What is significantly different now from 1970s is that the communications are multi-way and can reach a potentially infinite number of people across a wide range of jurisdictions almost without restriction (Wall, 2003). More importantly, cyberspace challenges many of the principles upon which people's conventional understanding of crime and policing are based. In this sense, continued work in technoethics is of major importance and in great need in uncovering what impacts individuals online activities and psychologies that lead to unethical consequences. According to Luppicini (2008), "technoethics deals with human processes and practices connected to technology, which are becoming embedded within social, political, and moral spheres of life" (2008, p. 4). The case of Ghostnet provided a good example of a technoethical inquiry into controversial cyber behaviours drawing on the theoretical perspective of ANT. It highlights the fact that almost all technology use for information exchange and communications is intermediated by human agency, along with the social and ethical (and unethical) values of participating agents that need to be taken to be a top priority within developing and deploying ICT's within society.

REFERENCES

Armin, J. (2009). *Hacker deploys cloud to smash passwords*. Retrieved from http://news.hostexploit.com/cyber-security-news/4739-hacker-deploys-cloud-to-smash-passwords.html

Bocij, P. (2006). *The dark side of the Internet*. Westport, CT: Greenwood Publishing.

Bunge, M. (1977). Towards a technoethics. *The Monist, 60*(1), 96–107. doi:10.5840/monist197760134

Callon, M. (1986). The sociology of an actor-network: The case of the electric vehicle. In Callon, M., Law, J., & Rip, A. (Eds.), *Mapping the dynamics of science and technology* (pp. 19–34). Basingstoke, UK: Macmillan Press.

Clarke, R. (1999). Internet privacy concerns confirm the case for intervention. *Communications of the ACM*, 60–67. doi:10.1145/293411.293475

Cooper, H. (2010). *Research synthesis & meta-analysis: A step-by-step approach*. Thousand Oaks, CA: Sage.

Cornish, P. (2009). *Cyber security and politically, socially and religiously motivated cyber attacks*. London, UK: European Parliament.

Creswell, J. (2007). *Qualitative inquiry & research design: Choosing among five approaches*. Thousand Oaks, CA: Sage.

Creswell, J. (2009). *Research design, qualitative, quantitative, and mixed methods approaches* (3rd ed.). Thousand Oaks, CA: Sage.

Dartnell, M. (2003). Weapons of mass instruction: Web activism and the transformation of global security. *Millennium: Journal of International Studies*, 477-499.

F-Secure. (2003). *Iraq War and information security*. Retrieved from http://www.fsecure.com/virus-info/iraq.shtml

Furnell, S., & Ward, J. (2006). Malware: An evolving threat. In Kanellis, P., Kiountouzis, E., Kolokotronis, N., & Martakos, D. (Eds.), *Digital crime and forensic science in cyberspace* (p. 357). Hershey, PA: IGI Global. doi:10.4018/978-1-59140-872-7.ch002

Hoven, V. D. J. (2000). The Internet and varieties of moral wrongdoing. In Langford, D. (Ed.), *Internet ethics* (p. 257). New York, NY: St. Martin's Press.

Information Warfare Monitor. (2009). *Tracking GhostNet: Investigating a cyber espionage network*. Toronto, ON, Canada: Information Warfare Monitor.

Information Warfare Monitor & Shadowserver Foundation. (2010). *Shadows in the cloud: Investigating cyber espionage 2.0*. Retrieved from: http://shadows-in-the-cloud.net

Johnson, D., & Post, D. (1997). The rise of law on the global network. In Kahin, B., & Nesson, C. (Eds.), *Borders in cyberspace*. Cambridge, MA: MIT Press.

Kurokawa, T. (2007). Ontology for cross-organizational communication. *The Quarterly Review*, 13–20.

Latour, B. (1991). Technology is society made durable. In Law, J. (Ed.), *A sociology of monsters: Essays on power, technology and domination* (pp. 196–233). London, UK: Routledge.

Law, J., & Law, J. (1999). After ANT: Complexity, naming and topology. *The Sociological Review, 46*, 1–14. doi:10.1111/1467-954X.46.s.1

Luppicini, R. (2008). Introducing technoethics. In Luppicini, R., & Adell, R. (Eds.), *Handbook of research on technoethics* (pp. 1–18). Hershey, PA: IGI Global. doi:10.4018/978-1-60566-022-6.ch001

Luppicini, R. (2009). Technoethical inquiry: From technological systems to society. *Global Media Journal, 2*(1), 5–21.

MaHring, M., Holmstrom, J., Keil, M., & Montealerge, R. (2004). Trojan actor-networks and swift translation. *Information Technology & People*, *17*(2), 210–238. doi:10.1108/09593840410542510

Malinowski, C. (2006). Training the cyber investigator. In Kanellis, P., Kiountouzis, E., Kolokotronis, N., & Martakos, D. (Eds.), *Digital crime and forensic science in cyberspace* (pp. 311–333). Hersey, PA: IGI Global. doi:10.4018/978-1-59140-872-7.ch014

Mason, R. O. (1986). Four ethical issues of the information age. *Management Information Systems Quarterly*, *10*(1), 5–12. doi:10.2307/248873

Micro, T. (2009). *Data-stealing malware on the rise—solutions to keep businesses and consumers safe*. Retrived from http://us.trendmicro.com/imperia/md/content/us/pdf/threats/securitylibrary/data_stealing_malware_focus_report_-_june_2009.pdf

Mills, E. (2010). *Report unearths targeted attacks on oil firms*. Retrieved from http://news.cnet.com/security/?keyword=espionage

Nagaraja, S., & Anderson, R. (2009). *Snooping dragon: Social malware surveillance of the Tibetan movement* (Tech. Rep. No. 746). Cambridge, UK: University of Cambridge.

Spinello, A. R. (2002). *Regulating cyberspace: The policies and technologies of control*. Westport, CT: Quorum Books.

Stair, R. M. (1996). *Principles of information systems: A managerial approach*. Florence, KY: Thomson Publishing.

Stephens, K., & McKee, L. (2010). *Cyber espionage: Is the United States getting more than its giving?* Smithfield, VA: National Security Cyberspace Institute.

Stevens, S. (2010). *Google studies*. Retrieved from http://www.discoverdecaturil.com/

Wall, D. (2003). *Cyberspace crime*. London, UK: Dartmouth Publishing.

Williams, P., Shimeall, T., & Dunlevy, C. (2002). Intelligence analysis for Internet security. *Contemporary Security Policy*, *23*(2), 1–38. doi:10.1080/713999739

Wilson, C. (2009). Botnets, cybercrime, and cyberterrorism: Vulnerabilities and policy issues for Congress. In Jacobson, G. V. (Ed.), *Cybersecurity, botnets, and cyberterrorism* (p. 71). New York, NY: Nova Science Publishers.

Yin, R. K. (2003). *Case study resarch: Design and methods* (2nd ed.). Thousand Oaks, CA: Sage.

This work was previously published in the International Journal of Technoethics, Volume 2, Issue 2, edited by Rocci Luppicini, pp. 66-78, copyright 2011 by IGI Publishing (an imprint of IGI Global).

Chapter 10
The Impact of Context on Employee Perceptions of Acceptable Non–Work Related Computing

Troy J. Strader
Drake University, USA

Suzanne R. Clayton
Drake University, USA

J. Royce Fichtner
Drake University, USA

Lou Ann Simpson
Drake University, USA

ABSTRACT

Employees have access to a wide range of computer-related resources at work, and often these resources are used for non-work related personal activities. In this study, the authors address the relationship between employee's utilitarian ethical orientation, the factors that create the context that influences their ethical perceptions, and their overall perceptions regarding the level of acceptability for 14 different non-work related computing activities. The authors find that time and monetary cost associated with an activity has a negative relationship to perceived acceptability. Results indicate that contextual variables, such as an employee's supervisory or non-supervisory role, opportunity, computer self-efficacy, and whether or not an organization has computer use policies, training, and monitoring, influence individual ethical perceptions. Implications and conclusions are discussed for organizations and future research.

INTRODUCTION

Information technology today is ubiquitous. Professional employees in business and government are provided with a variety of information technology tools for communication and

DOI: 10.4018/978-1-4666-2931-8.ch010

productivity improvement and their use of these technologies is often unsupervised. As with many new information technology implementations, there are a wide range of expected benefits, but there are also some unanticipated consequences. Distributed information technology creates a number of opportunities for misuse and one of the most common examples is use of organizational computing resources for personal activities while working. Some of the terms used to describe this misuse of computer resources are cyberloafing, cyberslacking, or non-work related computing (NWRC). Throughout this paper we will refer to these activities as NWRC.

Regardless of how these activities are characterized, it appears to be a significant problem for many organizations. In a recent survey, nearly 64% of employees claimed to use the Internet for personal activities during work hours (Young, 2010). It has also been reported that, in one month in 2010, 57 million people visited social-networking sites from a work computer, spending an average of 15 minutes on them per day (Needleman, 2010). These activities can put a strain on network resources, reduce productivity, create potential legal liabilities, as well as a variety of other negative consequences. Organizations have a wide range of technical and non-technical solutions available to them, but underlying the problem is the need to understand the factors associated with these behaviors.

Technoethics is an interdisciplinary research area that involves the study of ethical issues associated with the development, use, and impact of technology on individuals, organizations, and society. This study focuses on employee perceptions of information technology use, and misuse, in organizations, as well as the factors that impact these perceptions. Within the broad field of technoethics research, this study would address issues related to technoethics and work, professional and organizational technoethics, computer and Internet technoethics, and digital property ethics.

In this study we address these issues from the perspective of individual employee's perceptions regarding the level of acceptability for certain NWRC activities given their ethical orientation. We address the following questions. The first two questions are included to replicate an earlier study of these issues (Strader et al., 2010). The final two questions are the primary focus of this study.

1. What ethical orientation do individual employees employ when determining the extent to which they perceive non-work related computing activities as ethical?
2. If individuals are utilitarians, which factors are associated with their assessment of activity related benefits and negative consequences?
3. What additional contextual factors impact employee perceptions?
4. Beyond the contextual variables, do demographic factors have an impact on ethical perceptions?

The answers to these questions will provide organizations with a better understanding of how their employees perceive these activities and why. These are particularly important questions to answer because existing countermeasures such as appropriate use policies or filter/monitoring systems that are intended to reduce misuse do not appear to work (Lee et al., 2007).

The paper is organized as follows. The theoretical background for this study is presented in the next section including discussion of existing research in this area. Study hypotheses are identified along with the theoretical basis for their inclusion in this study. The next section presents the research methodology and findings. The final section outlines the overall conclusions along with managerial and research implications.

THEORETICAL BACKGROUND

The digital world enabled by computers and other information technologies has produced numerous benefits for individuals, organizations, and society, but it has also created new and unique ethical issues that are not fully understood. From a very broad perspective, ethics issues arising in the digital world have been addressed by a number of previous studies. Examples of issues addressed include the cultural lag between rapid technology development and slower develop of new ethical guidelines (Marshall, 1999), the influence of computer guidelines and the belief in universal moral rules on ethical intentions regarding the use of computers in the workplace (Peterson, 2002), the varying ethical views on appropriate ethical uses for software and other information technologies (Calluzzo & Cante, 2004), the relationship between intellectual property and privacy attitudes, Machiavellianism and Ethical Ideology, and experience as a programmer on acceptance of unethical information technology practices (Winter et al., 2004), the increase in ethical problems as technological revolutions progress toward and into the power stage of their development (Moor, 2005), and the ethics of information transparency (Turilli & Floridi, 2010).

The studies that address information technology ethics from a broad philosophical perspective provide background for studies that utilize an empirical methodology to address more specific practical contexts and the resulting individual behaviors and perceptions. In this study we address a specific instance of unethical behavior, and perceptions regarding these behaviors, where employees use an organization's computer resources for their own personal activities while they are working.

Non-Work Related Computing Research Studies

Non-work related computing (NWRC) has been studied in a variety of contexts and the studies have been based on several different theoretical foundations. Most studies are based on the premise that personal use of organizational computing resources while working has negative consequences and should be inhibited through some form of monitoring and/or punishment. The following provides a summary of recent studies in this research area including a discussion of the factors that influence NWRC attitudes and behavior. The

unique contextual variables utilized in this study are described at the end of this section.

To some extent, all of the following studies consider some form of positive and/or negative consequences when attempting to explain NWRC behavior. Some of the studies also provide more specific definitions and categories for describing various forms of NWRC activities.

Henle and Blanchard (2008) studied the impact of role ambiguity, role conflict, role overload, and organizational sanctions on cyberloafing behavior. They found that the behavior was a response to work stressors. They found that employees were more likely to cyberloaf when they perceived role ambiguity or role conflict, and they were less likely to do this when they perceived role overload. However, they were more likely to cyberloaf in response to these stressors when they perceived that the likelihood that they would be punished by the organization was low.

A study by Mastrangelo, Everton, and Jolton (2006) explored definitions and motivations for personal use of work computers. They describe two distinct forms of NWRC – counterproductive versus nonproductive. Nonproductive computer use is the much more common form. They find that nonproductive computer use occurs among employees who have had an Internet connection longer, whose Internet connection is faster at work than at home, and who concurrently work more jobs. Counterproductive computer use is more prevalent when Internet access at work is new. It appears that individual experience with the technology infrastructure impacts behavior.

NWRC behavior can also be categorized as either minor cyberloafing or serious cyberloafing (Blanchard & Henle, 2008). Similar to the previous study, they find that the minor form of cyberloafing is much more common. Based on their study they find that employees' perceptions of the norms of their reference group were related to minor cyberloafing behavior. It appears that the employees do not consider minor cyberloafing to be inappropriate, or perhaps they do not perceive significant negative consequences. Employees' perceptions of norms were not related to serious cyberloafing. Employees do not feel that their coworkers or supervisors would approve, but they also do not sense that anyone will catch them in their inappropriate behavior. Negative consequences would be due to bad luck, not just the behavior itself.

The impact of task characteristics and organizational culture are addressed in another study (Bock et al., 2010). They found that when employees performed tasks that were primarily non-routine, NWRC control mechanisms were less effective in reducing the inappropriate behavior. In bureaucratic organizational cultures, punitive discipline systems were more satisfactory to employees while In innovative and supportive organizational cultures the employees were more satisfied with positive discipline systems.

A study by Chun and Bock (2006) addressed perceived benefits and costs more directly. They found that perceived benefits (time saving, fun and enjoyment, learning, team building and facilitating communication) were positively associated with NWRC behavior while perceived costs (fines, loss of status in organization, termination) were negatively related. They also found that subjective norms (peer group, supervisor, culture) had an impact on NWRC behavior. An additional finding of interest was that habit may overrule other rational decisional factors when attempting to explain behavior.

Bock and Ho (2009) investigated the relationship between times spent on offline non-work related activities, non-work related computing, and time spent on work after working hours, on job performance. They found that NWRC does reduce employee productivity while time spent on offline non-work related activities positively affects job performance which was contrary to their expectations.

Explaining NWRC can also be done by comparing alternative models (Pee et al., 2008). In this study, models derived from the Theory of Interpersonal Behavior (TIB) and the Theory of Planned Behavior (TPB) was compared to identify which perspective best explained NWRC behavior. They found that the TIB-based model had higher explanatory power then the TPB-based model. They found that social factors and perceived consequences (convenience, warnings/reprimands, and decrease in productivity) significantly influenced employees' intentions to do NWRC, while this intention, habit, and other facilitating conditions (existence of productivity measurement systems) led to the behavior. As with the previous study, it appears that a key to reducing inappropriate behavior is to prevent NWRC from becoming a habit for an employee.

A study by Garrett and Danziger (2008) emphasizes that expected outcomes of Internet use most effectively explain behavior. Expected outcomes include positive perceptions of the Internet's utility, routinized use of computers, job commitment, and organizational restrictions on computer use, and each of these factors is a significant predictor for NWRC behavior.

The ethics of NWRC, and individual ethical orientations, is included in a recent study. A study by Lee et al. (2007) investigated how people's perceptions of moral dimensions contribute to non-work related use of the Internet for personal purposes during work hours. Denial of responsibility and moral obligation are considered as predictors of personal Web use intention. Denial of responsibility is related to rationalizing the consequences of one's behavior. Moral obligation represents an individual's perception of responsibility to perform or refuse to perform a certain behavior and it therefore has a relationship to human ethical decision making. Interestingly, denial of responsibility did affect personal Web usage, but moral obligation did not. These results point to the need for further study of the impact of individual ethical orientations, in combination with other factors, and how it may impact employee's NWRC behavior and perceptions.

In summary, recent studies of NWRC behavior all appear to incorporate some form of positive and negative consequences for explaining intentions and behavior, as well as outcomes such as job performance. It is apparent that there is no clear consensus on the impact of NWRC, but there is interest in understanding the ethical dimension. This study is based on the utilitarianism theory that explains individual ethical perceptions based on an overall assessment of positive and negative consequences associated with activities. In addition, contextual variables are considered because they may alter consequence perceptions. The following section describes the study's hypotheses and the rationale for their inclusion.

STUDY HYPOTHESES

Individuals use two distinct forms of reasoning when assessing whether an act is ethical or not. These forms of reasoning are described as formalist and utilitarian (Brady, 1985; Alder et al., 2007). The formalist orientation assesses ethics based on a set of rules or principles for guiding behavior. The utilitarian orientation would take into account the relative positive and negative consequences of the action and judge actions as ethical that create the greatest overall good. Most studies in this area have focused on psychological and sociological factors and their impact on NWRC behavior. Assessing an individual's ethical orientation as a basis for understanding employee computer resource misuse is less common, but has been addressed by a few studies. The unique contribution of this study is that it focuses on employee perceptions rather than behavior, and it addresses the impact of a NWRC activity's consequences as well as additional contextual variables that may impact individual employee perceptions. Given that employees continue to do NWRC while at work, it appears that they utilize a utilitarian ethical orientation and feel that in many instances the positive benefits associated

with these activities outweighs any negative consequences to the individual or organization. This leads to the first hypothesis.

H1: Individual employees use a utilitarian orientation to determine the extent to which they perceive a NWRC activity to be acceptable.

If support is found for H1, then the next question is to identify which factors are considered by individuals when they are assessing the negative consequences associated with a NWRC activity. A previous study identified two negative consequences for an organization that are characteristics of each individual NWRC activity – monetary cost (H2) and time (H3) (Strader et al., 2010). We test these first two hypotheses to replicate the previous study and to establish a common background for continuing this study and identifying additional contextual variables that may impact individual ethical perceptions.

H2: Individual employees consider the organization's monetary cost when determining whether a NWRC activity is acceptable.
H3: Individual employees consider the time involved in an activity when determining whether a NWRC activity is acceptable.

In addition to the characteristics of the activity itself, additional factors that create a decision making context are considered in the following hypotheses. The first contextual factor is the role of the individual employee within the organization regarding whether or not they supervise other employees. A study focused on profiling cyberslackers in the workplace found that young executives are the most likely to cyberslack due to the pressure of their jobs, but years of service and pay status did not have significant influence on an individual's degree of cyberslacking (Ugrin et al., 2007). Supervisory role was not specifically included in this study, but there seems to be some conflicting findings related to what we would typically associated with supervisor status – executive status, pressure, larger number of years of service, and higher pay. Thus, there is some uncertainty regarding whether supervision of others has an impact on NWRC behavior or perceptions of others misbehavior. For this study we assume that supervisors have negative perceptions of other's NWRC behavior because they are evaluated on the performance of their group and thus would be concerned about behavior that is counterproductive. Therefore, H4 is stated as follows.

H4: Individuals who supervise other employees view NWRC as less acceptable than non-supervisor employees.

A study by Lee et al. (2007) considered some contextual variables when attempting to understand and predict NWRC behavior in the workplace. In this study we extend the current model to include five of these factors that influence the context in which employees assess the level of acceptability for certain NWRC activities. These contextual variables include opportunity for NWRC, computer self-efficacy indicating an employee's level of ability related to using computer resources, and three components of an organization's culture regarding computer use (policies, training on these policies, and monitoring of computer use).

One requirement for NWRC is access to a computing device and/or the Internet. If the opportunity does not exist at work then misuse is far less likely and perhaps impossible. For many employees, though, computers and Internet access devices are readily available. Employees may perceive lower levels of negative consequences for NWRC because they have easy access and opportunity to do certain computer-related personal activities while at work. Therefore, H5 is stated as follows.

H5: The more opportunity an employee has to use a computer while working, the more acceptable they perceive NWRC activities.

Similar to opportunity, higher levels of computer self-efficacy may also reduce an employee's perceptions that there is significant effort associated with NWRC. They may feel that doing computer-related personal activities is easy which affects their negative consequence perceptions leading to H6.

H6: The higher the level of an employee's computer self-efficacy, the more acceptable they perceive non-work related computing activities.

The next three factors are components of an overall organizational culture related to computer use at work. Organizations may have computer use policies, they may train their employees on these policies, and, in addition, they may monitor their employee's computer use. Some companies may do all three, while others do none. If companies do all three they are being very restrictive which indicates higher negative consequences for misuse. This will influence employee's perceptions and the three associated hypotheses (H7, H8, and H9) are shown.

H7: Organizations with policies describing acceptable computer use at work create an environment where employees perceive non-work related computing activities as less acceptable.

H8: Organizations with programs to train employees on acceptable computer use at work create an environment where employees perceive non-work related computing activities as less acceptable.

H9: Organizations that monitor their employees' computer use create an environment where employees perceive non-work related computing activities as less acceptable.

Finally, gender and age are added to the previous models to test whether they influence perceptions regarding the level of acceptability for certain NWRC activities. In this research area there is no strong support for why these factors should alter perceptions so the expectation is that neither gender nor age will be significant predictors. Therefore, hypotheses are not provided, but the relationship will be tested.

Figure 1 provides an overall summary for the study hypotheses, hypothesis numbers, and expected relationships (positive or negative).

METHODOLOGY

To test the hypotheses, a survey was developed that included 14 examples of NWRC activities. The first page of the survey is included in Appendix A. Most previous studies focused on misuse of Internet access, but the examples included in this survey are expanded to encompass a wider range of activities including various forms of viewing information online, use of output devices, use

of digital storage space and bandwidth, sending e-mail, playing games online and online shopping. The population of interest for this study is employed, educated, individuals who have at least some opportunity to access computer resources while working at an office or offsite. The survey was distributed to approximately 120 people in four graduate classes in business administration and public administration. Ninety usable surveys were returned for a response rate of 75%. The characteristics of the survey sample are shown in Table 1.

Table 1. Survey sample characteristics

Demographic	Categories	Number	%
Gender	Male	47	52.2%
	Female	43	47.8%
Age	18-20	1	1.1%
	21-24	32	35.6%
	25-29	34	37.8%
	30-39	13	14.4%
	40-49	8	8.9%
	50+	2	2.2%
Supervise other employees	Yes	26	28.9%
	No	64	71.1%

Figure 1. Summary of study hypotheses

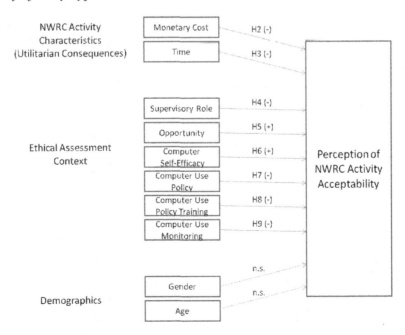

For each of the 14 tasks, respondents were asked to indicate the degree to which they agreed with a statement that a certain activity is acceptable using a seven point scale ranging from strongly disagree (1) to strongly agree (7). This was followed by questions to assess their perceptions of the monetary cost and time associated with each activity. Seven point scales were used for these perceptions with monetary costs ranging from no cost (1), a few cents (2), 25-50 cents (3), $1 (4), a few dollars (5), 10-50 dollars (6), and $50+ (7) for a single instance of the associated activity, and time ranging from no time (1), a few seconds (2), 1-2 minutes (3), 5-10 minutes (4), 10-30 minutes (5), one hour (6), and several hours (7). Additional questions asked respondents to report whether they supervise employees or not, their level of self-efficacy related to computer use, percentage of a typical work day that they have access to some form of digital computing and/or Internet access device, and three items related to organizational culture regarding computer use and misuse (acceptable use policies, training, and monitoring). Measures of self-effi-

cacy, opportunity (availability), and attributes of organizational computer use policies and monitoring, were adapted from a study of individual Web usage in organizations (Lee et al., 2007). Measurement items are summarized in Appendix B. Demographic information (gender and age) was also collected for each respondent.

RESULTS

Hypothesis 1 states that employees are utilitarians whose views on the level of acceptability for various personal uses for computer resources at work depend on their perceptions of the overall positive and negative consequences associated with the activities. Table 2 summarizes the overall level of acceptability found for each of the 14 activities included in the survey.

The results support H1. Perceptions of activity acceptability range from viewing weather information online (6.29, highly acceptable) to using an organization's CDs to burn copies of personal files (1.60, highly unacceptable). If

Table 2. Employee perceptions of acceptability for computer resource uses at work

Non-Work Related Computing Activity	Perceived Acceptability (1=low, 7=high)
9. View weather information online	6.29
14. Read newspaper online	5.53
2. Do Internet searches for personal interests	4.67
3. Send e-mail using work e-mail address	4.50
11. Scan personal documents to create PDF files	4.04
6. Print personal documents on office printer	4.02
4. Make copies of personal documents	3.97
7. Use spreadsheet to analyze personal investments	3.57
10. Shop online for personal items	3.08
1. Store music files on office computer	2.78
12. View YouTube streaming videos on office computer	2.51
8. Send and receive Twitter messages	2.23
5. Play games online	1.74
13. Use organization's CDs to burn music file copies	1.60

employees were not utilitarians, then they would be expected to judge each of these activities as unacceptable because they are all examples of using organizational computer resources for personal activities while working which has potentially negative consequences for the organization. The findings support the idea that employees apply a utilitarian approach to evaluating the ethics for each of these activities. Some activities are more acceptable than others. The next step is to identify the factors that alter these perceptions.

To test the remaining hypotheses, a series of regression models were tested using data from the surveys. Results for these regression models are summarized in Table 3. Model 1 is used to test H2 and H3. Correlations between model constructs are summarized in Table 4.

H2 states that the perceived level of activity acceptability will be influenced by the perceived monetary cost for that activity. H3 is a similar hypothesis, but it is related to the perceptions of the time involved in the activity. As in the previous study of these two factors (Strader et al.,

2010), the regression results summarized in Model 1 support both H2 and H3. The greater the monetary cost and time an activity takes, the less acceptable that activity is perceived to be. Negative consequences for the organization appear to influence employee perceptions of NWRC activity acceptability.

While the activity characteristics (cost and time) influence perceptions, the factors that influence the decision making context are also expected to alter perceptions. One issue is whether being a supervisor creates different views on the negative consequences of NWRC activities which will then change perceptions of acceptability. H4 looks at whether supervisors have significantly different perceptions regarding NWRC activity acceptability than do non-supervisor employees. The data support H4. Employees who supervise other employees regard each activity as being relatively less acceptable.

Another contextual issue is the opportunity to use computer resources for personal activities at work. The survey asks - On average, for what

Table 3. Parameter estimates for models to test hypotheses

Predictor	Model 1		Model 2		Model 3	
	b	t	b	t	b	t
Intercept	5.33	28.44***	4.01	11.00***	3.95	10.45***
Activity Monetary Cost (COST)	-0.29	-9.29***	-0.28	-9.10***	-0.28	-9.15***
Activity Time (TIME)	-0.19	-4.59***	-0.23	-5.57***	-0.23	-5.56***
Supervise Other Employees (SUPER)			-0.32	-2.57**	-0.33	-2.54**
Access Opportunity (OPP)			0.01	5.66***	0.01	5.58***
Computer Self-Efficacy (CSE)			0.16	3.65***	0.16	3.62***
Use Policy (POLICY)			-0.32	-2.24**	-0.32	-2.23**
Use Training (TRAIN)			0.37	3.07***	0.36	2.88***
Use Monitoring (MONITOR)			-0.38	-3.12***	-0.39	-3.17***
Gender					0.11	0.96
Age					0.02	0.30
R-Square	0.100		0.122		0.149	

***p <.01; **p <.05; *p <.10

Table 4. Construct correlation matrix

	COST	TIME	SUPER	OPP	CSE	POLICY	TRAIN	MONITOR
COST								
TIME	0.256							
SUPER	0.027	-0.118						
OPP	-0.120	0.016	-0.242					
CSE	0.018	-0.012	-0.010	-0.015				
POLICY	0.041	-0.041	-0.026	0.196	0.192			
TRAIN	0.081	0.028	-0.035	0.047	0.072	0.427		
MONITOR	-0.061	-0.107	0.077	0.232	0.028	0.367	0.249	

percentage of your work time do you have easy access to the Internet (either through a personal computer or other device)? A higher percentage implies a greater opportunity to do NWRC activities. This will alter the context in which the employees perceive these activities because high opportunity may lead to a lower expected negative consequence associated with the activities. Little effort is required to access computer resources because they are easily available for much of the day. The results support H5. Greater opportunity is associated with higher perceptions that NWRC activities are acceptable.

Computer self-efficacy is another factor that will alter perceptions regarding the negative consequences associated with NWRC. If someone can use a computer without much effort or without anyone else's help then they may feel that there are less negative consequences for NWRC activities. The data support H6. Employees with higher levels of computer self-efficacy perceive NWRC activities as more acceptable.

The final three contextual factors include three issues related to an organization's computer use policies, training on these policies, and monitoring of computer use. Respondents were asked to report whether each of these exist in their organization. The assumption is that organizations that have all three will give an employee the impression that negative consequences for NWRC are higher,

while organizations that do none of these things will lower negative consequence expectations. The data support H7 and H9. Computer policies and monitoring of computer use create a context in which employees perceive NWRC activities as less acceptable. The counter-intuitive result is that the test for H8 is significant, but the impact is opposite of what was expected. We find that training on policies is associated with perceptions that NWRC activities are more acceptable. There is no clear rationale for this result and more study of this issue is needed.

Finally, demographic factors (gender and age) were added to the model and results are shown in Model 3. As expected, basic demographic differences for gender and age are not significant predictors for differing employee perceptions regarding NWRC. It is apparent that perceptions are impacted by deeper psychological factors and ethical orientations.

These results confirm that a variety of contextual variables have a significant impact on employee perceptions of the level of acceptability for specific NWRC activities. Ethical orientation, consequences, and decision making context shape perceptions and thus should be considered when developing organizational computer use policies and designing future research studies. The following provides a discussion of implications and conclusions resulting from this study's findings.

IMPLICATIONS AND CONCLUSION

We conclude by discussing managerial and research implications. The findings from this study have implications for both the organizations where employees have access to computer resources as well as the employees themselves. When organizations develop computer use policies, and training and monitoring programs, they need to recognize that their employees have perceptions regarding NWRC activities that are impacted by a number of factors. The employees do not perceive all NWRC activities to be equally egregious. Activities that take little time and do not directly cost the organization any money are viewed as acceptable by someone with a utilitarian orientation. In addition, contextual factors also impact people's perceptions. Supervisory role, the extent to which there is opportunity to use computer resources, self-assurance in their ability to effortlessly use computers, and the existing organizational culture surrounding NWRC activities all impact perceptions. The implication for computer use policies is that they need to be flexible. Certain activities should be allowed because employees do not recognize a high degree of negative consequences associated with those activities. For other activities, the employees will feel that they are unacceptable and thus can be listed as something that is not allowed. Obviously, more study needs to be done in this area, but it appears that there is a common finding – employee perceptions regarding negative consequences associated with various NWRC activities impact their perceptions, attitudes, intentions, and behavior. If flexibility in computer use is possible for a particular job type, then it should be allowed. There will still be instances where any form of NWRC is unacceptable because of the nature of the position.

Based on a review of the literature in this area, and the findings from this study, there are several directions for future research. First, additional contextual variables should be considered in studies of NWRC ethical perceptions or user behavior. The method used to evaluate a job could be considered. Jobs where performance is based on output may provide an opportunity for a flexible computer use policy. For example, it does not matter when an author does their writing, only that they complete the assignment by a deadline. On the other hand, for a customer service job, that employee must be attentive to the service task at all times which gives little latitude for flexibility on NWRC. The industry in which a company operates may also impact perceptions and behavior. Industries where computer activities are tracked more closely – government, legal, medical – may have strict limits on NWRC while other industries where innovation and creativity are paramount may result in employees being more open to NWRC.

Another direction for future research could be to investigate other information technology ethical questions using a similar methodology to that employed in this study. Examples include individual perceptions of online information privacy or digital copyright infringement. Do similar contextual factors impact individual perceptions of acceptability as they do for views on NWRC?

Finally, the positive consequences for NWRC need much more study. Most studies focus only on the negative. There appear to be some situations where letting people cyberslack may actually improve productivity and job satisfaction. Identifying these situations will provide additional valuable knowledge regarding appropriate computer use policies and theoretical understanding of ethics in a world of ubiquitous information technology.

REFERENCES

Alder, G. S., Schminke, M., & Noel, T. W. (2007). The impact of individual ethics on reactions to potentially invasive HR practices. *Journal of Business Ethics*, *75*, 201–214. doi:10.1007/s10551-006-9247-6

Blanchard, A. L., & Henle, C. A. (2008). Correlates of different forms of cyberloafing: The role of norms and external locus of control. *Computers in Human Behavior, 24,* 1067–1084. doi:10.1016/j. chb.2007.03.008

Bock, G.-W., & Ho, S. L. (2009). Non-work related computing (NWRC). *Communications of the ACM, 52*(4), 124–128. doi:10.1145/1498765.1498799

Bock, G.-W., Shin, Y., Liu, P., & Sun, H. (2010). The role of task characteristics and organizational culture in non-work-related computing: A fit perspective. *ACM SIGMIS Database, 41*(2), 132–151. doi:10.1145/1795377.1795385

Brady, E. N. (1985). A Janus-headed model of ethical theory: Looking two ways at business/society issues. *Academy of Management Review, 10,* 568–576.

Calluzzo, V. J., & Cante, C. J. (2004). Ethics in information technology and software use. *Journal of Business Ethics, 51*(3), 301–312. doi:10.1023/B:BUSI.0000032658.12032.4e

Chun, Z. Y., & Bock, G.-W. (2006). Why employees do non-work-related computing: An investigation of factors affecting NWRC in a workplace. In *Proceedings of the Tenth Pacific Asia Conference on Information Systems* (pp. 1259-1273).

Garrett, R. K., & Danziger, J. N. (2008). Disaffection or expected outcome: Understanding personal Internet use during work. *Journal of Computer-Mediated Communication, 13,* 937–958. doi:10.1111/j.1083-6101.2008.00425.x

Henle, C. A., & Blanchard, A. L. (2008). The interaction of work stressors and organizational sanctions on cyberloafing. *Journal of Managerial Issues, 20*(3), 383–400.

Lee, Y., Lee, Z., & Kim, Y. (2007). Understanding personal Web usage in organizations. *Journal of Organizational Computing and Electronic Commerce, 17*(1), 75–99.

Marshall, K. P. (1999). Has technology introduced new ethical problems? *Journal of Business Ethics, 19*(1), 81–90. doi:10.1023/A:1006154023743

Mastrangelo, P. M., Everton, W., & Jolton, J. A. (2006). Personal use of work computers: Distraction versus destruction. *Cyberpsychology & Behavior, 9*(6), 730–741. doi:10.1089/cpb.2006.9.730

Moor, J. H. (2005). Why we need better ethics for emerging technologies. *Ethics and Information Technology, 7*(3), 111–119. doi:10.1007/s10676-006-0008-0

Needleman, S. E. (2010, May 20). *A Facebook-free workplace? Curbing cyberslacking.* Retrieved from http://online.wsj.com/

Pee, L. G., Woon, I. M. Y., & Kankanhalli, A. (2008). Explaining non-work-related computing in the workplace: A comparison of alternative models. *Information & Management, 45*(2), 120–130. doi:10.1016/j.im.2008.01.004

Peterson, D. K. (2002). Computer ethics: The influence of guidelines and universal moral beliefs. *Information Technology & People, 15*(4), 346–361. doi:10.1108/09593840210453124

Strader, T. J., Simpon, L. A., & Clayton, S. R. (2010). Using computer resources for personal activities at work: Employee perceptions of acceptable behavior. *Journal of International Technology and Information Management, 18*(3-4), 465–476.

Turilli, M., & Floridi, L. (2010). The ethics of information transparency. *Ethics and Information Technology, 11*(2), 105–112. doi:10.1007/s10676-009-9187-9

Ugrin, J. C., Pearson, J. M., & Odom, M. D. (2007). Profiling cyber-slackers in the workplace: Demographic, cultural, and workplace factors. *Journal of Internet Commerce, 6*(3), 75–89. doi:10.1300/J179v06n03_04

Winter, S. J., Stylianou, A. C., & Giacalone, R. A. (2004). Individual differences in the acceptability of unethical information technology practices: The case of Machiavellianism and ethical ideology. *Journal of Business Ethics*, *54*(3), 275–296. doi:10.1007/s10551-004-1772-6

Young, K. (2010). Killer surf issues: Crafting an organizational model to combat employee Internet abuse. *Information & Management*, 34–38.

APPENDIX A: SURVEY

Individual Perceptions of Acceptable Activities

Table A1. For questions 1-14 please circle the appropriate number indicating the extent to which you disagree or agree with the following statements

It is acceptable for an employee to:	Strongly disagree						Strongly agree
1. store personal music files on the computer in their office.	1	2	3	4	5	6	7
2. use their office computer to do Internet searches for personal interests.	1	2	3	4	5	6	7
3. send a personal e-mail message from work using their work e-mail account.	1	2	3	4	5	6	7
4. use an office photocopier to make copies of personal documents.	1	2	3	4	5	6	7
5. play games online.	1	2	3	4	5	6	7
6. print personal documents using a printer at work.	1	2	3	4	5	6	7
7. analyze personal investment information using spreadsheet software on an office computer.	1	2	3	4	5	6	7
8. send and receive Twitter messages (tweets) at work to communicate with friends	1	2	3	4	5	6	7
9. view weather information online prior to driving home after work.	1	2	3	4	5	6	7
10. use their office computer to shop online for personal items.	1	2	3	4	5	6	7
11. use an office scanner to create PDF copies of personal documents.	1	2	3	4	5	6	7
12. view YouTube streaming videos using their office computer.	1	2	3	4	5	6	7
13. use the organization's CDs to burn personal copies of music files.	1	2	3	4	5	6	7
14. use their office computer to read an online newspaper.	1	2	3	4	5	6	7

APPENDIX B: MEASUREMENT ITEMS

Supervisory Role (SUPER)

Does your current position require supervision of other employees? ___ Yes ___ No

Opportunity (OPP)

On average, for what percentage of your work time do you have easy access to the Internet (either through a personal computer or other device)? (Write a number between 0 and 100)

Computer Self-Efficacy (CSE)

Strongly Strongly
   disagree agree
   I could use a computer for personal 1 2 3 4 5 6 7
activities without spending much effort.
I could use a computer for personal 1 2 3 4 5 6 7
activities without others' help.

Organizational Computer Use Written Policies (POLICY)

Does your organization have a written policy describing acceptable and/or unacceptable computer use while at work? ___ Yes ___ No

Organizational Computer Use Policy Training (TRAIN)

When you started working for your current organization, did you receive training related to their policies related to acceptable/unacceptable computer use while at work? ___ Yes ___ No

Organizational Computer Use Monitoring (MONITOR)

Is your computer use monitored (by a person or computer) while you are working? ___ Yes ___ No

Gender

Your gender: ___ Male ___ Female

Age

Your age: _____ 18-20 years _____ 21-24 years _____ 25-29 years _____ 30-39 years _____ 40-49 years _____ 50-59 years _____ 60+ years

This work was previously published in the International Journal of Technoethics, Volume 2, Issue 2, edited by Rocci Luppicini, pp. 30-45, copyright 2011 by IGI Publishing (an imprint of IGI Global).

Chapter 11
Chicken Killers or Bandwidth Patriots?
A Case Study of Ethics in Virtual Reality

Kurt Reymers
Morrisville State College, USA

ABSTRACT

In 2008, a resident of a computerized virtual world called "Second Life" programmed and began selling a "realistic" virtual chicken. It required food and water to survive, was vulnerable to physical damage, and could reproduce. This development led to the mass adoption of chicken farms and large-scale trade in virtual chickens and eggs. Not long after the release of the virtual chickens, a number of incidents occurred which demonstrate the negotiated nature of territorial and normative boundaries. Neighbors of chicken farmers complained of slow performance of the simulation and some users began terminating the chickens, kicking or shooting them to "death." All of these virtual world phenomena, from the interactive role-playing of virtual farmers to the social, political and economic repercussions within and beyond the virtual world, can be examined with a critical focus on the ethical ramifications of virtual world conflicts. This paper views the case of the virtual chicken wars from three different ethical perspectives: as a resource dilemma, as providing an argument from moral and psychological harm, and as a case in which just war theory can be applied.

THE ETHICAL IMPLICATIONS OF VIRTUAL REALITY

Throughout history new approaches to ethics have emerged along with social change. With the development of agriculture and the attending emergence of early civilizations, new ethical guidelines guided group life within the context of urban settlement (some of which became the basis for foundational religious texts). As the development of capitalism emerged, ethical frameworks were developed to deal with contractual law and property ownership. With the industrial revolution driving the growth of the applied sciences and technologies in the twentieth century, ethics

DOI: 10.4018/978-1-4666-2931-8.ch011

have become focused upon the impact of science and technology upon individual and collective subjects. As we move into the twenty-first century, an information revolution is transforming social organization in ways that demand a return to new ethical considerations. Technoethics, the interdisciplinary research area related to the moral and ethical aspects of technology in society, helps to embrace the questions that stem from such technological developments, as it can with the recent adoption of virtual reality as a place of residence, business, entertainment and education for millions of people.

The ethical implications of virtual reality are only beginning to be taken in a serious and scholarly way. This is perhaps due to the fact that many popular virtual reality interfaces involve video gaming and similar entertainments, relegating them to the sphere of play and separating them categorically from the realm of the serious. For instance, World of Warcraft, now the most popular interactive, three-dimensional computerized virtual reality interface, involves a fantasy role-playing architecture through which twelve million users, more or less, interact (as of October 7, 2010, www.worldofwarcraft.com). Yet it is considered merely a game by most. While the ethics of interaction in these types of virtual worlds would be subject to as intense a scrutiny as possible if it were to take place in face-to-face reality, because such a virtual reality is "only a game," the impacts are typically minimized: such games are deemed irrelevant outside of the context of the virtual world and their feedback into the face-to-face world is deemed by many to be negligible. However, there is evidence to suggest that the relevance of virtual worlds to face-to-face interaction may be growing.

The most popular human-computer virtual reality interfaces today are known as MMOR-PGS (massively multiplayer online role-playing games), or simply MMOs. Examples of such MMOs beyond World of Warcraft include Eve Online, Blue Mars, Entropia Universe, AlphaWorld,

There.com, D&D Online, EverQuest, Lineage, Final Fantasy, Runescape, Asheron's Call, Cybertown, WorldsAway, and many others. Today, the development of virtual worlds has gone beyond the video gaming genre and has become a unique form of social interaction. With the emergence of avatar-built virtual realities the user has gained control over elements of the development of their online three-dimensional experience, following the limitations of the virtual reality laid down by the software they are using (for example, gravity, sun movement, and other universal, if simulated, forces). This introduces to virtual worlds the quality of *frontier genesis*, or social spaces where new boundaries are being developed, new norms created, new status arrangements negotiated and new territories contested. With only a loose set of rules provided by the developers of the software, these virtual frontiers can be seen as sociological laboratories, giving glimpses into the continual process of the (re)development of society and the inevitable ethical choices made by participants. As Malaby (2009, p. 132) puts it, "What should command [our] attention…is the way in which it is now possible to build, with the help of game design and other techniques, complex spaces designed to be spaces of possibility but without the conventional boundaries that mark games. This generates a remarkable opportunity for us to explore issues such as creativity, governance, ethics, and many others… Institutions, it seems, may be changing in their ability to govern themselves and others, and the advent of virtual worlds is at the forefront of this transformation."

This study will investigate a case of contested boundaries in the virtual world of Second Life. Second Life was developed by San Francisco's Linden Research, Inc. (aka Linden Labs) and released to the public in June 2003. Several books have recently been published about Second Life and Linden Labs (Boellstorff, 2010; Malaby, 2009; Au, 2008; Castronova, 2008; Meadows, 2008), analyzing core issues of virtual reality such as personal identity, the organization of governance, economic behavior and creative capitalism. Interaction in Second Life is driven by avatars, a three-dimensional representation of a human, animal, hybrid, or any other of a number of representations limited only by the imagination and Linden Scripting Language (the computer code that makes objects look and act as they do). What is clear is that the participants of Second Life have signed up in great numbers. As of October 20, 2010, Linden Lab reports just over twenty million signups. The actual number of users is more difficult to establish due to the existence of alternate avatars, or "alts," created by the same user (Linden Lab, 2010). At any given time "in-world" (in the virtual reality of Second Life), there are typically over 50,000 avatars being animated by users behind the other side of the computer screen (Linden Lab, 2010). These avatars are typically used to allow users to bond with one another and become committed to shared cultures and communities that range from the healing to the perverse (not unlike face-to-face reality). Building virtual objects is made possible by anyone in Second Life, though it requires the development of some skill in manipulating such objects in a three-dimensional virtual environment using the familiar mouse and keyboard input devices. Further involvement can come by adding coded scripts to objects to allow them to perform certain functions (such as firing a virtual gun). A few users have maximized the utility of their second lives, turning them into a means for making a profit in face-to-face life.

Take, for example, Second Life avatar Anshe Chung. In *I, Avatar: The culture and consequences of Second Life* (2008), Mark Stephen Meadows describes the rush of businesses into virtual worlds and the following disappointment. In May 2006, Chung appeared on the cover of *BusinessWeek* magazine. She was known in the real world as Ailin Graef, a woman who had multiple avatars in multiple game and social worlds. In Second Life, she moved into the virtual real estate market and, as the value of land in Second Life increased,

she earned more than US$100,000 and had assets totaling over L$1,000,000 (the unit of currency in Second Life, known as the "Linden"). Seeing Second Life as a new market opportunity, many companies (including Microsoft, MTV, NBC, AOL, BBC Radio, Fox, Reuters, Sony, Popular Science, Playboy, Mercedes-Benz, Nissan, Pontiac, Toyota, BMW, AMD, Dell, Coca-Cola, Sears, Adidas, Reebok, and many others) rushed to acquire a presence in Second Life (Meadows 2008: 64-65). Nonetheless, traditional marketers still saw Second Life as too risky for mainstream business. After the *BusinessWeek* article ran, Allison Fass of *Forbes.com* concluded that Second Life was not a healthy place for business. Her article (titled "Sex, Pranks, and Reality") finishes with a quote from Erik Hauser, creative director of Swivel Media, Wells Fargo's agency: "Going into Second Life now is the equivalent of running a field marketing program in Iraq" (Meadows, 2008, p. 65).

Many users choose not to create and they participate primarily for the gratification of the social interaction itself. One of the predominant institutions in Second Life today is the dance club. A huge hit upon their development in 2004, the number grew exponentially in a short time (Malaby, 2009, p. 112). With a huge number of themes (beach club, formal dance, jazz, honky-tonk, space themed, etc.), they continue to attract large numbers of users who socialize and virtually dance the night away. Other pastimes include playing virtual bingo-like games called Tringo (and now its successor, Zyngo), as well as other variations, often socially. Role-playing is a common endeavor for users of virtual worlds and a number of communities ranging from fantasy to science fiction, steampunk to post-apocalyptic themes have emerged in Second Life. Like social groups and communities which emerged via past computer technologies (for example, bulletin board systems, AOL, the WELL, USENET, IRC, etc.), Second Life offers the same opportunity for any interest group willing to invest time and energy into building the social relationships that define that community. In 2008, a newly released product in Second Life, the "virtual chicken," spurred the organization of a new group affiliation: cyber-pastoralists who self-identify as the "New World Virtual Farmers." Particularly, chicken farming has become very popular. The repercussions of this choice provide the focus of this case study of ethics in Second Life.

Chicken Farming in Second Life

In 2008, a virtual creature called the *sionChicken,* which is designed to be lifelike in certain aspects of its simulation, was developed by a Second Life resident named Sion Zaius (Figure 1). In face-to-face reality, Zauis is "a young college student whose first language is not English," according to Nika Dreamscape (2010), a journalist in Second Life. "He's described as "painfully shy" by those who know him."

The virtual chicken he created required food and water to survive, was vulnerable to physical damage, and could reproduce (Figures 2 and 3).

Though created as a project for friends, he soon found that they were popular in the virtual realm of Second Life and began a business selling chickens and associated products in the various marketplaces of Second Life (Dreamscape, 2010).

This development led to the creation of chicken farms and the mass sale and distribution of these curious virtual entities, a kind of minimally intelligent "bot," or virtual robot.

Second Life is organized economically as a free-market space within which developers are privy to profit from their creations, and the incentive for distributing these virtual chickens was fiduciary on *all* levels, from producer to consumer. The consumer's stake involves virtual reproduction. When chickens lay their eggs, the color scheme is important for determining their age - scarce eggs are worth more on the egg-trading market (Figure 4). These markets determine the value of eggs and, ultimately, the flock that one has accumulated. Trading in this way, people have accumulated thousands of Lindens, the virtual world currency of Second Life, which are then transferable into real-world currency.

Figure 1. Second Life avatar Sion Zaius (Source: Mistral, 2009)

Figure 2. Virtual chickens and feed (Source: Mistral, 2009)

Figure 3. Virtual chickens and feed (Source: Mistral, 2009)

Figure 4 Virtual chicken color schemes (Source: Reymers, KJJewell Chicken Farm, Second Life, 2010)

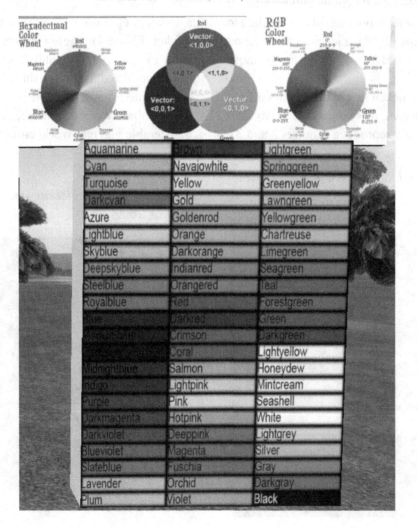

Like other aspects of cultural development in Second Life, chicken farming was hardly neutral in its impact on residents of the virtual world. Chicken breeding became an investment activity and one estimate put the number of virtual chickens at over 100,000 (Davison, 2009). The virtual chicks, once hatched, would roam around their pens (or, if unpenned, eventually walk away and disappear "off grid") and bump into each other and the walls of the pens. Each collision needed to be tracked by the computer server running the simulation (or sim), and if enough chickens were present, the combined calculations slowed the performance of the sim significantly. Many chicken farm neighbors in the same sim complained of lag, the generic label for the experience of slow simulation performance which reduces the synchronicity of the virtual experience and makes it frustrating to interact. The code which allows the chicken's greater accuracy in their virtuality compromised neighboring experiences and affected the social and experiential realism which sustains meaning in the virtual world. As a result, the reaction of some was violent. Neighboring residents took to kicking the offending chickens to death, or even shooting them with virtual guns (Figure 5).

Many virtual realities, including Second Life, have "combat systems" which allow users to track the "health" of their avatars – if they are struck by a virtual bullet (or crossbow bolt, or stone, or laser, etc.), the system keeps track of the "damage" done to the avatar (damage which is not lasting and does not truly compromise the "life" of the avatar). The virtual chickens work in a similar matter and are susceptible to such "physical" damage, though unlike a user's avatar, they can only "heal" with the aid of first-aid kits purchased from Sion Zaius and can sometimes "die" outright. Some particularly devious programmers began a kind of "bio-warfare" campaign against chickens in Second Life, creating an object in the virtual world called the sionChicken Killer (Figure 6) that acted as a food decoy, but when "eaten" would provide no sustenance - the chickens would die of starvation after a few days.

This prompted avatar journalist Pixeleen Mistral, author of an "in-world" newspaper article in the *Alphaville Herald* about the sionChicken war, to muse on the economic nature of Second Life in asking "Will this then create a market for chicken killer detectors?" (Mistral, 2010a). The opportunity for an arms race in the chicken wars was ripe. The text of the sionChicken Killer product reads as follows:

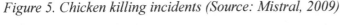

Figure 5. Chicken killing incidents (Source: Mistral, 2009)

Figure 6. sionChicken killer product (Source: Davison, n. d.)

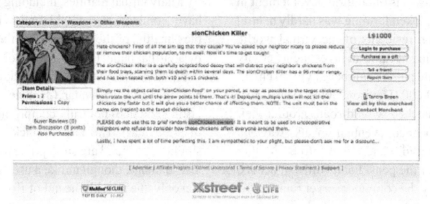

sionChicken Killer

Hate chickens? Tired of all the sim lag that they cause? You've asked your neighbor nicely to please reduce or remove their chicken population, to no avail. Now it's time to get tough!

The sionChicken Killer is a carefully scripted food decoy that will distract your neighbor's chickens from their food trays, starving them to death within several days. The sionChicken Killer has a 96 meter range, and has been tested with both v12 and v11 chickens.

Simply rez the object called "sionChicken food" on your parcel, as near as possible to the target chickens, then move the unit until the arrow points to them. That's it! Deploying multiple units will not kill the chickens any faster but it will give you a better chance of affecting them. NOTE: The unit must be in the same sim (region) as the target chickens.

PLEASE do not use this to grief random sionChicken owners! It is meant to be used on uncooperative neighbors who refuse to consider how these chickens affect everyone around them.

Lastly, I have spent a lot of time perfecting this. I am sympathetic to your plight, but please don't ask me for a discount... Cost: L$1000 (Mistral 2010a).

At this point, the conflict came down to software versus software. The point made in the sionChicken Killer description about "griefing" random sionChicken owners is an important one, as it helps to distinguish between two different types of "chicken killers": those who were attempting to eliminate lag in their regions, and those who were playing a game, intentionally killing chickens for the sheer sport of it. The most famous of these griefers in the chicken wars of Second Life became known as the "Soviet Woodbury" faction, apparently tied to Woodbury College in Montpelier, Vermont. According to Davison (n. d.), "Realizing that (a) there are a group of chicken owners who care about their pets, and (b) their [sic] are chicken-killing weapons available for purchase, a group of griefers, allegedly younger residents allied with Woodbury College begin killing "innocent" chickens on purpose." Some of these griefers had their accounts banned and the Woodbury College presence in Second Life was eliminated by Linden Lab on April 20, 2010, citing only that "this decision [is] based on historical and recent events that constitute a breach of the Second Life community standards and terms of service. We ask that you please respect the decision and do not take part in the Second Life platform in the future" (Young, 2010).

In an attempt to remediate the lag issue, creator Sion Zaius updated the version 11 and 12 chickens to reduce the lag that came from collisions. Furthermore, he released a new version of a device used to ensure the safety of one's eggs and keep them from breaking, called the "Proteggtor." The new version was programmed to delete all of the old, lagging eggs so that old laggy chickens would not hatch from them. Unfortunately, the initial release deleted ALL eggs, including newer, "lagless" versions. Chicken farmers were angry and demanded compensation. After repairing and updating the new Proteggtor, Zaius claimed to have no responsibility for lost or broken eggs related to his unannounced "Proteggtor" update. This infuriated many of the New World Virtual Farmers, enough to threaten real-world lawsuits against Zaius's real-life driver. A noticeable backlash against Zaius also appeared in blogs and news websites related to chicken farming specifically, and Second Life generally (Dreamscape, 2010; Mistral, 2010a; Davison, n.d.). "Keep in mind that Sion Zaius created a lot of bad will in his customer base due to what appeared to be abusive business practices" (Mistral, personal communication, October 25, 2010).

As a result of the chicken debacle, "petable" turtles and bunnies known as "Ozimals" are becoming more popular as of 2010. New products are being developed by innovative Second Life residents who have recognized the niche opening in this very popular viral market of cyber-pastoralism. In some cases, they too have been targeted for termination, albeit to a far lesser degree than were the now infamous sionChickens.

This case reveals several ethical dilemmas that are unique not in their substance but in their context: virtual worlds. These dilemmas could be characterized in three ways. First, the conflict revolves around a resource allocation dispute, which in the context of the virtual world could be considered a resource dilemma (ultimately ending in a "tragedy of the commons"). Secondly, as a conflict borne out of the behavior of psychologically engaged actors, it could be examined from the point of view of moral hazard and psychological harm. In terms of ethical theory, this case lends itself to examination through deontological and consequentialist perspectives. Thirdly, inasmuch as the conflict took on the characteristics of a "war," it could be considered through the lens of just war theory. To generate further thought about ethics in virtual reality, I will briefly examine the case from each of these perspectives in the following sections.

A Tragedy of the Commons

The case of the chicken wars could be examined as a resource dilemma, also known as a "tragedy of the commons" (Hardin, 1968). This ethical dilemma exists when actors behave only out of rational self-interest, depleting the shared resources available for their collective well-being, resulting in conflict. In this case the capacity for the computer that is creating the simulated environment to calculate collisions was the shared resource. When stressed with too much information, the server slows down, creating a phenomenon called *latency* (also known as *lag*). "Second Life sends packets to your computer. There are about 10-30 pipes in between you and the grid servers, depending on where you're located. The amount of time it takes to traverse that length is called *latency* or *ping time*. More latency means things are just that little bit less responsive" (Nino, 2006). More latency also means greater "packet loss" from the information being sent to your computer from the Second Life server that controls the simulation within which one's avatar is acting. The impact that this lack of responsiveness can have on the experience of avatar embodiment can be disturbing. "Packet loss is generally pretty nasty. It causes all sorts of weird side-effects. 'Rubber-banding' is one, where your avatar is walking or flying a short distance and suddenly snaps back... Slow rezzing of textures (or not rezzing at all). Avatars that are invisible except for their attachments. Being *Ruthed* (suddenly changing to the default female shape [named Ruth]) or seeing someone else as *Ruthed*. All sorts of odd things like this are caused by packet-loss" (Nino, 2006).

The property of latency is related to *bandwidth*, a measure of data transmission rates. The rational self-interest of the chicken farmer is to grow their livestock for fun and to profit from the unique eggs that are created. The rational self-interest of the neighbors of chicken farmers may be to dance, role-play, educate, or take on some other form of socialization. In a resource dilemma it is when the property is overtaxed and the shared resource runs out that conflict emerges. The analogy in virtual reality occurs when the "lag meter" spikes.

Questions inevitably arise. Is lag simply a technical error, or are phenomena that impact the virtual environment, such as slow server response time and bandwidth packet loss, ecological variables of virtual reality? If so, do these ecological variables present the basis for a shared "reality" that constitutes a "commons" which can be tragically overused? And who is to bear the cost of the developer's programming error? Is the developer responsible for chickens killed out of the frustration of neighbors whose resource was being depleted by the chickens? Or is this responsibility solely on those performing the chicken slaying? These questions point out the fact that ethical dilemmas are at hand, ones that involve common commodities (server time and bandwidth). In fact, the resource dilemma is resolved by sharing the cost, though admittedly much of it is borne by the chicken farmers, due to a lack of regulation and enforcement over personal property in Second Life.

Historically, the eighteenth century provided much of the foundation for governance over personal property and market economies which fostered the development of capitalism. John Locke, Jean-Jacques Rousseau, Adam Smith and many other philosophers, economists and political thinkers established treatises and ethical guidelines that lead to the social contracts that govern over economic and property relationships today. Constitutional democracies and contractual law are the form of governance deemed to keep the

world's nations politically stable and economically prosperous today. Enter virtual reality.

In Second Life, a virtual world not even a decade old, governance is generally maintained by a small group of developers at Linden Lab who have neither time nor interest in every small dispute that goes on in Second Life, nor provides an established contract beyond their Terms of Service. Like a wild frontier, the sheriffs are few and far between and the arm of the law is not so long. Posses and vigilantes frequently decide the outcome of property dilemmas under such conditions. And the vigilantes killing chickens were not the only offenders. Chicken farmers who had little regard for neighbors sometimes placed a large number of chickens in a small area, deeply intensifying the lag problem. These small, populous virtual chicken breeding operations could be compared to the concentrated animal feeding operations (CAFOs) of the food factory systems common today in industrialized countries (Gurian-Sherman, 2008). Here, the ethics of animal breeding intersect with virtual chickens, but I will leave that to another discussion. What of the ultimate source of the chickens themselves? What responsibility does the developer of the virtual chicken, Sion Zaius, hold? Zaius did attempt to remediate the problem, but in some cases (as with the "ProtEggtor" product) the solution was worse than the problem.

Few involved seemed to act within a context of enlightened self-interest, seeking compromise and negotiation, but rather resorted to the most expedient solution: violence in the case of neighbors, retrenchment in the case of farmers, or disclaiming responsibility on the part of the developer. No infrastructure of rules or ability to enforce existing community standards was evident. As an experiment in near-anarchy, this case is illustrative of anomic communities, or social groups within which the rules of law and social norms break down. Sociologically speaking, at least some parts of Second Life resemble ethically the state of moral decay experienced in some of America's inner cities (Anderson, 1999).

However, since the human ecology of Second Life typically involves far fewer people than an urban environment, perhaps it may be more illustrative to compare this example to one found in a face-to-face confrontation over chickens in a small town near the college where I teach. In this case, an eccentric resident living within, but near the edge of, the village limits sought to raise chickens on his property and was granted that right (according to him) by a town councilman prior to a recent turnover in administrations. When a new administration sought to limit that right, the resident balked, and the town council sanctioned a new law banning the raising of chickens and other farm animals within the village limits. The resident protested (Figure 7) and ultimately the case went to court. After the local court upheld the law, the resident appealed to the county court which upheld the citizen's right to own chickens. One side of this story can be read at the resident's website, www.EarlvilleChickens.com.

This case highlights two facets of difference between the face-to-face world and the virtual

Figure 7. Earlville Chickens: "Honk for Earlvil-leChickens.com" (Source: Reymers, 2008)

world. The first difference is the established and often intransigent political and legal process by which the case proceeded and the availability of courts of law to mediate the dispute. The second difference is the avenues of attention the resident had at his disposal to protest what he considered to be an unfair demand by the village, ranging from wearing a sandwich board in the town center to establishing a (rather bizarre) website about his case. These are the positives and the negatives of an established face-to-face community that many residents of Second Life seek to avoid. Nonetheless, Second Life does have its own boundaries, fuzzy and indeterminate though they may be.

One of the chief methods Malaby (2009) identifies for establishing order and governance both within the virtual world of Second Life and that of its maker, Linden Lab, involves the transition from traditionally executed norms and cultural rules for the workplace (negotiated through traditional media) to the use of what has been called "code/space" (Kitchin & Dodge, 2006). Institutionally, this meant the turn from memos, face-to-face meetings, and other traditional vertical, "top-down" bureaucratic methods of communication at Linden Lab to a software product called *Jira*. Jira is an open-ended communication tool "designed to help a group of people keep track of the development of a software product and [it] allows for the relatively straightforward coding of further tools that can be layered onto its software to make use of the information it tracks" (Malaby, 2009, p. 68). Individually, this transition in the workplace led to the use elements of contingency and indeterminacy – games, in fact -- to negotiate the new media platforms used to convey the tasks and strategies necessary to create the virtual world, which itself is a kind of game. Much of the interface of Second Life is predicated on the developers intimate knowledge of video games (frequently played within the Linden Labs office, which is set up almost exclusively around this type of play) of all varieties: race games, arcade games, combat games, strategy games, "God" games, first-person shooters, and so on. Offline games, such as "Nerf battle" were common as well. "Games," says Malaby (2009, p. 85), "are socially constructed by a shared commitment to their legitimacy as contrived spaces where indeterminate outcomes can unfold." This injects an intentionally irrational element of unpredictability into the business of Linden Lab and its product. During his time at Linden Lab, Malaby describes the place as constantly teetering between success and failure.

This irrationality (in the Weberian sense) and unpredictability, however, harnessed the creativity of the masses and pushed Second Life in successful directions that would not have been otherwise known. For instance, the dance clubs and Zyngo parlors, castles, spaceports and, of course, chicken farms, were all outcomes of the nearly complete freedom given to users to create their own world. The sionChicken developed by Sion Zaius was one such "game" that represented both a rational and successful business strategy, but one that created an irrationality in the form of a dilemma within the framework of the resources important to the community. The debate about whether the experience simply was in fact, "just a game," also was central. "[This] debate asks if chicken killing is a legitimate form of gameplay in a player-created game, and whether meta-gamers should be able to hijack the narrative of others. In practical terms, one has to assume that not everyone will share your notions of fair play and either find a way to exclude those people or incorporate their narrative into your own. In other words, either lock the griefers out, or include the idea of "bad guys" in your game's narrative structure" (Mistral, personal communication, October 25, 2010). Without the authoritative structure of law that we have constructed in the "real" world our virtual counterpart can expect this condition of unpredictability and irrationality to unleash more such dilemmas. What is an interesting new development is the role which cooptation takes in

negotiating the boundaries of acceptability: if you incorporate the bad guys into your narrative, and they cease to be able to do any harm! Is narrative the replacement for law in the virtual world?

Moral Hazard Perspectives

Another ethical dilemma present in the sionChicken case involves the treatment of objects in simulated environments and the moral hazards involved therein. Is it ethically permissible to kill a virtual chicken in a premeditated fashion? Does the moral hazard of doing so lie solely on the infringement of property rights, or is there a deeper reason why simulations of living things be regarded with ethical pause? Is there an argument from moral development or from psychological harm? Does something need to be "real" to be subject to moral hazard? And to what extent is the degree of accuracy of a virtual experience related to its believability, its impact on our psyche as a real experience, and thus its condition as an entity deserving of moral consideration? Here we enter a *Blade Runner* world of postmodern definitions of reality.

In a now classic article on virtual reality in the 1990s titled "A Rape in Cyberspace," Julian Dibbell (as cited in Vitanza, 2005) anticipated the tension between simulated and actual experience. In that article, Dibbell describes a MUD, or multi-user domain (a kind of text-based only virtual reality that preceded three dimensional graphic user interface virtual reality) named LambdaMOO. In the MUD, one user manipulated the computer code so as to force another user's character to "perform" (within the context of their shared reality, which is merely the text description that defines the MUD) certain unsavory sexual practices upon the character of the perpetrator. The afflicted user reportedly experienced emotional trauma not unlike that of an actual rape victim. Reid (1995, p. 165, as cited in MacKinnon, 1997) writes, "Users treat the worlds depicted by MUD programs as if they were real. The illusion of reality lies not in the machinery itself but in the user's willingness to treat the manifestations of his or her imaginings as if they were real." The line between the virtual and the real is thinner than common sense might allow us to believe.

In two further anticipatory articles that were published prior to the emergence of most well-subscribed virtual worlds, researchers have noted that in the ethics of representation and action in virtual reality the degree of realism is important. "[Virtual reality] applications differ in the kinds of *reality claims* they make, i.e., the implicit or explicit promises about the realism of (features of) the virtual environment," says Brey (1999, p. 12; original emphasis). He continues, "When certain reality claims are made, the application can be expected to live up to certain *standards of accuracy*." In the case of Second Life, the features of the chickens that created their realism (their "physicality," need for food, shelter, etc.) was also the feature that created the resource dilemma and allowed neighboring residents to kick or shoot the chickens to death. In this case, the degree of realism has led to the creation of certain ethical choices not anticipated by the users, choices that mirror ones made necessary in the world of non-virtual chickens and non-virtual neighbors, but ones without the institutional infrastructure and support of established law and developed social norms. Brey (1999, p. 13) notes that it is "the developers [that] should hold the responsibility to take proper precautions to ensure that modeling mistakes do not occur, especially when the stakes are high... [and] the responsibility to inform users if such mistakes do occur and are difficult to correct." From this perspective, Sion Zaius would clearly be taking much of the blame for the creation of the lag problems and subsequent exterminations. To his credit, creating new, "lagless" chickens (versions 11 and 12) provided a satisfactory resolution to the dilemma. But with them came a strict End User License Agreement removing any future liability to him due to the behavior of his product, about which many virtual chicken owners

were quite unhappy. Virtual chicken owner Nika Dreamscape wrote a story on *The Chicken Blog* ("The Saga of Sion," April 1, 2010) describing the perspective of the owners.

Suddenly there was a license which they [the chicken owners] felt left them defenseless against any maligned business practices. It was one-sided and bound the buyer completely... As a result, people left. They moved on to the new AI pets popping up in the marketplace that promised bigger and better things... While the group was maintained by his two customer service reps, Sion seemed to vanish, and has since rarely been seen or heard from. His chickens have not seen an update or new content in half a year now. I was one of Sion's most challenging critics. I wrote, in depth, about some of his most fumbling missteps. I highlighted his lack of communication with his community... I'm not asking if, in specific incidents, Sion was right or wrong. I'm asking if perhaps it was to [sic] much too soon for a single, unas-suming young student in college to readily endure success, demands, public scrutiny, public service with complete rationale [sic]. I don't think I would have been able to [sic].

This owner, for one, could sympathize with the plight of Sion Zaius in recognizing his limited ability to cope with the problem. But she also held this person clearly responsible during the conflict, holding his conduct up to public scrutiny and demanding restitution for investment losses.

Interestingly, a Second Life estate owner, Intlibber Brautigan, also initially took offense to the scripting problem of the old chickens, but saw an opportunity to help the cause rather than create conflict. A transcript of Brautigan's work with Zaius was recorded by journalist dana vanMoer of the *Daily SL News*:

Today I had an IM from Intlibber Brautigan, he said he had a story for me and asked if I knew about 'Sion chickens'...

IntLibber Brautigan: they lag the crap out of sims, physics and script lag. They've been spreading like an infection across the estate - I spent last night dealing with lag complaints all over the estate. 10 chickens make 1 ms of script lag but each chicken makes 150 potential collisions!

He was so infuriated he considered an AR [Abuse Report] against the creator citing griefing issues but instead decided to speak to the designer and point out the problems.

Sion Zaius worked with Intlibber to fix the issues and had this to say:

Sion Zaius: yes, there have been problems with sion chicken causing physical lag. I'm working on the update which is called "version 11 lagless"

Originally the chickens were updated from the feed but this was changed about a week ago so I asked Sion what people needed to do as I understand you can't just pick them up and re-rez them.

Sion Zaius: since you cannot pick up chickens or eggs into your inventory without breaking them, you have to use chicken transport boxes or proteggtors to do so, if you use those objects, your chicken/egg will be updated automatically.

Sion Zaius: then, people would have to box up living chickens once, and free them again lag-improvement should then occur instantly - its awesome, this chicken has as collision score of 3.4... it had 140 before

IntLibber Brautigan: We will be requiring all of our residents who have chickens to update them if they want to keep them. The lag improvement is just tremendous, its going to improve sim life for everybody.

This shows the possibility of cooperation rather than competition as one of the resolutions to the ethical dilemma of resource competition. The commons need not be trampled by the self-interested masses: enlightened self-interest empowered a cooperative strategy in this case, resolving the problem for many by reducing the need for the resource at the source.

Brey also notes the function of meaning in virtual environments. "VR simulations of objects may approach the perceptual complexity and interactive richness of everyday physical objects, and may for this reason more easily generate belief in their veracity and objectivity than other sorts of representations" (1999, p. 13). According to Ford (2001, p. 118), who expands on Brey's article to include multi-user environments, "people often become emotionally invested in online personae within the context of a community." The important question then is, "Can virtual chickens be considered personae?" They exhibit many of the characteristics of real chickens. They are valued by their owners, who must actively care for them in order to benefit from them. They can be terminated, like a real chicken. The term *personae* comes from the Greek word meaning *mask*. Is the mask of virtuality enough to argue that the chickens are no more than pixels and/or electrons and deserve no more rights than their constituent parts? Or is their status as "bot" enough to confer even minimal ethical considerations? Do the feelings of the chicken owners or the chicken killers count, despite the virtual nature of the chickens themselves? Does a representation or a simulation have a claim to any moral rights? These are certainly difficult questions to tackle theoretically. The answers to most of these questions in the actual case of the sionChickens were resoundingly negative. In response to the question "What were the reasons given by the chicken killers (or bandwidth patriots, if you like) for their behavior?" Pixeleen Mistral says "there were two rationales: the griefers said it was just part of the gameplay, and those concerned about the degraded performance of sims said they were trying to keep the chicken farms from slowing the sims down for everyone (i.e., the chicken farmers were taking more than their share of the sim resources)." Regardless of the fact that these chicken-bots were seen to be a "real" part of the Second Life community, by the New World Virtual Farmers at least, it is hard to believe that those involved in the "slaughter" paused to consider if the bots themselves had any natural right to exist. After all, in the minds of many this was only a game and it was simply necessary to redistribute resources appropriately so that all players could have a lag-free experience: the right-to-bandwidth clearly trumped the chickens' right-to-life.

Of course, lives are not at stake in this case. Life, or more particularly the premeditated end of it, predisposes moral discussion. While it is clear that there is an exchange value for virtual chickens, from a deontological perspective the question remains: is there any *inherent* value in simulated objects? Plato denied this possibility, privileging the ideal and the real over the "sham," or virtuality (Vitanza, 2005, p. 1). An ethical thinker of this ilk, critical of consequentialism, might ask if the termination of simulated life could lead down a slippery slope to a position whereby some people, unwilling to strip simulation from nature, fantasy from reality, allow the impulses of the lawless virtual world to drive their behavior in the face-to-face world. The application of virtuality ethics in our face-to-face world, from this point of view, might have profound negative repercussions. It would be equivalent to the most hardened and harsh forms of libertarianism. Furthermore, the position of truth suffers when one merely need change ones narrative in order to deal with deviance. From a deontological perspective this condition is untenable. Inherent truths are indeed masked by the personae of virtuality and the entire game of the New World Virtual Farmers can be seen as a sham which need not gain our consideration, until the impacts of such behavior are encountered in the face-to-face reality that contains the inherent truths of life.

A more utilitarian consideration, on the other hand, would allow for such virtual and face-to-face crossovers provided they could be evaluated in terms of serving the greater good. This might follow a more classical liberal logic, that of the laissez-faire free market. Allow developers to create virtual chickens; expose these chickens to the marketplace; if they are valued they will succeed and if they create social problems which the market cannot bear, they will fail. The greater good served is to enhance the social relationships of those participating in the virtual world; the market externalities are the latency issues. And, if what Dreamscape (2010) says is true, the turn to a new product (Ozimals, for instance) may signal the demise of the offending sionChicken and the end of any resource conflicts and ethical controversy. The invisible hand of the market both creates a greater social value and mitigates ethical problems. From this point of view, because they are driven by completely economic rules (with all of the attending commodities ultimately taking the form of bits and bytes), virtual worlds are a perfect case to employ consequentialist ethics.

The common "runaway train dilemma" has little bearing on virtual reality, for instance. In this dilemma, a runaway train is heading toward a group of people stuck in a car broken down on the tracks. You stand at a switch for the tracks and have the ability to divert the train onto a path where only one pedestrian is standing. Is it ethical to take the action of switching the train's path, despite the fact that your actions would in turn lead to the death of the pedestrian? Consequentialists argue that saving the lives of many counterbalances the loss of the life of the few (or the one). In the case of virtual reality, however, such actions are moot because all action is an illusion of the mind. Let the train hit the car, then just restart to begin anew. "No harm" defenses of violence in video games have been mounted from a similar perspective, particularly by evolutionary psychologists (Ferguson, 2010; Ferguson & Rueda, 2010).

Objections to such cold and calculating ethics necessarily involve the examination of the term "virtual." Brey (2008, p. 4) notes that one meaning of the term does not mean "unreal," but rather "practically but not formally real." For example, the statement "buffalo are virtually extinct" is not meant to indicate they are the opposite of real, but that they have been practically wiped out. Likewise, although avatars can fly and not die in Second Life, they are driven by human intent, emotion and a deep, immersive psychological engagement. They are the personae of human actors. Avatars thus deserve a similar respect as their corresponding drivers, or "meatpuppets," to use the language of cyberpunk. Social constructionists would approve of such a definition. This perspective suggests that traces of "reality" do intermingle with virtual worlds people inhabit, which could disarm a purely calculative utilitarian ethics of virtual reality.

Just War Theory

A third ethical perspective from which this "lag war" could be examined from the point of view of Aristotelian virtue ethics, in particular focusing upon the cardinal virtue of justice. Upon examination of the issue of justice in the framework of a truly positive pacifism, Carhart (2010) connects virtuous character to the act-based ethical framework of just war theory. "By recognizing that the use of force may sometimes be permissible or even necessary as part of the pursuit of peace, the positive pacifism described above departs from the consistent ideal of absolute pacifism and may be identified with a cautious version of the just war theory."

The classic just war tradition of medieval philosophers dismisses normative agreements regarding rules of war agreed upon by each party in a conflict. Rather, rules of justice defined within the context of virtuous behavior should define the terms of engagement. Since the Second Life chicken conflicts were taking place in a new frontier,

only loosely bound by convention and social norms and where few significant rules of conduct and combat existed, this seems a valuable approach. The principles of just war are commonly held to be: "having just cause, being a last resort, being declared by a proper authority, possessing right intention, having a reasonable chance of success, and the end being proportional to the means used" (Moseley, 2009). These principles cover a broad range of ethical perspectives including elements of virtue ethics, teleological ethics, Kantian duty ethics, and consequentialism.

Before hashing out each of these principles in the Second Life chicken war case, let us establish once again that the "field of battle," so to speak, is the virtual world, a shared experience mediated through the computer and its input and output mechanisms. The world is broken down into regions called "sims," each controlled by a unique computer server. The establishment of the accuracy of the simulation is the chief public good being fought over. Technically, this resource can best be exploited when bandwidth is maximized and lag is minimized. And the contestants in this fight involved avatars, or virtual personae driven by users. Now, recalling these conditions, can principles of just war be applied in this case?

First, was there just cause for the neighbors of the chicken farmers to terminate the chickens without negotiation? Certainly they felt so. The culprits striking the first blow, by their account, were the agents responsible for placing the chicken-bots in the sim, thereby creating what was deemed to be an excessive amount of lag. They were only reacting to an untenable situation introduced by the chicken farmers. But what if the farmers were unaware of the impact their chickens were having on the sim? In fact, this condition was recognized by the developers of the sionChicken Killer (the food distracter "bio-weapon"), when they wrote on the product advertisement "PLEASE do not use this to grief random sionChicken owners! It is meant to be used on uncooperative neighbors who refuse to consider how these chickens affect everyone around them" (Mistral, 2009). Nonetheless, the need to create such a plea was an indication that some of the sionChicken killing had gone beyond the boundaries of just war and had become "griefing," or "activities designed to make another player's life or experience in Second Life unpleasant" (Second Life Wiki, 2010).

Second, was chicken killing a last resort for residents? In some cases, chicken killing may have seemed a last resort when chicken farmers did not respond to neighbor's requests to reduce lag, and the larger governing authority in Second Life (Linden Labs, to whom one can report abuse of the Terms of Service contract) was unresponsive. And while the costs of teleporting (the main form of travel in Second Life) to another sim which was unladen by virtual chickens was nil, for one who invests in renting or buying property in Second Life and takes the time and energy to create one's own personal space, the idea of moving due to a neighboring avatar's behavior can be undesirable and costly. Property, in other words, takes on vestiges of reality in Second Life, and that boundary of civic engagement was as real as the boundary in the EarlvilleChickens.com case.

Third, the declaration of war by a "proper" authority was impossible, due to the fact that such an authority does not exist in virtual reality. Governance is at the dictatorial behest of Linden Labs, who (in following the generalized ethic found in the origins of the Internet) created a virtual world where freedom and communitarianism are centrally valued and rules and roles left up to the individual (Reymers, 2004). As in most virtual communities, griefing (alternatively called "flaming," "phishing," or "trolling," though each term has a specific nuance) is more than happenstance and often has a central place in the definition of the community (it is often absorbed into the shared narrative of the actors in Second Life). This is precisely due to the fact that no central authority exists and laws and norms are left to each unique community to establish. Occasionally Linden Labs will mediate a dispute, but they

certainly did not sanction the chicken killing that was going on. As Davison (2009) suggests, it was precisely due to the collapse of boundaries between normative expectations in different communities that the sionChicken incidents occurred. The third principle for just war in this case is not clearly definable due to the fact that there is no agreement on the propriety of authority in Second Life.

Fourth, did the chicken killers possess the right intention in perpetrating their violent solution to the lag problem? It seems that if the chicken farmers were provided warning that such solutions would be forthcoming, and they did nothing to respond, the intention would have been clear. But was it the *right* intention? Given that bandwidth is the lifeblood of virtual reality, the air that avatars breathe, anything that compromises this valuable resource may be said to compromise the very existence of the interactive interface itself. Therefore, if one considers interaction (with the virtual environment) to be central to both the purpose and the capacity of one's presence in Second Life, and accuracy of interaction is a direct result of bandwidth availability, protecting bandwidth availability is akin to protecting one's existence. From this perspective, "chicken killers" may be seen more heroically as "bandwidth patriots." The right intention principle seems to be met. However, one can see this principle spiraling out of control. Would it be acceptable in virtual worlds, for instance, to somehow disable (or "kill") another user's avatar if you deemed they were using up too much bandwidth (as a result of wearing heavily scripted objects, such as fancy hair, shiny jewelry, etc.)? Certainly a balance needs to be struck between stylizing ones avatar with "bling" (or ones land with virtual chickens) and bandwidth considerations. Griefers who indiscriminately killed chickens without warning and without the purpose of reducing the lag problem would not have been acting with the right intention (and thus might be considered a kind of "chicken war criminal" – although there is no equivalent of The Hague at which these avatars could be tried for their crimes, unless you consider the developers at Linden Labs to be playing that role, as in the Soviet Woodbury case).

Fifth, the chances for the reasonable success of exterminating the problem chickens was relatively high, due to the fact that sims are isolated from one another and eliminating the chickens from one sim, while not addressing the more global problem of chicken lag, solves the problem from the point of view of the "bandwidth patriot." Furthermore, such local "raids" drew attention to the fact that a more global problem existed; thereby creating the kind of awareness that ultimately motivated Sion Zaius to alter his product (virtual chickens) into a "lagless" version. In retrospect, these "bandwidth patriots" did succeed in their objective to rid Second Life of most of the laggy initial versions of the sionChicken.

Finally, the consequentialist question comes up again: did the ends justify the means? Or, more accurately, were the means used to protect bandwidth proportional to the value of the bandwidth obtained? How *bad* a problem was the lag issue, and did chicken owners *deserve* to have their flocks culled? Was there a greater good produced by the chicken wars? The problem of the extent of lag can be answered technically, as it was in the following response to an article written about the chicken wars in the *Alphaville Herald*, Second Life's online newspaper:

With the Version 11 Lagless model, the collisions produced by sion chickens have dropped from 150-250 down to 3-30 under most circumstances. The script lag still runs from 0.1-0.3 ms per chicken. For this reason I recommend that estate owners impose a covenant limit of no more than 1 chicken per 1k sqm of land. In our Ancapistan estates, we've found that these limitations keep sims healthy and enjoyable for the most part, however we do recommend an upper limit of about 50-60 chickens per sim ONLY if these are the only objects in the sim. The more prims in the sim, the more collisions that will happen, and other scripted objects

of course will take up some of the total script time (Intlibber Brautigan, as cited in Mistral, 2009).

Whether or not the chicken farmers deserved to have their flocks terminated is a value-based question whose answer varies depending on the supplicant. From the point of view of the "bandwidth patriot," laggy chickens earned their just desserts: death. From the point of view of the New World Virtual Farmer, there was no honor, integrity or merit in the unilateral decision of the chicken killers to exterminate their flocks – an extreme view might have held that it was a "bot genocide" imposed upon them from the outside, a kind of "electronic ethnic cleansing." The very culture of the virtual farmers was at risk. A less extreme view might simply acknowledge the loss of investment property that cost, in some cases, a significant amount of real money. And from the estate owners' point of view, it became necessary to hold a balance between the interests of the many renters upon their estates and the economic advantage that chicken farming brought. The resident above continues:

When these chickens started spreading across our estate, we considered banning them, but considering the economic activity they stimulate, we sought instead to help the makers at Sion Labs reduce lag as much as possible. Provided these tips are followed, theres no reason why chickens cannot be permitted in any sim in the numbers specified.

Estate owners IMHO should look at the chicken craze as a means of expanding occupancy, particularly by imposing limits in numbers of chickens per 1k sqm, this obviously means a chicken farmer needs a lot of land to farm a lot of chickens (Intlibber Brautigan, as cited in Mistral, 2009).

The consequences of the chicken war were variable. Was a greater good achieved? Perhaps for Sion Zaius, who continues to reap great profits from his chickens and the associated products needed for them to survive. Perhaps there was a greater good served to the Second Life community at large, for whom the growing problem of laggy chickens was resolved through the actions of a few "bandwidth patriots." However, a greater good was not likely to be seen by the aggrieved chicken farmers who lost valuable (even if virtual) livestock as a result of the conflict. And it is uncertain without further research, but perhaps a greater good was not served for the face-to-face communities of the users behind the avatars, whose first lives may have been influenced in their Second Life roles as patriotic chicken killers, victimized farmers, or grief mongers. A just war minimizes casualties, and the casualties of this virtual war, while physically negligible, may have taken a psychological toll.

CONCLUSION

No ethical dilemma has a simple solution or one that satisfies all parties. Nonetheless, the structure of the community within which such dilemmas arise influences the outcome for all. Where resources are limited, communities come into conflict around the behaviors that exploit those local resources. A resource dilemma, or tragedy of the commons, challenges those trapped within it to take the allegedly irrational step of bowing to a group interest over one's own perceived self-interest, but as many have discussed (Axelrod, 1984; Sagan, 1998), this strategy can be more lucrative (and thus rational) for all individuals involved in benefitting from a common environmental resource. In this case, lag and bandwidth are the environmental resources being taxed, and so an ethically sound position advocates eliminating the elements of the virtual reality that put an undue burden on this resource on a shared basis. The definitions of "undue" and "shared," however, are the sticky points left to be negotiated by the actors involved within the context of their circumstances. One of the difficulties in

deciding where these lines should be drawn are the multiple sources of lag and the lack of good information about what exactly is slowing one's Second Life experience.

From a consequentialist perspective, then, if the achievement of the greatest good for the greatest number of people requires the extermination of the offending laggy chickens by those other than their owners, so be it. However, this argument offends the more principled ownership argument regarding said chickens. We would certainly not feel this to be the right course of action if the terminated chickens were our own, particularly when a technological fix was right around the corner. The extermination of laggy chickens in an arbitrary and capricious (as it was in many cases, such as the Soviet Woodbury attacks and others), could not be justified using this argument. Nevertheless, from a consequentialist argument, the chicken-killers were in the right – the end of the chicken wars in Second Life (brought about by a software upgrade introduced by the creator), overall, justified the means by which this conflict was resolved.

A second ethical dilemma regarding the virtual entities known as sionChickens involves the degree to which they are virtual. This dilemma stems from the relationship between simulation/belief and agency. With the representation of behavior in avatars, it is assumed that users have enough self-coherence to willingly suspend their disbelief. But is this a safe assumption? As the degree of accuracy of the representation provides a virtual world that more and more closely approximates the parameters of the face-to-face world, it can be expected that phenomena outside of the control of the designers of the virtual world – let's call them environmental contingencies – will lead to many different reactions and consequences that must be dealt with on a social level. Malaby (2009) goes so far as to suggest that Second Life is explicitly designed to allow the development of such contingencies (yet just as explicitly avoids governing over the consequences of users' creations – it is left to users to hash this out on their own). The degree of responsibility that Linden Lab took in the chicken lag problem was minimal. Rather, the mantle of responsibility was passed to Sion Zaius, the creator of the offending chickens. A lack of responsiveness on his part clearly evoked the violent and negative responses from residents that were fed up with lag. In market terms, the situation took care of itself. But this still leaves the question of psychological harm unanswered.

Third, the notion that conflict can be based on principles of just war leads to the conclusion that cyberwar is one of the complexities of virtual reality that challenges the bedrock notions of our culture. While questions of just cause, last resort, intentionality and having a reasonable chance of success are as clear, or fuzzy, as the case may be, as in face-to-face situations of conflict, the chief difference in the case of Second Life involves the place of authority within the structure of disputes. Without a distinct political authority and system of comprehensive rules (beyond the TOS, or Terms of Service provided by Linden Lab) the establishment of authority is weak or non-existent. Occasionally, Linden Lab bans users or whole groups (as in the case of Woodbury College), but typically Linden Lab stays out of a conflict if there is not a clear violation of the TOS. This may be a result of earlier conflicts in which Linden Labs was involved, as in the case of the Jessie Wall described by Au (2008, pp. 103-117). Estate owners can establish a covenant and ban those who break it, but the offending parties can return in mere moments after creating a new Second Life account (with a different name, but with a near equal ability to harass). This basic anonymity and powerlessness to prevent griefers is what typifies most online communities. The sole effective form of resistance against such griefing, flaming, trolling or phishing is to ignore the offending parties until they go away, or to incorporate them into one's online narrative. The establishment of law in virtual reality is unlikely, except where existing face-to-face world law (such as the Digital Mil-

lennium Copyright Act) may impede on behavior in the virtual world.

This intrinsic lack of a coherent social contract beyond custom can be taken as a symptom of a much larger movement involving a greater crisis of metaphysics. In connecting virtual reality to modern philosophy, Colin Beardon argues that the artifice of virtual realities and simulation are such a symptom:

The idea that modern philosophy is in crisis is not new. Some postmodernists express this by saying that we are at the end of the project that began with the Enlightenment (Dews, 1989). Laufer (1991) has shown how philosophy has moved through three stages since the Enlightenment: the first (from 1790 until 1890) was dominated by Newtonian science and Kantian philosophy; the second (from 1890 until 1945) was dominated by Comtean positivism and what we would call "modernism"; and the third (the period since 1945) is the period of deepening philosophical confusion and the emerging concept of the "artificial".

If this analysis is correct, then the emergence of virtual reality at this point in time is a reflection, not just of technical, economic and political developments (which are of course also very important) but of the fact that our traditional philosophical system is now collapsing at its most central point - metaphysics. Our concern with the ethics of virtual reality is therefore doubly difficult. Ethics has been severely attacked and has been in a state of confusion for at least fifty years (Ayer, 1936), and virtual reality is a reflection of deep philosophical confusion (Beardon, 1992, p. 4).

Though he doesn't name it, Beardon eludes to the postmodern condition of Lyotard (1979) and many others who might claim that in the new era, one of pastiche and simulation, the interactive virtual world experience is the end of the grand narrative of broadcast dominance and the beginning of something new. The events and technologies that shape what is new – virtual reality in this case – do in fact reveal cracks in the façade of traditional ethics. Those cracks may eventually be filled in by some future Locke or Rousseau, but for now they represent holes in the fabric of our socially constructed virtual worlds.

If it is the case that the simulations created in virtual realities reflect a shift to a postmodern era, we can expect that traditional and modern modes of ethics involving enlightened self-interest, individual responsibility, ends and means, and arguments from moral development to be less than apt for describing and anticipating the actions and reactions of free agents in Second Life. In fact, even free agency comes under scrutiny in postmodernism. The case of the virtual chicken wars in Second Life seems to confirm these expectations of Beardon (1992). The crisis of philosophy is intensified as new, postmodern virtual realities bear down upon our old, modern face-to-face reality. This crisis leads to one of Beardon's final questions, "What is the nature of the responsibilities one has when offering a new version of reality?" In Second Life, and increasingly in cyberculture generally, this crisis is realized through the modality of the construction that people collectively create as they compete for resources in virtual worlds, share social bonds, create communities and factions, and work out amongst themselves what it means to kill a virtual chicken.

REFERENCES

Anderson, E. (1999). *Code of the street: Decency, violence and the moral life of the inner city.* New York, NY: W.W. Norton & Co.

Au, W. J. (2008). *The making of Second Life: Notes from the new world.* New York, NY: HarperCollins.

Axelrod, R. (1984). *The evolution of cooperation.* New York, NY: Basic Books.

Beardon, C. (1992). The ethics of virtual reality. *Intelligent Tutoring Media, 3,* 23–28.

Boellstorff, T. (2008). *Coming of age in Second Life: An anthropologist explores the virtual human.* Princeton, NJ: Princeton University Press.

Brey, P. (1999). The ethics of representation and action in virtual reality. *Ethics and Information Technology, 1,* 5–14. doi:10.1023/A:1010069907461

Brey, P. (2008). Virtual reality and computer simulation. In Himma, K., & Tavani, H. (Eds.), *Handbook of information and computer ethics.* New York, NY: John Wiley & Sons. doi:10.1002/9780470281819.ch15

Carhart, R. (2010). Pacifism and virtue ethics. *The Lyceum, 11*(2).

Castronova, E. (2007). *Exodus to the virtual world: How online fun is changing reality.* New York, NY: Palgrave Macmillan.

Davison, P. (2009). *Sion chicken.* Retrieved from http://www.nextnature.net/2009/12/sion-chicken

Davison, P. (n. d.). *On chickens: Sion and Second Life* [Web log]. Retrieved from http://00pd.com/post.php?id=2

Dreamscape, N. (2010). *The saga of Sion: SL chicken blog.* Retrieved from http://slchickenblog.wordpress.com/2010/04/01/the-saga-of-sion/

Ferguson, C. J. (2010). The modern hunter-gatherer hunts aliens and gathers power-ups: The evolutionary appeal of violent video games and how they can be beneficial. In Koch, N. (Ed.), *Evolutionary psychology and information systems research* (pp. 329–342). New York, NY: Springer. doi:10.1007/978-1-4419-6139-6_15

Ferguson, C. J., & Rueda, S. M. (2010). The hitman study: Violent video game exposure effects on aggressive behavior, hostile feelings and depression. *European Psychologist, 15*(2), 99–108. doi:10.1027/1016-9040/a000010

Ford, P. (2001). A further analysis of the ethics of representation in virtual reality: Multi-user environments. *Ethics and Information Technology, 3*, 113–121. doi:10.1023/A:1011846009390

Gurian-Sherman, D. (2008). *CAFOs uncovered: The untold costs of confined animal feeding operations.* Retrieved from http://www.ucsusa.org/food_and_environment/sustainable_food/cafos-uncovered.html

Hardin, G. (1968). The tragedy of the commons. *Science, 162*, 1243–1248. doi:10.1126/science.162.3859.1243

Kellner, D. (n. d.). *The media and social problems.* Retrieved from http://gseis.ucla.edu/faculty/kellner/essays/medsocialproblems.pdf

Kitchin, R., & Dodge, M. (2006). Software and the mundane management of air travel. *First Monday, 7.*

Linden Lab. (2009). *XML statistics.* Retrieved from http://secondlife.com/xmlhttp/secondlife.php

Lyotard, J.-F. (1979). *The postmodern condition: A report on knowledge.* Minneapolis, MN: University of Minnesota Press.

MacKinnon, R. (1997). Virtual rape. *Journal of Computer-Mediated Communication, 2*(4).

Malaby, T. (2009). *Making virtual worlds: Linden Lab and Second Life.* Ithaca, NY: Cornell University Press.

Meadows, M. S. (2008). *I, Avatar: The culture and consequences of having a Second Life.* Berkeley, CA: New Riders.

Mistral, P. (2009, April 7). *Bio-warfare attacks on second life chicken farms.* Retrieved from http://alphavilleherald.com/2009/07/biowarfare-attack-on-second-life-chicken-farms.html

Moseley, A. (2009). *Just war theory.* Retrieved from http://www.iep.utm.edu/justwar

Nino, T. (2006). *Under the grid - Other causes of lag.* Retrieved from http://www.secondlifeinsider.com/2006/12/26/under-the-grid-other-causes-of-lag

Reymers, K. (2004). Communitarianism on the internet: An ethnographic analysis of the Usenet newsgroup tpy2k, 1996-2004 (Doctoral dissertation, Morrisville State College). *WorldCat Dissertations* (OCLC: 57415279).

Sagan, C. (1998). *Billions and billions: Thoughts on life and death at the brink of the millennium.* New York, NY: Ballantine Books.

Second Life Wiki. (2010). *Griefing.* Retrieved from http://wiki.secondlife.com/wiki/Griefing

Vitanza, V. (2005). *Cyberreader* (Abridged ed.). New York, NY: Pearson Education.

Weinberger, D. (2002). *Small pieces loosely joined: A unified theory of the web.* New York, NY: Perseus Publishing.

Young, J. (2010). *Woodbury U. banned from Second Life, again.* Retrieved from http://chronicle.com/blogs/wiredcampus/woodbury-u-banned-from-second-life-again/23352

This work was previously published in the International Journal of Technoethics, Volume 2, Issue 3, edited by Rocci Luppicini, pp. 1-22, copyright 2011 by IGI Publishing (an imprint of IGI Global).

Section 3
Morality and Techno–Mediated Violence

Chapter 12

Structural and Technology–Mediated Violence:
Profiling and the Urgent Need of New Tutelary Technoknowledge

Lorenzo Magnani
University of Pavia, Italy, & Sun Yat-sen University, China

ABSTRACT

A kind of common prejudice is the one that tends to assign the attribute "violent" only to physical and possibly bloody acts – homicides, for example – or physical injuries; but linguistic, structural, and other various aspects of violence – also embedded in artifacts – have to be taken into account. The paper will deal with the so-called "technology-mediated violence" taking advantage of the illustration of the case of profiling. If production of knowledge is important and central, this is not always welcome and so people have to acknowledge that the motto introduced in the book Morality in a Technological World (Magnani, 2007) knowledge as a duty has various limitations. Indeed, a warning has to be formulated regarding the problem of identity and cyberprivacy. The author contends that when too much knowledge about people is incorporated in external artificial things, human beings' "visibility" can become excessive and dangerous. Two aims are in front of people to counteract this kind of technological violence, which also jeopardizes Rechtsstaat and constitutional democracies: preserving people against the various forms of circulation of knowledge about them and building new suitable "technoknowledge" (also to originate new "embodied" legal institutions) to reach this protective result.

DOI: 10.4018/978-1-4666-2931-8.ch012

1. INDIVIDUAL AND STRUCTURAL/ TECHNOLOGY-MEDIATED VIOLENCE

In my recent book *Understanding Violence. The Intertwining of Morality, Religion, and Violence: A Philosophical Stance* (Magnani, 2011) I have stressed the attention to a kind of common prejudice which is the one that tends to assign the attribute "violent" only to physical and possibly bloody acts – homicides, for example – or physical injuries; but linguistic, structural, and other various aspects of violence – also embedded in artifacts – have to be taken into account. However, even homicide itself is more complicated than expected, in fact recent research on the legal framework of homicide and on its biological roots shows many puzzling aspects. Brookman (2005) contends that it is not appropriate to think of homicide just as a kind of criminal or violent behavior, because the phenomenon is complex and socially constructed for the most part. Those who kill do so for very different reasons and under different circumstances, and some structurally originated killing is often hidden and disregarded, even from the legal point of view, such as in the case of multiple homicides perpetrated by artificial technostructures such as corporations. He illustrates various explanations of homicide: psychoanalytic and clinical approaches, accounts from evolutionary psychology or social and cognitive psychology frameworks, sociological and legal aspects and the biological roots of killing.

The extension of the meaning of the word violence brings up the need to reconsider our concepts of safety, ethics, morality, technology, law, and justice. May be the philosophy-violence connection (with the help of other related disciplines, both scientific and cognitive) will generate novel ideas and suggestions. Of course people and intellectuals are clearly aware that drug, alcohol, revenge, frustration, and mobbing behaviors are related to violent events, and so many aspects of linguistic and structural violence are acknowledged, but this acknowledgment is almost always fugitive

and superFcial. In sum, we see the spectacle of violence everywhere, but, so to say, the violence always out there, involves *other* human beings and we can stay distant from the theme of violence by adopting a simple and familiar – but practically empty – view of it. Indeed, it is implicit that, if we know for certain that we are the possible target of violent behavior, we *a priori* think of ourselves, spontaneously, as exempt from any contamination, in our supposed purity and immunity. We just hope not to become victims, but this is usually considered just a question of good luck, and, above all, we are not interested (even if we are, or think of ourselves as, philosophers) in analyzing the possible existence and character of *our* own (more or less) violent behavior. It is better not "to problematise our confidence in, or familiarity with, 'what violence is'" (Catley, 2003).

Familiarity with violence involves a trivial and simple *sense* of violence as interpersonal, physical, and illegitimate, which can be clearly seen in the case of workplace violence as a:

[...] Deviant set of behaviours to be eradicated through a series of familiar strategic interventions. Workplace violence becomes reduced to a technically rational set of "procedural issues about workforce selection, early detection of potential troublemakers, adequacy of liability insurance, risk management and effective exclusion of potential as well as actual offenders" [...]. And it is this familiarity that erases questions about the constitution of violence that might lead us to ask other critical questions about the organisation of work and the work organisation beyond individual pathography. Arguably, the familiarity of violence as interpersonal and illegitimate has encouraged explanations of workplace violence to focus on the individual and the eradication of such deplorable behaviour. In these explanations, the focal point has tended around the exposition of the personality and motivations of the "perpetrator", typically with a view to profiling the violent individual (Catley, 2003).

The passage is clear and eloquent. Violence tends to become an easy matter for psychologists, sociologists, etc., who make – sometimes vane – efforts to depict the structural aspects of situations and artifactual institutions and how they change (families, markets, nations, prisons, workplaces, races, genders, classes, and so on). However, what really matters is to describe how individuals respond in a dysfunctional way because of their personality, motivation, and (lack of) rationality. The full violent potential of structures, artifacts, institutions, cultures, and ideologies is marginalized and disregarded. For example we rarely acknowledge the fact that contemporary humans can – with scarce awareness – damage the future as well as the present causing future populations to be our "unborn victims", thanks to the consumption of energy, the pollution of air and water, or the demolition of cultural traditions. It is only the "abnormal" individual that is seen as prone to suddenly perpetrating violence, and only (right wing?) repressive policies that are considered extinguishers of the violent anomaly.

To summarize: it is the individual that is violent, but the inherent violence of structures (where for example injustice and relations of dominations are common, and violent attitudes are constantly distributed and promoted) is dissimulated or seen as a "good" violence, in the sense that it is ineluctable, obvious, and thus, acceptable. Every day in Italy at least three workers die in their workplace, and this seems natural and ineluctable. The potential harmfulness of organizational practices and workplace management leads to violent events and are rarely seen as causing them. When they are seen as violence-promoters, this consideration is a kind of "external" description which exhausts itself thanks to a huge quantity of hypocritical rhetoric and celebration. Again, we tend to focus on the familiar idea of violence as basically physical and we ignore or belittle other aspects of violence which neither leave their mark on the body nor relate back to it.

In his recent book Žižek (2009) reveals "the hypocrisy of those who, while combating *subjective violence*, commit *systemic violence* that generates the very phenomena they abhor" (p. 174). Language and technological media in modern life continually show a spectacle of brutality accompanied by a kind of urgent (but "empty") "SOS violence", which de facto systematically serves to mask and to obliterate the symbolic (embedded in language) and structural-systemic violence (often embedded in artifacts) like that of capitalism, racism, patriarchy, etc. Žižek provides the following, amazing example of the intertwining between charity and social reproduction of globalized capitalism: "Just this kind of pseudo urgency was exploited by Starbucks a couple of year ago when, at store entrances, posters greeting customers pointed out that almost half of the chain's proFIts went into health-care for the children of Guatemala, the source of their coffee, the inference being that with every cup you drink, you safe a child's life" (p. 5). The instantiation of this hypocrisy is clear when analyzing the philanthropic attitude of what Žižek calls the *democratic liberal communist capitalists*, who absolve themselves as systemic violent monopolizers and ruthless pursuers of proFIt thanks to an individual policy of charity and of social and humanitarian responsibility and transcendental gratitude. Žižek also warns us that a cold analysis of violence typical of the media and human sciences somehow reproduces and participates in its horror; also empathising with the victims, the "overpowering horror of violent acts [...] inexorably functions as a lure which prevents us from thinking" (p. 3).

Another aspect of structural violence can be traced back from the analysis of the commodification of sexuality in our rich capitalist societies. The urgency of sex tends to violate the possibility of love and the encounter with the Other:

Houellebecq (2006) depicts the morning-after of the Sexual Revolution, the sterility of a universe dominated by a superego injunction to enjoy [...]

sex is an absolute necessity [...] so love cannot flourish without sex; simultaneously, however, love is impossible precisely because of sex: sex, which "proliferates as the epitome of late capitalism's dominance, has permanently stained human relationships as inevitable reproductions of the dehumanizing nature of liberal society; it has essentially ruined love". Sex is thus, to put it in Derridean terms, simultaneously the condition of the possibility and of the impossibility of love (p. 9).

To make another example, harassing, bullying, and mobbing are typical violent behaviors that usually involve only language instead of direct physical injury: that is, blood is not usually shed. Bullying can also be ascribed to such a category, but in a somewhat minor dimension since bullies often resort to both physical and linguistic threats. Of course, the sophisticated intellectual and philosopher – but not the average more or less cultured person – says "certainly, mobbing involves violence!", but this violence is *de facto* considered secondary, at most it is only worth considering as "real" violence when it is in support of actual events of physical violence. It is well-known that people tend to tell themselves that the mobbing acts they perform are just the same as cases of innocuous gossiping, implementing a process of dissimulation and self-deception.

2. VICTIMS AND TECHNOLOGICAL MEDIA

A strategy that aims at weakening serious attention to violence is often exploited by the media (and is common in everyday people). The mass media very often label episodes of violence, for example terrorism, as something pathological: "the terrorists kill for the sheer pleasure of killing and dying, the joy of sadism and suicide", "they constitute a pathological mass movement". Following Wieviorka (2009, p. 77) we can say that these considerations only serve to obscure any

hidden causes for the "horrifying attacks" and represent a form of medicalization or *psychiatrization* of a behavior that comes to be dismissed as simply insane.[1]

Another strategy that aims at playing down the problem of violence consists in labeling some violent episodes as "new", in a kind of "ahistorical individual pathology" where organizations and collectives are considered simply "affected" by violence, and certainly not part of it. Here is an eloquent example: "To represent workplace violence as a 'new' event serves to deny and evacuate any historical relationship between violence and work. This is not to suggest that the 'object' of workplace violence has remained constant throughout time, but to inoculate against any imagined sense that violence has not been present in the organisation of work" (Catley, 2003). We know that this last statement is patently false just like in the case of bullying in schools – often marked as "new". Catley nicely concludes: "Consequently, the result of accounts of workplace violence that distance themselves from viewing workplace violence as historical and sociological in favor of an ahistorical individual pathology is a continuance of the representation of the organization as being *affected* by violence rather than possibly *contributing* to, constructing or reproducing violence" (ibid.)[2]

A meaningful recent analysis of the so-called "deadly consensus", which is at the basis of UK workplace safety is provided by Tombs and White (2009). The authors document the regulatory degradation as a result of its compatibility with neoliberal economic strategy. The degradation is justified by political and economical reasons, which rapidly become the *moral* grounds at the level of the individuals' legitimacy in disregarding safety. A subsequent analysis of empirical trends within safety enforcement reveals a virtual collapse of formal enforcement, as political and resource pressures have taken their toll on the regulatory authority. Furthermore, an increasing impunity with which employers can actually kill and injure (even if they do not *mean* to, as the end of their action) is also observed.

In sum, we can say that in a common sense (as it emerges, for instance, in the mass media) the representation of violence is really partial and rudimental. In our current technological global-ized world media and violence live in a kind of symbiosis (Wieviorka, 2009, p. 68) – just think about terrorism – this presents two aspects: the first is that victims are continuously exhibited and described, and this is positive, because they are not simply disregarded as collateral, like in the past, but are increasingly represented in the violence that they have suffered and which has damaged their moral and/or physical integrity. One can at least think "that the thesis that violence is func-tional loses its validity" (p. 160) since an effective depiction of violence halts the belief that violence has a (more or less) just reason and a reasonable scope. The second aspect is concerned with what the victims are like, so to speak, the "words" the journalists depict them with, and the fact that the subjective side of the victims is rarely presented, favoring empathy and the sense of the audience's psychological incorporation of the actuality of violence. Thus a culture of fear is favored but, at the same time, violence may be unintentionally promoted and reinforced.

In intellectual settings, with the exception of literature and arts, and of other – so to say – more or less "technical" disciplines (theory of law, religious studies, some branches of psychology, sociology, psychoanalysis, paleoanthropology, evolutionary psychology, evolutionary biol-ogy, and criminology), the lack of curiosity for violence is widespread and violence is basically *undertheorized*. In the case of "technical" disci-plines an interest in violence is very recent. As interestingly noted by Gusterson (2007), when – starting from the 1980s – "new wars" and local processes of militarization with high civil-ian casualty rates came about in Africa, Central America, the former Eastern block, and in South Asia, anthropologists showed a greater interest

in various kinds of violence: a kind of "cultural turn", Gusterson says. They started to analyze terror, torture, death squads, ethnic cleansing, gue-rilla movements, fear as a way of life, permanent war economy, the problem of military people as "supercitizens", war tourism, the memory work[3] inherent in making war and peace, and also the problem of nuclear weapons and American militarism. They also stressed the role of current military apologetics in the USA, which favored and still favors degraded popular culture saturated with racial stereotypes, aestheticized destruction, and images of violent hyper-masculinity. In this perspective of a new widespread militarism (as a source of so much suffering in the world) it can be said that the "war on terror" has provided the occasion for a new intellectual commitment to the theme of violence debating the merits of military anthropology versus critical ethnography of the military. However, we have to say that all too often, especially in psychology and sociology, the analysis of violent behavior is exhausted by the description of the more or less vague characters of various psychopathological individuals or groups, where empirical data and practical needs are the central aim, but deeper structural and theoreti-cal aspects are disregarded, while they could be crucial to a fuller understanding and subsequently to hypothesize long-term solutions.

3. THE URGENT NEED FOR A PHILOSOPHY OF VIOLENCE

In the paper "Philosophy – The luxurious supple-ment of violence", Catley (2003) says

In many of the growing number of accounts of workplace violence there exists a particular sense of certainty; a certain confidence in what violence "really" is. With these accounts, philosophy ap-pears unnecessary – and even luxurious – in the face of the obvious and bloody reality of workplace violence. [...] one outcome is an absence of a sense of curiosity about the concept of violence in many typical commentaries on workplace violence. Through a turn to philosophy it is suggested, we might possibly enquire into other senses of violence that may otherwise be erased. However, weary of simply "adding" philosophy, this paper also begins to sketch out some possible consequences a philosophy-violence connection might have for doing things with philosophy and organisation.

Workplace violence is only a marginal aspect of the widespread violence all over the world but I agree with Catley that philosophers usually do not consider this theme; he says they lack a "curiosity" about violence.

Similarly, in a recent book about the relationships between phenomenology and violence, which mainly addresses the problem of war, Dodd (2009, p. 2) says:

It might strike one as a strange point of departure for a reflection on the obscurity of violence to raise the question of its properly philosophical character. Does not virtually any obscurity, not to mention profound questions of human existence, by definition invite philosophical reflection? This already begs the question. For perhaps it is instead the case that the problems of violence are not, in the end, all that obscure, even if they may be difficult to understand.

Furthermore, the existence of a conviction about a kind of stupidity of violence is illustrated:

In its barest form, the stupidity of violence principle states that violence is and can be only a mere means. As a mere means, pure violence remains trapped, according to the principle, within the confines of a very narrow dimension of reality defined by the application of means. Violence as such is thus blind; when taken for itself it is ultimately without direction. The practices of violence, however traumatic and extreme, fade into indefinite superficialities unless supported

by a meaningful cause or end. To be sure, the stupidity of violence does not detract from the seriousness of the consequences of violence, the damage it inflicts – the shredded flesh, the famine and disease, the pain both physical and spiritual, and the shocking number of corpses that it leaves in its wake. [...] violence is stupid, in that it involves nothing more significant than what can be captured and organized in a technical fashion. [...] Likewise, the morality of war does not find, according to this principle, any chance of being expressed, so long as we only follow along the passage of the event of violence that has been so tightly reduced to this line of cause to effect (Dodd, 2009, pp. 11-12).

The lack of curiosity is accompanied by a sense of certainty about "what violence is", Catley says, and I agree with him. There is a sense of certainty that suppresses its historical variability and that only leaves a hypocritical sense of it as a merely deplorable outbreak on an otherwise peaceful scene. "Philosophy, it would seem, appears unnecessary – and even luxurious – in the face of the obvious and bloody reality of workplace violence" (Dodd, 2009, pp. 11-12).

In sum, we do not only need philosophers of science or moral philosophers but also "philosophers of violence"

I strongly believe that the problem of violence is not less important (and worth of philosophical attention) than the problem of morality: furthermore, they cannot be separated one from the other. Indeed, I devoted the entire book *Understanding Violence* to illustrate this complex and yet omni-pervasive intertwining between morality and violence.[4]

I strongly agree with Catley that philosophy can become a "catalyst" for "energizing" our thinking about the multifaceted aspects of violence. However, we have to remember that of course "philosophy in the forms of its philosophical institutions and practices may too exert its own forms of violence. Rather than standing in opposition to violence, philosophy may well reproduce that which it seeks to condemn. The faithful love of ideas harbors an affair with violence" (Catley, 2003). It is obvious that philosophy can afford to explore the problem of violence only if choosing a side to stand on; philosophy has to be aware that, given the inherent violent nature of knowledge, language, and technology, it is not immune to it, and that every option chosen has the chance of producing violent effects. For example "Here one might think of the institutionalization of philosophy, where what is included and excluded in the canon is decided. Or the silencing effects of truth, where 'truth' is mobilized to close down dissenting positions. Philosophy becomes complicit in the sorts of domination enacted when the arbitrary takes on the status of the natural to close down alternative ways of knowing, courses of action, and subject positions in order to preserve dominant relationships of power" (Catley, 2003): to make an example, how can we forget that often the presentation of a sophisticated philosophical argument about "creation" is experienced and "perceived" as a tremendous violence by some religious and/or uncultured people?

As I have already said in the preface to my book *Understanding Violence* (Magnani, 2011), philosophy has always displayed a tendency to "dishonor violence" by disregarding it. Step by step the philosophical intellectual commitment to violence has become more and more objective and abstract and is being transformed into an autonomous subject of philosophical reflection. Now we can say that philosophy has learnt to *respect* violence and – so to speak – its "moral dignity" as a philosophical topic.

An initial theoretical step is to acknowledge that, often, structural technological violence is seen as *morally* legitimate: we have to immediately note that when parents, policemen, teachers, and other agents inflict physical or invisible violence on the basis of *legal* and/or *moral reasons*, these reasons do not cancel the violence perpetrated and violence does not have to be condoned. In

chapter three of my book (Magnani, 2011) I have described how a kind of common psychological mechanism called "embubblement" is at the basis of rendering violence invisible and condoning it. Human beings are prisoners of what I call *moral bubbles*, which systematically disguise their violence to themselves: this concept is also of help in analyzing and explaining why so many kinds of violence in our world today are treated as if they were something else. Such a structural violence, which is in various ways legitimated, leads to the central core of my commitment to the analysis of various relationships between morality and violence.

It seems to me there is an increasing interest in the intertwining between morality and violence in concrete research concerning human behavior. For example Smith Holt and colleagues (Smith Holt *et al.*, 2009, p. 4) contend that there is a common thread that emerges in various recent studies about killing: "[...] something about moral code, a religious or ethical belief enmeshed within a cultural context, determines one's stance on various types of killing and, indeed, on inhibitors to killing".

A philosopher I recently found to explicitly (seemingly, at least) acknowledge the strict link between morality and violence is Allan Bäck. He first of all contends that aggression is not necessarily a destructive or morally bad act:

In contrast to aggression, I shall thus take "violence" in the basic sense to signify a certain sort of aggression, namely an aggressive activity to which judgments, of being good or bad, apply. In many moral and legal theories, such judgments require, among other things, considering the conscious of the moral agent(s). According to such views, if you are aware of what you are doing, will to do so, and could do otherwise, you are morally responsible for those acts. You might, then, be able to do something without being morally responsible, if you did not will the act, or were not aware of what you are doing, or were in a mental state of extreme duress of emotion.

At any rate, I shall suppose that normal cases of human intentional action are subject to moral judgment and are chosen (Bäck, 2009, p. 369).

From the perspective of the perpetrator an action is violent when it is aggressive, it is "chosen" and when the victim would not want to suffer that harm (i.e., "the action unjustly violates the rights of the victim, where 'rights' signifies the morally ideal set of entitlements that the recipient of the action – typically a person – ought to have"). In such a way violence "contains" a moral component that "is associated with choosing to engage in actions that harm another person and attempting to force that person to act as you want". In this basic sense

1. Violence, from the perspective of the perpetrator, is the fruit of moral deliberation (always endowed with its "idealistic" and "pure" halo - Baumeister, 1997), but

2. Usually, when we judge a human act to be violent we mean something *pejorative*, we need to add the condition that the selected choice is morally wrong, the agent ought not have made that choice

3. It is clear that this judgment of violence is of second order and derives from a moral judgment that usually does not belong to the perpetrator; indeed I strongly think the case of a perpetrator that consciously performs violence, by disobeying to *his own* moral conviction, is a rare case of a real Mephistophelian behavior. It is clear that the second meaning of violence tends to *obliterate* the first one: we always forget that violent actions are usually performed on the basis of moral deliberations.

Taking advantage of a both philosophical and interdisciplinary outlook, my aim is to present a fuller picture of what I call *technology-mediated violence*. I will try to avoid reproducing the coldly intellectual/reductionist atmosphere that certain disciplines (such as the so-called "ethics of technology") evoke when dealing with technologies. These approaches usefully stress how contemporary technologies, and the related knowledge, are often inadequate or even dangerous from an ethical point of view, but the "gentle" neutrality exhibited by the intellectual analysis in the classical terms of "moral philosophy" ends up in a systematic lack of attention to the violent effects that are actually perpetrated – presently and actively – by certain technologies: this kind of analysis results in violence remaining "covered". In the following section, I will explore issues relating to a wide range of scenarios – including ethical, juridical, and cognitive ones – related to the problem of violence perpetrated through technological artifacts, at the expense of privacy, identity, *Rechtsstaad*, and constitutional democracy. To anticipate the content taking advantage of a kind of motto, I can say: "when technologies powerfully distribute habits and/or moral norms, they enact a parallel distribution of violence causing harm and wounds".

4. INTRODUCING TECHNOVIOLENCE: THE CASE OF PROFILING

Even though producing knowledge is an important goal, as I have stressed in my book *Morality in a Technological World. Knowledge as Duty* (Magnani, 2007), in actually doing so in certain circumstances is not always a welcome prospect: we must be cautious when dealing with issues of identity and cyberprivacy, where the excessive production and dissemination of knowledge can be dangerous, at least in the perspective of our current morality of freedom and of the politics of constitutional democracies.

Recent cognitive science research[5] contends that our consciousness is formed in part by representations about ourselves that can be stored outside the body (narratives that authenticate our identity, various kinds of data), but much of consciousness also consists of internal representations about ourselves (a representation of the body and a narrative that constructs our identity) to which other people do not have access. We rarely transfer this second type of representation onto external mediators, as we might with written narratives, concrete drawings, computational data, and other configurations. Such internal representations can be also considered a kind of information property (moral capital) that cannot be disseminated – even to spouses, since the need for privacy exists even between husband and wife. There are various reasons for withholding such representations: to avoid harm of some kind, for example, or to preserve intimacy by sharing one's secrets only in love or friendship. But perhaps the most deleterious effect of the loss of privacy is its impact on free choice. The right to privacy is also related to the respect for others' agency, for their status as "choosers"; indeed, to respect people is also to concede that it is necessary to take into account how one's own decisions may affect the enterprises of others.

Distributing Violence through Technological Profiling

First of all, technological profiling can be seen as the latest sophisticated powerful tool for intensively scapegoating a person or a group of persons: the scapegoat is a typical moral/religious mechanism of ancient groups and societies, where a paroxysm of violence would tend to focus on an arbitrary victim and a unanimous antipathy generated by "mimetic desire" (and the related envy) would grow against him. Following Girard (1977, 1986) we can say that in the case of ancient social groups the extreme brutal elimination of the victim would reduce the appetite for violence that had possessed everyone just a moment before, leaving the group suddenly appeased and calm, thus achieving equilibrium in the related social organization (a sacrifice-oriented social organiza-

tion may be repugnant to us but is no less "social" just because it is rudimentary violent).

When we lose ownership of our actions, we lose responsibility for them, given the fact that collective moral behavior guarantees more ownership of our destinies and involves the right to own our own destinies and to choose in our own way. When too much of our identity and data is externalized in distant, objective "things," for example through automated profiling (knowledge discovery in database, KDD of which data mining,[6] DM, is a part), the possibility of being reified in those externalized data increases, and our dignity/autonomy subsequently decreases, and refined *discriminations* are also favored. Indeed, the resulting profiling is often applied to a person because her data match the profile, so diagnosing her life styles, preferences, dispositions to risk, earning capacities and medical aspects. As I have anticipated, I think that, in principle, profiling furnishes the latest most sophisticated hidden way of socially intensively scapegoating a person (or a group of persons), for example through continuous unfair discrimination and masking: "[…] in the case of KDD the hypothesis emerges in the process of data mining and is tested on the population rather than a sample. […] when trivial information turns out to correlate with sensitive information, an insurance company or an employer may use the trivial information to violently exclude a person without this being evident" (Hildebrandt, 2008, p. 55).[7] It is also because of the opacity of the computational processes that leave us completely unaware, that some of our interests can be at stake.

The ostracism and stigmatization of individuals and minorities could increase and generate an explicit loss of freedom. Moreover, less choice may exist because of an implicit loss of freedom: for example, technology may one day lead a person to say "I am no longer free to simply drive to X location *without leaving a record*". Less privacy could also mean the loss of protection against insults that "slight the individual's ownership of himself" (as happens in the case of prisoners or of slaves): consequently, people will say *mine* with less authority and *yours* with less respect. Indeed when we lose ownership of our actions, we lose responsibility for them, given the fact that morality involves the right to decently possess our own destinies and to choose in our own way. Finally, the risk of infantilizing people increases, as does the likelihood of rendering them conformist and conventional, and they are, as a result, faced with a greater potential for being oppressed. Given these possibilities, it is apparent that privacy is politically linked in many ways to the concepts of liberalism and democracy (Reiman, 1995, pp. 35-44).

By protecting privacy, however, we protect people's ability to develop and realize projects in their own way. Even if unlicensed scrutiny does not cause any direct damage, it can be an intrusive, dehumanizing force that fosters resentment in those who are subjected to it. We must be able to ascribe agency to people – actual or potential – in order to conceive of them as fully human. It is only when our freedom is respected and we are guaranteed the chance to assume responsibility that we can control our lives and obtain what we deserve.

Guessing General Hypotheses from Private Data

Cyber-warfare, cyber-terrorism, identity theft, attacks on abortion providers, holocaust revisionism, racist hate speech, organized crime, child pornography, and hacktivism – a blend of hacking and activism against companies through their web sites – are very common violent cases on the Internet (Van den Hoven, 2000, p. 127). As a virtually uncontrollable medium that cuts across different sovereign countries, cyberspace has proved to be fertile ground for new moral problems and new opportunities for violence and wrongdoing, mainly because it has created new moral "ontologies" that affect human behavior by generating conflicts and controversies.

Beyond supports of paper, telephone, and media, many human interactions are strongly mediated and potentially recorded through the Internet. At present, identity acquires a wide meaning: it must be considered in a broad sense: the amount of data about us as individuals is enormous, and it is all stored in external things/means. This repository of electronic data for every individual human, at least those in rich countries, can be described as an external "data shadow" that, together with the biological body, forms a kind of cyborg that identifies us or potentially identifies us. The expression "data shadow" was coined in Sweden in the early 1970s: it refers to a true, impartial image of a person from a given perspective (Gotterbarn, 2000, p. 216).

More than in the past, much of the information about human beings is now simulated, duplicated, and replaced in an external environment. The need of the modern state for the registration of its citizens – originally motivated by tax obligations and by subscription in the national army – appears excessive. Moreover, corporate and global governance demand new sophisticated means for identification:

Supposedly justified by an appeal to security threats, fraud and abuse, citizens are screened, located, detected and their data stored, aggregated and analysed. At the same time potential customers are profiled to detect their habits and preferences in order to provide for targeted services. Both industry and the European Commission are investing huge sums of money into what they call Ambient Intelligence and the creation of an "Internet of Things". Such intelligent networked environments will entirely depend on real time monitoring and real time profiling, resulting in real time adaptation of the environment (Hildebrandt, 2008, p. 55).

Indeed, AmI (Ambient Intelligence) is supposed to turn the offline world online.[8] We will not have to provide a deliberate input, because we are read by the environment that monitors our behavior, diminishing human intervention as far as possible.

The activity of profiling is a massive *abductive/inductive* activity of new knowledge production, basically performed by machines, and not by organisms.[9] These machines are computational programs – which take advantage of various abductive[10] and inductive inferential procedures – "trained" to recover unexpected correlations in masses of data aggregated in large databases and aimed at diagnosing individual profiles. These machines not simply permit the making of queries in databases, summing up the attributes of predefined categories, but provide an abductive or inductive generation of suitable knowledge about individuals, that is extracted from data. It is extremely important to note that we cannot reflect upon the way that profiling impacts our actions because we have no access to the way they are produced and used (Hildebrandt, 2008, p. 58).

Of course abducing/inducing from private data through profiling is fruitful. For example "Business enterprise is less interested in a foolproof registration of the inhabitants of a territory. Its focus is on acquiring relevant data about as many customers and potential customers as possible as part of their marketing and sales strategies" (Hildebrandt, 2008, p. 56). Customers have to be persuaded and to this aim they are interested in a refined type of categorization more than in the identification of a particular customer.

Technological Frameworks Jeopardizing Constitutional Democracy and *Rechtsstaat*

Hildebrandt usefully contends that there is a deep impact of profiling "on human identity in constitutional democracy". In this perspective we can say that the protection of democracy is equivalent to the need of promoting what is sometimes called cyberdemocracy. This impact jeopardizes the so-called *rule of law*, which instead protects human identity as fundamentally *under*determined: "This

requires us to foster a legal-political framework that both produces and sustains a citizen's freedom to act (positive freedom) that is the hallmark of political self-determination, and a citizen's freedom from unreasonable constraints (negative freedom) that is the hallmark of liberal democracy" (Hildebrandt, 2008, p. 57). Rule of law, Hildebrandt reminds us, relates to a *text* depending on writing ("an affordance of the printing press"), and so it is mediated by a very simple and ancient artifact: it is well-known that rule of law aims both at protecting human rights and at limiting governments. Profiling challenges the interplay between negative and positive freedom, which, to be preserved, need that transparency which solely can provide those knowledge contents which renders individuals' actions and deliberations based on conscious reflections, autonomously, adequately, and responsibly done. Unfortunately, in presence of the current profiling activities, we do not have any access to knowledge that conditions those choices. The following are some obvious consequences: 1) we think we are alone but we are watched by machines (and this will be more than evident in the case of the possible outcoming effects of AmI – Ambient Intelligence) (Hildebrandt, 2007, 2009b); 2) we "think we are making private decisions based on a fair idea of what is going on, while in fact we have no clue as to why service providers, insurance companies or government agencies are dealing with us the way they do" (Hildebrandt, 2008, p. 61).

Various technological progresses will create an even worse situation, and not only in the perspective of identity and privacy. What will happen when a silicon chip transponder, once surgically implanted in a human being, can transmit identifying data like one's name and location through a radio signal? Of the many, many possibilities, one of the more benign is that it may render physical credit cards obsolete. And what will be the result when a transponder directly linked with neural fibers in, say, one's arms, can be used not only to generate remotely controlled movement for people with nerve damage, but also to enhance the strength of an uninjured person? For the blind, extrasensory input will merely be compensatory data, but for those with normal sight it will be a powerful "prosthesis" that turns them into "super-cyborgs" with tremendously powerful sensory capabilities. What about the prospect of "hooking up" a nervous system to the Internet? It will revolutionize medicine if it is possible to download electronic signals that can replace disease-causing chemical signals. It is worth remembering that the human nervous system is, by nature, electro-chemical – part electronic, part chemical (Warwick, 2003, p. 135).

When delegating information or tracing identity, human beings have always used external mediators such as pictures, documents, various types of publications, and fingerprints, etc., but those tools were not simultaneously electronic, dynamic, and global, as is the case with the Internet and other technological devices, and at their level it was always possible to delete records, because they were not strongly shared across contexts. Where we exist cybernetically in cyberspace is no longer simple to define, and new technologies of telepresence and ambient intelligence will continue to complicate the issue.

New Moral/Violent Ontologies

In general, this complex new "information being" involves new moral ontologies. We can no longer apply old moral rules and old-fashioned arguments to beings that are both biological (concrete) and virtual, physically situated in a three-dimensional *local* space but potentially "existing" also as "globally omnipresent" information-packets (Schiller, 2000). It is easy to materialize cybernetic aspects of people in three-dimensional local space, even if it is not quite as spectacular as in old *Star Trek* episodes, when Scottie beamed the glimmering forms of Captain Kirk and Mr. Spock from the transporter room to the surface of an alien planet. Human rights, the notion of care, obligation and

duty, have to be updated for these cybernetic human beings. These notions are usually transparent when applied to macro-physical objects in local spatio-temporal coordinates. They suddenly become obscure, however, in cyberspace:

What is an agent? What is an object? What is an integrated product? Is that one thing or two sold together? What is the corpus delicti? If someone sends an e-mail with sexual innuendo, where and when does the sexual harassment take place? Where and when it is read by the envisaged addressee? Or when it was typed? Stored on the server? Or as it was piped through all the countries through which it was routed? And what if it was a forwarded message? (Van den Hoven, 2000, p. 131)

In the era of massive profiling, privacy cannot be reduced to the mere non disclosure of personal data, as a mere private good, because personal data remain out there, hidden in the cybernetic space. Privacy has to be also considered a democratic good: in this perspective citizens' freedom does not have to be silently constrained thanks to the structural/hidden moral (and/or legal) normative effects induced by profiling technologies. The common view about the efficacy of simply "hiding personal data" is mistaken: this does not protect from group profiling and its "candid" effect on our choices.

Moreover, a loss of privacy also implies a loss of security, given the fact that individual identity becomes more vulnerable. I have to stress the fact that a loss of privacy also generates an unequal access to information, which can promote a market failure, like Hildebrandt usefully notes: a market segmentation where companies can gain profit from a specific knowledge and/or unequal bargaining positions, and consent based on ignorance. Indeed, people do not have access to the group profiles that have been inferred abductively or inductively by a mass of data gathered and do not have any idea on how these profiles can impact our chances in life. Hildebrandt nicely concludes: "As for profiling, privacy and security both seem to revolve around the question 'who is in control: citizens or profilers?' But again control is often reduced to hiding or disclosing personal data and this does not cover privacy and security as public values" (Hildebrandt, 2008, p. 63). As I will illustrate below, new profiling technologies, fruit of a gigantic abductive/inductive increase of "knowledge" about human beings, paradoxically threaten to increase our *moral ignorance*. I have previously pointed out that the idea of "knowledge as a duty" is relevant to scientific, social, and political problems, but it is even more important – as I will soon illustrate, cf. the following section – when we are faced with reinterpreting and updating the ethical intelligibility of behaviors and events that can generate dangers. Obviously, information has an extraordinary instrumental value in promoting the ends of self-realization and autonomy, but it can jeopardize or violate various aspects of our individual and social life which, for example, we morally or politically extremely value. For example we value "democracy", and at the same time we clearly see that profiling could threaten it. The question is, do we have an appropriate knowledge able to describe how profiling menaces democracy so that we can also build and activate suitable counterbalances?

In general, the new moral issues of the cyberage are well-known. Some problems, like financial deception and child abuse, were perpetrated for years via traditional media and have become exacerbated by the Internet. Others issues arose as a result of new technology – problems like hacking, cracking, and cooking up computer viruses, or the difficulties created by artificial intelligence artifacts, which are able to reason on their own (Steinhart, 1999). Also needing attention is the proliferation of spam. The utility of spamming clashes with the time spent in data trash and garbage of the recipient: in this sense it can be said that spamming externalizes costs. Another problem related to spam, it is what I called

super-expressed knowledge (Magnani, 2007) - huge, expensive cache of data and information available on the Net and on our computers that is overused, unused, and/or not useful. Interestingly, a survey commissioned some years ago by the U.S. Air Force (AF) estimated that AF had as much as 50% duplication in stored files across the enterprise - something like two petabytes of data, a giant information wasteland whose maintenance cost several hundred million dollars every year (Corrigan & Sprehe, 2010).

Yet another challenge is the huge problem of the digital divide, for while millions of people are active cyborgs, a gulf has opened between them and others who number in the billions. Members of the latter group come from both rich and poor countries, and they have not attained cyborg status. A further split, at least inside the "cyborg community," occurred because of the way technology dictates the speed, ease, and scope of communication; the dominance of certain text features creates an unfair playing field, and the disparity between computer users with the most sophisticated equipment and those with more modest systems indicates that we are far from the ideal of an unmediated Internet. And the list continues: professional abuses, policy vacuums, intellectual property questions, espionage, a reduced role for humans – all these things create further moral tangles. Finally, if we consider the Internet as a global collection of copying machines, like Mike Godwin contends, huge copyright problems arise.[11]

In light of my motto "knowledge as a duty," the Internet and the new computational tools make available many goods that are critical for human welfare by increasing access to information and knowledge regarding people's biological needs: food, drink, shelter, mobility, and sexuality. It also serves their rational aspects by providing information, stimulating imagination, challenging reason, explaining science, and supplying humor, to cite a few examples (Van den Hoven, 2000). If the Internet's power, fecundity, speed, and efficiency in distributing information are universally acknowledged, moral problems immediately arise, related to distortion, deception, inaccuracy of data, copyright, patents, trademarks abuses.[12]

Let us come back to the problem of privacy. We must safeguard freedom of information and, at the same time preserve privacy, the right to private property, freedom of conscience, individual autonomy, and self determination.[13] The problem is, how can traditional laws of our constitutional democracies protect people's privacy, given the fact that profiling technologies – as I will soon explain – cannot be touched by laws as we traditionally intend them? Harm to others through Internet and other computational tools is often simply invisible (or it cannot be tracked) and we do not possess neither knowledge about its ways of occurring nor of course laws able to limit it in a more efficacious way.

Even in the complicated world of cyberspace, however, Mill's principle still establishes the moral threshold, in this case in the sense that only the prevention of harm to others justifies the limitation of freedom (but, how to weigh and balance harms or freedoms?): "The only freedom which deserves the name, is that of pursuing our own good in our own way, so long as we do not attempt to deprive others of theirs, or impede their efforts to obtain it" (Mill, 1966, p. 18). Mill also observes that truth emerges more easily in an environment in which ideas are freely and forcefully discussed. I will address the problem of morally and legally contrasting abuse at the level of privacy in the following sections.

How New Knowledge Can Violate Privacy

In the previous subsection I have explained how new profiling technologies, fruit of a gigantic abductive/inductive increase of "knowledge" about human beings, paradoxically threaten to increase our *moral ignorance*. I have previously pointed out

that the idea of "knowledge as a duty" is relevant to scientific, social, and political problems, but it is even more important when we are faced with reinterpreting and updating the ethical intelligibility of behaviors and events which can generate dangerous consequences. In sum, profiling technologies are in tune with my motto "knowledge is a duty", because they increase knowledge thanks to the abductive/inductive capacity of rendering visible the invisible, but they also violate autonomy, security, freedom, etc. How can we contrast these effects we do not like and we perceive as violent and dangerous? My answer is: thanks to a further production of knowledge able to reinterpret and update the ethical intelligibility of those violent processes and events, a new knowledge that can also furnish the condition of possibility of effective legal remedies.

An organism and its environment co-evolve each other: profiling is thus a crucial general *cognitive* sign of life, because it consists of a repeated identification of risks and opportunities by an organism in its environment (Hildebrandt, 2008): "Interestingly enough, such organic diagnostic profiling is not dependent on conscious reflection. One could call it a cognitive capacity of all living organisms, without thereby claiming consciousness for an amoeba. One could also call it a form of intelligence based on the capacity to adapt: monitoring and testing, subsequent adaptation and repeated checking is what makes for the difference between the living and the inorganic world. This is what allows any organism to maintain its identity in the course of time, detecting opportunities to grow and spread as well as risks that need to be acted upon" (Hildebrandt, 2008, p. 58).

In sum, profiling is a diagnostic cognitive activity which produces important knowledge contents, made available in various ways to organisms. The externally distributed resources – like for example identity marks – have culturally specific different degrees of stability and so various chances to be re-internalized, varying from the very stable and reliable, like words and phrases of "natural" language and certain symbols, to others which are more evanescent and transient. Only when they are stable can we properly speak of the establishment of an "extended mind", like Wilson and Clark contend. Finally, internal and external resources (that is neural and environmental) are not identical but complementary – in this sense human beings can be appropriately considered "cyborgs" (Clark, 2003).

The "cyborg" comprises a vast quantity of external information and that it is possible to "materialize" many of our cybernetic aspects – at home, I can, in principle,[14] print out a picture of you as well as gather information about your sexual, political, and literary preferences based on records of your online purchases. Consequently, those aspects are not only available to the person they concern; because they are potentially "globally omnipresent," everyone can, in theory, see them. We must note that computer technology also distances the human body from the actions performed, and because it is an automating technology, it has a great potential to displace human presence.

The current problem is that profiles have no clear status: the situation in western countries presents many puzzling cases, which depict a general lack of awareness of their real ways of affecting our lives: "[…] today's technological and organisational infrastructure makes it next to impossible seriously to check whether and when the directive is violated, creating an illusion of adequate data protection" (Hildebrandt, 2008, p. 65). This implies that technological and organizational infrastructure de facto implement *moral axiologies* constituting new frameworks, which affect human habits, regulate behaviors, and at the same time these technologies *cannot* be controlled by the current democratic legal systems just because of their own structure, which is characterized by impenetrability. Technologies invite certain behaviors and inhibit others, or even enforce certain behaviors while prohibiting others.

I am convinced that current political powers in western countries already started to exploit this despotic aspect of the new technologies almost everywhere. Here an example, even if not related to the problem of privacy: the last Berlusconi Italian government uses various informational technologies – for example at the level of Ministry of University, which implement centralized complicated web systems worth of a police system – to enforce its own government policies, because the government probably sees the technological tools as proper extensions of its exercise of power. De facto the government embeds legal norms in technological devices trying to sidestep the checks and balances supposedly inherent in legal regulation. In sum "legal" normative effects embedded in technology go beyond what has been legitimately adopted through the democratic decision of the Parliament. Those embedded norms *are not legal* but just the product of more or less aggressive and incompetent Rome governmental bureaucracies, which produce relevant normative effects without the real approval of the Parliament and without the awareness of both Parliament and citizens.

The Need of New Technoknowledge to Counteract Technoviolence

To avoid becoming victims of the normative effect of new technologies my moral motto "knowledge as duty" is still applicable. New intellectual frameworks have to be built, as soon as possible. For example it is necessary to acknowledge that legal norms aiming at defending privacy have to be "incorporated" in the profiling technological tools themselves: "We are now moving into a new age, creating new classes of scribes, demanding a new literacy among ordinary citizens. To abstain from technological embodiment of legal norms in the emerging technologies that will constitute and regulate our world, would mean the end of the rule of law, paving the way for an instrumentalist rule by means of legal and technological instruments. Lawyers need to develop urgently a hermeneutic of profiling technologies, supplementing their understanding of written text. Only if such a hermeneutic is in place can we prevent a type of technologically embodiment of legal norms that in fact eradicates the sensitive *mélange* of instrumental and protective aspects that is crucial for the rule of law" (Hildebrandt, 2008).

New technical and scientific knowledge, together with an improvement of ethical and institutional capacity, are simply urgent. It is only through the acknowledgement of the necessity of embodying legal norms in new technological devices and infrastructures, that we can avoid "the end of law as an effective and a legitimate instrument for constitutional democracy" (Hildebrandt, 2009a, p. 443).

Hildebrandt illustrates some new interesting perspectives. The so called RFID technologies should include the "adoption of design criteria that avoid risks to privacy and security, not only at the technological but also at the organisational and business process levels" (Hildebrandt, 2009a, p. 443). Constructive technology assessment (CTA) represents a case for "upstream" involvement in technological design, i.e., not installing ethical commissions after the technology is a finished product but getting involved at the earliest possible stage of technological design. I have repeatedly observed in *Morality in a Technological World* that designers' good intentions do not determine the actual consequences and affordances of a technology. Nevertheless, privacy-enhancing technologies (PETs), aimed for example at hiding of data (anonymization) and use of pseudonyms, could moderate privacy risks. Finally "Countering the threats of autonomic profiling citizens will need more than the possibility of opting out, it will need effective transparency enhancing tools (TETs) that render accessible and assessable the profiles that may affect their lives" (Hildebrandt, 2008, p. 67). When detected, the actual unexpected negative consequences and affordances of technology can be "intentionally" counterbalanced.

This new cognitive/ethical potential has to elevate cognition from the implicit and unperceived to the explicit, which will allow us to understand technological processes: once they are organized into an appropriate hierarchy, they can be more readily managed. In doing so, the implementation of PETs and TETs still requires intentionality and free will and, consequently, "moral responsibility."

CONCLUSION

To conclude, I have just said that to avoid becoming victims of the normative effect of new technologies devoted to violent activities of profiling the moral motto "knowledge as duty" is still applicable. New techno-intellectual frameworks have to be built, urgently, for example by incorporating legal norms aiming at defending privacy in the profiling technological tools themselves. By embodying legal norms in new technological devices and infrastructures, we can counteract "the end of law", that law which is one of the fundamental aspects of a constitutional democracy. I certainly think that the mediated intentionality expressed in large and complex artifactual cognitive systems reminds us that we should not evade moral responsibility, but I would further clearly emphasize that circumventing such responsibility arises from a lack of a commitment to producing suitable knowledge able to manage their negative violent consequences. As a result, modern people are often jeopardized by the unanticipated, global-scale outcomes of localized individual intentions, and this gap between local intentions and global consequences derives from using systems (technologies, for instance) that are just energized by that intentionality itself, like in the case of profiling technology. Fully understanding functions and effects of our various technological cognitive systems will help us to avoid damage to the earth and to the human beings, which is why I strongly support the motto "knowledge as a duty."

REFERENCES

Bäck, A. (2009). Thinking clearly about violence. In Bufacchi, V. (Ed.), *Violence: A philosophical anthology* (pp. 365–374). Basingstoke, UK: Palgrave Macmillan.

Baumeister, R. (1997). *Evil: Inside human violence and cruelty*. New York, NY: Freeman/Holt.

Boddy, C. R. (2010). Corporate psychopaths, bullying and unfair supervision in the workplace. *Journal of Business Ethics*, *100*(3), 367–379. doi:10.1007/s10551-010-0689-5

Boddy, C. R., Ladyshewsky, R. K., & Galvin, P. (2010). The influence of corporate psychopaths on corporate social responsibility and organizational commitment to employees. *Journal of Business Ethics*, *97*, 1–19. doi:10.1007/s10551-010-0492-3

Brookman, F. (2005). *Understanding homicide*. London, UK: Sage.

Bynum, T. W., & Rogerson, S. (Eds.). (2004). *Computer ethics and professional responsibility*. Malden, MA: Blackwell.

Catley, B. (2003). Philosophy – the luxurious supplement of violence? In *Proceedings of the Third International Critical Management Studies Conference on Critique and Inclusivity: Opening the Agenda*, Lancaster, UK.

Clark, A. (2003). *Natural-born cyborgs: Minds, technologies, and the future of human intelligence*. Oxford, UK: Oxford University Press.

Corrigan, M., & Sprehe, T. (2010). Cleaning up your information wasteland. *Information & Management*, *443*, 27–31.

Damasio, A. R. (1999). *The feeling of what happens*. New York, NY: Hartcourt Brace.

Davis, N. D., & Splichal, S. L. (2000). *Access denied: Freedom of information in the information age*. Ames, IA: Iowa State University Press.

Dodd, J. (2009). *Violence and phenomenology*. New York, NY: Routledge.

Eckstrand, N., & Yates, C. S. (Eds.). (2011). *Philosophy and the return of violence*. New York, NY: Continuum.

Girard, R. (1977). *Violence and the sacred*. Baltimore, MD: Johns Hopkins University Press. (Original work published 1972)

Girard, R. (1986). *The scapegoat*. Baltimore, MD: Johns Hopkins University Press.

Godwin, M. (1998). *Cyber rights: Defending free speech in the digital age*. Toronto, ON, Canada: Random House.

Goldman, A. (1992). *Liaisons: Philosophy meets the cognitive and social sciences*. Cambridge, MA: MIT Press.

Gotterbarn, D. (2000). Virtual information and the software engineering code of ethics. In Langford, D. (Ed.), *Internet ethics* (pp. 200–219). New York, NY: Macmillan.

Gusterson, H. (2007). Anthropology and militarism. *Annual Review of Anthropology*, *16*, 155–175. doi:10.1146/annurev.anthro.36.081406.094302

Hildebrandt, M. (2008). Profiling and the rule of law. *Identity in the Information Society*, *1*, 55–70. doi:10.1007/s12394-008-0003-1

Hildebrandt, M. (2009a). Technology and the end of law. In Keirsbilck, B., Devroe, W., & Claes, E. (Eds.), *Facing the limits of the law* (pp. 1–22). Heidelberg, Germany: Springer-Verlag. doi:10.1007/978-3-540-79856-9_23

Hildebrandt, M. (2009b). Profiling and AmI. In Rannenberg, K., Royer, D., & Deuker, A. (Eds.), *The future of identity in the information society* (pp. 273–310). Heidelberg, Germany: Springer-Verlag. doi:10.1007/978-3-642-01820-6_7

Houellebecq, M. (2006). *The possibility of an island*. New York, NY: Knopf.

Johnson, D. G. (1994). *Computer ethics* (2nd ed.). Upper Saddle River, NJ: Prentice Hall.

Langford, D. (Ed.). (2000). *Internet ethics*. New York, NY: Macmillan.

Magnani, L. (2007). *Morality in a technological world: Knowledge as duty*. Cambridge, UK: Cambridge University Press. doi:10.1017/CBO9780511498657

Magnani, L. (2009). *Abductive cognition: The epistemological and eco-cognitive dimensions of hypothetical reasoning*. Heidelberg, Germany: Springer-Verlag.

Magnani, L. (2011). *Understanding violence: The intertwining of morality, religion, and violence: A philosophical stance*. Heidelberg, Germany: Springer-Verlag.

Magnani, L. (in press). Abducing personal data, destroying privacy. Diagnosing profiles through artifactual mediators. In Hildebrandt, M., & de Vries, E. (Eds.), *Privacy, due process and the computational turn. Philosophers of law meet philosophers of technology*. London, UK: Routledge.

Mawhood, J., & Tysver, D. (2000). Law and internet. In Langford, D. (Ed.), *Internet ethics* (pp. 96–126). New York, NY: Macmillan.

McCumber, J. (2011). Philosophy after 9/11. In Eckstrand, N., & Yates, C. S. (Eds.), *Philosophy and the return of violence* (pp. 17–30). New York, NY: Continuum.

Mill, J. S. (1966). On liberty. In Mill, J. S. (Ed.), *On liberty, representative government, the subjection of women* (12th ed., pp. 5–141). Oxford, UK: Oxford University Press. (Original work published 1859)

Reiman, J. H. (1976). Privacy, intimacy, personhood. *Philosophy & Public Affairs, 6*(1).

Reiman, J. H. (1995). Driving to the Panopticon: a philosophical exploration of the risks to privacy posed by the highway technology of the future. *Computer and High Technology Law Journal, 11*(1), 27–44.

Rowlands, M. (1999). *The body in mind*. Cambridge, UK: Cambridge University Press. doi:10.1017/CBO9780511583261

Schiller, H. J. (2000). The global information highway. In Winston, M. E., & Edelbach, R. D. (Eds.), *Society, ethics and technology* (pp. 171–181). Belmont, CA: Wadsworth.

Shoeman, F. (Ed.). (1984). *Philosophical dimensions of privacy: an anthology*. Cambridge, UK: Cambridge University Press. doi:10.1017/CBO9780511625138

Smith Holt, S., Loucks, N., & Adler, J. R. (2009). Religion, culture, and killing. In Rothbart, D., & Korostelina, K. V. (Eds.), *Why we kill: Understanding violence across cultures and disciplines* (pp. 1–6). London, UK: Middlesex University Press.

Stallman, R. (1991). Why software should be free. In Stallman, R. (Ed.), *Free software, free society: The selected essays of Richard M. Stallman*. Boston, MA: Free Software Foundation.

Steinhart, E. (1999). Emergent values for automations: ethical problems of life in the generalized internet. *Journal of Ethics and Information Technology, 1*(2), 155–160. doi:10.1023/A:1010078223411

Tombs, S., & White, D. (2009). A deadly consensus. *The British Journal of Criminology, 20*, 1–20.

Van den Hoven, J. (2000). The Internet and varieties of moral wrongdoing. In Langford, D. (Ed.), *Internet ethics* (pp. 127–157). New York, NY: Macmillan.

Warwick, K. (2003). Cyborg morals, cyborg values, cyborg ethics. *Ethics and Information Technology, 5*, 131–137. doi:10.1023/B:ETIN.0000006870.65865.cf

Wieviorka, W. (2009). *Violence: A new approach.* London, UK: Sage.

Winston, M. E., & Edelbach, R. D. (Eds.). (2000). *Society, ethics, and technology.* Belmont, CA: Wadsworth/Thomson Learning.

Žižek, S. (2009). *Violence.* London, UK: Profile Books.

ENDNOTES

[1] On the relationships between psychiatry and violence cfr. chapter five, section "Pure evil", of my book (Magnani, 2011).

[2] Amazingly, empirical research showed that in Australia around 26% of workplace bullying can be attributed to the 1% of the employee population, representing "corporate psychopaths". Corporate psychopaths, like toxic leaders, cause major harm to the welfare of others, trigger organizational chaos, and are a major obstacle to efficiency and productivity (Boddy *et al.*, 2010; Boddy, 2010).

[3] A collective mobilization of memory about past injuries.

[4] However, a new focus of attention on violence in philosophy seems to be "in the air". Very recently McCumber (2011), in "Philosophy after 9/11", has stressed the need to reshape research in philosophy (and its related community) from various perspectives, for example defending Enlightenment, beyond the mere use of the "[...] standard arsenal of argument forms, of which our Analytical colleagues make excellent though wrongly exclusive use" (p. 28). The paper is part of a collection with the eloquent title *Philosophy and the Return of Violence* (Eckstrand & Yates, 2011).

[5] Cf., for example, Damasio (1999).

[6] Risk assessment, credit scoring, marketing, anti-money laundering, criminal profiling, customer relationship management, predictive medicine, e-learning, and financial markets all thrive on data mining techniques which reflect a type of knowledge construction based on various types of computation.

[7] On the effect of artifactual mediators on profiling, privacy, and identity, cf. Magnani (2012).

[8] "If your 'Ambient Intelligent Environment' caters to your preferences before you become aware of them, this will invite or even re–enforce certain behaviours, like drinking coffee that is prepared automatically at a certain hour or going to sleep early during week days (because the central heating system has lowered the temperature). Other behaviours may be inhibited or even ruled out, because the fridge may refuse a person another beer if the caloric intake exceeds a certain point" (Hildebrandt, 2009a, p. 452).

[9] On abduction and induction from a cognitive, epistemological and computational perspective, cf. the recent Magnani (2009, Ch. 2).

[10] Abduction refers to that cognitive activity accounting for the introduction of new explanatory hypotheses in science or in diagnostic reasoning (for example in medicine). Abduction is the process of *inferring* certain facts and/or laws and hypotheses that render some sentences plausible, that *explain* or *discover* some (eventually new) phenomenon or observation; it is the process of reasoning in which explanatory hypotheses are formed and evaluated. There are two main epistemological meanings of the word abduction: 1) abduction that only generates plausible hypotheses ("selective" or "creative") and 2) abduction considered as inference "to the best explanation", which also evaluates hypotheses. To illustrate from the field of medical knowledge, the discovery of a new disease and the manifestations it causes can be considered as the result of

a creative abductive inference. Therefore, "creative" abduction deals with the whole field of the growth of scientific knowledge. This is irrelevant in medical diagnosis where instead the task is to "select" from an encyclopedia of pre-stored diagnostic entities. A full illustration of abductive cognition is given in Magnani (2009).

[11] Godwin (1998). Cf. also the classical Johnson (1994) and Stallman (1991).

[12] Goldman (1992) and Mawhood and Tysver (2000). On the Internet's retention of information in corporations, governments, and other institutions, cf. Davis and Splichal (2000).

[13] The Internet and the so-called "Intelligent Vehicle Highway Systems" (IVHS) have also created "external" devices that challenge our established rights to privacy, making it necessary to clearly define and shape new moral ontologies. Recently, ECHELON, the most powerful intelligence-gathering organization in the world, has garnered attention as it has become suspected of conducting global electronic surveillance – mainly of traffic to and from North America – which could threaten the privacy of people all over the world. The European Union parliament established a temporary committee on the ECHELON interception system (1999–2004) to investigate its actions. The 2001 draft report "on the existence of a global system for the interception of private and commercial communications (ECHELON interception system)" can be found at http://www.europarl.eu.int/tempcom/echelon/pdf/prechelon_en.pdf

[14] Of course if I have the resources to invest in the relevant software, like in the case of large companies.

This work was previously published in the International Journal of Technoethics, Volume 2, Issue 4, edited by Rocci Luppicini, pp. 1-19, copyright 2011 by IGI Publishing (an imprint of IGI Global).

Chapter 13
Fairness and Regulation of Violence in Technological Design

Cameron Shelley
University of Waterloo, Canada

ABSTRACT

The purpose of this article is to explore how the design of technology relates to the fairness of the distribution of violence in modern society. Deliberately or not, the design of technological artifacts embodies the priorities of its designers, including how violence is meted out to those affected by the design. Designers make implicit predictions about the context in which designs will perform, predictions that will not always be satisfied. Errors from failed predictions can affect people in ways that designers may not appreciate. In this article, several examples of how artifacts distribute violence are considered. The Taylor-Russell diagram is introduced as a means of representing and exploring this issue. The role of government regulation, safety, and social role in design assessment is discussed.

1. INTRODUCTION

Hybrid cars are applauded primarily because of their fuel efficiency and the reduced amount of pollution that they produce. Besides these advantages, hybrid cars such as the Toyota Prius are also quieter than conventional cars. This difference results from their reliance on electric instead of gas or diesel engines. In the absence of internal combustion and the vibrations that it causes, cars like the Prius can move through city streets in relative silence.

The quiet operation of hybrid and electric cars might be thought of as an added bonus. Mostly, it is. However, it has been noted that quiet cars like the Prius could pose a danger to the blind. That is because, when blind persons walk near cars or cross streets in traffic, they use the sound made by

DOI: 10.4018/978-1-4666-2931-8.ch013

the engines of the cars in traffic to estimate their speed, location, and direction of travel. Since the Prius is so quiet, this information is diminished. Without this information, blind pedestrians might be unable to detect an oncoming hybrid and so step out in front of one, resulting in a collision.

This situation illustrates a significant connection between technological design and violence: The design of a piece of technology can, in effect, act to regulate the distribution of violence within a society. The violence, in this case, consists of collisions between pedestrians and cars, in addition to the health threats posed by pollutants, greenhouse gases, and noise. The design of the hybrid car affects how these harms are distributed between the two constituencies involved here, namely blind pedestrians and residents of areas where the cars operate.

This problem raises an issue of *fairness*. Here, fairness refers to what is termed "distributive justice" (Aristotle, 1998), that is, the distribution of burdens and benefits amongst the members of a society. Shelley (in press) shows that fairness is a common issue relating technological design to society. That is, the design of technologies often has implications for the distribution of benefits or harms between different constituencies within a society. In the case of quiet hybrid cars, the harms are threats of injury through collision versus through emissions or noise, and the constituencies are blind pedestrians and local residents, respectively. The concern becomes problematic because the design of the cars effectively pits the two constituencies against one another: The more that blind pedestrians face a threat from Priuses, the less that residents face a threat from emissions and noise pollution.

Of course, the danger is not yet acute because electric hybrids and electric cars are relatively rare so far. However, the threat is being taken seriously. The US government recently passed the Pedestrian Safety Enhancement Act of 2010, requiring the Secretary of Transportation to study methods for protecting vulnerable pedestrians. In response to this Act, Nissan has decided to design its electric Leaf car to emit recorded sounds when it is put in gear (Eaton, 2010). In that way, it is hoped that blind pedestrians will be alerted to the approach of the car. In assessing this law, and Nissan's response to it, we must address the issue of how fairly this design treats the competing interests of pedestrians and residents.

The purpose of this article is to explore further the role of fairness in the regulation of violence in technological design, as illustrated by the electric car example. In particular, I wish to demonstrate how such issues can be identified and clarified through the use of the *Taylor-Russell diagram* (T-R). In the remainder of this paper, I provide an overview of the T-R diagram and how it applies to further examples in which the design of technologies serve to regulate the distribution of violence in societies. The impact and resolution of fairness issues can then be addressed.

2. THE TAYLOR-RUSSELL DIAGRAM

The Taylor-Russell (T-R) diagram was invented by the psychologists Taylor and Russell to help analyze the validity of scholastic aptitude tests, such as the modern SAT (Taylor & Russell, 1939). This section provides a brief overview of the T-R diagram using the analysis of the SAT as an illustration.

The basic purpose of a SAT is to predict which applicants would successfully complete their degrees if admitted to university. Prediction is important in this context because there are a fewer places available than there are applicants for them, so that it is not possible to offer admission to all. The best solution would be to know in advance which applicants would be successful. However, since such foreknowledge is hardly possible, the next best solution is to generate a prediction. In the United States, predictions are made on the basis of standardized tests.

The result of testing and then educating students is, among other things, a pair of data points. The first point is the student's score on the SAT while the second point is the students graduating average. These points can be paired and plotted in a scatterplot, such as those in Figure 1. In each plot, the predictions, that is, the SAT scores, lie along the x-axis. The actual events or outcomes lie along the y-axis. When enough data points are accumulated, the plots begin to reveal the accuracy of the prediction system. In the first plot (a), the prediction system is not especially accurate; the points fall across a wide area of the scatterplot. The width of the smallest ellipse, with its major axis along the main diagonal of the plot, illustrates how broadly the points fall in the plot. The correlation between predictions and actual events is low, roughly 0.2. In Figure 1(b), the accuracy of the system is higher, with a correlation of 0.5, and the width of the ellipse is smaller. In Figure 1(c), the accuracy of the system is high, with a correlation of 0.8, and the width of the ellipse is narrow. If the predictive accuracy of the system were perfect, then every point would fall on the main diagonal, where x=y, and the ellipse would be completely flat. Of course, no non-trivial system can be predicted with perfect accuracy.

The ineradicable uncertainty that remains in such predictive systems creates a problem for designers of the SAT: The test will always give rise to some errors. These errors are illustrated in Figure 2. The scatterplot in this figure is divided into four quadrants by a horizontal and a vertical line. The horizontal line represents the point on the y-axis that divides success (graduation) versus failure (drop out). This line may be called the *design threshold*. The vertical line represents the point on the x-axis that divides admissible SAT scores from inadmissible ones. This line may be called the *prediction cutoff*.

These lines give a unique meaning to each quadrant as a different kind of outcome of the SAT:

1. The upper-right quadrant contains those points that are *true positives*, that is, students who exceeded the cutoff for admission (tested "positive") and who successfully graduated (the prediction was "true").

2. The lower left quadrant contains those points that are *true negatives*, that is, students who fell short of the cutoff for admission (tested "negative") and who would have failed to graduate.

3. The lower right quadrant contains those points that are *false positives*, that is, students who exceeded the cutoff for admission but

Figure 1. Three scatterplots representing prediction/event correlations of (a) 0.2, (b) 0.5, and (c) 0.8

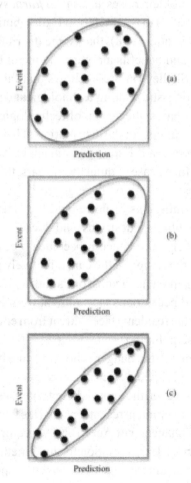

Figure 2. A T-R diagram, showing the relationships between predictions versus actual events

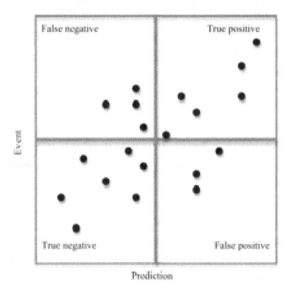

Prediction

3. FAIRNESS IN DESIGN

The SAT system generates a problem of fairness because it helps to determine the distribution of a social good, namely admission to university. Designers of the system have to determine how to trade off each sort of error against the other. Would it be better to have a higher cutoff and reject more students who would have graduated, or would it be better to have a lower cutoff and accept more students who will not end up graduating? As noted by Hammond (1996), social constituencies that have a stake in each kind of error will tend to advocate for the cutoff that best satisfies their interests, values, or policy goals. When two or more groups find themselves on opposite sides of such a dispute, a problem of fairness is present.

Consider the state of university admissions in the United States. Currently, US universities have a low cutoff, meaning that they admit a large number of students, many of whom do not graduate. At present, little more than half of the students admitted proceed to obtain a bachelor's degree (Smith, 2011). In other words, the prediction cutoff is set to produce a preponderance of false positives. Libertarians, such as the venture capitalist and founder of PayPal, Peter Thiel, regard university as not much more than a scam or a bubble. It is a waste of both private and government capital to fund a system with such a high drop-out rate. Thiel would like to see many fewer students admitted to university and has begun a fellowship program to pay promising entrepreneurs to drop out (Weisberg, 2010).

More collectivist-minded observers look on the false negative rate with more concern. If the system were more restrictive, then more students who would proceed to graduate would be denied admission. This could stunt the intellectual growth of these young people and, by pushing them into private business earlier, redirect their energies from altruistic pursuits into the pursuit of individual wealth.

who did not proceed to graduate (the prediction was "false").

4. The upper left quadrant contains those points that are *false negatives*, that is, students who fell short of the cutoff for admission but who would have successfully graduated.

In some cases, there may be missing data, e.g., the graduation averages of students who were not admitted. This data can be estimated using statistical models and comparisons with universities having open admissions policies.

For current purposes, the T-R diagram in Figure 2 provides a clear understanding of the errors that a prediction system gives rise to, and how these errors are related:

1. The system produces both false positives and false negatives, and
2. The incidence of each kind of error is determined by overall accuracy of the system and the location of the prediction cutoff.

The issue of fairness depends crucially on the second point.

Assume that the accuracy of the SAT system cannot be greatly increased. Perhaps the admission testing problem is simply too noisy for further improvement. In that case, the only remaining aspect of the system's design open to dispute is the setting of prediction cutoff (Hammond, 1996). If the libertarian constituency succeeds, then the cutoff must be moved rightward, to minimize the number of false positives. If the collectivist constituency succeeds, then the cutoff must remain to the left, to minimize the number of false negatives. The goal of policy makers, in this case, is to decide which setting of the prediction cutoff is the most fair to the legitimate interests of both constituencies.

In the following sections, the problem of fairness and violence in the design of technology will be addressed through a number of examples. These examples are divided into three categories: (1) designs that commit violence, (2) designs that respond to violence, and (3) designs that inadvertently contribute to violence. T-R diagrams are used to show that the problem of fairness and violence has the same structure in each sort of instance.

4. REGULATION OF THE COMMISSION OF VIOLENCE

a. Military Robots

Some technological systems are designed to allow their operators to deal out violence. The current use of armed ground robots and aerial vehicles in Afghanistan and Iraq provide good examples of this use of technology. The TALON robot, for example, may be armed with rifles, machine guns, grenade launchers or anti-tank rocket launchers. It is operated remotely by a human operator and was introduced into Iraq by the US Army in 2007. Unmanned areal drones have been used in both theaters by the US Air Force and the CIA. These drones may carry air-to-surface missiles and are

flown remotely by pilots who may be stationed hundreds or thousands of kilometers away (Singer, 2009).

At some point, weapons systems of these types will be designed to operate autonomously. That is, military ground robots and drones will be able to fire their weapons—at human targets—without prior explicit and direct authorization from a human operator. The imperative for designing autonomous drones is straightforward: It might be advantageous for a drone to take action in a situation in which it is out of communication with its controllers, perhaps due to an equipment failure or jamming of communications by the enemy. Imagine that the drone locates an enemy leader while its communications system is not functioning. Its controllers on the ground would likely prefer that the system should attack the enemy leader even without explicit authorization.

Such systems will be controversial because they are pieces of technology designed to commit violence without direct human mediation. Thus, the regulation of this form of violence is a growing concern for potential designers, their clients, and their potential victims.

One concern for designers of autonomous drones will be fairness. One of the most obvious problems facing a drone considering an attack will be the problem of false positives. That is, the drone's software will have to weigh the possibility that the target it has identified as an enemy combatant is, in fact, not an enemy combatant and, therefore, ought not to be attacked (Sharkey, 2008). Slightly less obvious but equally significant is the problem of false negatives, that is, enemy combatants who are spotted but misclassified as non-combatants. After all, the main motivation for designing drones to be autonomous is to reduce the possibility of this sort of error. The difficulty identified in this situation is represented in the T-R diagram given Table 1.

It is not hard to see the constituencies that would take opposed views of the two sorts of error described here. In the case of drones in

Table 1. Fairness in autonomous drone attacks

		Prediction: Appearance of target	
		Low threat	*High threat*
Event: Significance of target	*High*	A significant target is ignored	A significant target is attacked
	Low	A non-combatant target is ignored	A non-combatant target is attacked

Afghanistan, citizens of NATO nations with troops exposed to danger there will prefer to minimize the false negatives, that is, significant targets that remain unattacked. In this way, more significant targets will be attacked, before they can carry on to attack the NATO soldiers in their area. Afghan citizens will prefer to minimize the false positives, that is, insignificant (e.g., non-combatant) targets that are attacked by automated robots. Indeed, the Afghan government has repeatedly expressed this preference regarding NATO air attacks there (Coghlin, 2008). Were autonomous drones deployed there, the same issues would have to be resolved not by human operators but by the drones themselves.

b. Intelligent Co-Drivers

The quest for autonomous vehicles is also central to many civilian research programs. Google, for example, has invested millions of dollars and thousands of hours into cars that will drive themselves along conventional roadways that are shared with cars driven by human drivers (Markoff, 2010). However, it is unlikely that fully autonomous cars will be licensed for regular use on the roads anytime time soon (Shepardson, 2011).

In the nearer term, we can probably look forward to cars that feature computerized driving assistance. Already, several cars come with self-parking systems that parallel park the car without steering input from the driver (Trillin, 2007). For the next step, MIT engineers are working on co-driving systems that use sensors on the

car to track the car and its surroundings with the aim of warning the driver of impending problems or even taking action to avoid collisions (Finn, 2011). Similar systems are being developed by other groups.

It is quite conceivable that such a system could develop to the point where it would save lives overall. However, taking driving decisions out of the hands of drivers raises difficult issues of various sorts, including fairness. The immediate aim of co-drivers appears to be to save the human driver of the car in the event of a crash, e.g., by adjusting seatbelts and airbags (Pauzie & Amditis, 2011, p. 19). Yet, it is not clear that saving the driver will always produce the best outcome. Imagine a situation in which a head-on collision is impending. The instinct of the driver might be to swerve so as the place the passenger side of his car between himself and the oncoming vehicle. However, this maneuver exposes the passengers to greater danger. In such a situation, should the co-driver system countermand the driver's steering and potentially sacrifice the driver to save the passengers? Should the car swerve to avoid nearby pedestrians? This scenario is reminiscent of the notorious Trolley Problem, in which people face the dilemma of pushing a man under an oncoming trolley in order to save a group standing further down the tracks (Foot, 1967). Whether or not to push the innocent man is a controversial issue, but any sophisticated co-driving system will have to be programmed to solve that sort of problem, and quickly.

A key setting in a co-driving system will be the willingness to sacrifice the driver to save other people with a stake in any violent situation. The effect of this willingness on the outcomes of potential accidents is displayed in Table 2.

If the willingness to sacrifice drivers is high, then drivers will be endangered even in situations where doing so turns out not to be necessary to save others. A driver might find his car edging towards a telephone pole on the roadside because

Table 2. Fairness in autonomous co-drivers

		Prediction: Willingness for driver sacrifice	
		Low	*High*
Event: Overall casualties	High	Causalities increased	Casualties minimized
	Low	Driver is spared from danger	Driver is excessively endangered

the co-driver is concerned about an oncoming school bus near the center line of the road, for example. This would constitute a false positive. Drivers will likely disapprove of purchasing cars configured in this way. Now consider a system with a low willingness to sacrifice drivers for the sake of others. A false negative would concern a situation in which the system saves a driver at the expense of endangering others. Such a system would be resisted by constituents who frequently walk or cycle on or near the roadways, for example. It would be similar to the situation that have existed in some cities, e.g., Bogota, in which upper-class, car-driving citizens routinely parked on or drove over sidewalks in poor areas of town, thus placing children there in danger (Peñalosa, 2003). Wealthier car owners would have cars equipped with co-drivers whereas less well-to-do users of the roadway would be without and consequently be exposed to violence on behalf of the car owners.

5. REGULATION OF RESPONSES TO VIOLENCE

Military drones and civilian co-drivers illustrate how issues of fairness arise in the regulation of violence that is within the control of automated systems. Regulation of such violence involves explicitly programming considerations of fairness into the operation of these designs. However, similar considerations apply to systems that do not control violence but rather control responses to violence. Often, violent incidents call forth violent responses, involving such designs indirectly in the commission of violence.

a. ShotSpotter

ShotSpotter is a system designed to detect and locate gunshots in urban areas, and to coordinate police responses. The system consists of a set of microphones, distributed within a community, that are connected to a central computer through a network. The microphones relay sounds that they pick up to the central computer, which compares them to sound profiles in order to determine which sounds are gunshots. A match between sounds and gunshot profiles results in a report sent to a police dispatcher. The report identifies the type of gunshot, presents its location on a map, and offers the opportunity for the dispatcher to play back the sound as stored in an electronic sound file. The dispatcher then decides on the appropriate response, e.g., sending a cruiser to check out the scene.

Here is a description of one such incident as described by Watters (2007):

On February 3, 2007, three sensors picked up gunfire at 1:49 am. A red dot on the map appeared just blocks from the Eastmont station, and officer Martin Ziebarth, who was there filling out paperwork, rushed out into the night. A few minutes later, he came upon 21-year-old Addiel Meza walking down the street with the butt of a pistol sticking out of his pants pocket. Ziebarth stopped his car, jumped out, and yelled at Meza to stop. Meza began to run and pulled out his gun.

At 1:57 am, ShotSpotter showed a fresh red dot on Foothill Boulevard. The shots recorded at that moment came rapidly, five staccato blasts in less than two seconds.

As the echoes faded, Meza ran east on Foothill and then north on 61st. Before reaching the end

Table 3. Fairness in gunshot detection

		Prediction: Sound matches gunshot profile	
		Poorly	*Closely*
Event: Gunshot occurs	*Yes*	Gunshots not reported	Gunshots reported
	No	Non-gunshots not investigated	Non-gunshots investigated

of the block, he slowed and then collapsed to the sidewalk. He died that night at Highland Hospital.

This incident illustrates a number of issues concerning the application of ShotSpotter (Shelley, in press). No system works perfectly, and ShotSpotter may make mistakes. The nature of these mistakes is represented in Table 3.

In some cases, non-gunshots may be misclassified as gunshots, e.g., firecrackers, automobile backfires, or recordings of gunshots being played back. In such cases, police are likely to be sent to investigate a non-incident. Besides being a waste of police resources, such incidents could cast suspicion on innocent parties. ShotSpotter is deployed in areas that suffer from heavy gun use, so finding a young man there with a gun in his pocket, and an adversarial attitude towards police, may not be uncommon. Since police arrive expecting to find a shooter, a presumption of guilt will prevail and an unnecessary exchange of gunfire may result. Proponents of individual liberties will likely object to such cases.

In other cases, gunshots may be misclassified as other sounds. In that event, someone guilty of illegal shooting may go unpursued, and any victim unaided. Advocates of community safety will presumably object to these outcomes. The police themselves, perhaps wanting to confirm the usefulness and cost-effectiveness of their decision to deploy ShotSpotter, would also have an incentive to minimize the number of false negatives. The result is that ShotSpotter can create an issue of fairness in which the interests of advocates for

individual liberties are pitted against the interests of advocates for civil order. Published statistics for the accuracy of ShotSpotter are limited to an early study suggesting that ShotSpotter then had a false negative rate of 19%, and without a determination of its false positive rate (Watkins, Mazerolle, & Rogan, 2002). Thus, it is difficult to say where the system falls in this conflict of interests.

(In fact, a just-published survey of police dispatchers suggests that ShotSpotter has a false positive rate of about 33%, whereas the occurrence of false negatives is described as "very rare" (Selby, Henderson, & Tayyabkhan, 2011).)

In any event, the ShotSpotter example illustrates how technologies designed to guide the response to violence can shape the commission of violence.

b. Arsonist Detection

A somewhat similar problem with a possible, high-tech solution is the detection of criminals such as arsonists. It seems that arsonists tend to linger at the scene of their crimes, to observe the fruits of their labors. Thus, their faces might be visible in video recordings of crowds observing a fire from the street. In the case of serial arsonists, their faces would occur in several such videos. Anyone whose face turned up in multiple videos might be a person of interest to the police. The problem of *questionable observer detection* is the problem of mining video recordings of crowds at the scene of criminal incidents to detect serial perpetrators (Barr, Bowyer, & Flynn, 2011). The system is not advanced enough for practical application yet, although the developers hope to interest the FBI and the Intelligence Advanced Research Projects Agency in the US (Sandhana, 2011).

The possible outcomes of such as system are represented in Table 4.

Here, a false positive consists of an innocent person who is identified as being in a crowd observing a fire on suspiciously many occasions. A

Table 4. Fairness in arsonist detection

		Prediction: Repeated detection	
		Low	*High*
Event: Arsonist present	*Yes*	Arsonist is not identified	Arsonist is identified
	No	Innocent person is not suspected	Innocent person is under suspicion

false negative consists of an arsonist who is not detected as present in enough crowd scenes to cause an alarm report.

Barr, Bowyer, and Flynn discuss the technical trade-offs between these kinds of errors. They note that a false negative is harder to correct than a false positive because of the way in which their detection algorithm works. This analysis does not, of course, relate to the problem of fairness present here. As in the ShotSpotter case, a false positive may result in a strong presumption of guilt on the part of police or, later on, a jury. A false negative also brings a burden, as it means that a serial arsonist may evade detection and thus be in a position to inflict further damage on the community. As with the case of student testing discussed by Hammond, this case presents a conflict between the interests of an individual, that is, a suspected arsonist, with the interests of a social group, that is, the victims of arson. As in the case of ShotSpotter, it may be expected that the authorities who deploy such a system might prefer to minimize false negatives in order to be seen to protect the community. Also, it may be assumed that people falsely cast into suspicion through the system would be acquitted upon further investigation.

6. REGULATION OF INADVERTENT VIOLENCE

In the systems discussed above, the problem of violence is overtly present in the purpose of the system. The fact that violence is obviously in-

volved makes the importance of fairness obvious. In other cases, however, violence is not an overtly present as a design issue. Instead, violence may be an inadvertent result of how the design responds to circumstances.

a. GPS Navigation

GPS navigation systems are a result of online map navigation aids developed in the mid 1990s. The system MapQuest, for example, allowed users to specify an origin and a destination site and receive a set of driving directions, accompanied by maps, specifying how to navigate from the origin to the destination. These directions could be printed out and taken in the car to be read en route by the driver or a passenger.

With the advent of inexpensive and accurate GPS location electronics, it became possible for such directions to be spoken to the driver by a device that tracked the car's location in real time. This is the basic function of the current GPS navigation system.

Stories of errors resulting from the use of GPS navigation systems are commonplace in the news. Some are largely amusing, as in the examples of motorists driving into rivers instead of over bridges when following the instructions of the GPS unit (Prigg, 2010). Some stories are more serious and involved fatalities, such as the driver who drove into a reservoir and drowned while following navigation instructions late at night (Tremlett, 2010).

The reasons for these incidents remain unclear. However, in some situations, it may be that the use of a GPS navigation system tends to detach the driver's attention from their surroundings (Shulman, 2011). This detachment, in turn, tends to decrease the driver's situational awareness. Thus, when navigational directions given by the unit are erroneous, the driver may not be in position to identify and make the appropriate course correction. The interaction of navigational accuracy

Table 5. Fairness in GPS navigation

		Prediction: Navigational inaccuracy	
		Low	*High*
Event: Situational awareness	High	Anticipation for correction wasted	Correction anticipated
	Low	Appropriate compliance	Inappropriate compliance

and driver situational awareness are displayed in Table 5.

The false positive in this case is the danger of an accident caused by an inappropriate maneuver, such as making a turn into a river at the command of the system. This sort of error is the one that dominates current discussion of the issue. The false negative in this case is the expenditure of attention by the driver on the situation in anticipation of a correction that turns out to be unneeded. Since the GPS navigation system is accurate in this case, the effort is wasted. This sort of error has not received much public discussion but is an important component in understanding the issue created by GPS navigation systems.

Attention is an important resource for people, especially in an era of increasing distractions. Demands for the attention of a driver are plentiful. There are the established demands of the roadway, the radio, and passengers. There are also new demands in the form of cell phones and in-car entertainment systems. With the limited supply of attention available and increasing demands, allocation of attention for the driver is becoming more problematic. Thus, a trade-off arises among the possible objects of attention.

One noteworthy feature of this situation is that it pits the interests of the individual against the interests of others. The false negative error consists of attention allocated by the driver for anticipating course corrections that prove to be unnecessary. That attention could be considered wasted when there are more rewarding tasks on which to spend it. The false positive error creates the danger of

a violent outcome not only for the driver but for anyone else in the vicinity, such as other drivers or pedestrians. Attention to others on the roadway is not a priority for the GPS navigation system. So, in cases where it is not a high priority for the driver, the effect may be to externalize the risk of violence on people not involved in using the GPS system in question.

b. Wrap Rage

So-called "clamshell packaging" consists of transparent, rigidly formed plastic shells that comprise the packaging of many small to medium-sized products as consumers encounter them in stores. This packaging has become a mainstay of industry as it offers several advantages. First, it allows items within the package to be displayed attractively. Second, it helps to prevent damage to the products during shipment. Third, and most importantly, it helps to protect the product against theft once it is on display in stores (Friess, 2006).

The third consideration is important because shoplifting is a major source of loss for retailers. The US National Association for Shoplifting Prevention states that more than $13 billion worth of goods are stolen from US retailers each year (National Association for Shoplifting Prevention, 2006). Clamshell packaging helps to defeat shoplifting by making goods difficult to access—the packaging must often be removed with a knife and considerable effort—and too large for easy concealment. Also, small radio frequency identification tags can be affixed to the packaging so that it will set off an alarm if the goods are removed from the store with the packaging in place.

Unfortunately, the attributes that make the packaging useful for security purposes also make it frustrating or even dangerous for consumers. Cutting the packaging with a sharp implement can leave sharp edges that cut the consumer's hands when they try to grasp the product inside. Also, because the packaging is often simply difficult to open, the consumer may become frustrated and attack the packaging with a knife, scissors, or screwdriver. This phenomenon is sometimes known as *wrap rage*. The result can be a potentially serious, self-inflicted injury (Friess, 2006):

... Emergency room doctors say they're slammed the week after Christmas with such injuries and see them regularly all year. Dr. Christian Arbelaez, a Boston-area ER physician, sees about a case a week, some as serious as tendon and nerve damage that require orthopedic surgeons to repair. "I would definitely like to tell (manufacturers) that serious hand injuries are occurring because of this packaging," said Arbelaez, a member of the Trauma Care and Injury Control National Committee of American College of Emergency Physicians.

It is clear, then, that the decision to use clamshell packaging as a part of packaging design involves a trade-off between security and violence to bodily integrity. The issue of fairness involved in this trade-off can be captured in Table 6.

One interesting feature of this issue of fairness is that it pits a risk of injury on the one hand against an economic loss on the other. That is, the false positive is a risk of personal injury, whereas the false negative is a risk of loss to shoplifting.

Table 6. Fairness in product packaging

		Prediction: Adopt clamshell packaging	
		No	*Yes*
Event: Threat of shoplifting	*High*	Merchants fall prey to shoplifting	Merchants protected from shoplifting
	Low	Customer able to open packaging easily	Customer frustrated/injured by packaging

7. VIOLENCE, FAIRNESS, AND THE STATE

Hobbes (1985) famously argued that violence lies at the foundation of a legitimate state. That is, there exists a social contract on which the citizens give up to the state their right to use violence in self-defense. In exchange for this monopoly on violence, the state undertakes to use it to create conditions of peace and good order in the land. For this purpose, the state might provide a police force to deal with criminals and a military to deal with predatory foreigners.

Some of the examples described may be viewed as manifestations of such a social contract. The regulation of military robots, for example, clearly illustrates the concern of the US military for dealing with violent threats against American interests. Similarly, the examples of ShotSpotter and arsonist detection illustrate the concern, or potential concern, of the state in ensuring peace and good order within its own territory.

These examples also emphasize the importance of fairness as a consideration for the government's deployment of violence. Consider the conduct of war. On Hobbes's view, the state uses violence as a means of corporate self-defense against the predations of other states, much as an individual uses violence as a means of self-defense against attackers when there is no government to perform this function. This analogy suggests that the state may apply any violent action that it deems necessary in order to preserve itself.

The debate over autonomous military robots suggests that this view is not held by the US military. Discussion of the programming of autonomous military robots includes the issue of fairness, that is, the trade-off between successful attacks on enemies and "collateral damage" to non-combatants. Of course, this concern may be primarily prudential. That is, fairness may be a concern because achieving it or, at least, the appearance of it, would maximize the overall effectiveness of robotic warfare for the achievement of American aims. Whether or not fairness enters the debate about autonomous military robots for its own sake or its usefulness for some other purpose, fairness is clearly important in a way that Hobbes's discussion of the social contract seems to overlook.

Interestingly, the illustrations given above suggest that the same level of concern does not apply to the state's monopoly on violence within its territory. ShotSpotter can, and should, be viewed as a means that the state has of enforcing its monopoly on violence amongst its own citizens. There has been very little, if any, public discussion of the fairness of ShotSpotter's design (Shelley, in press). The sensitivity of the system appears to be left in the hands of the police who, however well intentioned, have a vested interest in the situation. That is, their interests lie largely in minimizing false negatives, gunshots that go unreported. Whether or not this emphasis is most appropriate for the citizenry as a whole has not been publically debated. Given that ShotSpotter is deployed in many areas dominated by ethnic minorities, its use raises issues akin to those of racial profiling. Of course, this lack of transparency was no concern for Hobbes, who viewed the sovereign's power as absolute, so long as peace and public order are maintained. However, in a democratic society, the design of a public surveillance system would normally be the subject of public discussion. Similar comments would apply to an arsonist detection system in the event that one is put in place.

8. VIOLENCE, FAIRNESS AND SAFETY

Some of the cases discussed above do not show concern by the state to monopolize the use of violence by its citizens. In the case of quiet electric cars, for example, there is no attempt by the US government to reserve for itself the right to run over blind pedestrians. Instead, the thrust of the Pedestrian Safety Enhancement Act is to create conditions wherein blind citizens can expect to cross the street without undue risk of stepping out in front of a stealth car. In short, these examples conform to Hammond's (1996) characterization of fairness as a conflict between groups within society who have opposed interests.

The examples discussed above display some interesting contrasts. The US government was swift to act on a potential threat to blind pedestrians. After all, the Toyota Prius has been on the streets for years without reportedly having caused a rash of casualties among the blind (Masnick, 2009). This situation stands in stark contrast to the situation of GPS navigation systems, which seem to have contributed to thousands of accidents but have not yet been the subject of government regulation (Shulman, 2011). The same could be said of people who have suffered violence due to wrap rage.

Perhaps the difference between the two sorts of situation is the victims of the violence in question. In the first case, the victims are a clearly disadvantaged part of the population, that is, people who are blind. In the latter cases, the victims have no similar disadvantage. Indeed, the fact that they have suffered these forms of violence often suggests to observers that they are stupid or too ready to conform to orders (in the case of GPS navigation). Government investment in education might be thought to discharge the state's responsibility to these victims. That is, by providing a general education, the state has done what it can to prepare people to watch the road appropriately, or attend to their own safety while using scissors or kitchen knives.

This argument is open to question, particularly in the case of driving. Modern governments issue driving licenses only after applicants pass a test to assess their skills. Nowadays, GPS navigation systems are becoming almost as common as steering wheels and gear shifts. In these circumstances, skill in using GPS systems could be considered as basic as skill with a steering wheel. Thus, drivers should be educated about, and tested on, proper use of GPS navigation systems, while manufacturers should be required to design the devices to be more responsive to facts about human attention.

The problem of autonomous co-drivers is more likely to attract attention when they become available. People may be less likely to blame the victim when an autonomous co-driver causes the death of a human being behind the wheel. (That is, unless the driver drove carelessly due to over-reliance on the co-driver for safety.)

Previous experience suggests that the state has a role to play in establishing a fair trade-off between the interests of drivers and others. For example, many jurisdictions have adopted *red-light cameras*, that is cameras mounted at traffic intersections that photograph cars that run a red light. The object is to discourage drivers from running red lights and to punish those who do, thereby reducing the risk of accidents and casualties. According to a US Department of Transportation Study (US Federal Highway Administration, 2005), these cameras can reduce the casualties from accidents overall. The cameras do discourage drivers from entering intersections on the red light, when they are likely to be struck (or to strike) from the side by cars entering at a right angle. At the same time, drivers are more likely to brake suddenly to avoid running the red light, which increases their risk of being rear-ended by a car following them. Since rear-end collisions tend to cause fewer casualties that side-on collisions, the overall casualty rate is reduced.

A co-driving system occupies a similar role to the red-light camera. That is, it is in a position to trade off one sort of violence against another when the circumstances arise. Of course, co-driving systems would not be deployed by the state but by private car-makers. Nevertheless, they play the part of traffic cop to some extent and thus warrant government oversight.

9. VIOLENCE, FAIRNESS AND STATUS

When considering fairness in how technology distributes violence in society, we have to consider the trade-off represented by the design of the technology, and the role of the state in its regulation. In addition, we have to consider how the violence in question affects the dignity or status of the people involved. Whether or not violence is fairly distributed may depend upon who is the recipient or the originator of it.

To see the importance of status in issues of fairness, consider the case of Casey Martin. As Sandel (2009, pp. 203-207) explains, Martin was a golfer with a disability that prevented him from walking properly. Martin asked the Professional Golfer's Association (PGA) for permission to ride on a golf cart during tournaments. The PGA replied that golf carts were not permitted, whereupon Martin took the case to court, eventually ending up in the US Supreme Court. Martin argued that requiring him to walk amounted to discrimination against him as a disabled person. The PGA argued that the endurance needed to walk the course during a tournament was part of the sport. The Justices ruled in Martin's favor.

In Sandel's view, the issue for the PGA came down to the status of the game as a sport and its competitors as athletes. Although golf clearly demands skill, it does not involve running, jumping, or other vigorous activity, and the ball stands still while in play. The need to walk the course during a tournament is the most obvious, sustained physical activity that golf involves. If players could excel at golf while riding in a cart, then its status as a sport might be diminished, along with the dignity of its top players.

Similar considerations might apply to some of the examples of violence and technological design discussed above. One could imagine, for example, blind people arguing that the relative silence of electric cars constitutes a kind of discrimination against them as pedestrians. The counterargument analogous with the Casey case would be that quietness is an inherent part of city living for residents. Of course, such an argument would be hard to sustain in the face of the fact that cities have always been somewhat noisy places to live.

The case of intelligent co-drivers is more telling. Safety features in cars have generally been for enhancing the safety of the occupants, and drivers are expected to act to preserve their own lives in the event of a sudden accident. However, as automatic systems take over more and more of the task of driving, less control and prestige attach to the role of driver. In that case, the driver becomes more like a passenger and, for that reason, may not have legitimate claim to the indulgence that drivers otherwise enjoy. In ceding control of their cars to computerized co-drivers, drivers may be ceding any special consideration in the weight their safety is given when an accident happens.

Perhaps this sort of demotion may be observed in the case of GPS navigation. Inadvertently, drivers using GPS navigation systems have already accepted a reduction in their status as vehicle operators in exchange for the convenience of freeing some of their attention from the needs of way-finding. When accidents result from this lack of attention, drivers are frequently ridiculed for having forgotten that they, and not their GPS units, are the bosses of their cars. On this view, it seems as though someone who operates a car but is not fully its boss should not complain about an increase in their risk of suffering violence. Someone taking this view would not find it unfair that GPS navigation systems regularly rely on false or outdated information and tend to remove the driver's attention from the road.

10. CONCLUSION

Although it may sound strange, violence is something that is distributed throughout a society, just like any other benefit or burden. As such, it makes sense to talk about the fairness of this distribution. Also, since the design of technological artifacts is a means by which violence gets distributed, it makes sense to investigate the fairness with which this occurs. In this article, the Taylor-Russell diagram has been used to clarify ways in which design mediates fairness and thus distributes violence.

Three ways in which design mediates violence have been discussed, namely initiating violence, responding to violence, and inadvertent provoking of violence. Each mode of violence can be elucidated by the same method of analysis. The key issue of fairness in each case concerns the trade-off that the technology is designed to bring about. Violence is often suffered by some party to the design in order that another party may realize a benefit, often an increase in freedom from violence. Even in cases where overall violence is reduced, the problem of whether or not the resulting distribution of violence is fair remains.

Problems of fairness have various causes. In some cases, problems may arise from the need for regulation of the appropriate kind. The prospect of armed and autonomous military robots has attracted much attention and public debate, suggesting that appropriate regulation may be in the works. The presence of ShotSpotter has not provoked much attention and, perhaps for that reason, has not yet been the subject of public debate or regulation.

Of course, public scrutiny is no guarantee that issues of fairness will be given appropriate consideration. In the case of quiet electric cars, the potential threat to blind pedestrians, a group that rightly enjoys broad public sympathy, created a regulatory response without much investigation into the magnitude of the threat. In the case of errors by drivers using GPS navigation, the public is clearly less sympathetic even in the face of evidence of persistent and violent incidents. It may be that the public is unsympathetic because its concept of *driver* has not kept up with the current realities of vehicle operation.

This article has touched on some initial issues where the distribution of violence through technological design is concerned. Hopefully, further exploration of these issues will follow.

REFERENCES

Aristotle,. (1998). *Nicomachean ethics* (Ross, D., Ackrill, J. L., & Urmson, J. O., Trans.). Oxford, UK: Oxford University Press.

Barr, J. R., Bowyer, K. W., & Flynn, P. J. (2011, January 5-7). Detecting questionable observers using face track clustering. In *Proceedings of the IEEE Workshop on Applications of Computer Vision* (pp. 182-189).

Coghlin, T. (2008, July 7). *Afghan inquiry into American bombing of 'wedding party'*. Retrieved September 21, 2011, from The Sunday Times website: http://www.timesonline.co.uk/tol/news/world/asia/article4281078.ece

Eaton, K. (2010, December 13). *Senate Approves Bill Requiring Silent EVs Like Nissan's Leaf to Make Noise*. Retrieved September 20, 2011, from FastCompany.com website: http://www.fastcompany.com/1709381/senate-approves-bill-making-noisy-electric-cars-mandatory-as-nissan-ships-first-leaf-ev

Finn, E. (2011, June 14). *'Smart cars' that are actually, well, smart*. Retrieved June 17, 2011, from MIT News website: http://web.mit.edu/newsoffice/2011/smart-cars-0614.html

Foot, P. (1967). Abortion and the doctrine of the double effect. *Oxford Review, 5*, 5–15.

Friess, S. (2006, June 22). *Tales from packaging hell*. Retrieved September 20, 2011, from Wired.com website: http://www.wired.com/science/discoveries/news/2006/05/70874

Hammond, K. R. (1996). *Human judgment and social policy: Irreducible uncertainty, inevitable error, unavoidable injustice*. New York, NY: Oxford University Press.

Hobbes, T. (1985). *The Leviathan*. London, UK: Penguin Classics. (Original work published 1651)

Markoff, J. (2010, October 9). *Google Cars Drive themselves, in Traffic*. Retrieved June 17, 2011, from The New York Times website: http://www.nytimes.com/2010/10/10/science/10google.html?_r=1

Masnick, M. (2009, September 22). *Nissan to Add Futuristic Sound Effects to its Electric Car to keep it from Hitting Unaware Pedestrians*. Retrieved September 23, 2011, from TechDirt website: http://www.techdirt.com/articles/20090921/0141396258.shtml

National Association for Shoplifting Prevention. (2006). *Shoplifting statistics*. Retrieved September 20, 2011, from Shopliftingprevention.org website: http://www.shopliftingprevention.org/WhatNASPOffers/NRC/PublicEducStats.htm

Pauzie, A., & Amditis, A. (2011). Intelligent driver support system functions in cars and their potential consequences for safety. In Barnard, Y., Risser, R., & Krems, J. (Eds.), *The safety of intelligent driver support systems: Design, evaluation and social perspectives* (pp. 7–26). Farnham, UK: Ashgate.

Peñalosa, E. (2003). Parks for livable cities: Lessons from a radical mayor. *Places, 15*(3), 30.

Prigg, M. (2010, February 25). *Satnav is to blame, says driver rescued from swollen river*. Retrieved September 23, 2011, from London Evening Standard website: http://www.thisislondon.co.uk/standard/article-23809822-satnav-is-to-blame-says-driver-rescued-from-swollen-river.do

Sandel, M. (2009). *Justice: What's the right thing to do?* New York, NY: Farrar, Straus and Giroux.

Sandhana, L. (2011, January 24). *Crime-Fighting Technology Spots Lingering Arsonists in the Crowd*. Retrieved September 20, 2011, from FastCompany website: http://www.fastcompany.com/1719341/questionable-observer-detector-spots-lingering-arsonists-at-the-scene-of-the-crime

Selby, N., Henderson, D., & Tayyabkahn, T. (2011). *Shotspotter gunshot location system efficacy report.* Retrieved November 10, 2011, from http://csganalysis.files.wordpress.com/2011/08/shotspotter_efficacystudy_gls8_45p_let_2011-07-08_en.pdf

Sharkey, N. (2008). Grounds for discrimination: Autonomous robot weapons. *RUSI Defense Systems, 11*(2), 86–89.

Shelley, C. (in press). Fairness in technological design. *Journal for Science and Engineering Ethics.*

Shepardson, D. (2011, June 13). *NHTSA chief skeptical of Google's driverless vehicles.* Retrieved June 17, 2011, from The Detroit News website: http://www.detnews.com/article/20110613/AUTO01/106130418/1361/NHTSA-chief-skeptical-of-Google-s-driverless-vehicles

Shulman, A. N. (2011). GPS and the end of the road. *The New Atlantis: A Journal of Technology and Society.* Retrieved September 16, 2011, from http://www.thenewatlantis.com/publications/gps-and-the-end-of-the-road

Singer, P. W. (2009). *Robots at War: The New Battlefield.* Retrieved September 21, 2011, from The Wilson Quarterly website: http://www.wilsonquarterly.com/article.cfm?aid=1313

Smith, D. B. (2011, May 1). *The university has no clothes.* Retrieved May 27, 2011, from New York Magazine website: http://nymag.com/news/features/college-education-2011-5/

Taylor, H. C., & Russell, J. T. (1939). The relationship of validity coefficients to the practical effectiveness of tests in selection: Discussion and tables. *The Journal of Applied Psychology, 23,* 565–578. doi:10.1037/h0057079

The Economist. (2008, October 23). *Surveillance technology: if looks could kill.* Retrieved from The Economist website: http://www.economist.com/node/12465303?story_id=12465303

Tremlett, G. (2010, October 4). *GPS directs driver to death in Spain's largest reservoir.* Retrieved September 23, 2011, from The Guardian website: http://www.guardian.co.uk/world/2010/oct/04/gps-driver-death-spanish-reservoir

Trillin, C. (2007, January, 26). *Park, He Said.* Retrieved June 17, 2011, from The New York Times website: http://www.nytimes.com/2007/01/26/opinion/26trillin.html?_r=1

US Federal Highway Administration. (2005). *Safety Evaluation of Red-Light Cameras.* Retrieved September 23, 2011 from Federal Highway Administration website: http://www.fhwa.dot.gov/publications/research/safety/05049/

Watkins, C., Mazerolle, L. G., & Rogan, D. F. (2002). Technological approaches to controlling random gunfire: Results of a gunshot detection system field test. *Policing: An International Journal of Police Strategies and Management, 25*(2), 345–370. doi:10.1108/13639510210429400

Watters, E. (2007). *ShotSpotter.* Retrieved June 24, 2011, from Wired Magazine website: http://www.wired.com/wired/archive/15.04/shotspotter.html

Weisberg, J. (2010, October 16). *Hyper-libertarian Facebook billionaire Peter Thiel's appalling plan to pay students to quit college.* Retrieved September 20, 2011 from Slate website: http://www.slate.com/id/2271265/

This work was previously published in the International Journal of Technoethics, Volume 2, Issue 4, edited by Rocci Luppicini, pp. 20-36, copyright 2011 by IGI Publishing (an imprint of IGI Global).

Chapter 14
Unintended Affordances as Violent Mediators:
Maladaptive Effects of Technologically Enriched Human Niches

Emanuele Bardone
University of Pavia, Italy, & Tallinn University, Estonia

ABSTRACT

This paper outlines a theoretical framework meant to help people understand the emergence of violent mediators in human cognitive niches enriched by technology. Violent mediators are external objects that mediate relations with the environment in a way facilitating – not causing – the adoption of violent behaviors. In order to cast light on the dynamics violent mediators are involved in, the author illustrates the role played by what is called unintended affordances. In doing so, the author presents three specific examples of unintended affordances as violent mediators: multitasking while driving, desultory behavior and cyberstalking. The last part of the paper presents the notion of counteractive cognitive niche as a possible, yet partial, solution to the problem concerning the emergence of unintended affordance.

1. INTRODUCTION

Technological improvements give the illusion that – once a specific technology is available – a problem or a class of problems can quickly and easily be solved. If in certain cases that might be true, in some others we would better slow down to avoid or, at least, minimize side effects or unexpected consequences. Technologies that actually solve one problem may cause another to appear.

For one reason or the other, the violent and detrimental effects of a technology are hidden or just removed from sight because of the benefits that a particular technology is bringing about with respect to specific problems. For example, GPS devices are extremely useful at helping us interact

DOI: 10.4018/978-1-4666-2931-8.ch014

with places and travel between them. Especially when navigating through unfamiliar places, GPS is just what one needs to avoid losing time, missing an important appointment, etc. However, GPS devices might create new problems. For instance, driving is not just driving. But it is driving and navigating. Without GPS, these two acts are not separated and cannot be. For instance, while one is driving, she or he should pay attention to other cars, landmarks, barriers, pedestrians, curves. All these activities involve driving as well as navigating skills. However, if navigating is completely delegated to GPS, it is very likely to adopt dangerous, antisocial, and narcissistic behaviors like, for instance, changing direction suddenly or failing to notice the presence of pedestrians. Who is to blame?

Generally speaking, technology may create moral dilemmas. Moral dilemmas emerge, as designers and engineers face the challenge of accommodating different, heterogeneous, and sometimes conflicting values. When successfully identified, those moral dilemmas may lead to design trade-offs (Shelley, 2011, in press). With such trade-offs we face what Kuran called "moral overload" (Kuran, 1998). Moral overload emerges when an agent is overloaded by different obligations, which cannot be all fulfilled at the same time. The outcome of moral overload is what is called "moral residue" (Van den Hove, Lokhorst, & Poel, in press). Moral residue is the feeling we have when we have not fulfilled a duty or a value commitment has not been met.

Van den Hoven et al. (in press) posit that moral residue is not necessarily a bad thing, because it gives us an incentive to avoid moral overload in the future by using technology itself as part of the solution. More precisely, it brings up a second-order principle that helps us drive technological innovation. Accordingly, ethics would no longer be a source of constraints, but an active partner in finding innovative solutions.

I am quite sympathetic with Van den Hoven and colleagues' approach. Ethics is not necessarily a source of constraints. What I disagree on is about the meaning of the moral residue Van de Hoven and colleagues talk about. To me a moral residue is also residual of something that is intrinsic to us as moral beings that cannot be designed away: violence. Magnani (2011) in his Understanding Violence offers a deep and insightful account about the relationship between violence and morality, which may help us clarify the point. In a nutshell, he argues that our morality is what often gives us license to be violent by providing us with what we perceive as overwhelming reasons and/or emotions for doing or not doing something. Such overwhelming reasons and emotions conceal our violence. Conversely, what we commonly call violence is any other violent action or behavior that our morality cannot justify as good or right. This philosophical stance leads Magnani to contend that the problem of violence in technology, on the one hand, and the problems related to the so-called "ethics of technology" (or design ethics), on the other, are two different domains, and the former has priority over the latter (Magnani, in press).

In this paper I adopt Magnani's approach in a specific context: I will try to outline a theoretical framework meant to help us understand the emergence of violent mediators in human cognitive niches enriched by technology. Violent mediators mediate our relations with the environment in a way facilitating – not causing – the adoption of violent behaviors. More specifically, violent mediators mediate the maladaptive impact that human cognitive niches may happen to have.

In order to develop this idea, I will rely on the concept of affordance. I will argue that violent mediators result from the emergence of what I call unintended affordances. In my view, unintended affordances are not features of a particular object as, for instance, sitting is for a chair or climbing is for stairs. Conversely, I will try to show how unintended affordances emerge as opportunities for action provided by a cognitive niche or part of it, considered as a whole.

The paper will proceed as follows. In the first part I will present the notion of affordance originally introduced by James Gibson. I will refer to a particular interpretation of the notion: I will argue that the detection of an affordance is the result of an abductive process, in which an agent infers what his or her environment may offer in terms of cognitive chances.

Relying on this particular interpretation of affordance, I will introduce the case of unintended affordance by contrasting it with that of failed affordance. This conceptual analysis will bring us in the third part of paper to illustrating three specific examples of unintended affordances as violent mediators. The three examples are related to: 1) the emergence of anti-social behaviors due to multitasking when driving (especially with relation to GPS devices); 2) the emergence of desultory behavior when using information and communication technologies; 3) the phenomenon of cyberstalking.

The last part of the paper is an attempt to illustrate at a more theoretical level how to counterbalance the emergence of unintended affordances. I will present the notion of counteractive cognitive niche as a possible, yet partial, solution to the problem concerning the emergence of unintended affordances. More specifically, I will claim that the construction of counteractive cognitive niches is related to the identification of appropriate "cognitive firewalls" to counterbalance the maladaptive effects of our cognitive niches.

2. AFFORDANCES AS ABDUCTIVE ANCHORS

Gibson defines "affordance" as what the environment offers, provides, or furnishes. For instance, a chair affords an opportunity for sitting, air breathing, water swimming, stairs climbing, and so on. By cutting across the subjective/objective frontier, affordances refer to the idea of agent-environment mutuality. Gibson did not only provide clear examples, but also a list of definitions (Wells, 2002) that may help us have an idea about what affordances are:

1. Affordances are an opportunity for action, not a property of an object;
2. Affordance are the values and meanings of things which can be directly detected;
3. Affordances are ecological facts;
4. Affordances imply the mutuality of perceiver and environment.

I contend that the Gibsonian ecological perspective achieves two important results. First of all, human and animal agencies are somehow hybrid, in the sense that they strongly rely on the environment and on what it offers. Secondly, Gibson provides a general framework about how organisms directly perceive objects and their affordances. His hypothesis is highly stimulating: "[…] the perceiving of an affordance is not a process of perceiving a value-free physical object […] it is a process of perceiving a value-rich ecological object", and then, "physics may be value free, but ecology is not" (Gibson, 1979, p. 140).

As just mentioned, Gibson defines affordance as what the environment offers, provides, or furnishes. For instance, a chair affords an opportunity for sitting, air breathing, water swimming, stairs climbing, and so on. But what does that exactly mean? What happens when somebody is afforded by something?

Within an abductive framework introduced by Magnani and Bardone (2008), that a chair affords sitting means we can perceive some clues (for instance, robustness, rigidity, flatness) from which a person can more or less tacitly understand that he or she can engage the environment in a specific way: he or she can sit down. Now, suppose the same person has another object O, and she or he can only perceive its flatness. He or she does not know if it is rigid and robust, for instance. Anyway, he or she decides to sit down on it and he/she does that successfully. Is there any difference between the two cases?

I claim the two cases should be distinguished: in the first one, the cues the person comes up with (flatness, robustness, rigidity) are highly diagnostic to know whether or not he or she can sit down on it, whereas in the second case he or she eventually decides to sit down, but we do not have any precise clue about. How many things are there that are flat, but one cannot sit down on? A nail head is flat, but it is not useful for sitting. That is to say, that in the case of the chair, the signs the person comes up with are highly diagnostic. Accordingly, affordances can be related to the variable (degree of) abductivity of a configuration of signs: a chair affords sitting in the sense that the action of sitting is a result of a sign activity in which we perceive some physical properties (flatness, rigidity, etc.), and therefore we can ordinarily "infer" (in Peircean sense) that a possible way to cope with a chair is sitting on it.

According to our perspective, the original Gibsonian notion of affordance deals with those situations in which the signs and clues we can detect prompt or suggest interacting with the environment in a certain way rather than others. In this sense, I maintain that detecting affordances deals with a (semiotic) inferential activity (Windsor, 2004). Indeed, we may be afforded by the environment, if we can detect those signs and cues from which we may abduce the presence of a given affordance.

There are a number of points we should now make clear. They will help us clarify our conception of affordance. I claim that an affordance can be considered as a hypothetical sign configuration where signs or clues are the information specifying the affordance. It is hypothetical in the sense that signs as such do not have value, but they might be symptomatic for an abductive agent. For I contend that an affordance informs us about an environmental symptomaticity, meaning that through a hypothetical process we recognize that the environment suggests for us and, at the same time, enables us to behave a certain way. We can do something with the environment, we can have a certain interaction, and we can exploit the resources ecologically available in a certain way. This is related to how to transform the environment from a source of constraints to a source of resources. The human agent (like any other living organism) tries to attain a stable and functional relationship with his or her surroundings. I now claim that affordances invite us to couple with the environment by informing us about possible symptomaticities. In doing so, affordances become anchors transforming the environment into an abductive texture that helps us establish and maintain a functional relationship with it (Bardone, 2011).

The second point worth mentioning is related to how the human agent regulates his relationship with his environment. I have partly answered to this question. My idea is that the human agent abductively regulates his relationship with the environment. That is, the human agent is constantly engaged in controlling his or her own behavior through continuous manipulative activity. Such manipulative activity (which is eco-cognitive one) hangs on to abductive anchors, namely, affordances that permit the human agent to take some part of the environment as local representatives of some other. So, the human agent operates in the presence of abductive anchors, namely, affordances, that stabilize environmental uncertainties by directly signaling some pre-associations between the human agent and the environment (or part of it).

The third point is that an affordance should not be confused with a resource. I have just argued that an affordance is what informs us that the environment may support a certain action so that a resource can be exploited. Going back to the example of the chair, I contend that an affordance is what informs us that we can perform a certain action in the environment (sitting) in order to exploit part of it as a resource (the chair). This is coherent with the idea introduced by Gibson and later developed by some other authors who see an affordance as an "action possibility".

The idea that an affordance is not a resource but rather, something that offers information about one, allows it to be seen as anything involving some eco-cognitive dimension. That is, insofar as we gain information on environmental symptomaticity to exploit a latent resource, then we have an affordance.

3. WHY AND WHEN WE ARE NOT AFFORDED (AS WE WOULD LIKE TO BE)

In the previous section I illustrated how an abductive framework is particularly fruitful in understanding the process of being afforded. The aim of this section is, conversely, to shed light on those cases in which people are not afforded, and then to explain why they are not. The interest in spelling out the reasons why people might be not afforded will turn out of great importance for introducing the idea of unintended affordances as violent mediators.

I have already pointed out that an affordance is a symptomatic configuration of signs informing a person about a way of exploiting the environment, meaning that the environment enables her for cognitive coupling. I shall distinguish two main ways in which a person or a group of persons may be not afforded as we would like them to be. The case of the so-called failed affordances and that of unintended affordances. Let us start with the former one.

3.1. Failed Affordances

An affordance might not be detected and thus properly exploited, because it is poorly designed so that a person can hardly make use of it, meaning that the sign configuration is poorly symptomatic. In this case we have what we call failed affordances. Norman introduced the distinction between potential affordances and perceived affordances precisely because he wanted to draw a line of demarcation between merely potential chances and those that are fully exploited by the user (Norman, 1999). He argued that only in the second case we do have affordances. However, it is worth noting that it is nearly impossible to predict whether or not a user will detect or perceive the affordances constructed by the designers.

Indeed, the cooperation between designers and users can be enhanced so as to allow designers, for instance, to characterize and thus predict the most likely user reaction. However, like any other activity involving a complex communicative act, there are always design trade-offs (Sutcliffe, 2003) between user needs and environmental constraints identifying not the optimal solution, but the satisfying one. Design trade-offs simply indicate that some affordances intended by the designer cannot be optimally constructed, meaning that an affordance, as a sign configuration, might be ambiguous to the user. If so, then the distinction between potential affordances and perceived affordances starts blurring. Let us see how our conception of affordance may be helpful to solve this problem.

As already pointed out, an affordance informs us about environmental symptomaticity. That is, an affordance is a sign configuration that is totally or scarcely ambiguous so that we promptly infer that an environmental chance is available to us. Therefore, the ambiguity of a sign configuration can be a valuable indicator to demarcate the boundaries of what can be called an affordance and what cannot. However, it is worth noting that on some occasions the user fails to detect an affordance as it was intended in the designer's mind simply because of design trade-offs. For I propose to use the term failed affordances to indicate those situations in which a sign configuration might favor some misunderstanding between designers and users as resulting from design trade-offs. Notwithstanding the fact that they are ambiguous – to a certain extent, failed affordances are still affordances, as their ambiguity is more a result of design trade-offs than of the absence of symptomaticity.

Failed affordances may become violent mediators. However, I argue that failed affordances are more about situations in which we cannot really exploit all the potentialities of a designed object. To make a very simple example, it is like constructing a hammer, which is not very good for hammering. So, to us failed affordances are not fully enabling to do what they were supposed to enable us to do – at least in the designer's mind. This last consideration is very important for introducing the idea of unintended affordance. As I will illustrate in the following, unintended affordances precisely point at those situations in which we are enabled to do something, which is however 1) unpredicted and 2) it bears negative/violent consequences for somebody. It is important to stress here that when we refer to violent behaviors, we always do that by adopting a particular and situated morality. As already mentioned, morality has a sort of cognitive priority with relation to violence, as it is what makes violence conceivable/cognizable and, at the same time, it is also what covers it up (Magnani, 2011). I will come back to this issue again in the end of the paper.

3.2. Unintended Affordances

In his Ecological Approach to Visual Perception Gibson claimed that to perceive an affordance is not to classify an object. He mentioned a number of interesting cases for illustrating it. Consider for instance an elongated object of moderate size and weight. It affords wielding. Such an object, however, may be used for different purposes. For instance, it can be used as a hammer. It may also be used as a rake to pull in things that are beyond one's reach. Or it may be used as a lever. Another example is a graspable and rigid object of moderate size and weight. An object as such affords throwing, so it may be used as a missile. But it may also be a ball and so affording playing.

The list of objects that may afford different possibilities is potentially endless. It is also amazing how some people manage to solve problems by using objects in a way that we would never think it was even imaginable. However, there are certain objects that clearly are not meant to let the user be too much creative. Technological devices are indeed thought to afford us in rather specific ways. Take the example of a trackpad in a laptop. Each gesture the user can perform on a trackpad affords a specific function. It is very unlikely – actually impossible – that dragging a file to the trash bin can also be accomplished by a four-finger pinch, for example. The affordances of a technological device are usually made as much predictable as possible by the designer. So, to a large extent, the designer has the power of constraining how a certain object can be used. As already said in the previous subsection, it can be the case that the designer fails to design easy-to-detect affordances. In such a case, the user might waste time and energy in accomplishing his or her task with that object.

There is, however, another case – our main focus here – in which technological devices provide what I called unintended affordances, that is, affordances that emerge in a specific cognitive niche as unintended or unplanned chances for behavior, which, in turn, have violent consequences with respect to a given morality.[1] Our major claim is that unintended affordances are not directly resulting from design failure – meaning that they are not failed affordances, but they are related to the intrinsic limitations affecting any cognitive niche. As already mentioned, my claim is that unintended affordances are not due to a particular object or device, but they are maladaptive consequences emerging in a particular cognitive niche. Let us try to clarify our claim.

As already mentioned, Gibson argued that to perceive an affordance is not to classify an object. That means that giving a name to an object and/or listing its main features are not necessary conditions to perceive its affordances and/or learn how to use it. Our claim is that Gibson's contention may also be extended to the identification of the boundaries of an object. Indeed, objects have boundaries. That is, one object is clearly distinguishable from one another. A laptop is distinguishable from a mug, a bench from a mobile phone, and so on and so forth. They have different functionalities, different affordances. However, to perceive an affordance does not necessarily imply to perceive/detect such boundaries. That is, affordances may emerge from the simultaneous use of distinct objects that are available in one's cognitive niche. In this case, it is not just an object affording us, but it is a cluster of objects as belonging to a cognitive niche. I argue that unintended affordances may emerge precisely from this: a cluster of objects that are part of a cognitive niche.

Coming back to our idea of affordance as a sign configuration that is symptomatic to somebody, I claim that in the case of unintended affordances the sign configuration is not delivered by a particular object or device. That is, we cannot draw a linear causal chain back to a particular design flaw of a device or object. Conversely, it is delivered ecologically by a portion of a cognitive niche, which goes beyond the boundaries of objects. It is a set of objects as belonging to a cognitive niche that, interacting with each other, may favor the emergence of unintended possibilities for action. As I will show in the next section, unintended affordances mediate the maladaptive dimension of a cognitive niche.

4. UNINTENDED AFFORDANCES AS VIOLENT MEDIATORS: THREE CASES IN POINT

This section will be devoted to illustrating three particular cases in which unintended affordances emerge as violent mediators. The examples are not meant to be complete case studies. Rather, they are meant to be more explorative than demonstrative. The first example is related to multitasking

and driving. It is supposed to make clear that unintended affordances are not related to design failure. Or, at least, the design of an object is not the only variable to take into consideration, as I call for a more holistic approach. The second example is a bit more complicated, and it deals with the emergence of desultory behavior. In this case the holistic component is much more evident, as desultory behavior seems to be an intended consequence related to the abundance of stimuli generated by the progressive virtualization of our sociality. The last example is about the so-called cyberstalking.

4.1. Multitasking and the Adoption of Anti-Social Behaviors

As mentioned in the beginning of the paper, GPS devices are meant to provide instructions to navigate from one place to another in a way that it relieves the driver of knowing exactly where he or she is going to. I have already mentioned that driving is actually composed by two activities (Iaria, Petrides, Dagher, Pike, & Bohbot, 2003). One is driving and the other is navigating. When we know the way, because we are familiar with the place or because we have created a sort of map in our mind, for instance, we usually set out to take the next exit safely so that we do not obstruct other drivers and we successfully avoid any other obstacles that happen to be on our way. GPS does not afford us creating a mental map, but it also does not necessarily afford us forgetting about other drivers. In fact, GPS may be used in conjunction with a map or with the assistance of another person.

However, in particular circumstances GPS may drastically contribute to the emergence of unintended affordances. For instance, car mirrors are fundamental aids for navigation, as they extend the drivers' field of vision in different ways. Interior and external mirrors perform different functions and their positioning has a major impact on our driving performance. Bad positioning and an excessive reliance on GPS can drastically decrease the quality of our driving and, at the same time, affording anti-social behaviors. It seems that unintended affordances do not necessarily arise because GPS is badly designed. Unintended affordances might emerge from the interaction among several objects (including the GPS device) in one's cognitive niche.

Other elements belonging to the driver's cognitive niche can impact on his or her performance. For instance, talking on the phone, texting, switching radio station, and the human body itself may have a negative impact on our ability to deal with GPS devices, for example, driver fatigue, which slows down our response to situations on the road. More generally, unintended affordances in this context are likely to appear and facilitate the adoption of violent behaviors (also as consequences of a car crash like yelling, insulting or even worse) whenever the driver is involved in multitasking. There is overwhelming evidence to prove that multitasking degrades driving performance when a secondary task is added (Just, Keller, & Cynkar, 2008). It is worth to note here that multitasking is not due to a particular object. Conversely, multitasking is afforded by a particular niche. More precisely, it is the totality of the driving experience that is affected, when the driver's particular cognitive niche overstimulates him or her by offering a wide range of secondary tasks (as mentioned above, talking on the phone, texting, listening to music, switching radio station, etc.). In fact, the driving environment has been progressively transformed into something more than a cognitive niche assisting the driver for rather specific tasks (namely, driving and navigating). In the last decades the driver's niche has been constantly re-shaped to resemble more and more one's living room: a comfortable armchair and several sources, controls of pleasure providing both cognitive and emotional stimulation (Urry, 2006). That, of course, did come at a cost: the emergence of unintended affordances.

4.2. Self-Violence and Desultory Behavior

Another interesting example in which unintended affordances emerge is what we may refer to as "desultory behavior", which is a concept introduced by blogger and university professor Timothy Pychyl.[2] In a nutshell, Psychyl claims that social media like Twitter and Facebook create desultory behavior. That is, the tendency to jump from one activity to another in a hasty, erratic, superficial and causal way, say, without being really involved in any of them. The major consequence of desultory behavior would be disconnecting people from one's own long-term goals as a sort of self-violence.

According to Pychyl, interruptions play a fundamental role in creating desultory behavior. Using our terminology, it seems that the combination of various objects with different affordances – simple mobile phones, smart phones, social software like Facebook and Twitter – affords the user to interrupt whatever he or she is doing just to do something else in a sort of endless loop. Interruptions are afforded for instance by different forms of notifications like sounds or banners appearing on the screen of one's device. Such notifications are useful for managing one's incoming information flow. One does not need to check his or her mailbox every now and then. In this sense, notifications help us delegate part of the task.

However, especially when they come from different devices or programs (mail client, Skype, Facebook, etc.), notifications may interfere with our work flow tremendously.[3] The perceptual nature of notification may easily create what I call gaps of inactivity, to which we respond by stopping doing what we are doing, direct our attention to something else, and then try to go back to what we were doing. More specifically, notifications work as attention traps, which are hard to resist. Our perceptual system is designed so as to identify as an object of interest everything that makes available a certain amount of information at initial exposure (Schulkind, 2000). That is what notifications do. In plain English, notifications attract our attention. However, they do that regardless the relevance of the incoming email, text, or whatever. That is, we do not switch from one activity to another for a reason that is somehow relevant to what we are doing.

It is worth adding that we may be engaged in desultory behavior without necessarily being interrupted by notifications, say, by an external trigger. Being desultory might become a habit of our mind as a maladaptive consequence brought about by our technologically enriched niche. That is, in a way we learn how to be erratic by exploiting unintended affordances that happen to be eco-cognitively available.

From a more general point of view, it is interesting to note that gaps of inactivity can occur without any technological device. Consider, for instance, a quite unknown phenomenon called mind wandering (Smallwood, McSpadden, & Schooler, 2008). Mind wandering is basically drifting away from what we are doing, whether it is reading a book, writing an essay, participating in a conversation, etc. It involves losing contact with what is happening around us. Speculations have recently arisen about the function of mind wandering. One interesting hypothesis suggests that mind wandering is not a sign of a lazy mind. Contrary to popular belief, neurological studies have shown that it is related to switching to thinking of goals that are temporally displaced, say, in the future (Christoff, Gordon, Smallwood, Smith, & Schooler, 2009). So, what we do is disconnecting from our short-term goals or tasks – reading a book, for example – to literally immerse ourselves into thinking of long-term goals.[4]

Desultory behavior seems to be related to mind wandering. However, it is just apparently so. I may argue that desultory behavior interferes with mind wandering, which is actually important for the self, as I just noted. We may say that the desultory mind does not even start to wander.

Conversely, it just turns its attention to something else, which, however, has nothing to deal with one's long-term goals.

4.3. Inter-Personal Violence and Desultory Behavior

Desultory behavior may be self-destructive but also affording anti-social, uncooperative or simply impolite behaviors. For example, it is becoming more and more frequent that in the middle of a conversation somebody would wipe out his or her smart phone and check his or her email, Facebook or whatever. In this case, I posit that desultory behavior affords creating gaps of social inactivity, which temporarily disconnect the person from his or her immediate social environment, connect to a more remote one, and back. That may be perceived as violent and, at the same time, lead to the paradox of being absent from those who are around us – our neighbors – in order to be present in another community that is, however, composed of purely interactive beings who "are spending more and more time in parallel universes, disconnected from reality" (Brey, Innerarity, & Mayos, 2009, p. 28). In other words, we are alone together (Turkle, 2011).

There are two main consequences worth mentioning here with relation to the emergence of anti-social behaviors. The first is the most obvious one and it regards what comes out of desultory behavior in one's real life. Embarrassment, confusion, self-blaming, irritability are among the most common responses from those who are "victims" of social desultory behavior. They may perceive desultory behavior as just violent and/or particularly annoying. The second consequence is more related to the kind of community that desultory behavior affords. Our take is that a community of purely interactive beings – which is clearly not representative of all virtual communities one may be part of – amplifies some aspects of human relations that are widely recognized as violent, such as narcissism and superficiality, which, in turn, degrade or even jeopardize one's social capital.

By definition narcissism denotes those people who are incapable of building long-term bonds with other persons so as to reach interpersonal intimacy and reciprocal social exchange. What is interesting here is that narcissists are extremely successful insofar as they can rely on skills related to initiating relationships and subsequently exploiting them for becoming popular in the short term (Buffardi & Campbell, 2008). I posit that communities of purely interactive beings afford narcissism to emerge. The desultor who wipes out the phone and check Facebook, for instance, while having a conversation is not motivated by the need or desire of communicating, but by the need of filling a temporary (and temporal) gap that might be not necessarily connected with boredom, but with a maladaptive habit as mentioned above. The consequence is that any connection established as a response to such a temporary gap can easily correlate to shallow activities characterized by short-term rewards. This sets up the stage for narcissism to emerge.

4.4. Cyberstalking and the Violence of Meaningful Bits: When Men are the Victims

Another example of unintended affordances is related to cyberstalking. Here again we see the same pattern as above. There is no single device or technology providing "bad" affordances for action. But it is a complex and quite complicated network of devices, namely, a cognitive niche or part of it, which produces unintended chances facilitating violent behaviors to emerge.

Generally speaking, stalking is defined as unwanted, malicious, highly obstructive, and obsessive attention that an individual receives from another individual. Cyberstalking is commonly considered a different deviant behavior from traditional stalking, as it is usually pursued by a person by means of information and com-

munication technologies. Indeed, it is not our aim here to give an exhaustive overview about such a complicated and still evolving matter. As far as we are concerned here we mention one particular case, in which unintended affordances can be easily seen at work. Let us see what it is.

As just mentioned, cyberstalking is regarded as a specific deviant and harassing behavior that is not entirely overlapping with traditional stalking. For instance, cyberstalking is usually considered physically less threatening as it lacks the physical contact between the stalker and his or her victim. Interestingly, a recent British research reported that men are considerably more at risk from cyberstalking than from face-to-face stalking (Maple, Short, & Brown, 2011). I argue that this particular case might be seen as resulting from the emergence of unintended affordances.

Specifically, what we call the "cyberspace" affords certain women to be more aggressive. That is due to two features characterizing the cyberspace as a disembodied cognitive niche. First of all, interpersonal violence cannot be perpetrated by relying on those parts of the human body affording to harm and threat a person. For instance, hands, elbow, eye contact, etc. This puts men and women on a par. That is, when violence is mediated by language, women are not certainly weaker than men. Besides, the disembodied nature of the cyberspace grants anonymity, which, in turn, distorts or even impairs the mechanisms of inhibition (Suler, 2004). As Suler argued, the loss of inhibition may be correlated to other phenomena like dissociative anonymity, invisibility, asynchronicity, solipsistic introjection, dissociative imagination, and minimization of authority. All of those phenomena – though affecting both men and women – may encourage women to be more aggressive in the absence of direct physical contact.[5]

Another important aspect to mention here is related to language and its relationship with violence. We tend to think that violence equates with physical violence, say, when blood is shed.

However, words are sometimes just swords of a different kind.[6] The cyberspace as a disembodied cognitive niche has enormously potentiated the transmission of words as meaningful bits. Interestingly, the British research also tells us that the most severe psychological consequences are reported by those men who have been victims of multiple modes of harassment, that is, involving several and different means for communication, i.e., cell phone texts, social networking websites, instant messaging, etc. This finding seems to support the idea that unintended affordances cannot be tracked back to a single device or software. In fact, in the case of cyberstalking it is the systematic and coordinated use of different "object" that affords the stalkers to create in their victim's mind the claustrophobic feeling that they cannot escape, a sort of jail made out of written words as meaningful bits.

5. VIOLENCE, COUNTERACTIVE NICHES, AND COGNITIVE FIREWALLS

In this last section, I will try to present a possible response to the emergence of unintended affordances. It is worth stressing here that our proposal is meant to cast some light on the problem, rather than to solve it. More specifically, I will point at the construction of counteractive niches as a necessary step for counterbalancing unintended affordances. In the second part of the section I will come back to the problem of violence and its relations with morality I mentioned in the introduction of the paper.

5.1. Luck, Violence, and Counteractive Niches

Quite often problems are solved by luck. That is, new and interesting perspectives emerge by luck. To us there is nothing magical or irrational in giving luck credits for something important that has

happened to us. In our view, acknowledging the importance of luck – especially in problem-solving and decision-making – means that there is always something that we do in and with our environment that we do not completely control and, yet, has an impact on us (Magnani & Bardone, 2011).

Our take is that luck is nothing but a sign that our cognition is distributed (Magnani & Bardone, 2008). I claim that luck is an ecological event, meaning that it comes about in the environment. It follows that if one wants to reduce as much as he could the impact of luck that would also mean shrinking from distributing our cognition. If interpreted in this way, luck may have some interesting connection with the notion of affordance that should be brought to light.

Indeed, luck cannot be planned or predicted, but, as long as it is interpreted as an ecological event, we may try to partially domesticate it by manipulating our environment so that the chances we have to prosper may increase (Magnani & Bardone, 2011). That is, we are engaged in a process in which we try to select those aspects of the environment that are affording certain activities rather than others with relation to our goals and aims.

Indeed, such an activity of environmental selection, namely, niche construction (Odling-Smee, Laland, & Feldman, 2003), does not exhaust all the cognitive possibilities that are offered by the environment. Indeed, humans as eco-cognitive predators have no rivals. Human eco-cognitive dominance is a fact (Flinn, Geary, & Ward, 2005). But eco-cognitive dominance does not imply eco-cognitive exhaustion. Already Gibson (1979) pointed out that the environment is a sort of "global market" that provides us with potentially unlimited possibilities. That is true also in relation with environments that are profoundly altered by human activity of human construction. Now, I claim that luck is nothing but a sign pointing at the impossibility of exhausting and so controlling the unlimited supply of possibilities that the environment offers, in good but also, as far as we are concerned here, in bad. That means that luck – good as well as bad – is like a software running in the background: it never stops affecting us. The last contention is very important to shed light on the dynamics promoting unintended affordances, namely, the emergence of maladaptive cognitive niches – independently from our intention.[7] In order to clarify this point a brief evolutionary excursus is needed.

In any human cognitive niche, the selective pressures of the local environment are drastically modified in order to lessen the negative impacts of all those elements which they are not suited to. This new perspective constitutes a radical departure from traditional theory of evolution introducing a second inheritance system called ecological inheritance system (Odling-Smee et al., 2003). The evolutionary impact of such an ecological inheritance system is not related to the presence of replicators – like in the case of natural selection, but the persistence of certain ecological modifications rather than others.

Now, what is interesting to stress here is that natural selection as a blind selective process promoting the survival of the fittest is not halted by niche construction (Odling-Smee et al., 2003). That means – to use our terminology – that luck, though partially domesticated, is still operating and affecting the chances humans have to live long enough to reproduce and prosper. One major consequence that we may draw from this is the impossibility of ruling out the emergence of un-intended affordances as a direct consequence of the impossibility of fully domesticating luck or, more technically, of exhausting the environmental supply of affordances.

In all the three examples I have illustrated, unintended affordances mediate those maladaptive effects characterizing a cognitive niche – as a set of affordances – by administrating, once detected, violence of different kinds. A cognitive niche enriched by GPS does help us avoid to get lost, but it might compromise our ability of navigating due to multitasking, which, in turn, provide

unintended affordances for adopting anti-social or careless behaviors. In the second example, the cognitive niche enriched by multiple sources of information and stimuli may empower a person to know more and to tighten his or her social connections; but, at the same time, it may lead to the adoption of desultory behavior affecting both his or her goals and social cooperation. In the third and last example, the cognitive niche in which words between people can freely travel at light speed from one place of the globe to another provides the chance to harass somebody without any physical mediation, but only by means of language – usually written language.

I posit that violence is intrinsically connected with technology. More precisely, violence is structural to any activity in which humans exhibit their potentialities as eco-cognitive engineers. Here, I consider violence as a sort of primitive behavior arising from the very fact that humans are surviving machines acting in cognitive niches that happened to be increasingly human-tailored. Building on Thom's analysis of language (1980) and Magnani's (2011) philosophical stance on violence, we may specify our claim by saying that human cognitive niches have become in the course of human evolution what mediates our relations with vital pieces of information about the fundamental biological opponents (life/death, good/bad). Therefore, the construction or modi-fication of cognitive niches is intrinsically and essentially connected with violence.

Acknowledging the inevitability of violence is not in itself a justification for adopting violent behaviors. It does not lead to indifference. Magnani clearly points out that acknowledging that we are intrinsically violent beings is a way to be – at least – responsible violent beings (Magnani, 2011). Respecting violence might even become a sort of primeval and unconditional duty. That makes a big difference when one comes to face up to the challenge of counterbalancing unintended affordances as violent mediators. Let us see how.

As already mentioned, natural selection is not halted by niche construction. Selective pressures are modified, but still constraining our chances for prospering as individuals and members of society. Whether or not changes in selective pressures are driven or caused by human eco-cognitive activities, humans may experience some adaptive lag, namely, a mismatch between current selection pressures and profitable behavioral options that are available to the human agent (Laland & Brown, 2006). Unintended affordances will be more likely to come out, when such a mismatch becomes wider and wider. Interestingly, Odling-Smee and colleagues posit that more sophisticated niche constructors – humans, in our case – exhibit a greater capacity for building what they call "counteractive niche" (Odling-Smee et al., 2003). That is, greater eco-cognitive engineering can better counteract those independent changes that have emerged in a niche and that have widened the mismatch between current selection pressures and the available behavioral repertoire.

In case of humans, I claim that counteractive cognitive niches have the major role of facing up to the emergence of unintended affordances. This is an important point that needs to be clarified. As repeatedly noted in this paper, negative unintended affordances cannot be merely attributed to the faulty design of a particular object. That is the case of failed affordances, which usually have smaller impacts if any. Conversely, unintended affordances are mediators – usually producing intra- and inter-personal violence – emerging at the eco-cognitive level, meaning within a complex network of interactions. For the proper response can only be identified in counteracting the particular niche which has been intoxicated, so to say.

5.2. Cognitive Firewalls as Counteractive Affordances

Our claim is that counteractive cognitive niches counterbalance the emergence of unintended affordances by creating suitable cognitive firewalls.

According to Cosmides and Tooby, cognitive firewalls are "systems of representational quarantine and error correction" (Cosmides & Tooby, 2000, p. 105), which have a specific role: to avoid inference propagates errors due to wrong information given as input (see also Magnani, 2011; Magnani & Bertolotti, 2011). Cosmides and Tooby posit that cognitive firewalls are of that importance because of the massive amount of information that humans should handle and, at the same time, they could exploit in enriched cognitive niches. The chance to extract and exploit information ecologically delivered does come at a cost. The cost is related to the ceaselessly activity of locating, monitoring, and updating relevant pieces of information in order to avoid maladaptive outcomes. Within a cognitive framework, which acknowledges the distributed nature of human cognition, cognitive firewalls may be characterized in a slightly different way. We may consider cognitive firewalls as counteractive affordances, which redistribute our cognitive and moral effort when dealing with those parts of the cognitive niche that is more fragile. Let us clarify this point.

In Section 1 I argued that an affordance is a sign configuration from which we may easily infer that we can exploit the environment a certain way. In the case of unintended affordances, the sign configuration becomes available as an unintended consequence in our cognitive niche, and it becomes symptomatic to some people, who – it is important to stress – do not necessarily intend to be violent or harassing. Cognitive firewalls as counteractive affordances are effective insofar as they block the inferential process from being triggered. In doing so, they avoid a cognitive niche to be irreparably intoxicated due to the emergence of unintended affordances. That is, cognitive firewalls lessen or try to minimize the maladaptive effects a cognitive niche inevitably has.

Given their distributed and ecological nature, cognitive firewalls may be of different types encompassing several dimensions, namely, technological, but also epistemological, social, political, and moral one. Besides, they can be embedded directly into technological devices, but also into artifacts of a completely different kind like institutions. For instance, both Magnani (in press) and Hildebrandt (2009) acknowledge and actively support the re-shaping of present democratic institutions for protecting their core values. Magnani also pointed at what he called rounded knowledge as a key factor for creating new moral ontologies in order to face the problems related to our technologically enriched niches. This is how he described such kind of knowledge: "knowledge built only from scientific fields has limited value, as would knowledge drawn only from humanistic fields [...]. It must be a well-rounded, varied body of information drawn from many disciplines, and even then it must be useful, available, and appropriately applied" (Magnani, 2007, p. 106). That is clearly pointing at the construction of cognitive firewalls.

As a concluding remark for this section, the construction of counteractive cognitive niches may be characterized by a pronounced discontinuity with the past, since it may involve a reconfiguration of the available cognitive as well as moral repertoire. It is worth noting that such a reconfiguration still has to do with violence, since it is ultimately driven by a more or less tacit moral set of principles inspiring it. Say, we want things to change. This last consideration is in line with what Magnani argues on a more philosophical level about the intertwining between morality and violence, as I have repeatedly noted all along the paper. As he fairly points out, "[h]uman beings are prisoners of what I call moral bubbles, which systematically disguise their violence to themselves" (Magnani, 2011). Indeed, the kind of violence involved in counteractive niche construction do not involve what is usually labeled as individual violence (i.e., somebody punches or kills somebody else), but structural one, which, as noted again by Magnani, is usually seen as morally legitimate and thus concealed or hidden on the basis of reasons. For example, a stalker finds morally legitimate to

intrude into his victim's life. Actually, he would not label himself as an intruder. He might just say that his victim is inhumane, that is, lacking empathy and compassion for him. If one adopts the cyberstalker's morality, anti-stalking measures are indeed perceived as violent, thus immoral. For instance, a cyberstalker in a cognitive niche de-potentiating the chance of cyberstalking by the presence of suitable cognitive firewalls would not be able to discharge his psychic energy to get some psychological relief.

6. CONCLUSION

In this paper I have tried to shed light on a specific problem related to the relationship between violence and technology. That is, when part of our technologically enriched cognitive niches may afford us to be violent. I have supported this claim by relying on the notion of unintended affordance. I argued that unintended affordances arise in presence of a cluster of objects, not just a single one. In this sense, the kind of violence I have been talking about is eminently structural. Then, I have presented several examples in which the emergence of unintended affordances as violent mediators is more visible. In the last part of the paper, I have tried to cast some light on a possible response to the problem of unintended affordances based on the notion of counteractive niche. Far from being an exhaustive and complete treatment of the matter, I hope that the present one would give a contribution in the direction of making us a bit more responsible violent beings.

ACKNOWLEDGMENT

This research was supported by the Estonian Science Foundation and co-funded by the European Union through Marie Curie Actions.

REFERENCES

Bardone, E. (2011). *Seeking chances: From biased rationality to distributed cognition.* Heidelberg, Germany: Springer-Verlag.

Bertolotti, T., Bardone, E., & Magnani, L. (2011). Perverting activism: Cyberactivism and its potential failures in enhancing democratic institutions. *International Journal of Technoethics, 2*(2), 19–29. doi:10.4018/jte.2011040102

Brey, A., Innerarity, D., & Mayos, G. (2009). *The ignorance society and other essays.* Barcelona, Spain: Infonomia.

Buffardi, L. E., & Campbell, W. K. (2008). Narcissism and social networking web sites. *Personality and Social Psychology Bulletin, 34*(10), 1303–1314. doi:10.1177/0146167208320061

Christoff, K., Gordon, A. M., Smallwood, J., Smith, R., & Schooler, J. (2009). Experience sampling during fMRI reveals default network and executive system contributions to mind wandering. *Proceedings of the National Academy of Sciences of the United States of America, 106*(21), 8719–8724. doi:10.1073/pnas.0900234106

Cosmides, L., & Tooby, J. (2000). Consider the source: the evolution of adaptations for decoupling and metarepresentation. In Sperber, D. (Ed.), *Metarepresentations: A Multidisciplinary Perspective* (pp. 53–115). Oxford, UK: Oxford University Press.

Csikszentmihalyi, M. (1996). *Creativity: Flow and the psychology of discovery and invention.* New York, NY: Harper Perennial.

Flinn, M., Geary, D., & Ward, C. (2005). Ecological dominance, social competition, and coalitionary arms races. Why humans evolved extraordinary intelligence. *Evolution and Human Behavior, 26*(1), 10–46. doi:10.1016/j.evolhumbehav.2004.08.005

Gibson, J. J. (1979). *The ecological approach to visual perception.* Boston, MA: Houghton Mifflin.

Hildebrandt, M. (2009). Technology and the end of law. In Keirsbilck, B., Devroe, W. W., & Claes, E. (Eds.), *Facing the limits of the law* (pp. 1–22). Heidelberg, Germany: Springer-Verlag. doi:10.1007/978-3-540-79856-9_23

Iaria, G., Petrides, M., Dagher, A., Pike, B., & Bohbot, V. (2003). Cognitive strategies dependent on the hippocampus and caudate nucleus in human navigation: Variability and change with practice. *The Journal of Neuroscience, 23*(13), 5945–5952.

Just, M. A., Keller, T., & Cynkar, J. (2008). A decrease in brain activation associated with driving when listening to someone speak. *Brain Research, 1205,* 70–80. doi:10.1016/j.brainres.2007.12.075

Killingsworth, M., & Gilbert, D. (2010). A wandering mind is an unhappy mind. *Science, 330,* 932. doi:10.1126/science.1192439

Kuran, T. (1998). Moral overload and its alleviation. In Ben-Ner, A., & Putterman, L. (Eds.), *Economics, Values, Organization* (pp. 231–266). Cambridge, UK: Cambridge University Press.

Laland, K., & Brown, G. (2006). Niche construction, human behavior, and the adaptive-lag hypothesis. *Evolutionary Anthropology, 15,* 95–104. doi:10.1002/evan.20093

Magnani, L. (2007). *Morality in a technological world: Knowledge as duty.* Cambridge, UK: Cambridge University Press. doi:10.1017/CBO9780511498657

Magnani, L. (2011). *Understanding Violence: The Intertwining of Morality, Religion and Violence: A Philosophical Stance.* Heidelberg, Germany: Springer-Verlag.

Magnani, L. (in press). Abducing personal data, destroying privacy. Diagnosing profiles through artifactual mediators. In M. Hildebrandt & E. Vries de (Eds.), *Privacy, Due Process and the Computational Turn: Philosophers of Law Meet Philosophers of Technology*. London, UK: Routledge.

Magnani, L. (in press). Structural and technology-mediated violence. Profiling and the urgent need of new tutelary technoknowledge. *International Journal of Technoethics*.

Magnani, L., & Bardone, E. (2008). Sharing representations and creating chances through cognitive niche construction. The role of affordances and abduction. In Iwata, S., Oshawa, Y., Tsumoto, S., Zhong, N., Shi, Y., & Magnani, L. (Eds.), *Communications and Discoveries from Multidisciplinary Data* (pp. 3–40). Heidelberg, Germany: Springer-Verlag. doi:10.1007/978-3-540-78733-4_1

Magnani, L., & Bardone, E. (2011). From epistemic luck to chance-seeking. The role of cognitive niche construction. In A. König, A. Dengel, K. Hinkelmann, K. Kise, R. Howlett, & L. Jain (Eds.), *Proceedings of the 15th International Conference on Knowledge-Based and Intelligent Information and Engineering Systems, Part II* (LNCS 6882, pp. 486-494).

Magnani, L., & Bertolotti, T. (2011). Cognitive bubbles and firewalls: Epistemic immunizations in human reasoning. In *Proceedings of the 33rd Annual Conference of the Cognitive Science Society* (pp. 3370-3375). Austin, TX: Cognitive Science Society.

Maple, C., Short, E., & Brown, A. (2011). *Cyberstalking in the United Kingdom: An analysis of the echo pilot survey*. Bedfordshire, UK: University of Bedfordshire National Centre for Cyberstalking Research.

Norman, D. (1999). Affordance, conventions and design. *Interaction, 6*(3), 38–43. doi:10.1145/301153.301168

Odling-Smee, F., Laland, K., & Feldman, M. (2003). *Niche Construction: A Neglected Process in Evolution*. Princeton, NJ: Princeton University Press.

Schulkind, M. (2000). Perceptual interference decays over short unfilled intervals. *Memory & Cognition, 28*, 949–956. doi:10.3758/BF03209342

Shelley, C. (in press). Fairness and regulation of violence in technological design. *International Journal of Technoethics*.

Shelley, C. (in press). Fairness in technological design. *Science and Engineering Ethics*.

Smallwood, J., McSpadden, M., & Schooler, J. (2008). When attention matters: The curious incident of the wandering mind. *Memory & Cognition, 36*, 1144–1150. doi:10.3758/MC.36.6.1144

Suler, J. (2004). The online disinhibition effect. *Cyberpsychology & Behavior, 7*(3), 321–326. doi:10.1089/1094931041291295

Sutcliffe, A. (2003). Symbiosis and synergy? Scenarios, task analysis and reuse of HCI knowledge. *Interacting with Computers, 15*, 245–263. doi:10.1016/S0953-5438(03)00002-X

Thom, R. (1980). *Modèles mathématiques de la morphogenèse* (Brookes, W. M., & Rand, D., Trans.). Paris, France: Christian Bourgois.

Turkle, S. (2011). *Alone Together: Why we expect more from technology and less from each other*. New York, NY: Basic Books.

Urry, J. (2006). Inhabiting the car. *The Sociological Review, 54*, 17–31. doi:10.1111/j.1467-954X.2006.00635.x

Van den Hove, J., Lokhorst, G.-J., & Van Poel, I. (in press). Engineering and the problem of moral overload. *Science and Engineering Ethics*.

Wells, A. J. (2002). Gibson's affordances and Turing's theory of computation. *Ecological Psychology, 14*(3), 141–180. doi:10.1207/S15326969ECO1403_3

White, B. (in press). Infosphere to ethosphere. Moral mediators in the nonviolent transformation of self and world. *International Journal of Technoethics*.

Windsor, W. L. (2004). An ecological approach to semiotics. *Journal for the Theory of Social Behaviour, 34*(2), 179–198. doi:10.1111/j.0021-8308.2004.00242.x

ENDNOTES

Aristotle used the term "Hamartia" to refer to behaviors that are not due to a character flaw, but flawed judgment intrinsically affecting humans as bounded rational beings. See White (in press).

[2] http://www.psychologytoday.com/blog/dont-delay/200904/twitter-commentary-desultory-behavior. On desultory behavior see also Bertolotti, Bardone, and Magnani (2011).

[3] On the cognitive importance of flow for creativity and self-fulfillment, see Csikszentmihalyi (1996) and the concept of flow.

[4] It should be noted that no all mind wandering related phenomena are positive. For more information see also Killingsworth and Gilbert (2010).

[5] To avoid possible misunderstanding on this point, I am not claiming that cyberstalkers are exclusively women. Or, even worse, that women are not brave enough to confront another men in real life situations. Sadly to say, women are still more at risk from cyberstalking than men as Maple and colleagues reported in their research.

[6] Magnani is quite clear in pointing out that "the sophisticated intellectual and philosopher – but not the average more or less cultured person – says certainly 'mobbing involves violence!', but this violence is de facto considered secondary, at most it is only worth considering as 'real' violence when it is in support of actual events of physical violence. It is well-known that people tend to tell themselves that the mobbing acts they perform are just the same as cases of innocuous gossiping, implementing a process of dissimulation and self-deception" (Magnani, 2011, p. 5).

[7] Generally speaking, maladaptive cognitive niches are those bringing about a wide spectrum of negative consequences. For instance, they increase the chance of having conflicts, intra- and inter- personal violence, an unequal distribution of costs and benefits, resentment among the population, and so on and so forth.

This work was previously published in the International Journal of Technoethics, Volume 2, Issue 4, edited by Rocci Luppicini, pp. 37-52, copyright 2011 by IGI Publishing (an imprint of IGI Global).

Chapter 15

Infosphere to Ethosphere:
Moral Mediators in the Nonviolent Transformation of Self and World

Jeffrey Benjamin White
KAIST, Korea

ABSTRACT

This paper reviews the complex, overlapping ideas of two prominent Italian philosophers, Lorenzo Magnani and Luciano Floridi, with the aim of facilitating the nonviolent transformation of self and world, and with a focus on information technologies in mediating this process. In Floridi's information ethics, problems of consistency arise between self-poiesis, anagnorisis, entropy, evil, and the narrative structure of the world. Solutions come from Magnani's work in distributed morality, moral mediators, moral bubbles and moral disengagement. Finally, two examples of information technology, one ancient and one new, a Socratic narrative and an information processing model of moral cognition, are offered as mediators for the nonviolent transformation of self and world respectively, while avoiding the tragic requirements inherent in Floridi's proposal.

1. SELF, POETRY, AND INFORMATION

Without adequate reasoning, even well-intentioned moral actions may fail – or, worse still, cause harm – and the best way to facilitate adequate reasoning is to confront problems with flexible and well-fed minds - Lorenzo Magnani (2007, p. 165)

DOI: 10.4018/978-1-4666-2931-8.ch015

Luciano Floridi has "defended a view of the world as the totality of informational structures dynamically interacting with each other", in which the world as information ecosystem, the "infosphere", and all individual inhabitants are essentially interconnected "informational entities" with any given thing's "least intrinsic value" "identified with its ontological status as an information object" as determined by an "analysis of *being* in terms of a minimal common ontology, whereby human beings as well as animals, plants, artifacts and so forth are interpreted as informational entities" (Floridi, 2011b, p. 564, 2002, p. 287, 2006, p. 33), "To be is to be an informational entity" (Floridi, 2008, p. 199). On Floridi's picture, living things are also informational entities, "inforgs", informational organisms, with human beings nestled amongst them as "interconnected informational organisms among other informational organisms and agents, sharing an informational environment" (at 4:53), and with that informational environment characterized by its degree of entropy (Floridi, 2011a, 2010, 2006). Entropy is important to Floridi's information ethics as it is central both to the role of human beings in the informational environment, and to the problem of evil, understood as increasing entropy (Floridi, 2006, 2002, 1999).

Floridi understands human beings as *homines poietici* - selves whose defining capacity is to create order, from which derives the relationship between entropy and evil, for example the identification of evil with the destruction of order (i.e., violence), and with each self emerging "as a break with nature, not as a super connection with it" (Floridi, 2006, 2011b, p. 560). Moreover, *homo poieticus* is unique in her/his reflexive awareness of this distinctive status, as an island of order within a sea of disorder, and with this recognition he/she is able to identify similar entities:

Selves are the ultimate negentropic technologies, through which information temporarily overcomes its own entropy, becomes conscious, and is finally able to recount the story of its own emergence in terms of a progressive detachment from external reality. There are still only informational structures. But some are things, some are organisms, and some are minds, intelligent and self-aware beings. Only minds are able to interpret other informational structures as things or organisms or selves. And this is part of their special position in the universe (Floridi, 2011b, pp. 564-565).

The "progressive detachment from external reality" is a process undertaken by minds, human beings, as they understand their place amongst other informational entities, a process aided and accelerated by information communication technologies (ICTs). Floridi holds that ICTs are making possible a "fourth revolution" in human self-understanding, following Copernicus, Darwin, and Freud. Note both that these first three plots an inward trend, from humans understanding their place amongst the stars, to their position within evolutionary time on Earth, to their selves as products of physical and metaphysical relationships, and that this fourth revolution follows this trend, as ICTs aid humans not only in understanding themselves, but in actively designing and constructing them. Floridi writes that:

ICTs are, among other things, egopoietic technologies or technologies of self construction, significantly affecting who we are, who we think we are, who we might become, and who we think we might become (Floridi, 2011b, p. 550).

As part of their "special position in the universe" (note: not some limited domain therein), detached from reality and from this perspective, through technology, increasingly empowered to both study and shape it, humans have a "vocation for responsible stewardship in the world" (Floridi, 2006, p. 34).

The more powerful homo poieticus becomes as an agent, the greater his duties and responsibilities become, as a moral agent, to oversee not only the development of his own character and habits but also the well-being and flourishing of each of his ever expanding spheres of influence, to include the whole infosphere (Floridi, 2006, p. 32).

Accordingly, Floridid stresses not only the *egopoietic*, self-constructing potential afforded by ICTs, but also the world transforming, *eco-*

poietic potential, where "The term "*ecopoiesis*" refers to the morally-informed construction of the environment, based on an ecologically-oriented perspective." (Floridi, 2006, p. 31)

On this basis, Floridi extends a principled ethics resting on entropy "meant to be a macroethics for creators not just users of their surrounding 'nature,'" in which, "fighting information entropy is the general moral law to be followed, not an impossible and ridiculous struggle against thermodynamics" (Floridi, 2006, p. 33, 2002, p. 300):

The duty of any moral agent should be evaluated in terms of contribution to the sustainable blooming of the infosphere, and any process, action or event that negatively affects the whole infosphere – not just an informational object – should be seen as an increase in its level of entropy and hence an instance of evil (Floridi, 2006, p. 32).

But, how are we to understand "entropy" and "evil" consistently throughout our "ever expanding spheres of influence", including the "whole infosphere", so that we might not only become responsible stewards of the world but also become morally "good" human beings?

First, we may wonder how we come to know ourselves as distinct selves in the first place. Floridi volunteers "what may be a fruitful approach to start understanding the construction of personal identities" centered on three terms: "egology", the development of selves, proceeding first by individualization and then by (self-)identification, "*self-poiesis*", self-creation in the first-person through progressive detachment from the world (the not-self), punctuated by "*anagnorisis*", the tragic realization that one's self is not the self (the not-not-self) one thought one's self to be:

In Aristotle, the phenomenon of anagnorisis refers to the protagonist's sudden recognition, discovery, or realization of his or her own or another character's true identity or nature. Through anagnorisis, previously unforeseen character information is revealed. Classic narratives in which anagnorisis plays a crucial role include Oedipus Rex and Mac-Beth. More recently one may mention The Sixth Sense, The Others, or Shutter Island. I shall not spoil the last three, if the reader has not watched them. Generalizing, one may say that, given an information flow, anagnorisis is the information process (epistemic change) through which a later stage in the information flow (the acquisition of new information) forces the correct reinterpretation of the whole information flow (all information previously and subsequently received). For this reason, I prefer to translate anagnorisis as realization (Floridi, 2011b, p. 564).

On this assay, one's self is most completely realized when the self one thought one was is discovered to not, in fact, be the self whom one is, compelling "the correct reinterpretation of the whole information flow." Moreover, this process occurs within narrative structures of our own creation, "through which we semanticize reality, i.e., through which we make sense of our environment, of ourselves in it, and of our interactions with and within it" (Floridi, 2011b, p. 564).

Notably, *anagnorisis* is not a stand-alone notion. In Aristotle's *Poetics*, his analysis is essentially tied with that of *hamartia*, a tragic error in judgment, from which disaster results (Aristotle, 1995, *Poetics*, books 13 and 14). *Hamartia* is not a character flaw, but flawed judgment due to emotion over-riding incomplete information. An agent acts toward an end, misses, and effects the opposite of those intentions, falling to tragic irony, realizing himself in terms not only opposite, but inverted, to those presupposed, and is thus presented the option to reconcile the difference, transcending prior limitations, becoming wiser, or close to it, only thereby expressing a flawed character, vicious ignorance (White, 2006).[1] So clarified, it is only to the original point of error, *hamartia*, that the information flow is re-cast, not the "entire" information flow as Floridi speculates. For example, the Occupy movement in the USA generally calls for repeal of the Federal Reserve act of 1913, to the point where the people feel a tragic error had been made, and not for the revision of all of Western history (Kincaid, 2011). Moreover, this illustration of *hamartia* in historical narrative reveals a rather disturbing fact about the "entire information flow" – we create it on the fly. Often enough where we end up is the product of a mistake, and often enough due to inadequate information. Finally, it is what we do at this point, fully informed of possibilities otherwise, that determines our characters, good or evil.

Two problems must be addressed. First, entropy. Entropy in information theory does not exactly match that of the physical sciences, with the former understanding entropy as the probability of the presentation of a message, and with the latter understanding entropy as a direct measure of disorder relative potential energy. Floridi is a "nonreductionist", supposing information of different forms to be qualified by exclusive accounts and measured by specific means, thereby admitting multiple sorts of entropy (Floridi, 2004). This position, however, stands against his "view of the world as the totality of informational structures dynamically interacting with each other", a position demonstrated by his own direct application of the "entropy method" from the artificial to the "standard domain" in assessing the moral significance of a broken windscreen (Floridi, 1999). One cannot have things both ways – either the world is coextensive with the domain of information, as implied when information counts as the fundamental ontological dimension, or information is a sub-domain of the world, as is implied when informational entropy is otherwise than natural entropy. We shall realize an informational view of self and world grounded on a single form in the next section.

Second, the problem of value. Raphael Capurro counters Floridi, holding that all value derives from human evaluation, as there is a human being – a "mind" - at the center of any claim to moral significance. Leveling the field of being on the basis of a common information ontology, while at once disengaging from the natural world that governs human being, creates a conceptual environment that dislocates the source of valuation from selves to some realm of ideals, "out there", diminishing both the human role in the production of value through work and its social value assessed through empathic mirroring of similarly embodied conditions, thereby inviting a slippery slope whereby things become more valuable than the human beings who engage with them (Capurro, 2008). This is problematic, recalling Plato, because as things gain value over people, we move from a healthy world to one with a "fever". Indeed, as recent Occupy movements around the world demonstrate, it seems that we already live in such a world, illuminating two important issues. One, as asserted on OccupyWallSt.org, the movement depends on ICTs to coordinate actions, by the so-called "Arab Spring tactic to achieve our ends and encourage the use of nonviolence to maximize the safety of all participants." This fact cements the moral significance of information. And, following Capurro, it also affirms that this value derives from human beings using information to construct a just social order. Two, there is no doubt that *anagnorisis* results in self-realization, for better or for worse. However, it is this potential for ending up worse that questions its fruitfulness as a strategy for advancing self-understanding. Certainly, at tragic ends already, it is a useful model for moral reasoning, but the people behind OccupyWallSt.org would certainly rather that the Federal Reserve Act had failed in the first place, just as Oedipus may have benefitted by DNA testing.

This is indeed a critical period in human development, even a "fourth revolution" of self-understanding catalyzed by information technologies and serving the potential construction of healthy selves and world. But, we must proceed with caution, else invite further tragedy. In this spirit, consider Lorenzo Magnani:

While I deeply believe that creating and acquiring new knowledge is critically important, even I must admit that all the information in the world is meaningless unless we can use it effectively: the principles and ways of reasoning that allow us to put new ethical knowledge to work are just as important as the knowledge itself (Magnani, 2007, p. 162).

It is not enough to recognize that all things are potential sources of information, to commit to promote "the flourishing of informational entities as well as of the whole infosphere" by limiting entropy within it. "Coherence in moral behavior is not necessarily good in itself" (Magnani, 2007, p. 145). And, order for the sake of order is not necessarily good in itself, either, as Occupy Wall-Streeters stand in testament. At least until the moral significance of entropy is clarified, commitments to avoid it are nothing we can "put to work".

We must conceptually transform Floridi's infosphere into the "ethosphere". But, with due caution, how should we proceed? To complete this task will require two things. For one, as it is inevitably some self, or selves, responsible for *hamartia,* piloting this transformation sans tragedy requires first of all a conception of self up to the task, "a conception of self that may be more easily shared (though of course, pluralistically) among the larger globe" (Ess, 2008, p. 167). An informationally modeled self is indeed easily shared, bereft as it is of culturally specific baggage, and so indeed the fourth revolution in self-understanding is well-presented as an informational one. Additionally, we require the tools and methods necessary to effect the transformation of the world of informational objects into one of moral value, infosphere to ethosphere. According to Charles Ess,

[...] we are in need of developing notions of distributed responsibility in an ethics of distributed morality - i.e., notions better suited to our realities as informational agents and patients, who are inextricably interwoven with one another via computer and other networks (Ess, 2008, p. 161).

2. MAGNANI, MEDIATION, AND METHOD

In one word, science (critically undertaken and methodically directed) is the narrow gate that leads to the true doctrine of practical wisdom, if we understand by this not merely what one ought to do, but what ought to serve teachers as a guide to construct well and clearly the road to wisdom which everyone should travel, and to secure others from going astray. - Immanuel Kant (1898, p. 174)

Lorenzo Magnani has written systematically on the role of abduction both in "scientific and everyday reasoning", "typical reasoning in presence of incomplete information", understood from Peirce as "inference that involves the generation and evaluation of explanatory hypotheses" and "inferring certain facts and/or laws and hypotheses that render some sentences plausible", present even in simple instances of visual perception, and the form of reasoning most applicable to moral problems (Magnani, 2005, 2011, p. 61; Magnani & Nersessian, 2002, p. 306). One Peircean illustration of this form of reasoning is offered in the distinction between two types of cloth by comparing the sensations of their textures, "but not immediately", requiring one "to move fingers over the cloth" (Magnani, 2007, p. 185). This is an example of "manipulative abduction", "thinking through doing" which, from evolutionary and developmental psychological perspectives eventuates in increasingly sophisticated sentential and symbolic reasoning through a process that Magnani designates the "disembodiment of mind" (Magnani, 2009).

Epistemically, by this process, the space of action is patterned in "construals" - effectively embodied and situated knowledge how. As these construals stabilize through repetition, they begin to pre-figure experience and rise to explicit awareness. Formalized, they become "parts of a memory system that crosses the boundary between person and environment", forming the basis of ritual, conventional law, guiding religious narratives, and theories – such as information ethics - with predictive force and moral value, all "disembodied" aspects of distributed cognition (Magnani, 2007, p. 242, p. 232). Accordingly, Magnani's work has also focused on "model-based reasoning", indicating "the construction and manipulation of various kinds of representations, not mainly sentential and/or formal, but mental and/or related to external mediators", following Peirce in understanding that all thought, "being minds", is mediated by "signs" - phenomena "available to interpretation" guide our actions "in a positive or negative way" and "become signs when we think and interpret them", including feelings, e.g., "somatic markers" (Magnani & Nersessian, 2002, p. 308; Magnani, 2009, p. 490; Damasio, 1999). "Being a mind" means "to be absorbed in making, manifesting, or reacting to a series of signs" - recalling Floridi, "able to interpret other informational structures as things or organisms or selves" - while being also "an external sign" for self and others (Magnani, 2009, pp. 489-490). Most importantly, "the "person-sign" is a future-conditional", "not fully formed in the present but depending on the future destiny of the concrete semiotic activity (future thoughts and experience of the community) in which s/he will be involved" (Magnani, 2009, p. 490). Thus, cognition is essentially evolved, developed, embodied, self-determining and distributed, "extended" through time and space, "consisting of the person plus the external physical representation" that facilitates and potentially changes the way that one reasons, as is the case with "epistemic mediators" like microscopes, telescopes, pen and paper, and information technology (Magnani & Nersessian, 2002, pp. 312-313).

Magnani locates morality in "all those situations in which humans have to manage problems related, for instance, to making decisions or policies that may have a moral concern and impact on our lives", including "coming up with new ideas that can solve old problems and even create new moral concerns towards new moral entities, such as animals and things" (Magnani & Bardone, 2008, p. 104). This is ethics as a creative enterprise, "creating ethics" through moral reasoning, which can "be viewed as a form of "possible worlds" anticipation, a way of getting chances to shape the human world and act in it", in this way "creating the world and its directions, in front of different (real or abstract) situations and problems" (Magnani & Bardone, 2008, p. 99). It "involves coming to see some aspects of reality in a particular way that influences human acting in shaping and surviving the future", and through this process "many external things, usually inert from the moral point of view, can be transformed into what we describe as "moral mediators", "moral tools" "distributed in 'external' objects and structures which function as ethical devices" (Magnani & Bardone, 2008, p. 100). Magnani takes inspiration from Edwin Hutchins' "mediating structures", emphasizing behavioral mediation, objects which direct, constrain, and guide action along culturally specific lines, contributing to cultural coherency, referring "to various external tools that can facilitate cognitive acts of navigating in both modern and 'primitive' settings" (Magnani, 2007, p. 237). Furthermore, in exhibiting "care" for and through these devices, we demonstrate for others a "conscientious" pattern exhibiting "a fundamental kind of moral inference and knowledge", by which "even a lowly kitchen utensil can be considered a moral mediator" (Magnani, 2011, p. 133). Thus, moral mediators range from objects to agents to actions to systems including binding narratives, through which we "make sense" of it all.

As a tool for "creative abduction", making sense of self and world in light of incomplete information, a moral mediator is a special kind of representation. As "distributed morality", it is a "task-transforming representation" able to "transform difficult tasks into ones that can be done by pattern matching" by performing what has been called "the "representational task" of representing moral problems so that their solutions are transparent" (Magnani & Bardone, 2008, p. 103).

Moral mediators represent a kind of redistribution of the moral effort through managing objects and information in such a way that we can overcome the poverty and the unsatisfactory character of the moral options immediately represented or found internally (for example principles, prototypes, etc.) (Magnani & Bardone, 2008, p. 105).

By "pattern matching", one quickly imagines metaphor; and, of course, metaphor has been recognized as central to cognition, being a capacity to recognize patterns of relationships amongst vastly different arrays of objects (Pinker, 2007). Here, metaphor and "pattern matching" are best understood as a form of abduction, model-based reasoning, in which Magnani and Bardone capture "a considerable part of the thinking activity" (cf. also Magnani & Nersessian, 2002; Magnani 2005).

Perhaps the most ubiquitous pattern of thought is the simple "possible worlds" counterfactual, "If... then..." Magnani and Bardone provide a poignant example of this pattern in a most ubiquitous form of ICT, a website, CostofWar.com. CostofWar.com presents a running tally of the US dollar costs of wars continuously waged by the USA since 2001. As of October, 2011, it shows a total *monetary* cost of all wars begun since 2001 (costs to the USA, begun by the USA) to be over 1.2 trillion, with the total cost of war (invasion and occupation) in Iraq to be over 800 billion. Alone, these are simply big numbers. They are not "immediately" morally significant, only becoming so when recast counterfactually as opportunities lost, for example "the number of children we could have insured, if..." Presenting alternative consequences, CostofWar.com "represents the same piece of information so that the problem we face, for instance, thinking about going to war or not, is completely changed", and qualifies as a moral mediator "because it mediates the task changing the representation we have of it, and making the solution more transparent" (Magnani & Bardone, 2008, p. 105). Thusly, information becomes eth-

ics, and we are on our way to transforming the infosphere into the ethosphere.

One product of Magnani's work that can help to illuminate the most difficult aspects of this transformation is his notion of the "moral bubble". Moral bubbles are "viscous" subdomains of the ethosphere that bind persons within sub-groups of humanity, and that "systematically disguise" the violence of their actions by making them unable to "incorporate"(empathize with) "the effect of [their] behavior on other human beings" or, in "bad faith" on themselves, while easing moral tensions by allowing them to "avoid the cognitive breakdown" "triggered by the constant appraisal" of every action inconsistent with given moral conviction (Magnani, 2011, p. 8, p. 59, p. 77). Magnani models the "glue-like" integrity of moral bubbles on the disposition of agents within epistemic bubbles to "resolve the tension between their thinking that they know *P* and their knowing *P* in favor of knowing that *P*", the difference being that, rather than "knowing" that something is *true*, "prisoners" of moral bubbles "know" that something is *right*, with this "knowledge" reinforced through gossip and narrative that "inform and disseminate the moral dominant knowledge of a group" (Magnani, 2011, p. 74, quoting John Woods, p. 79). Thus, a moral bubble is a type of "cognitive niche" in which persons are actively insulated both from the anxiety associated with admitting that "we know a lot less than we *think* we do", and from perceiving moral inconsistencies, feeling responsibility for their actions (Magnani, 2011, p. 74). As domains of action expand, prisoners of bubbles conflict with prisoners of other bubbles and even those within the same bubbles "whenever one's signaling does not conform with the 'standard' implications meant by the signal deployed." (Magnani, 2011, page 84) Even war is justified in a moral bubble as the "predominant view is that in a state of war, the act of killing is ruled by different moral principles from those that rule acts of killing in other contexts" (Magnani, 2011, p. 175).

"Embubblement" is facilitated through the violent use of the prototypical ICT, language, cutting humanity apart with some divided as means to others' ends, "exactly like a knife […] of biological origin", and as "the outward extension of an organic activity" (Magnani, 2011, p. 47). Violence is often justified, and the viscous integrity of the moral bubble maintained, through a "perverting of sympathy" tempering the cognitive dissonance arising when one "simultaneously holds two contrasting cognitions", and rather than reconcile them ends up "obliterating facts supporting one view and fabricating appropriate evidence justifying that suppression in favor of the other" (Magnani, 2011, pp. 103-104). In this way, embubblement creates conflict, "structural violence", that is "often diluted in the pervasive form of narratives", including "the fairytales that are told to children from early youth", as these inspire drives to champion the terms by which one thrives and strives for his/her own happy ending, thus presenting preservative actions as morally justified (Magnani, 2011, p. 123). Persons committed to the terms of such fairy tales pursue them with zeal, however violent, often "fabricating evidence" at the expense of others through deception, "bullshitting" (Magnani, 2011, p. 105). Magnani illustrates this fairy tale bullshit by way of recent "too big to fail" bank "bailouts" propagated on a "sacred myth" that at once, perversely, considers "government action of any sort beyond bare minimum" unnecessary while at once claiming that government financial support is necessary for the integrity of the "global" economic system, itself an over-inflated bubble, in a prime example of perverted sympathy "actively supported and encouraged not only by politicians, but also by intellectual elite" who leveraged expert testimony to undermine democracy and "dupe public opinion", thereby serving one bubble at the expense everyone else (Magnani, 2011, p. 105). Thus, even employed in preservation of order, information is a tool of violence, resulting in tragic injustice, recalling Occupy Wall Street, entropy, and evil.

Magnani's moral bubbles provide a model that helps to clarify the relationship between entropy and evil:

It is well-known that, from the point of view of physics, organisms are far-from-equilibrium systems relative to their physical or abiotic surroundings. Apparently they violate the second law of thermodynamics because they stay alive, the law stating that net entropy always increases and that complex and concentrated stores of energy necessarily break down(Magnani, 2011, p. 121).

They apparently violate the second law because they construct local orders - bubbles, niches, narratives, theories, and so on – maintain these orders, and pattern them in terms of modes of engagement that, ostensibly, increase survivability in the natural world, i.e., knowledge:

To create cognitive niches is a way that an organism (which is always smartly and plastically "active", looking for profitable resources, and aiming at enhancing fitness) has to stay alive without violating the second law: indeed it "cannot" violate it (Magnani, 2011, p. 121).

The second law is not violated because the niche is a local space of order created at entropic expense to whatever is other. This "entropic expense" translates into actions and attitudes ranging from violence to ignorance, in any case raising the level of energy necessary for systems external (logically, not necessarily physically) to the bubble in question to remain stable and retain integrity. Thus, increasing entropy in the "whole" infosphere, bubble-making is objectively "evil". Meanwhile, being the space of information in terms of which people live and die, these orders are essentially moral, with their creation and maintenance perceived, selfishly, "good".

Consider in this light Kantian moral theory, impelling persons to "Act so that the maxim of thy will can always at the same time hold good as a principle of universal legislation", while "The direct opposite of the principle of morality is, when the principle of private happiness is made the determining principle of the will" (Kant, 1898, p. 93, p. 96). Putting these together, a principle of immorality emerges: "Act so that the maxim of thy will is private happiness." The "propensity to evil" that this modified maxim represents is an inclination to "reverse the moral order" due to the weakness, impurity, and/or corruption of the heart, which for Kant is the seat of the "springs" of will, source of motivation, where the morally good agent acts solely by the spring of moral law, "alone an adequate spring", and "*Whatever is not done from this faith is sin*", while the bad agent acts selfishly by "deceiving itself as to its own good or bad dispositions" through "malignancy" that "extends itself then outwardly also to falsehood and deception of others which, if it is not to be called badness, at least deserves to be called worthlessness" (Kant, 1898, pp. 217-218, p. 223). Worthless, "the moral agent falls into a perfidious moral bubble, where the evil is not perceived", in what Magnani calls moral "disengagement" through which "The moral agent falls into a perfidious and self-deceitful reengagement in a new decisional framework where evil is simply supposed to be good, and so morally justified"[2] (Magnani, 2011, p. 181).

In illustration, imagine a single, universal plane of moral order representing a low-energy stable state of maximal potential energy. Given the natural tendency toward disorder, further imagine the spontaneous disintegration of this universal moral domain into local regions, bubbles that retain internal order at the accelerated disorder of whatever is external. Accordingly, "evil" is related with entropy as the tendency to exacerbate this disintegration, and "good" the tendency otherwise. However, recalling that organisms create bubbles as a way to "stay alive without violating the second law", so think it "good", subjective perceptions of "evil" differ from the preceding portrait. "Evil" is (subjectively) *idealized* as agents

and actions "that intentionally harm other people thanks to a transgression of a moral rule perfectly present and approved in the agent's mind", while *actual* "evil-doers" are "embedded in their "moral bubbles", understanding their actions "as totally or abundantly justified" with "evil" existing "only in the experience of the victim" (Magnani, 2011, p. 182).

Still, IF we want to become "good" persons, THEN by what method might we move in that direction? A demonstrated tendency to the good exists in non-violent movements for social change. Magnani writes that:

[...] nonviolence always promotes what I can call a "moral epistemology": new "regimes" of truth related to the "inessentiality of something, or its nothingness, in the form of illuminating its fragility and pursuing the orchestration of its collapse (Magnani, 2011, p. 142).

This description appears violent, and in the minds of the "victims" of non-violent movements for social justice, even "evil". However, in this orchestrated collapse of the inessential the proliferation of bubbles is contravened. Through non-violence, people *work* – use metabolic energy – to increase universal moral order. Granted, they increase disorder in target moral bubbles – *bubbles get popped* - and granted that global disorder is always increasing per the second law, so they may appear engaged in a "ridiculous struggle against thermodynamics". But, the resulting moral order, the ethosphere, is transformed otherwise. Consider Martin Luther King, Jr., on these points:

Nonviolent direct action seeks to create such a crisis and foster such a tension that a community which has constantly refused to negotiate is forced to confront the issue [...]. Just as Socrates felt that it was necessary to create the tension in the mind so that individuals could rise from the bondage of myths and half-truths to the unfettered realm of creative analyses and objective appraisal, so must we see the need for nonviolent gadflies to create a kind of tension in society that will help men rise from the dark depths of prejudice and racism to the majestic heights of understanding and brotherhood (King, 1963).

King's method is to force embubbled persons to overcome their cognitive dissonance by forbidding them to continue "obliterating facts supporting one view and fabricating appropriate evidence justifying that suppression in favor of the other." Thus, reducing the number of moral bubbles in the world, bringing people together, and increasing order. Tellingly, he points out this intention through an entropically significant physical metaphor - "majestic heights". He points up. And up is a good thing.

3. TECHNOLOGICALLY MEDIATED MORAL EMBUBBLEMENT

How we ought to live is the central problem of morality, and the way we reason dictates how we live. - Lorenzo Magnani (2007, p. 164)

Since King, however, we appear to have traveled in the other direction, with the "War on Terror", as articulated in GW Bush's infamous "doctrine" "Either you are with us, or you are with the terrorists", delivered along with a supporting narrative justifying the occupation of Afghanistan due to its alleged involvement in the deaths of more than 3000 US citizens, "They hate us for our freedom" (Bush 2001). This language split the world like a knife along an "Axis of Evil", "rogue states" Iraq, Iran, North Korea, and later Syria, Libya, and Cuba.[3] In Iraq alone, with "documented civilian deaths from violence" over 103,000 (Franks, 2011) while "the deliberate destruction of Iraq's water and sewage systems by US bombings has been the major cause (for a decade) of an outbreak of diarrhea and hepatitis, particularly lethal to pregnant women and young

children" (Hassan, 2005) and with "defects in newborns 11x higher than normal" due to depleted uranium munitions (Chulav, 2010) poison with a half-life measured in the billions of years (International Atomic Energy Agency, n. d.) Iraqis have suffered "irreversible injustice" due to "countless numbers of families, friends, refugees and orphans who have suffered losses" for which "the US and its allies are responsible", while "the denial by the governments of the US and its allies over the number of excess Iraqi civilian casualties is further adding to this injustice" (Karagiozakis, 2009). With the critical reason for the invasion, alleged "weapons of mass destruction", now acknowledged to have been founded on incomplete, faulty – if not bullshit – information (Schwarz & Cohen, 2011) it is no mystery why so many are questioning "the whole information flow".[4] We are at a tragic end, due to tragic error, and so many continue to suffer for it. Who is the terrorist, now? Before we ever look up, again, we - as future conditionals - must reconcile with the fact that we have become the opposite of who we set out to become.

As difficult as this process may be on its own, it is made all the more difficult through the employment of ICTs in the reinforcement of embubbling paradigms through a process that Magnani terms "cognitive hacking". This is nothing less than technologically mediated and targeted deception. Information, through technology, is deployed in a violent "semantic attack" targeting the technology users' minds and behaviors "by manipulating their perception of reality".

In this case information is used in a violent and sophisticated way – beyond the well-known violent effects of traditional mass-media propaganda – to influence and to affect the behavior of humans through computational tools: a full process of information warfare, as it is called (Magnani, 2011, p. 87)

Killing and poisoning so many people on the basis of *intentionally* bad information raises the issue of the relative moral status of persons and things. Certainly, deceiving a population in order to serve one's own selfish purposes is to treat those persons as a means to one's own ends, as a thing and not as an end in itself, and so is, at least from a Kantian perspective, patently immoral. However, this still leaves the relative moral *status* of persons and things unspecified, and deserving of immediate attention.

The distinction between persons and things is subtle, if not fuzzy, especially as we are embedded in a technological world, with technologies of all sorts increasingly mediating even the most mundane aspects of our lives. Lorenzo Magnani has argued that "advanced and more pervasive technology" has "blurred the line between humans and things" making it "increasingly difficult to discern where the human body ends and the non-human thing begins", "so that we delegate action to external things (objects, tools, artifacts) that in turn share our human existence with us", with this "hybridization" in turn necessitating "treating people as things", however stating that "instead of treating people as means, we can improve their lives by recognizing their part-thingness and respecting them as things" (Magnani, 2007, pp. x-xi). This position has created some controversy, to which Magnani has responded by clarifying that we must "respect [not treat] people as things" "only when things are already endowed with an ethical value, and sometimes endowed with much more ethical worth than humans" (Magnani, 2010, p. 162).

Also, consider that children are often killed and violated just because of the needs of "things" to which we have attributed too much ethical value (capitalistic profit, for example, to use an old-fashioned socialist idea). So I answer: we have to respect people at least to the same degree that we respect some "sacred" things. For now we are very far from this target (Magnani, 2010, p. 163).

Magnani simply holds that "human beings must be respected as we now respect many things; people deserve, at the very least, the same level of care we now lavish on library books" (Magnani, 2007, pp. 20-21). A sentiment more strongly represented by Martin Luther King, Jr.:

I am aware that there are many who wince at a distinction between property and persons – who hold both sacrosanct. My views are not so rigid. A life is sacred. Property is intended to serve life, and no matter how much we surround it with rights and respect, it has no personal being. It is part of the earth man walks on; it is not man (King, 2011, p. 58).

Things, property, technologies and narratives *serve the living*. That is their purpose, as given in their origins as construals and employment as moral mediators. Reminding persons of this fact is the rationale behind crafting the claim that we must respect persons (at least as well) as things. There is certainly nothing controversial here.

Rather, what is controversial is admitting that things *are* valued above living, breathing human beings. For instance, in war, immoral treatment seems justifiable, with the value of life relative things like "oil" and "economy" turned upside-down:

Wars compel cultures to acknowledge that they attribute greater value and respect to tanks and technological weapons (and the all-encompassing commodification of human needs) than to an intact natural community of living plants, animals, and human beings (Magnani, 2011, p. 232).

As we have seen, people tend to insulate themselves from such uncomfortable realizations, embubbled in justificatory language like "Axis of Evil" and narratives like "they hate us for our freedom." "War violently kills human beings, and this fact is too horrific for some people to accept when faced with it", so they maintain their ignorance by "building an emotional firewall" behind which they "regress into a sort of psychological refuge from the horrors of war", thereby keeping war "out there" (Magnani, 2011, p. 232). For direct agents of war, soldiers and mercenaries, this process is also aided by recent advances in technology, such as remotely controlled missile carriers which insulate the killer/operator from her/his violence behind a computer monitor within an air conditioned concrete bunker, "death by joystick". Magnani analyzes this condition in terms of "bad faith", wherein people dislocate the source of valuation from selves to some realm of ideals, valuing themselves in terms of narrative roles to which they give themselves over, for example transforming good democratic citizenship into blind support of the military and its dependent corporations or transforming military valor from courage under fire to faithful disregard for "collateral damage", thus insulating themselves from criticism, "and more importantly, from responsibility for actions undertaken *as* that role", by creating "a safe external entity that cannot participate in their own identity" (Magnani, 2007, p. 136). After an afternoon of air-conditioned murder, the killer/operator in the above example simply jumps into her/his SUV and goes home, an exercise in bad faith only possible through recent technological advances. Thus, aided by technology, persons make bubbles of themselves, become things, and so, rather than actively constructive *homines poietici,* become "mere objects of their own evaluation".

Consider in this light the ubiquitous distinction between public and private life whereby a person does things at work that he might not do otherwise. Classic examples include the politician war-criminal claiming that "everything we did was legal", and the foot-soldier that he was "only following orders". Contrast this with that example of a person who does what he loves to do, and fulfills that role as a life of art and practice. Socrates, for example, famously went home rather than follow orders to arrest a man on trumped up charges so that his wealth might be seized, and

the man executed. In the latter case, this person is not his role, but rather the role is himself. He is doing what needs to be done, and in this way is a means to others' happiness, to their ends. However, there is no bad faith. He has not made a bubble of himself, but rather refused to enter one with others. Thus, we may fully grasp the respect due to human beings *as* things, because, as this example illustrates, the thing - a role, job, office, etcetera - is itself a human being, and can only ever *be* a human being. "If we fail to associate external things' positive qualities with the person, if we do not 'respect' the human being as a means/thing, then we do not recognize that instrumental condition as incorporating ends in itself" (Magnani, 2007, p. 136). Indeed, respecting people as things is a moral necessity.

The "instrumental condition" is simply the fact that human beings are able to act towards the realization of constructive projects, create order. Where this inherent potential is leveled to the potential of any valuable thing, like a library book perhaps valuable but essentially not able to act in the realization of constructive goals, then the value of work is lost. The value of work does not reside in the product, but in the realization of the self as the work is done. *One's self is one's own greatest life's work.* Also the most difficult. Perhaps this is why so many prefer to live in bad faith, while others fail due to what Kant calls weakness of will, "To will is present with me, but how to perform I find not", ultimately the failure to gain the knowledge necessary to do the job that needs to be done, and why Magnani maintains that "appropriate ethical knowledge and proper moral reasoning are the basic conditions for maintaining freedom and taking responsibility for our actions", the foundations of moral life and ultimately why he holds that knowledge is a duty, especially the self-knowledge to overcome "the lack of knowledge" due to bad faith that exists as "a blind spot in one's knowledge of oneself" (Kant, 1898, p. 217; Magnani, 2007, p. 221, p. 249). This is also why models of self-construction are so important, to serve as technological *ego-* and *ecopoietic* mediators in the crucial yet difficult task of becoming "good" persons, perhaps even rendering the solution transparent.

In this direction, Luciano Floridi extends his "3C model", suggesting that it accounts for the deepest difficulties in reasoning about the self, namely that a self is experienced as one self over time, perduring through changes, retaining integrity, and is capable of self-reflection whereby it becomes an object of self-evaluation. The 3C model grounds the emergence of selves in corporeal processing (chemical), from which arises cognitive (sentience, awareness), and finally conscious processing (self-awareness). Each sphere of processing supervenes on the last, delimited by "membranes", as "the self emerges as a break with nature":

Each membrane, and hence each step in the detachment of the individual from the world, is made possible by a specific, auto-reinforcing, bonding force. The corporeal membrane relies on chemical bonds and orientations. The cognitive membrane relies on the bonds and orientations provided by what is known in information theory as mutual information, that is the (measure of) the interdependence of data (the textbook example is the mutual dependence between smoke and fire). And, finally, the consciousness membrane relies on the bonds and orientations provided by semantics (here narratives provide plenty of examples), which ultimately makes possible a stable and long-lasting detachment from reality. At each stage, corporeal, cognitive and consciousness elements fit together in structures (body, cognition, mind) that owe their unity and coordination to such bonding forces. The more virtual the structure becomes, the more it is disengaged from the external environment in favor of an autonomously constructed world of meanings and interpretations, the less physical and more virtual the bonding force can be (Floridi, 2011b, p. 560).

Crucially, in light of preceding discussion, though "each membrane contributes to the construction of the self", "inextricably mixed together to give rise to a self and its personal identity",

[...] this truism hides the fundamental fact that, once a membrane is in place, the particular inside that it detaches from the relevant outside becomes conceivably independent of the previous stages of development. It is correct to stress that there is no butterfly without the caterpillar, but insisting that once the butterfly is born the caterpillar must still be there for the butterfly to live and flourish is a conceptual confusion. There is no development of the self without the corporeal and the cognitive faculties, but once the latter have given rise to a consciousness membrane, the life of the self may be entirely internal and independent of the specific body and faculties that made it possible. While in the air, you no longer need the springboard, even if it was the springboard that allowed you to jump so high, and your airborne time is limited by gravity (Floridi, 2011b, p. 561).

Floridi is describing nothing less than self-embubblement in fairy tales, to which we may respond in like terms. Certainly, the butterfly is no longer a caterpillar in our eyes, but doesn't a single thread of physical information underwrite both entities? Certainly, airborne we forget the springboard, but failing to recall our origins in that weightless instant in the end transforms us only into a tragic mess in the dirt, *anagnorisis*. We can also certainly allow that the "consciousness membrane relies on the bonds and orientations provided by semantics", of which narratives are prime examples, but, morally speaking, do we really want our moral fables to encourage "a stable and long-lasting detachment from reality?" Where there is smoke there is (often) fire, but this is not for the sake of a message. It is simply a fact of matter.

4. NON-VIOLENT TRANSFORMATION OF SELF AND WORLD

Very well, replied Ivan, you need not become soldiers unless you wish to. - Leo Tolstoy, Ivan the Fool (1917, Ch. 11, para 17)

Floridi's approach invites a slippery slope dislocation of valuation hardly "suited to our *realities* as informational agents and patients" and hardly providing for "responsible stewardship in the world." Moreover, to forge an ethics on the disengagement "from the external environment in favor of an autonomously constructed world of meanings and interpretations" is only to proliferate bubbles. Disorder. We have seen how such a move is spontaneous, and why it may even seem appealing, entropically embodied as we are, but fairy tale happy endings exist, sadly, only in fairy tales, and ultimately, all returns to ground.

We are left with *anagnorisis* - which is to require that one first create a bubble of himself, only then have it unexpectedly popped, an experience perhaps fruitful but better off avoided - or Magnani's approach, creative abduction engaged with the world and with a "well fed mind", wary of embubblement. In the latter spirit, in my own work, I have developed the ACTWith model, inspired equally by bottom-up hybrid neural network models, complex/dynamic systems, traditional moral philosophy, and neurology – especially social cognitive neuroscience, empathy and mirroring (White, 2006, 2010, in press, forthcoming). Space forbids complete exposition, but it is detailed in other vehicles, so here will note only those ways in which the ACTWith (As-if Coming-to-Terms-With) approach, representing the bare minimum moral architecture, mediates *self-poiesis* sans tragedy. It is a situated, embodied and embedded information processing framework composed of a four-fold cycle - As-if (closed) coming-to-terms-with (closed); As-if (open) coming-to-terms-with (closed); As-if (closed) coming-to-terms-with (open); and As-if (open) coming-to-terms-with (open) - with the open "as-if" operations feeling a situation out, and the open "coming-to-terms-with" operations defining the situation accordingly, an affect-first self-as-future-conditional model of agency retaining practical wisdom as its limiting condition. As in Floridi's model, integrity is ultimately a matter of embodiment. But, two important differences bear highlighting in the present context. First, in order to properly evaluate mirrored/empathized information, information is fundamentally understood in natural energetic/metabolic terms. Second, these ontic grounds discourage embubblement and encourage "conscientiousness", habitual openness to self and others, as information crucial for solving life's most difficult problems (for example, the *ego-* and *ecopoietic* problems reviewed throughout this paper) is most efficiently collected thereby, and having done so, the risk of *hamartia* is reduced. Tragedy avoided.[5] Representing the essential, embodied social nature of moral cognition, emphasizing the guiding resource that is shared experience, the ACTWith model both reinforces the threads of moral philosophical tradition that bind ancient tradition with cutting-edge science and technology, and extends this technology in an accessible form for active self-construction. It does so by extending a pattern of information processing essential to moral cognition, and as agents match this pattern, it serves as a technological mediator of *self-poiesis*, thereby rendering moral solutions transparent. Thus, facilitating the non-violent constructive transformation of self, the ACTWith model is an example of ICTs "significantly affecting who we are, who we think we are, who we might become, and who we think we might become."

But, what of narratives in light of which each one of us makes sense of it all? As we have seen, narratives represent the ordered spaces of information in terms of which one lives, dies, succeeds or fails, and so provide crucial support in the realization of personal identity.

We usually see our own lives and those of others as a series of narratives, and we continually reinterpret and revise our narrative self-understanding. For example, we scroll through our cache of stories to find one that can best clarify the moral problem at hand and that we can reconcile with our self-representations and ideals (Magnani, 2007, pp. 177-178).

Our final problem is to find a narrative mediator for moral self-construction so that, fully informed of this potential, we can best determine our characters, good or evil. Scrolling through the cache of stories reviewed so far, we have already glimpsed such a narrative, in the illustration of a universal plane of moral order representing a low-energy stable state of maximal potential energy.

This picture is inspired by the narrative in terms of which Socrates makes sense of his environment, of himself in it, and of his interactions with and within it. In the *Gorgias*, Socrates maintains that language should be used to get rid of injustice (Plato, 1997, 480d) rather than to "scheme" for selfish enrichment at the expense of others (481b), that bad leaders use language in the former mode, making people evil and encouraging bubbles, while good leaders in the latter even if it means telling people what they don't want to hear, (503a-) Callicles – an upstart politician of the former mode - poignantly objects "won't this human life of ours be turned upside down, and won't everything we do evidently be the opposite of what we should do?"(481c) Socrates affirms this, noting that Callicles is in a state of internal "dissonance", bad faith, pretending it just to do injustice (482a-c). Callicles denies this, maintaining that it is natural for the more powerful to take from the less powerful (483d), and, foreshadowing the most tragic instance of affirming the consequent in the history of future-conditionals, that only a "slave" (prisoner) would not do similarly if able, "one who is better dead than alive" (483a,b). Socrates then forces Callicles to confess that doing for one's self at the expense of others is

unnatural, but that fairness is, and that we ought to be fair (488e-489a), thereby popping his moral bubble. At this perceived violence, Callicles notes that Socrates lives outside of the reigning moral bubble of political opinion, issuing the veiled threat that some evil man might drag him back in to be judged in those embubbled terms (521c), to be imprisoned by those terms, with Socrates expecting that, as he aims for what is best rather than making people happy (521e), "it wouldn't be all that strange if I were put to death" (521d). Finally, Socrates recounts the fable that brought him to this position. This fable establishes a beacon at the endpoint of life, a plane of moral judgement marked by the necessary three points for a geometric plane, represented as judges, from which two roads proceed, one down, to tragedy and *anagnorisis,* and one up (524a). On that field, all are judged outside of their moral bubbles according to evidence of their moral lives, whether seriously "warped" from life in a selfishly confined moral bubble, or "straight" from a life on a level plane of universal moral law, terms by which Socrates feels he will be judged favorably.

Socrates' chosen fable helps him become "the most just man in Athens" in "a stable and long-lasting detachment" from the political reality of bad faith because it sets out the terms from which he acts, to which he aims, and ultimately for which he dies, thus serving as a mediator of moral self-construction ensuring against *hamartia* and *anagnorisis*. Living as if on a level plane of justice, Socrates remains an inhabitant of the ethosphere writ large, resisting embubblement through work, philosophy, and thereby providing through fairy-tale example the support we need to effect the nonviolent transformation of the world. He does effectively affirm the consequent of his life as a future conditional. But, don't we all? And, if this is our greatest moral sin, it is also that which makes us most worthy.

5. CONCLUSION

One's self is one's most important life's work, but it is also the most difficult. In no small part this is due to the importance of the product, nothing less than one's only self. Working with the right information, towards the right ends, aided by technology, the product of a life's work is a just human being. This is a life worth living. Poorly informed and misdirected, one may suffer a tragic irony, to have acted with best intentions only become a very bad person, for it is truly terrible when one judges his entire life wrong. Moreover, as the world we build is as well the product of our actions, actions the effects of which are amplified by technologies designed to carry our intentions, to act on bad information is to risk a tragic end for not only one life, but for us all and everyone yet to come. Should we have no desire to suffer twice, both in our personal realities as well as in retrospect, looking back on the wreck made of the Earth through lifetimes of collective bad action, tools for the non-violent transformation of self and world must be our highest priority.

The conceptual tools necessary to effect constructive change to moral ends of both self and world was this paper's central focus. We began by briefly reviewing Luciano Floridi's information ethics. Four interconnected issues, arose – self-determination, moral valuation, entropy and narrative. We found that Floridi misplaces the source of moral valuation away from human beings, that the entropic base of his principled ethics is inconsistent in both assertion and application, and that his appointed pattern for effective self-realization is a path not only avoidable, but better off avoided. Granted that self, world, and their revolutions are best understood informationally, we found resources to understand Floridi's "infosphere" as an "ethosphere" in Lorenzo Magnani's cognitive niches, and tools to effect the transformation of infosphere to ethosphere in his moral mediators. With these came an understanding of human beings as entropically grounded creators of order whose fundamental tool (and weapon) in the creation and maintenance of said orders is the prototypical ICT, language. Through Magnani's work in moral bubbles and insights into non-violent social change, we came to realize the power of myth to shape our lives, as well as our active role in the creation, maintenance, and reformation of said myths, their limits, and the ends to which they bring us. Finally, then, we were able to see how Floridi's 3C model fails to adequately represent moral self-development in a way that serves to mediate moral self-determination in actual moral subjects. In the last section, we briefly reviewed a model of moral cognition that does not suffer the shortcomings of the 3C model, and took a fresh look at an old fable, one which, when freely chosen, sets terms to shape a moral life into a just man. With a just world in sight, we must only get there.

Though we have the capacity to construct our own selves, our world, and the myths in light of which self and world all make sense, choosing the right story, and living accordingly, is a most difficult task. Luckily, we have the resources in moral philosophers, moral exemplars, and their guiding narratives with which to begin, and are gifted with the technological accumen to design the tools to get this most important work done. Hopefully, the preceding paper helps to set us on the right path in both efforts. In the end, what we do at this point, fully informed of possibilities and fully capable of realizing them, will determine who we become, good or evil.

REFERENCES

Aristotle,. (1995). *Aristotle: Selections* (Irwin, T., & Fine, G. (Trans. Eds.)). Indianapolis, IN: Hackett.

Ayotte, B. J., Bosworth, H. B., Potter, G. G., Williams, H. T., & Steffens, D. C. (2009). The moderating role of personality factors in the relationship between depression and neuro-psychological functioning among older adults. *International Journal of Geriatric Psychiatry*, *24*(9), 1010–1019. doi:10.1002/gps.2213

Bush, G. W. (2001). *Address to a joint session of Congress and the American People*. Retrieved from http://georgewbush-whitehouse.archives.gov/news/releases/2001/09/20010920-8.html

Capurro, R. (2008). On Floridi's metaphysical foundation of information ecology. *Ethics and Information Technology*, *10*(2-3), 167–173. Retrieved from http://www.capurro.de/floridi.html doi:10.1007/s10676-008-9162-x

Chulov, M. (2010). *Research links rise in Falluja birth defects and cancers to US assault*. Retrieved from http://www.guardian.co.uk/world/2010/dec/30/faulluja-birth-defects-iraq

Damasio, A. R. (1999). *The feeling of what happens: Body and emotion in the making of consciousness*. New York, NY: Harcourt Brace.

Floridi, L. (2002). On the intrinsic value of information objects and the infosphere. *Ethics and Information Technology*, *4*(4), 287–304. doi:10.1023/A:1021342422699

Floridi, L. (2004). Open problems in the philosophy of information. *Metaphilosophy*, *35*(4), 554–582. doi:10.1111/j.1467-9973.2004.00336.x

Floridi, L. (2006). Information ethics, its nature and scope. *SIGCAS Computers and Society*, *36*(3), 21–36. doi:10.1145/1195716.1195719

Floridi, L. (2008). Information ethics: A reappraisal. *Ethics and Information Technology*, *10*(2-3), 189–204. doi:10.1007/s10676-008-9176-4

Floridi, L. (2010). *The Cambridge handbook of information and computer ethics*. Cambridge, UK: Cambridge University Press.

Floridi, L. (2011a.) *The fourth technological revolution*. Retrieved from http://www.youtube.com/watch?v=c-kJsyU8tgI

Floridi, L. (2011b). The informational nature of personal identity. *Minds and Machines*, *21*(4), 549–566. doi:10.1007/s11023-011-9259-6

Floridi, L., & Sanders, J. W. (1999). Entropy as Evil in Information Ethics. *Etica & Politica*, *1*(2). Retrieved from http://www2.units.it/etica/1999_2/floridi/node17.html

Franks, T. (2011). *Iraq body count*. Retrieved from http://www.iraqbodycount.org/

Hassan, G. (2005). *Living conditions in Iraq: A criminal tragedy*. Retrieved from http://www.globalresearch.ca/articles/HAS506A.html

Hoppenbrouwers, S. S., Farzan, F., Barr, M. S., Voineskos, A. N., Schutter, D. J. L. G., Fitzgerald, P. B., & Daskalakis, Z. J. (2010). Personality goes a long a way: An interhemispheric connectivity study. *Frontiers in Psychiatry*, *1*(140), 1–6.

International Atomic Energy Agency. (n. d.). *Depleted uranium*. Retrieved from http://www.iaea.org/newscenter/features/du/du_qaa.shtml

Kant, I., Abbott, T. K., & Liberty Fund. (1898). *Kant's critique of practical reason and other works on the theory of ethics*. London, UK: Longmans, Green. Retrieved from http://files.libertyfund.org/files/360/Kant_0212_EBk_v6.0.pdf

Karagiozakis, M. (2009). Counting excess civilian casualties of the Iraq War: Science or Politics? *The Journal of Humanitarian Assistance*. Retrieved from http://sites.tufts.edu/jha/archives/559

Kern, M. L., & Friedman, H. S. (2008). Do conscientious individuals live longer? A quantitative review. *Health Psychology*, *27*(5), 505–512. doi:10.1037/0278-6133.27.5.505

Kincaid, P. (2011). *Occupy Wall Street protestors demands the abolishing of the Federal Reserve.* Retrieved from http://presscore.ca/2011/?p=4637

King, M. L., Jr. (1963). *Letter from the Birmingham Jail*. Retrieved from http://mlk-kpp01.stanford.edu/index.php/resources/article/annotated_letter_from_birmingham

King, M. L. Jr. (2011). *The trumpet of conscience*. Boston, MA: Beacon Press.

Leopold, A., Schwartz, C. W., & Leopold, A. (1970). *A Sand County Almanac: with essays on conservation from Round River*. New York, NY: Ballantine Books.

Magnani, L. (2005). An abductive theory of scientific reasoning. *Semiotica, 2005*, 261–286. doi:10.1515/semi.2005.2005.153-1-4.261

Magnani, L. (2007). *Morality in a technological world: Knowledge as duty.* Cambridge, UK: Cambridge University Press. doi:10.1017/CBO9780511498657

Magnani, L. (2009). Beyond Mind: How brains make up artificial cognitive systems. *Minds and Machines, 19*(4), 477–493. doi:10.1007/s11023-009-9171-5

Magnani, L. (2010). Is knowledge a duty? Yes, it is, and we also have to "respect people as things", at least in our technological world: Response to Bernd Carsten Stahl's Review of Morality in a Technological World: Knowledge as Duty. *Minds and Machines, 20*(1), 161–164. doi:10.1007/s11023-010-9179-x

Magnani, L. (2011). *Understanding violence: The intertwining of morality, religion and violence: a philosophical stance*. Heidelberg, Germany: Springer-Verlag.Bottom of Form.

Magnani, L., & Bardone, E. (2008). Distributed morality: externalizing ethical knowledge in technological artifacts. *Foundations of Science, 13*(1), 99–108. doi:10.1007/s10699-007-9116-5

Magnani, L., & Nersessian, N. J. (2002). *Model-Based Reasoning: Science, Technology, Values*. Boston, MA: Kluwer Academic. doi:10.1007/978-1-4615-0605-8

Plato,. (1997). *Complete works* (Cooper, J. M., & Hutchinson, D. S. (Trans. Eds.)). Indianapolis, IN: Hackett.

Schwarz, G., & Cohen, T. (2011). *Rumsfeld: WMD issue was 'the big one' in Iraq invasion*. Retrieved from http://politicalticker.blogs.cnn.com/2011/02/20/rumsfeld-wmds-werent-only-reason-for-war-in-iraq/

Sharp, E. S., Pedersen, N. L., Gatz, M., & Reynolds, C. A. (2010). Cognitive engagement and cognitive aging: Is openness protective? *Psychology and Aging, 25*(1), 60–73. doi:10.1037/a0018748

Tolstoy, L., & Garnett, C. (1917). Ivan the Fool. In *The Harvard Classics Shelf of Fiction* (*Vol. 17*). New York, NY: P.F. Collier & Son.

White, J. (2006). *Conscience: Toward the mechanism of morality*. Columbia, MO: University of Missouri-Columbia.

White, J. (2010). Understanding and augmenting human morality: An introduction to the ACTWith model of conscience. *Studies in Computational Intelligence, 314*, 607–621. doi:10.1007/978-3-642-15223-8_34

White, J. (in press). An information processing model of psychopathy and anti-social personality disorders integrating neural and psychological accounts towards the assay of social implications of psychopathic agents. In *Moral Psychology*. New York, NY: Nova.

White, J. (in press). Manufacturing morality: A general theory of moral agency grounding computational implementations: the ACTWith model. In *Computational Intelligence*. New York, NY: Nova.

White, J. (in press). *The mechanism of morality* [Monograph].

ENDNOTES

[1] These ideas play a central role in self-determination via the ACTWith model as articulated in 2006, and are more fully developed in my forthcoming book *The Mechanism of Morality* (White, in press).

[2] Note the telling physical metaphor, "falls", as in entropically spontaneous.

[3] http://news.bbc.co.uk/2/hi/1971852.stm All of which sharing a common yet unstated condition, no private central bank, a condition since rectified in all but three.

[4] Moreover, the arguments deploying this bad information were themselves fallacious, by Magnani's analysis exhibiting fallacies of four forms "generated by a constitutive lack of any process of self-corrective cognition" - i) the appeal to authority (faulty), ii) the unquestioned referral to habits of thought (mixed with a lack of evidence and a barrier to alternative analysis), iii) flawed and merely persuasive reasoning (not seeing weapons seemed to provide evidence that Iraq had them), iv) minimal role of empirical evidence (further explained away as denial and deception or discounted because it did not support the habitual knowledge of a robust and active WMD capability) (Magnani, 2007, pp. 156-157).

[5] Here, it may also be noted that conscientiousness and openness to experience are associated with myriad positive facotrs, including for example higher cognitive ability (Sharp et al., 2010), longer life span in greater health (Kern & Friedman, 2008), and decreased depression (Ayotte et al., 2009), while greater prefrontal integration of brain hemispheres (represented in the ACTWith model's two basic operations in the open-open mode) is associated with greater agreeableness and sociability (Hoppenbrouwers et al., 2010).

This work was previously published in the International Journal of Technoethics, Volume 2, Issue 4, edited by Rocci Luppicini, pp. 53-70, copyright 2011 by IGI Publishing (an imprint of IGI Global).

Chapter 16
Facebook Has It:
The Irresistible Violence of Social Cognition in the Age of Social Networking

Tommaso Bertolotti
University of Pavia, Italy

ABSTRACT

Over the past years, mass media increasingly identified many aspects of social networking with those of established social practices such as gossip. This produced two main outcomes: on the one hand, social networks users were described as gossipers mainly aiming at invading their friends' and acquaintances' privacy; on the other hand the potentially violent consequences of social networking were legitimated by referring to a series of recent studies stressing the importance of gossip for the social evolution of human beings. This paper explores the differences between the two kinds of gossip-related sociability, the traditional one and the technologically structured one (where the social framework coincides with the technological one, as in social networking websites). The aim of this reflection is to add to the critical knowledge available today about the effects that transparent technologies have on everyday life, especially as far as the social implications are concerned, in order to prevent (or contrast) those "ignorance bubbles" whose outcomes can be already dramatic.

1. INTRODUCTION

Once upon a time, "rumor has it that…" was the typical way of introducing the latest gossip heard over the vineyard: today, the old adage seems about to be changed to "Facebook has it that…" since social networking websites are becoming the prime source for information about our social connections. That is, gossip.

DOI: 10.4018/978-1-4666-2931-8.ch016

Over the past few years, mass media increasingly identified many aspects of social networking with those of established social practices such as gossip. This produced two main outcomes: on the one hand, social networks users were described as gossipers mainly aiming at invading their friends' and acquaintances' privacy; on the other hand the potentially violent consequences of social networking were legitimated by referring to a series of referring to a series of studies in anthropology, evolutionary psychology and sociobiology stressing the importance of gossip for the social evolution of human beings.[1] In this paper, I mean to explore the differences between the two kinds of gossip-related sociability, the traditional one (which can also rely on technological artifacts as means of communication) and the technologically structured one (where the social framework is determined by the technological one and coincides with it, as in social networking websites). Following the approach to the relationship between technology and knowledge undertaken in Magnani (2007), the aim of this paper is to act seminally and foster new reflections about the effects that transparent technologies have on everyday life, especially as far as the social implications are concerned, in order to prevent (or contrast) those ignorance bubbles whose outcomes can be already dramatic.

2. GOSSIP AS SOCIAL COGNITION

The revaluation of gossip, which started in the past century, is indebted towards two main disciplines: anthropology on the one hand, evolutionary studies on the other hand. Anthropologists focused on gossip as a means of social regulation (Gluckman, 1963; Yerkovich, 1977), but the mechanism often maintained a sense of otherness, both geographical and chronological: gossip could be indeed more than mere idle talk, but that was true as far as other people, at a different stage of development, were concerned. Conversely, evolutionary psychology and sociobiology immensely boosted gossip's reputation (no pun intended) by showing its relevance as far as it concerns the dawn of language and sociality (Dunbar, 2004; Wilson, Wilczynski, Wells, & Weiser, 2002). Partially because of the hard-to-die naturalistic fallacy, which reverberates in the moral justification of whatever is hypothesized to have been part of our evolutionary heritage, these studies managed to obtain a massive dissemination in public opinion: the result was a widespread acceptance of the common utility of gossip as "social hygiene", which came with a complete obliteration of its violent nature.

Gossip and its Evolutionary Dignity

One of the most quoted and successful approaches on gossip is that of evolutionary psychologist Dunbar (2004). His main take is that gossip evolved along the lines of grooming to allow hominids the possibility to cope with life in large social groups: in his view, the necessity of gossip is strictly connected with the development of language itself.

According to Dunbar, *Homo Sapiens*'s close evolutionary relatives – anthropoid primate societies – do indeed display an uncommon degree of sociability, notwithstanding the fact that apes and monkey are not considered eusocial species (i.e., ants and bees living in colonies). Living in large groups gave to primates and hominids a great deal of benefits: the main ones can be individuated in hunting and foraging, and as far as the protection of the individuals is concerned.

One key factor underpinning this kind of organization is the use of something similar to trust and constraints to protect and ensure the functioning of the social relationships. Actually, for all the benefits, living in interactive groups larger than many of any other non-eusocial animal exposes primates and humans (and therefore hominids, too) to an increased number of stresses, likely to trigger hostile behaviors among members of the same group. These stresses include disturbances when feeding, harassment by stronger and eager-to-dominate individuals, and more simply the uncomfortable effects brought about by the need to coordinate each other's behaviors with a result that's often less than ideal for every individual.

Primates' solution to this issue, connatural with high sociality, is the constitution of alliances by structuring a "sense of obligation" between individuals. This is achieved with the practice of social grooming, which "consists mainly in one to one activities involving physical contact, making the receiver to release endorphins and thus feel a sensation of well-being" (Dunbar, 2004, p. 101).

Research suggests that primates can live in groups counting up to 80 individuals: this number requires each of them to spend about one fifth of the day in social activities, that is, chiefly grooming. Humans have an average group size of 150 individuals (Hill & Dunbar, 2003), which is nearly the double of the primates' one. To break the ceiling of group size humans must have developed a new and more effective bonding mechanism: Dunbar's answer to this issue is language, in the form of gossip.

Linguistic exchanges with a social content seem to meet perfectly this requirement, allowing a significant increase in the interaction group: as we said grooming is a one to one activity, whereas several people can engage in a conversation (typically up to four people, one speaker and three listeners, taking turns). Curiously, people spend the same time as primates in sociable interaction, with the advantage that speaking is not an exclusive activity, and it can be easily undertaken together with several others.

Grooming would create social obligation through the manipulations of endorphins: it can be argued that gossip, in spite of being often a pleasurable activity, does not usually induce the release of as much endorphin as grooming. Yet, it manages to convey a simple yet powerful message of social commitment, similar to "I consider it more important standing here talking to you than being over there with [anyone else]" (Dunbar, 2004, p. 102).

There is another characteristic of language that makes it valuable as a bonding mechanism for large social groups: it allows us to exchange information in order to overcome monkeys' and apes' big constrain and get to know more than what we see: thus we are not limited anymore to first person knowledge. "That, after all, is what language itself is basically designed to do. Its role in social bonding is that it allows us to keep track of what is going on within our social networks, as well as using it to service those relationships. As

important as the latter function is, it is in many ways the first that is especially important. Lacking language, monkeys and apes are constrained in what they can know about their networks" (Dunbar, 2004, p. 102).

Thanks to gossip, we can easily find out what has gone on *behind our backs*, and decide to share with our friends and allies information that might be interesting and relevant for them to know.

Where Social Networking Sites Kick In

If this background is accurate, social networking websites appeal to an extremely potent and long established human inclination. As pointed out in Tufekci (2008) most social networking websites are similar in as much as they present a series of profiles, which *correspond* to the website's users, and those profiles are linked by relationships of "friendship". Letting alone the moral considerations implicated by such a deflationary use of the word *friendship*, let us focus on the effects it produces: calling virtual acquaintances (who may or may not correspond to real-life acquaintances) "friends" configures the whole exchange in a different way than if – say – they are called "links" or "connections" or simply "contacts". Mimicking the rhetoric of friendship, social networking websites such as Facebook and Myspace can appeal by simulating the affordances of real-life human relationships, perhaps in an even easier way.

The word "friendship" possesses in real-life an amazingly rich range of hues and variations, while sooner or later every uses of social networking website experiences the inadequacy of the word "friend" as a label which is meant to cover roughly anyone from high-school sweethearts to colleagues, including relatives, partners, neighbors, VIPs and so on.

Such more or less intended confusion triggered by the linguistic abuse of a word can trigger a wide range of perversions, but in this reflection we are going to focus on the one affecting the complex phenomenon known as gossip: gossip is so relevant to group-level interactions that Wilson identified it as a signal of the actuality of groups as levels of selection (Wilson et al., 2002), but the effects of gossip go beyond that. It could in fact be defined as a "group-projecting" behavior (Bardone, 2011): referring to friends, it is both true to say that gossip is what we do with friends, and that friends are the persons we gossip with.

The gossip-related nature of social networking websites, though, is not immediately clear and therefore calls for a philosophical enquiry: on the one hand, common sense is perfectly aware about the *gossipiness* of social networking websites (to the point that they are renowned to raise severe privacy issues), on the other hand it is rather rare to see on somebody's wall a post such as "WHOA!! I saw Mary kissing John at the bar yesterday night!! but she hasn't broken up with Peter yet, has she!? What a...," maybe complete with tags to the involved individuals. The gossipiness of social networking websites is subtler since it assumes the peculiar configuration of an invitation towards "self-gossip", which often winks at narcissism and voyeurism (Rosen, 2007).

Following this insight, in the next section I will focus on the evolution of gossip's attention-grasping mechanism, while in the last one the attention will be brought over how the violent potential of gossip can be increased by the epistemic multimodality characterizing those activities mediated by structures that are typical of web 2.0.

3. WHEN RELEVANCE BECOMES OLD-FASHIONED

In this section I mean to concentrate on relevance, considered to be the main pillar of gossiping exchanges. An illustration circulating online defines relevance as *the intersection between what you want to say and what your interlocutor wants to hear*. This definition exemplifies to what extent the notion of relevance is contextual: contributions

to conversation are never relevant *a priori*, but according to the way they relate to the existing conversation, the speakers' interests and a number of other factors (for instance chrono-geographical settings) (Walton, 2004, Ch. 1). As a matter of fact, one piece of gossip can be received as thrilling and exciting in one conversation, and totally disregarded (or even deemed as inappropriate) in another.

As we will show in the following subsections, being relevant coincides with being a good gossiper and one whose company is appreciated and sought for: to achieve this goal, a work of constant fine tuning and modulation of information one wants to share is required. What happens in gossip over social networking websites embodies a rather different conception of relevance. I am far from claiming that traditional gossip is not violent, but usually the violence is indeed aimed at a precise target: a person, a norm to be restated and enforced.

Shoot to Kill: Modulating and Fine Tuning Information

According to anthropologist Sally Yerkovich, what sets the boundary between what is gossip and what is a simple sharing of information consists in this: "information, no matter how salient or scandalous, isn't gossip unless the participants know enough about the people involved to experience the thrill of revelation" (Yerkovich, 1977).

The thrill of revelation is probably the key to pleasant and gratifying sides of gossip: we might reasonably suggest that to share an unheard piece of news causes a release of endorphins (both in the speaker and the hearer) similar to what groomers experience, thus further validating Dunbar's theory. In fact, when gossiping in an active conversational exchange,[2] what a speaker wants to avoid is facing a "So what?" reaction in her fellow gossipers. This means that the piece of information has not been acknowledged as interesting or, more specifically, as relevant.

Relevance has been individuated as the cause of the hiatus between human language and animal communication (Dessalles, 2000), and relevance is certainly a metaphorical lighthouse guiding conversations: "Think of how easily one may become ridiculous when uttering a dull remark, especially if there is a large audience. People who repeatedly fail to make relevant points are likely to be considered mentally defective. Conversely, people having the ability to make sound statements or interesting moves on certain subjects are likely to become the focal point of the conversational group" (Dessalles, 2000, p. 63).

The importance of relevant utterances is fundamental for gossip: "be relevant" was one of the conversational maxims introduced by philosopher Paul Grice as corollaries of the *cooperative principle* (Grice, 1975). Yet, if we overhear a typical gossiping exchange, we notice a systematic violation of Grice's quantity, quality and manner maxims: gossip is often less informative than required (to encourage further enquiries from hearers), evidences are mostly questionable (but hardly questioned) and remarks are obscure and ambiguous. Should this lead one into affirming that an interaction as widespread (and social in its deepest nature) as gossip can be labeled as uncooperative? It is clearly not the case, since these deficiencies are balanced by an empowerment of what Sperber and Wilson define as the presumption of relevance: "The utterance is presumed to be the most relevant one compatible with the speaker's abilities and preferences, and at least relevant enough to be worth the hearer's attention" (Sperber & Wilson, 1995, p. 233).

Consequently, in the conversational game of gossip, relevance is the key to both what to say and what to comprehend (enacting a sort of charity principle, and thus assuming that the speaker is trying her best to be relevant). It is clear, yet, that relevance is not a characteristic that a sentence displays per se, but it depends on the context. In fact, "[although] the remarks may be appreciated on the surface for their wittiness, viciousness,

etc., by anyone who hears them, they can only be completely understood with a knowledge of the shared characterization" (Yerkovich, 1977, p. 196).

This shared characterization must not be confused with a person's reputation, which is intended to be more or less shared between a vast number of that person's acquaintances: the shared characterization Yerkovich refers to is extremely dependent on the local gossip party. A speaker may have to refer to different shared characterizations of the same person, according to who her gossiping partners are: this leads to the fact, experienced by anyone who has ever engaged in gossip, that a single piece of news is usually *modulated so to be fine-tuned to the particular gossiping occasion*, in order to maximize the possibility that it is met as relevant and to avoid the emergence of conflictual situations: the same gossip is hardly ever told twice the same way, but different characterizations can be privileged every time.

Relevance as a Brute-Force Attack

The description of relevance and the pivotal role it plays in gossiping interactions clearly has some relatively high costs. To be relevant in the sense I just described, one must achieve a certain knowledge as far as it concerns elements such as: 1) the functioning of other people's mind; 2) the functioning of one's interlocutor mind; 3) one's interlocutor acquaintances and their respective evaluation; and 4) what one's interlocutor knows or is willing to know about one of her acquaintance. To this respect, Magnani (2009) argues that both argument evaluation and argument selection are often the fruits of an *abductive appraisal;* that is, striving to be relevant is guided by abductive skills, which both the arguer and audience employ to select convincing arguments on the one hand, and to evaluate the competitive narratives at stake, on the other.

These elements are not fixed but they are the object of constant negotiation between speakers, since – for instance – those I labeled as (3) and (4) are likely to be affected by the gossiping conversation itself.

If we consider the framework of gossip afforded by social networking websites, the situation is drastically different. As we already suggested, what is perceived as a frenetic gossiping network appears to be – at closer inspection – a self-gossiping dimension, in which the production of information does not partake of the negotiation typically characterizing gossip (Solove, 2007).

Using a well known analogy, it could be said that the self-gossip typical of social networking websites is akin to what happens in 1935 Walt Disney's *Silly Symphony* cartoon entitled "Music Land": users, through their profiles and their status updates, are like separate islands playing their own music – more or less skillfully – hoping that some of their friends will stop and listen, and display some interest about what they are doing, thinking, planning etc. Such an interactional schema is perfectly in-tune with the tenet of Web 2.0 assuming that every user must also be a content producer (Keen, 2007): a potential perversion, though, arises when this system deludes the user into believing it can support an exchange traditionally based on negotiation, exchange and precise regulation of the conveyed information according to the party.

In dialectical gossiping exchange the reward for delivering a juicy gossip can be easily read in one's interlocutor answer, or expression, and this can further corroborate the speaker's relevance-seeking strategies. If we consider Facebook as the most popular example of social networking website, the acknowledgement that a bit of self-gossip was considered relevant, worthy of a *friend*'s attention, comes through the apposition of a "comment" or of a standard "like".

The way to achieve this recognition is clearly different from the strategies typical of dialectical gossip, which is an activity usually engaged in one-to-one or in relatively small groups, wherefore it is possible (if not relatively easy) to infer the correct characterization shared by the party and gossip consequently. Conversely, technology-mediated

self-gossip is a one-to-many discourse: the quest for relevance cannot be met by hoping to infer the correct characterization (besides, it would mean to infer the correct characterization about oneself) that is shared by as many contacts as possible, and gossip consequently. The quest for relevance, therefore, becomes more of a brute-force attack: publishing tweets, or status updates, even quite frequently, until one catches somebody's attention. Seeing one's post liked or commented might be said to correspond to the endorphin-releasing moment in traditional gossip, inasmuch as the producer of self-gossip is finally acknowledged as an interesting source.

By using the expression "brute-force", I mean to focus our reflection on the peculiar cost-benefits relationship as far as it concerns technology-mediated gossip. In traditional gossiping exchanges, brute-force attempts would be considered a suicidal heuristic: imagine a speaker addressing to a party a series of utterances concerning what she is about to do, what she likes, how much she enjoyed yesterday's dinner, her mood upon waking up, and imagine to observe the party meeting these contributions to the dialogue with obstinate lack of interest until an utterance of the unfortunate speaker hits a lucky note and the party erupt in high-sounding approval, to regain their apathy a little later. This little mental experiment should show how radically different mechanisms are at play in social networking websites with respect to traditional gossip, and yet they appeal to our natural taste for gossip.

4. THE EVOLUTION OF SELF-REGULATION AND SELF-REPUTATION

In this section I will introduce an anthropological conception of traditional gossip as a tool for conflict management, whose main characteristic is to embody a self-regulation mechanism. I will consequently analyze the difference of gossip mediated by social networking platforms with this respect.

Gossip as a Self-Regulatory Conflict Management System

By asserting the evolutionary dignity that gossip was endowed with by recent studies, we mentioned how living in large groups exposes individuals to stresses that could break into serious conflicts between individuals: just as grooming is not always successful in the animal kingdom, so gossip-as-grooming is not always enough to prevent social skirmishes. Nevertheless, gossip can prove to be an effective conflict management tool also at a higher cognitive level, to rationalize hostilities among individuals and cliques (competitively engaged against each other) and reassert once again the general values of the group. Gluckman states that: "[Gossip and scandal] control disputation by allowing each individual or clique to fight fellow-members of the larger group with an acceptable, socially instituted customary weapon, which blows back on excessively explosive users" (Gluckman, 1963, p. 313).

To make this point clearer, we might use a reasonable analogy: a few centuries ago, duels were a common method for settling arguments. Weaponry was not at the contenders' total discretion but was subject to a long-established regulation. Similarly, gossip can indeed be used as a social weapon; saying social, though, we do not only mean that it can be obviously used to damage other peers in the same coalition, but also that its very way of operating is determined by the coalition itself. Since the content of gossip is bounded by the evoked morals of the community, struggles are kept within the common set of shared norms, and any attempt to trespass these norms out of rage or indignation is discouraged by the very same mechanism. Thus, even violent internal struggles achieve the result of maintaining the group structured at a convenient degree, as shown by the persistence of societies based on the concept of *honor culture* (Magnani, 2011, para 1.2).

This conception allows an expansion of what Yerkovich suggested by referring the "moral characters": language affords the construction of a distal model, a mediating structure that reproduces the dynamics of the coalition as an essentially moral narration. This narration is clearly about moral characters and not individuals per se: keeping track of every act of every peer would clearly cause a cognitive overload for every member. This is not the case when only a few relevant traits are considered and developed into (moral) typifications, so that the information can be easily recalled and used by individuals.

Social Networking Websites Alienate the Gossiping Structure

What I just described corroborates the aforementioned conception of gossip as a *group-projecting* behavior: the simple fact of gossiping manages on its own to spark the *poiesis* of a social group and enforce its regulations. Every particular group of gossips, and hence the shared pieces of news and the way they are characterized, lasts as long as the gossiping interaction lasts. This is not to say that gossip do not propagate outside the circle of acquaintances they were meant for, but everyone with a mild experience in gossip knows that a gossip, to be shared within a new gossiping party, often needs to be introduced and so to say *translated* in order to make it understandable as gossip (thrilling, pleasurable, scandalizing etc.) by the new party. The cost of this computational effort contributes to the self-regulatory aspects of gossip: if the translated gossip could provoke a reaction of disapproval (e.g., "Don't be so harsh on her! She was a very sweet person until that horrible accident happened"), or be perceived as dull by the receiving party (e.g., "This seems a perfectly ordinary story to me, but I reckon I don't know Jeff personally so…"), one would rather abstain from divulging the news.

In technology-mediated gossip, for instance in social networking websites, the gossiping structure lacks this group-projecting dimension, and hence the self-regulatory dimension. Friendships,[3] to begin with, are not assessed by continuous practice as it happens in real life, but sanctioned by yet another IT-mediated labeling which is valid until revoked (Rosen, 2007).

Furthermore, the structure affording the distribution of information does not depend uniquely on the speakers, as in regular gossip. The exchange depends on a third-party agency which also mediates how the regulation is enacted: this concern parallels the usual one about social networking websites' privacy issues, but has a different philosophical meaning. Gossip, as I just suggested, is indeed about sharing information, most often about people who would much rather not have that same information divulged and spread, but what gossip is not about is copy and paste mechanisms.

When gossiping, information is always subject to ending up in the wrong hands – willing or unwilling troublemakers – and thus causes potentially unwanted violence. Still, one of the ambivalences of gossip that made it such a fundamental device is that, as if aware of its poor epistemic value, gossip can always be downplayed as being "mere" gossip, probably transmitted not even transmitted in its correct form.

Conversely, in technology-mediated gossip, gossipers end up being alienated from gossip itself, because on the one hand search algorithms may allow to trace back the original source, on the other hand copy and paste functions make sure that the information is transmitted with nearly total copy fidelity.

To make things even more complicate and dangerous, whereas information in regular gossip has usually been only verbal, at least as ordinary citizens were concerned,[4] in social networking websites gossip is enriched with images and movies which further increase its epistemic value: text and pictures posted on Facebook, for instance, can circulate and become source of nuisances or harassment in a way that cannot compare to traditional forms of gossip.

Reputation Defined: Towards a Multimodal Self-Violence?

In this last section, I will focus my attention on the sensitive concept of *reputation*, and try to grasp how it might be jeopardized by a translation of gossiping habits into the technological mediation.

Reputation, that is the common image that is generally held about somebody, can be defined the product of a distribution of knowledge constructed *by negotiation* around that very person. It is not the same as a moral characterization inasmuch as it displays a further shared nature, but it is surely influenced by the different characterizations at play in gossiping scenarios: gossip is in fact one of the main forgers of reputation (Hunter, 1990), and the accurateness of reputation does also depend on the amount and diversification of circulating gossip (Sommerfeld, Krambeck, & Milinski, 2008).

Traditionally, the distinction between reputation and identity has always been tacitly acknowledged: reputation is an *artifact* superimposed on an agent by others, with a chaotic dependance on the agent's own actions. There are no monotonic correlations between one's deeds and the effects they have on the person's reputation: the very same action, perceived from different outlooks or performed at different moments – even if the intention is the same – can affect positively or negatively one's reputation. The proof of this complexity is that, reputation-wise, we are often trapped in *catch-22* situations, knowing that one course of action might hinder our reputation with certain respect, but abstaining from it would damage it from another perspective.

With respect to social networking website, a different phenomenon might arise. First of all, we must keep in mind that, as we suggested, the form of gossip afforded by social network is in the first place "self-gossip": users think they are publishing only what they want and they might be unconsciously led into believing that their self-gossip might improve their reputation.

Indeed, self-gossip does not enter the gossiping networks unmediated, but becomes the object of gossip itself: gossip, in fact, is about the information being offered together with a sense of characterization, which is also what makes it relevant and interesting, as I suggested in the previous sections: therefore, if social networking website delude the user into believing that one's profile can be a simulacrum of her identity, and if this simulacrum is thought to be able to project one's reputation as the subject wants it to be, this process might very likely see failure as its ultimate end.

Mere rumors have always been able to seriously damage people's reputation, and hence their life, sometimes beyond repair: still, it is important to stress how there is a fundamental difference, if only at a psychological level, between the rumor being introduced by somebody else rather than by the object of the rumor herself.

This phenomenon can be individuated as emerging right now, inasmuch as many of the violent consequences of technology-mediated gossip originated with the subject sharing sensible information about herself.

There is already a discrete amount of literature, as shown in Solove (2007), dealing with the reason why users feel like sharing that much information about themselves on social networks, whereas they would not do it in real life. A particular bias towards Facebook and other social networking websites embed users in a kind of "It's just Facebook" state of mind: to paraphrase a common motto, the common feeling is that *what happens in Facebook stays in Facebook*. Research shows that, even after facing (mild) consequences caused by excessive divulgation of personal information, users tend to go back to being active on Facebook. The main idea is well exemplified by Debatin, Lovejoy, Horn, and Hughes (2009): "The technical strategy is to tighten the privacy settings; the psychological strategy is to integrate

and transform the incidents into a meaningful and ultimately unthreatening context. Knowing or at least believing to know who the perpetrator is creates a feeling of control and reassurance. The perpetrators are branded immature individuals or creeps, and the incidents tend to be minimized as pranks, exceptions, or events from a different world: 'In the end,' said Brian, reflecting on the hacking incident, 'it's just Facebook…There was a time before Facebook. You can do without it. It's ok'" (p. 100).

Coherently with my previous observation, it might be suggested that gossip mechanisms in social networking websites create one further paradox, blurring the division between *profiles* and *avatars*. The concept of avatar, etymologically linked to that of incarnation, represented the essence of virtual social environments (Meadows, 2008): the avatar is a partial embodiment of the operating user, and their identities can be told one from the other. Furthermore, the avatar has usually been a warrant of anonymity for the user, allowing the Internet to become a reenactment of Victorian morality, where everyone could pursue more or less licit perversions protected by the mask of their avatar. Of course, avatars have reputations as well, but unless data leaks are the case, they do not affect that of their *puppeteers*.

In fact, all is needed is the recourse to the *virtual self*, as Goffman (1961) put it. According to his definition, the idea of the virtual self describes the expectations about the character a person is supposed to have playing a given role. So, one is who he is, and who he is supposed to be when playing a given role in society. In a traditional virtual context one can stress the process of "role distance" as much as she wants: Goffman used that expression to refer to all those cases in which a person distances herself (or who she actually is) from the role she has been assigned to in a given context. This can safely happen in the virtual world as long as no one expects the role of the avatar to coincide with the role of the actual user: conversely, in a social networking website, the role of my profile is expected to be the same role I display in real-life interactions.

At this point, it could be argued that the perceived – and yet sometimes hard to define – concoction of appeal and discomfort induced by social networking websites is caused by their being situated at the crossroads of two different lines of mental dispositions.

The first, as we described in the first section, is the assumed evolutionary background, which makes us eager to engage in social grooming by means of narratives rich of thrilling information about our peers; conversely, the second is extremely vivid due to its actuality and consists in the Internet seen as the land of anonymity, where one's deeds are devoid of consequences in real life, and where anything can be said.

This phenomenon can indeed be individuated inasmuch as many of the perverted consequences of the extreme social connectivity are characterized first of all by a moment of *sudden realization of the threat* they had exposed themselves to Debatin et al. (2009) and Solove (2007): given the transparency of social networking websites, users are completely clueless about the consequences caused by the information they share, which does not amount only to compromising writings, but may also consist in images and movies.

What I am trying to delineate clearly resembles a *cognitive bubble* mechanism (Magnani & Bertolotti, 2011), more precisely a *moral embubblement* (Magnani, 2011): users do not activate the moral knowledge that would be required to avert the violence they are inflicting to themselves in a technology-mediated structure.

An example can make this further apparent: after teaching certain classes and seminaries at university, I happened to add and be added on Facebook by several students. On Facebook, I expected to see (and I can easily share) the pictures of their summer holidays, and thus enrich the gossip I could be doing about them with as many saucy images as I wanted. Yet, if I decided to publicly and directly ask a student, in a real-life, exchange to share some pictures displaying her in bikini she would probably frown at my request, and she would be all the more infuriated if she caught me handing prints of her pictures to my colleagues, as we dwell in sleazy comments.

Such duplicity seems to corroborate my previous claim about the intersection of gossip and virtual realities: virtual realities, populated by disembodied avatars, are meant to be a parallel world, and failures in appreciating this separation can be labeled as forms of addictions, lunacy or acquire pathological dimension. On the other hand, social networking websites are engineered to be transparent and to scaffold the (social) lives of their users, but at the same time they wink to the possibility of engaging the same behaviors they would engage in a virtual parallel life.

Like relevance, so the notion of *appropriateness* has an eminently local nature, and can often be judged only over the consequences: whereas, for instance, for an anonymous avatar it could be appropriate to forge her virtual identity by publishing sexually explicit images of herself in a dedicated board, an "innocent" bikini set shared on Facebook might encourage comments that the user would not foresee. This could heavily damage her real-life reputation, when all she meant was to display herself as a solar, energetic person.

In this case, the problem is the lack of peculiar "lamination" displayed by social networking websites. According to Goffman, when a certain activity is re-framed, or re-keyed, it is said that a lamination (or layer) is added to that activity – the so-called natural strip that is the base appearance and character (1986, p. 82). As Hample et al. (2010) put it, laminations cover behaviors and/ or activities "in the same way a good veneer can cover inexpensive wood". Basically a lamination is a sort of new appearance of the natural strip that completes the process of meaning transformation.

Indeed, the prerequisite for a lamination to be perceived is that the partners in conversation apply the appropriate frame. So, a lamination does not conflict or erase all the pre-existing meanings that a certain activity – for instance, sharing sexually explicit images of oneself over the internet – may have. But it somehow becomes a new layer that metaphorically laminates it providing new interpretations (for instance in a BDSM virtual community), but, at the same time, it also favors the emergence of dissonance, in case the participants are not able to successfully re-key the activity and hence do not perceive them as *appropriate*.

Therefore, the outcome of technology-mediated self-gossip is a potentially lethal mix, in which the user thinks she can affect her reputation by operating on the identity on her profile as if it was an avatar, but is in truth enacting a kind of self-violence for the benefit of her public of "friends", splitting herself as the subject and the object of gossip. To do this, the agent needs to switch agencies between her real self and the self that she feeds to her contacts as the object of gossip, and the social networking website is the structure affording the unwilling self-disgrace of the user.

5. CONCLUSION: SOCIAL NETWORKING WEBSITES AS SUPERNORMAL STIMULI

The reflection I just proposed does not mean to a hypocritical and fundamentalist attack on social networking websites, I am a user myself and it is sincerely difficult to imagine the Internet experience to go back to a pre-Facebook era, also because of the device-based (and application-based) fragmentation we are experiencing every day.

The issue at stake the necessity for users to gain a major awareness of the potential risks they exposed themselves to, deluded by the transparency of the technological structure aimed at scaffolding our social life: it goes without saying that any discrete reproduction (like the computer-based one is) of a continuous phenomenon (like human sociality and social cognition) has costs and benefits. The transparency and pervasiveness of social networking, though, tend to obliterate the presence of costs.

The ability of driving, for instance, endows the driver with benefits and risks. The awareness of those risks compelled society to institute a costly system of courses and licenses to make sure that every driver classifies above decency line in avoiding harmful risks. Yet, a wrong photo circulating on the internet could damage one's life far more than an ordinary car accident. Just as in driving we expect people to be able to rely on more specific knowledge than their mere common sense, so we should expect users of social networking websites to rely on something better than a mere technical capability to use those websites. As a matter of fact, it is an enriched philosophical and moral knowledge which is required, which transcends the mere technical ability. It is not a problem of abstractly knowing the consequences of our deeds in socially networking websites, but of being able to engage the correct moral framework (for instance, the social cognition one and not the virtual reality one): only a deep philosophical reflection, involving mental experiments, dichotomies, focus on analogies and differences between the different frameworks, could arm users with the necessary "weaponry" to avert many of the violent consequences of misplaced behaviors in social networking websites.

On a final note, one aspect emerging from biological studies might further stress the urgency of a major reflection on a topic which has already begun to affect our lives powerfully: the danger and incredible fortune of social networking website could ultimately rest in their being a kind of "supernormal stimuli". As I suggested, on a first analysis, the appeal of social networking websites resides in the hard wired module that makes us eager to engage in social cognition, and in the more recent, soft wired one about the benefits of behavior in virtual environments.

The sum of these two factors, though, might not seem enough to explain the incredible diffusion of social networking websites in the past two decades: this is where the concept of *supernormal stimuli* can help our understanding. Naturalist and nobel laureate Niko Tinbergen had coined the expression in the 1930s to indicate the imitations of natural signals that would induce in animals a stronger reaction than the natural thing would, even in presence of the natural signal itself: famous examples include fishes attacking unnatural-looking dummy fishes with brighter colors than their natural counterparts, and trying to mate with unrealistic female dummies with exaggerated fertility attributes, even in presence of a "normal" living female. Recently, Barrett (2009) comprehensively suggested how this very concept could be applied to human beings and explain the exponential emergence of phenomena such as the pursuit of "cuteness" in artifacts and people, the diffusion of pornography and sexual esthetic enhancements, and even human beings' fondness for shopping malls and wars that are always more destructive: in all of these cases, according to Barrett, artifactual or artificial stimuli engage our primeval dispositions in a far more convincing way than their natural counterparts, even if the latter is at hand.

Shall we then conclude that social networking is to gossip what pornography is to sex? The analogy might be bit provocative, but the ultimate aim of this reflection is to provoke further reflection. Consistently with the supernormal stimuli hypothesis, research shows that social networking websites are often perceived as overwhelming: "[…in] the interviews, a majority of the [social networking sites] users talked about how much they enjoyed learning about their friends' and even strangers' lives. The heavy users in particular expressed that their use of these sites was partially driven by their curiosity about how people from their pasts were doing or whether they had changed. *Some students also reported 'getting lost' in social networking, continually checking profile after profile, leaving message after message.* One student talked about how receiving messages made her feel good, so she tried to leave messages to extend the good feeling to her friends" (Tufekci, 2008, pp. 556-556, emphasis added). With this respect, social networking websites can indeed be seen as a *supernormal* artifacts imitating social cognition. More attractive, easier, perceived as a more efficient way to connect and advertise oneself, they are indeed able to override our natural appetite for social cognition and informational grooming. The price to pay in return is to become violent self-gossipers, like butchers of our own image, feeding our friends' evolutionary appetite for gossip with bits of our private life, in exchange of the easy popularity that makes us appreciate virtual sociality so much.[5]

REFERENCES

Bardone, E. (2011). *Seeking chances: From biased rationality to distributed cognition.* Berlin, Germany: Springer-Verlag.

Barrett, D. (2009). *Supernormal stimuli: How primal urges overran their evolutionary purpose.* New York, NY: W.W. Norton.

Debatin, B., Lovejoy, J. P., Horn, A. K., & Hughes, B. N. (2009). Facebook and online privacy: Attitudes, behaviors, and unintended consequences. *Journal of Computer-Mediated Communication, 15*(1), 83–108. doi:10.1111/j.1083-6101.2009.01494.x

Dessalles, J. (2000). Language and hominid politics. In Knight, J. H. C., & Suddert-Kennedy, M. (Eds.), *The evolutionary emergence of language: Social function and the origin of linguistic form* (pp. 62–79). Cambridge, UK: Cambridge University Press. doi:10.1017/CBO9780511606441.005

Dunbar, R. (2004). Gossip in an evolutionary perspective. *Review of General Psychology, 8,* 100–110. doi:10.1037/1089-2680.8.2.100

Girard, R. (1982). *Le bouc émissaire.* Paris, France: Grasset.

Gluckman, M. (1963). Papers in Honor of Melville J. Herskovits: Gossip and scandal. *The American Economic Review, 4*(3), 307–316.

Goffman, E. (1961). *Encounters: Two studies in the sociology of interaction*. Indianapolis, IN: Bobbs-Merrill.

Goffman, E. (1986). *Frame analysis: an essay on the organization of experience*. Boston, MA: Northeastern University Press.

Grice, H. (1975). Logic and conversation. In Sternberg, R., & Kaufman, K. (Eds.), *Syntax and semantics 3: Speech acts*. New York, NY: Academic Press.

Hample, D., Han, B., & Payne, D. (2010). The aggressiveness of playful arguments. *Argumentation, 24*(4), 245–254. doi:10.1007/s10503-009-9173-8

Hill, R., & Dunbar, R. (2003). Social network size in humans. *Human Nature (Hawthorne, N.Y.), 14*(1), 53–72. doi:10.1007/s12110-003-1016-y

Hunter, V. (1990). Gossip and the politics of reputation in classical Athens. *Phoenix, 44*(4), 299-325.

Keen, A. (2007). *The cult of amateur: How today's Internet is killing our culture and assaulting our economy*. London, UK: Nicholas Brealey.

Magnani, L. (2009). *Abductive cognition: The epistemological and eco-cognitive dimensions of hypothetical reasoning*. Heidelberg, Germany: Springer-Verlag.

Magnani, L. (2011). *Understanding violence: The intertwining of morality, religion and violence: a philosophical stance*. Heidelberg, Germany: Springer-Verlag.

Magnani, L., & Bertolotti, T. (2011). Cognitive bubbles and firewalls: epistemic immunizations in human reasoning. In *Proceedings of the 33rd Annual Conference of the Cognitive Science Society*, Boston MA.

Meadows, M. S. (2008). *I, avatar: The culture and consequences of having a second life*. CA: New Riders.

Rosen, C. (2007). Virtual friendship and the new narcissism. *The New Atlantis: A Journal of Technology and Society*, 15-31.

Solove, D. (2007). *The future of reputation*. New Haven, CT: Yale University Press.

Sommerfeld, R., Krambeck, H., & Milinski, M. (2008). Multiple gossip statements and their effect on reputation and trustworthiness. *Proceedings. Biological Sciences, 275*(1650), 2529–2536. doi:10.1098/rspb.2008.0762

Sperber, D., & Wilson, D. (1995). *Relevance: communication and cognition*. Oxford, UK: Blackwell.

Tufekci, Z. (2008). Grooming, gossip, Facebook and Myspace. *Information Communication and Society, 11*(4), 544–564. doi:10.1080/13691180801999050

Walton, D. (2004). *Relevance in argumentation*. Mahwah, NJ: Lawrence Erlbaum.

Wilson, D., Wilczynski, C., Wells, A., & Weiser, L. (2002). Gossip and other aspects of language as group-level adptations. In Hayes, C., & Huber, L. (Eds.), *The evolution of cognition* (pp. 347–365). Cambridge, MA: MIT Press.

Yerkovich, S. (1977). Gossiping as a way of speaking. *The Journal of Communication, 27*, 192–196. doi:10.1111/j.1460-2466.1977.tb01817.x

ENDNOTES

[1] Consider for instance a recent article published by the online version of the *Daily Mail* title "Secret of Facebook's success: Sharing gossip with friends is 'addictive and arousing'".

2   It is important to bear in mind that not all technology-mediated gossip coincides with that mediated by social networking websites: it is clearly possible to gossip "traditionally" using email, instant messaging clients, text messages and so on.

3   With this respect, it should be kept in mind that Facebook and other social networking websites, because of their very nature, act both as a standalone niche and as instrumental artifacts in another niche. Clearly, a dimension of actual group projection is displayed by technology mediated gossip in as much as it creates coalitions of liking and commenting friends: nevertheless, from a certain perspective this formation of coalitions and groups might be said to strongly depend on the mediating structure.

4   Tabloids have long since accustomed us with visual gossip, but this concerned only VIPs, celebrities, politicians and so on, and informs another conception of the word "gossip".

5   Gossip, and self-gossip as well, can be a powerful trigger for *scapegoat* mechanisms (Girard, 1982; Magnani, 2011), with potentially devastating consequences for the person who is the object of gossip. Gossip can in fact often turn into actual mobbing, and the enriched form of self-gossip – displaying pictures, perfectly reproducible texts etc. – could become a multi-medial trigger by which the victim implicitly "authenticates" herself as such.

Section 4
Current Trends and Applications in Technoethics

Chapter 17
Technoethics and the State of Science and Technology Studies (STS) in Canada

Rocci Luppicini
University of Ottawa, Canada

ABSTRACT

University degree programs in STS (Science and Technology Studies) represent a popular training ground for scholars and other professions dealing with advanced studies in science and technology. Degree programs in STS are currently offered at universities around the globe with various specializations and orientations. This study explores the nature of science and technology in Canada and the state of ethics within STS curriculum in Canada. STS degree programs offered under various titles at nine universities in Canada are examined. Findings reveal that ethical aspects of science and technology study is lacking from the core content of most Canadian academic programs in STS. Key challenges are addressed and suggestions are made on how to leverage STS programs within Canadian universities. This study advances the understanding of the developing field of STS in Canada from a technoethical perspective.

INTRODUCTION

What is TE?

In a world where technology progresses more rapidly than the ability of society to master it, social and ethical consequences have become core components of technological advancement.

The concerted effort to study the ethical aspects of technology in life and society have nurtured in the field of Technoethics. The term, technoethics was first coined by Mario Bunge in the 1970s (Bunge, 1977) when fighting for greater moral and social responsibility among technologists and engineers concerning their creations. This concern for ethical aspects of technology evolved into a field of study under this same name. According to the Handbook of Research on Technoethics:

DOI: 10.4018/978-1-4666-2931-8.ch017

Technoethics is defined as an interdisciplinary field concerned with all ethical aspects of technology within a society shaped by technology. It deals with human processes and practices connected to technology which are becoming embedded within social, political, and moral spheres of life. It also examines social policies and interventions occurring in response to issues generated by technology development and use. This includes critical debates on the responsible use of technology for advancing human interests in society. To this end, it attempts to provide conceptual grounding to clarify the role of technology in relation to those affected by it and to help guide ethical problem-solving and decision making in areas of activity that rely on technology (Luppicini, 2009, p. 4)

What is STS?

Questions surrounding the role of science and technology in society have also spurred academic interest and debate under the general umbrella of Science and Technology Studies (Bijker, 1995; Ellul, 1967; Winner, 1993). As an interdisciplinary field of academic study and research, Science and Technology Studies (STS) focus on science and technology issues. It deals with the interrelation of social, political, and cultural variables within the advancement of scientific and technical innovation (Barnes, 1974; Hackett, Amsterdamska, Lynch, & Wajcman, 2007; Pinch, 1986). As is the case of most (if not all) interdisciplinary academic fields, Science and Technology Studies (STS) has diverse intellectual roots emerging from various areas of academic scholarship. In the case of STS, intellectual roots are grounded in the history of science and technology (Mumford, 1934), philosophy of science (Kuhn, 1962), philosophy of technology (Mitcham, 1994; Hickman, 2001), technocritical and feminist studies (Feenberg, 1991; Haraway, 1991), and sociological studies of science and technology (Bijker, 1995; Woolgar, 1991). The roots of STS differ in terms of epistemological

assumptions, conceptions of what is valuable and interesting, and by the manner in which they approach academic practices. These differing intellectual roots and values can lead to schisms which can threaten the future of STS.

University degree programs and research are at the nexus of this academic field research and represent a training ground for future STS professionals. Degree programs in STS are currently offered at universities around the globe with various specializations and orientations. As will become apparent in this article, academic training on ethical aspects of technology study is lacking from the core of most Canadian academic programs in STS.

BACKGROUND

STS Roots

One main branch of STS has been traced to mid-20th century Philosophy (Philosophy of Science in particular) and the realization that science cannot be understood from a single disciplinary perspective. Seminal works, such as Thomas Kuhn's *The Structure of Scientific Revolutions* (1962), helped to broaden philosophical views about the limitations of studying science solely in terms of facts and logic. This led to the spread of programs in history and philosophy of science which eventually included the study of science and technology (Fuller, 2005). A second main branch of STS grew out of the Social Sciences and concerted efforts to make the study of science and technology useful to society (Bauer, 1990). This second branch was driven largely by developments in Sociology and the study of scientific knowledge, which eventually expanded in scope to focus on science and technology (Woolgar, 1991). In particular, the sociological study of science and technology nurtured the development of STS through sociological critiques of science and technology (Latour, 1996), and influential writings on applications of science and technology grounded in social constructivism (Matthews, 1998). The branch of STS based in Philosophy is highly theoretical in orientation, whereas, the branch grounded in the Social Sciences has more of an applied focus (Griere, 1993).

STS at a Crossroads in Canada?

The global interest in STS programs is mirrored in countries like Canada, where there is evidence of an emerging academic field of STS. Programs in STS have been created at nine Canadian universities with orientations various specialization areas. Despite the apparent importance of STS and adequate access opportunity, current knowledge of STS within Canadian universities is limited and no attempts have been made to examine the field of STS in Canada. As a result, there is a paucity of research available about these programs which appear under a variety of titles including: STS, and History and Philosophy of Science and Technology.

This study seeks to understand the study of science and technology in Canada. More specifically, the following question directs this study: What is state of ethics within STS curriculum in Canada? This study seeks to advance understanding of the developing field of STS in Canada by exploring the current state of STS degree programs within universities across the country. This study deals with Canadian STS programs only and does not focus on STS programs outside of Canada.

METHODOLOGY

This study draws on qualitative research methods and content analysis designed to gather information on STS programs course offerings at the following universities: University of Toronto, University of Alberta, University of Calgary, Dalhousie University, Carleton University, York University, St. Thomas University, University of

Regina and the Université du Québec à Montréal (UQAM).

Data collection focused on the content of STS curriculum. To contextualize STS in Canadian institutions of higher education, academic degree program documentation is collected in addition to course descriptions. In terms of study delimitations, only non-technical degree programs that contain "science and technology" or "technology" degree programs are included within this STS research paper. It does not include academic programs dedicated the mastery of specific technology types (e.g., educational technology) and applications (e.g., digital design) or technology focused areas within the sciences and applied sciences, which are a distinct fields in themselves (e.g., Engineering and Computer Science). Accessing program information from the current academic year is possible by checking the university calendars and university websites for academic programs. Program information collected includes type of degree, degree name, courses offerings, and ethics oriented courses. In terms of data collection, the study employs qualitative analysis strategies drawn from Krippendorff's Content Analysis (1994). The purpose of the analysis is to identify major themes and areas of STS program content coverage. Overall, data analysis strategies support the study by focusing on the state of STS academic programs in Canada.

FINDINGS

STS Program Context

A scan of the online *Directory of Canadian Universities Database* (AUCC, 2007) and Canadian university websites revealed STS degree programs offered under various titles at nine universities in Canada. STS programs appear in Table 1.

From Table 1, one can observe the diversity of STS programs in Canada in terms of program titles and degree types. The most popular program

Table 1. Canadian degree programs in science and technology

Institution	Program Title	Degrees Offered	Contact
Carleton University	Technology, Society, Environment Studies	BA (minor)	http://www.carleton.ca/tse/
Dalhousie University and University of Kings' College	History of Science and Technology	BA and BSc	http://ug.cal.dal.ca/HSTC.htm
St. Thomas University	Science and Technology Studies	BA	http://w3.stu.ca/stu/sites/sts/whatis.htm
Université du Québec à Montréal	Science, technologie et société (STS)	BA	http://www.sts.uqam.ca/accueil/ac-cueil.as
University of Alberta	Science Technology and Society	BA	http://www.ois.ualberta.ca/sts.cfm
University of Calgary	Science, Technology and Society	BA and BSc	http://www.ucalgary.ca/comcul/scitech
University of Regina	Science and Technology Studies	BA	http://pager.uregina.ca/gencal/ugcal/facultyofArts/ugcal_223.shtml
University of Toronto	History and Philosophy of Science and Technology	MA and PhD	http://www.hps.utoronto.ca/
York University	Science and Technology Studies	BA, BSc, MA, PhD	http://www.yorku.ca/sts/

titles for STS are Science and Technology Studies (STS) and Science, Technology and Society (STS). The History of Science and Technology, History and Philosophy of Science and Technology, and Technology, Society, Environment Studies were other variations in program titles observed. The degree types for STS included BA, BSc, MA, and PhD degrees.

A closer scan of Canadian university degree programs in STS revealed that five of the nine programs do contain some level of program content focused on ethics or ethical considerations about

technology. The type and level of this content, however, varied in terms of content specificity (implicit or explicit) and placement (program description, course offerings). The discussion below provides a breakdown of Table 2.

Qualitative content analysis results of website program descriptions and consultation of current academic calendars for Canadian STS programs revealed ethical content in STS programs offered at St. Thomas University, the Université du Québec à Montréal, York University, Carleton University, and the University of Regina.

Table 2. Canadian science and technology degree programs with an ethical orientation

Institution	Program Description	Course Offerings
Carleton University	Yes (implict)	0 courses
Dalhousie University and University of Kings' College	No	0 courses
St. Thomas University	Yes (explicit)	5 courses
Université du Québec à Montréal	No	3 courses
University of Alberta	No	0 courses
University of Calgary	Yes (implicit)	1 course
University of Regina	Yes (explicit)	1 course
University of Toronto	No	0 courses
York University	Yes (implicit)	2 courses

St. Thomas University offers a BA in STS with a strong emphasis on ethics and technology in both program orientation and course offerings. The STS program at St. Thomas is an interdisciplinary program that addresses the myriad of social, legal and ethical issues that accompanies the advancement of science and technology. As stated in the program description:

The twin forces of science and technology are pervasive, powerful and come with a complicated array of ethical and social dilemmas that affect both our daily lives and our world. A person, therefore, cannot be considered well-educated or participate as an informed citizen, without knowledge of how science and technology interact with society (St. Thomas University Science and Technology Studies, 2010).

Although the St. Thomas program does not offer courses on ethics and technology (or courses that explicitly mention ethics), they do classify the following courses under an area philosophical/ethical concentration:

STS 2103: Science, Technology and Society II
STS 2163: Contemporary Perspectives on Science and Religion
STS 2603: Animals: Rights, Consciousness and Experimentation
STS 3103: Feminist Critiques of Science
STS 3563: Philosophy of Science

The only French STS program in Canada is offered at the from Université du Québec à Montréal (UQAM). This STS program does not address any ethical considerations in their program description but does offers multiple courses focusing on ethical considerations. According to the program description from UQAM:

Science, technologie et société (STS), c'est un domaine d'études en sciences humaines qui s'intéressent aux aspects humains et sociaux des sciences et des technologies. Les changements climatiques, les grandes percées biotechnologiques comme le clonage, les organismes modifiés génétiquement (OGM) ont des origines et des impacts importants sur la société qu'il est important de comprendre. Le programme STS a été développé pour répondre aux problématiques entraînées par le développement rapide des sciences et des technologies (Science, technologie et société, 2010).

In terms of course offerings, the UQAM program offers two courses related to ethics and technology:

PHI4347: Éthique et bioéthique
Introduction critique aux divers discours bioéthiques, en particulier bioéthique des populations, éthique médicale, ainsi que les débats entourant les questions des manipulations génétiques et des nouvelles technologies. En plus d'initier les étudiants aux différents courants bioéthiques et à leurs principaux modes d'argumentation, le cours visera à mettre en lumière les liens de la bioéthique et de l'éthique philosophique. Il s'attardera aussi à l'analyse des modes d'institutionnalisation de la bioéthique et de son incidence politique.

MOR4131: Enjeux moraux de la science et de la technologie
Étude des rapports fondamentaux entre éthique, science et technique. Interrogation éthique sur le développement explosif des sciences et des technologies dans des domaines qui touchent directement à la vie humaine et à la santé. Étude des principes et des valeurs mis en cause, des conséquences au plan individuel et collectif. Analyse de quelques problèmes particuliers: la recherche portant sur les humains et les animaux, le génie génétique, les nouvelles techniques de la reproduction humaine, le prélèvement et la transplantation d'organes, l'industrie

des biotechnologies, l'utilisation de l'énergie nucléaire, la pollution et la dégradation de l'environnement.

PHI4347: Éthique et bioéthique

Provides students with a focus on bioethics and the role of new technologies in controversial areas of public concern such as genetic manipulation. With a slightly broader focus, MOR4131 Enjeux moraux de la science et de la technologie focuses on the ethical aspects of technology in life and society such as nuclear energy, pollution, environmental destruction, biotechnology, and organ transpants.

Carleton University offers a BA in Technology, Society, Environment Studies which highlights the role of industrialization, consumption, and population growth, in generating new problems that our generation must deal with. Their program emphasizes the need of specially trained individuals to ensure an acceptable level of environmental quality and life in the midst of continual technological advancement. The emphasis on ethics is implicitly mentioned in the program description which seeks to balance the need for technological development with "human and environmental values" (Carleton University Technology, Society, and Environmental Studies, 2010). Ethical content is also apparent in program description references to human responsibility:

The TSE courses reflect the need to modernize education by tailoring it to our future needs. Taking one or more TSE courses makes it possible for you to acquire an additional dimension of understanding, additional vision, and the sense of responsibility for the survival and importance of the whole that distinguishes the manager from the subordinate, and the citizen from the subject. (Carleton University Technology, Society, and Environmental Studies, 2010)

In terms of course offerings, the Carleton program does not offer courses on ethics of technology but does offer special topics courses (TSES 4009, TSES 4010), which would allow such courses to be taken.

In a similar vein, the University of Calgary offers a BSc in Science, Technology and Society which addresses ethical aspects of science and technology implicitly in their program description. This program views the study of science and technology as a cultural and political force that influences contemporary society along with our human values:

From the air we breathe and the food we eat to the conversations we share and the journeys we make, science and technology play a critical role in everyday life around the globe. In Science, Technology and Society you will study and gauge the effects of science on our quality of life. By extension, you will learn to assess the benefits and drawbacks of innovation. You will acquire insight into human culture, values and concerns along with how society shapes science's potential and technology's limitations (University of Calgary STS, 2010).

University of Calgary offers one course highlighting the role of ethical and legal considerations linked to informational technology. According to the University of Calgary STS program overview:

STAS 341: New Media, Technology, and Society

A study of the implications of information technology for political, social and economic organization, individual psychology, and concepts of knowledge. Historical, ethical and legal implications will be discussed (University of Calgary STS, 2010).

The University of Regina offers a minor BA in STS with an orientation towards both science and technology. The program description has a focus on social and ethical issues in science:

This program provides students with an appreciation of the historical, philosophical, social and ethical issues of science. The effects on changes in technology and the science form an important part of these courses. For students who are pursuing a degree in science, the minor should provide an important adjunct to their existing studies. For students outside the Faculty of Science, the program will provide an important degree of scientific literacy and appreciation. Courses in this program are open to students in any program within the University. (University of Regina BA in STS, 2010)

While no exclusive courses on ethics and science or ethics and technology are offered at the University of Regina, ethical considerations are covered within one introductory course on globalization:

STS 100: Science and Technology in Global Society

This course will explore the key social, cultural, *ethical and political issues* associated with the development of science and technology. The course will focus on sociological features of science and technology, the influence of these forces on contemporary society, including discussion of the ethical challenges posed by technological development, and the social shaping of scientific and technological activities, products and systems used to serve the various interests within society (University of Regina BA in STS, 2010).

York University offers BA, BSc, MA, PhD degrees in STS. Despite having the greatest degree options in STS among the nine universities examined, there is only an implicit reference to ethics as values (cultural) in the program description to ethical considerations:

Science and Technology Studies (STS) is an interdisciplinary program that offers courses of study leading to BA, BSc, MA or PhD degrees. Its purpose is to expand our understanding of science and technology by exploring their social, cultural, philosophical and material dimensions. To achieve that purpose, the program draws upon the disciplines of the humanities and social sciences to offer courses treating specific scientific ideas, as well as courses addressing broader topics such as science and gender, science and religion, and technology and *cultural values*. Students are encouraged to draw connections across traditional boundaries as they seek an intellectual appreciation for the sciences and technology as powerful means for understanding, embodying and shaping the world and ourselves. Students will learn to analyse complex ideas about science and technology, and to discover how to trace the origins and implications of events and patterns of thought in the past and present.

While York University course offerings at the undergraduate level do not cover ethical issues, there are 2 optional graduate level courses that do address ethical concerns:

STS 6100 3.0: Biomedicine and the twentieth century

Contemporary, industrialized medical practice traces its roots back to a series of transformations that unfolded over the course of the last century. While epidemics of infectious diseases waned, medical research and health care delivery emerged as a focal point of domestic policy. Biomedical practice and investigation fractured into specialties (immunology, radiology, sleep medicine, allergy), even as the accepted distinctions between the normal or the pathological seemed to dissolve. Disease as self-identity became a matter of course. *Ethics* became the province of experts. This course will examine these and other trends in an effort to the ascendancy of health as a scientific object over the past one hundred years.

STS 6201 3.0: Technologies of Behavior

Offers an examination and analysis of social scientists' efforts to develop technologies, (broadly defined as applications of basic research and theory) to change, engineer, and control human behavior at the individual and group levels. The social and political impact of these technologies will also be explored, particularly in terms of the *ethical issues* that emerge, and in relation to the cultural authority of the scientist. Numerous examples will be drawn from 19th and 20th century psychology (e.g., phrenology, psychotherapy, behavior modification) as well as sociology and anthropology. An analysis of the social scientist as social engineer will serve as an organizing framework.

STS 6100 3.0: Biomedicine and the twentieth century

Highlights the role for ethics in terms of professional expertise and training, whereas, *STS 6201 3.0 Technologies of Behavior* places ethical issues at the centre of social and political concern about technological developments within society.

Canadian STS degree programs without an ethical orientation include those offered by Dalhousie University (joint program with King's College), the University of Alberta, and the University of Toronto. BA in History of Science and Technology offered at Dalhousie University has a historical and philosophical orientation to the study of science and technology that examines the changing role of science and technology in Western thought from Ancient to contemporary times (Dalhousie University History of Science and Technology, (2010). The University of Alberta offers a BA in Science Technology and Society is an interdisciplinary program exploring historical, philosophical, and sociological perspectives of science and technology (University of Alberta STS Program, 2010). Finally, the University of

Toronto offers an MA and PhD in History and Philosophy of Science and Technology which highlights "the history of medicine and biological sciences, history of mathematics and physical sciences, history of technology, general philosophy of science, philosophy of biology, philosophy of medicine, philosophy of physics and philosophy of mathematics" (University of Toronto IHPST, 2010). None of these programs offers course with an ethical orientation.

INTERPRETATION AND DISCUSSION

Findings revealed a lack of ethical emphasis within the content of Canadian STS degree programs as reflected in program descriptions and course listings. No STS program identified ethics as a core program dimension and the majority of STS programs offered one course or less dealing with ethical aspects of science and technology. However, it was found that ethical content was implicit within the majority of degree program descriptions or courses offered in all but three STS degree programs.

Based on the study findings, the author has identified a three key variables which appear to hamper the development of the ethical component of STS programs within Canada: (1) inadequate coverage of ethical content within STS programs, (2) lack of STS graduate programs to help raise public awareness of advances levels of STS, and (3) reappropriation of STS content areas by other academic departments and programs.

First, it appears that there is an inadequate of ethical aspects of STS within programs in Canada. Given the rapid change of science and technology, ethical implications created by new technologies are expected to expand as well. Moor (2005) believed that the transforming effect of technology created a technological revolution that was connected to growing ethical problems. If this is the case, than ethical content should be at the core rather than at the periphery of STS programs which deal with science and technology. This need for augmented ethical content in technology oriented studies is mirrored in work by Cutcliffe (1990):

For the most part, gone are the days when spending a few hours in a non-credit, or at most a one credit, course discussing professional licensing bidding on contracts, and the ethics of engineer-client relationships was sufficient to turn out the "well-rounded" engineer. Today, a much more sophisticated understanding of the value implications of engineering is expected. But are discussions of public safety and of the ethics of whistle-blowing all there is to engineering ethics? Ideally, something more is needed (Cutcliffe, 1990, p. 366).

The second major challenge comes from the lack of institutionalization of higher level STS programs required to raise greater awareness of the breadth and complexity of what STS offers. Within Canada, the majority of STS programs are undergraduate degree programs at institutions that do not have STS graduate programs. This can lead to a narrow view of the potential richness and complexity of STS program possibilities. A similar gap was also noted in the STS degree programs emerging in the US over a decade ago (Durban, 1999).

The third major challenge identified concerned the reappropriation of STS content areas by other academic departments and programs. As an instructor of technology and ethics within a communication program, I can attest to the growing interest within the social sciences and humanities to offer STS type courses on ethics and technology. This trend has also been linked to STS program development in the U.S as stated by Durbin (1999),"But sociological and anthropological studies of science, technology, and medicine, in recent decades, have become a fertile field for STS students" (p. 137). This can be looked at in two ways, as the spread of STS as an interdisciplinary academic field or as the dis-

integration of STS as STS students go elsewhere for their studies. Unfortunately for STS, the latter case may be the most probably given the current level of STS institutionalization within Canadian university and pervasive challenges in cultivating strong interdisciplinary connections between STS and other fields.

However, the demise of STS as an academic field is far from being realized and there is a great opportunity among STS scholars in Canada to strengthen program integrity and expand the scope of STS program coverage. Based on a technoethical perspective, a number of options are available, namely, the development of more balanced integration of ethics into STS curriculum, building STS continuity within institutions, and building inroads between other programs and STS. First, more balanced integration of ethics into STS curriculum is required. One the strengths of the STS program offered at St. Thomas is the close integration between the close orientation between program description and course offerings dealing with ethical aspects of science and technology (St. Thomas University Science and Technology Studies, 2010). Second, building STS continuity within institutions provides another means of advancing STS programs. For instance, York University offers students the chance to pursue STS work within multiple degree types and levels (BA, BSc, MA, PhD). This helps entrench STS studies within the university at both undergraduate and graduate levels while offering students STS degree flexibility. Finally, there are opportunities for STS curriculum building both within existing programs and between programs. In some cases, the creation of STS programming depended on having an existing department to house, as is the case as the University of Toronto where MA and PhD level STS studies are offered under History and Philosophy of Science and Technology. This arrangement is not unlike some program orientations in other countries like the U.S. Durbin (1999) states, "Two of the oldest and most traditional fields, history and philosophy—in spite of their typical resolutely academic inbred focus—include subfields in which it is possible to do interdisciplinary work related to STS (p. 137). This type of opportunity is open to STS scholars seeking to establish or extend STS program offerings within a university. These are just a few options available to Canadian leaders in STS program development. How far STS program development can progress also depends on other factors that extend beyond recommendations discussed in this article (i.e., political, financial, cultural).

Given that this is the first study of Canadian STS in higher education (and the first study to examine the nature of ethical content within STS programs), there are many unresolved questions to address to move beyond our current understanding of the academic side of STS in Canada. For instance, what does STS offer do that other disciplines and fields (such as sociology and communications) do not? How should STS relate to those fields in terms of cross-curricular exchange? Is there a core set of course offerings dedicated to ethics that should be included in any undergraduate STS degree program? Is there a core set of course offerings on ethics that should be included in any graduate STS degree program? Should course on ethics within STS be a required component university level training for scientific and technical students in other programs?

To conclude, this study is significant for multiple reasons. First, No study has explored STS across Canada. This study should help to inform a wider audience about STS (STS) and enable leaders to develop better strategies for managing programs to better meet the changing demands of scientific and technological innovation in Canada, particularly where advances in science and technology lead to the emergence of ethical dilemmas and public debate. Second, this study contributes vital information on content coverage of ethics within Canadian STS programs that can be used to prioritize program planning goals for the better management of the STS programs within Canadian universities and elsewhere.

REFERENCES

Association of Canadian Community Colleges. (2007). *Programs Database*. Retrieved July 20, 2007, from http://www.aucc.ca/can_uni/search/index_e.html

Bauer, H. (1990). Barriers Against Interdisciplinarity: Implications for Studies of Science, Technology, and Society (STS). *Science, Technology & Human Values*, 15(1), 105–119. doi:10.1177/016224399001500110

Bijker, W. (1995). *Of Bicycles, Bakelite, and Bulbs: Toward a Theory of Sociotechnical Change*. Cambridge, MA: MIT Press.

Bunge, M. (1977). Towards a technoethics. *The Monist*, 60(1), 96–107.

Carleton University Technology, Society, and Environmental Studies. (2010). Retrieved March 1, 2010, from http://www.carleton.ca/tse/

Cutcliffe, S. (1989). The emergence of STS as and academic field. *Research in Philosophy and Technology*, 9, 287–301.

Cutcliffe, S. (1990). The STS Curriculum: What Have We Learned in Twenty Years? *Science, Technology & Human Values*, 15(3), 360–372. doi:10.1177/016224399001500305

Dalhousie University History of Science and Technology. (2010). Retrieved March 1, 2010, from http://ug.cal.dal.ca/HSTC.htm#2

Ellul, J. (1964). *The Technological Society*. New York: Vintage Books.

Feenberg, A. (1991). *Critical Theory of Technology*. New York: Oxford University Press.

Fuller, S. (2005). *The Philosophy of STS*. New York: Routledge.

Griere, R. (1993). STS: Prospects for an Enlightened Postmodern Synthesis. *Science, Technology & Human Values*, *18*(1), 102–112. doi:10.1177/016224399301800106

Hackett, E., Amsterdamska, O., Lynch, M., & Wajcman, J. (Eds.). (2007). *The Handbook of STS*. Boston: MIT Press.

Haraway, D. (1991). *A Cyborg Manifesto, Science Technology and Socialist-Feminism in the Late Twentieth Century, The reinvention of nature*. New York: Routledge.

Hickman, L. (2001). *Philosophical Tools for Technological Culture, Putting Pragmatism to Work*. Indianapolis, IN: Indiana University Press.

Kuhn, T. (1962). *The Structure of Scientific Revolutions*. Chicago: University of Chicago Press.

Latour, B. (2005). *Reassembling the social: an introduction to actor-network-theory*. Oxford, UK: Oxford University Press.

Luppicini, R. (2009). Introducing Technoethics. In Luppicini, R., & Adell, R. (Eds.), *Handbook of Research on Technoethics* (pp. 1–15). Hershey, PA: IGI Global.

Mitcham, C. (1994). *Thinking through Technology: The Path between Engineering and Philosophy*. Chicago: University of Chicago Press.

Mumford, L. (1934). *Technics and Civilization*. New York: Harcourt Brace.

Science, technologie et société. (2010). *Accueil*. Retrieved March 1, 2010, from http://www.sts.uqam.ca/accueil/accueil.as

St. Thomas University Science and Technology Studies. (2010). Retrieved March 1, 2010, from http://w3.stu.ca/stu/sites/sts/whatis.htm

University of Alberta Science Technology and Society. (2010). Retrieved March 1, 2010, from http://www.ois.ualberta.ca/sts.cfm

University of Calgary Science, Technology and Society Program. (2010). *Society*. Retrieved March 1, 2010, from http://www.ucalgary.ca/comcul/scitech

University of Regina BA in STS. (2010). Retrieved March 1, 2010, from http://pager.uregina.ca/gencal/ugcal/facultyofArts/ugcal_223.shtml

University of Toronto IHPST. (2010). *Institute for the History and Philosophy of Science and Technology*. Retrieved March 1, 2010, from http://www.hps.utoronto.ca/

Winner, L. (1993). Upon opening the black box and finding it empty: Social Constructivism and the Philosophy of Technology. *Science, Technology & Human Values*, *18*, 362–378. doi:10.1177/016224399301800306

Woolgar, S. (1991). The turn to technology in social studies of science. *Science, Technology, & Human Values*, *17*, 20-50.

APPENDIX

Table A1. Course listing of Canadian degree programs in science and technology

Institution	Courses
Carleton University	TSES 2006 - Ecology and Culture TSES 2305 - Ancient Science and Technology TSES 3001 - Technology-Society Interactions TSES 3002 - Energy and Sustainability TSES 3500 - Interactions in Industrial Society TSES 4001 - Technology and Society: Risk TSES 4002 - Technology and Society: Forecasting TSES 4003 - Technology and Society: Innovation TSES 4005 - Information Technology and Society TSES 4006 - Technology and Society: Work TSES 4007 - Product Life Cycle Analysis TSES 4008 - Environmentally Harmonious Lifestyles TSES 4009 - Special Topics TSES 4010 - Special Topics
Dalhousie University	HSTC 1200X/Y.06: Introduction to the History of Science. HSTC 2000X/Y.06: Ancient and Medieval Science. HSTC 2105.03: The Life, Science and Philosophy of Albert Einstein. HSTC 2120.03: Magic, Heresy and Hermeticism: Occult Mentalities in the Scientific Revolution. HSTC 2200X/Y.06: Introduction to the History of Science. HSTC 2202.03: The Beginnings of Western Medicine: the Birth of the Body. HSTC 2204.03: The Darwinian Revolution. HSTC 2205.03: Natural Knowledge and Authority ? Science and the State. HSTC 2206.03: Bio-Politics: Human Nature in Contemporary Thought. HSTC 2207.03: Ghosts in the Machine: Topics in the History of Science, Technology and Mind. HSTC 2340.03: The Origins of Science Fiction in Early Modern Europe. HSTC 2400.03: Science and the Media. HSTC 2500.03: Science Fiction in Film. HSTC 2602.03: Astronomy Before the Telescope. HSTC 3000X/Y.06: The Scientific Revolution. HSTC 3100.03: Aristotle's Physics. HSTC 3120.03: Distilling Nature's Secrets: The Ancient Alchemists. HSTC 3121.03: In search of the Philosopher's Stone: The History of European Alchemy. HSTC 3130.03: The Origins of Chemistry: From Alchemy to Chemical Bonds. HSTC 3200.03: Science and Religion: Historical Perspectives. HSTC 3201.03: Science and Religion: Contemporary Perspectives. HSTC 3205.03: Natural Knowledge, Human Nature and Power: Francis Bacon and the Renaissance. HSTC 3210.03: Environmental History I. HSTC 3212.03: The Biosphere: Global Perspectives in Science and Philosophy. HSTC 3310.03: Hidden Worlds: Microscopy in Early Modern Europe. HSTC 3320.03: Omens, Science and Prediction in the Ancient World. HSTC 3331.03: History of the Marine Sciences. HSTC 3402.03: History of Mathematics I, Greek Geometry. HSTC 3411.03: Feminism and Science. HSTC 3412.03: Hypathia's Daughters: Women in Science. HSTC 3430.03: Experiments in the Mind: Thought Experiments in Physics. HSTC 3501.03: The Nature of Time I. HSTC 3502.03: The Nature of Time II. HSTC 3610.03: Studies in Ancient and Medieval Science. HSTC 3611.03: Studies in Early Modern Science (1500-1800). HSTC 3615.03: Studies in Science and Nature in the Modern Period: History of the Environment. HSTC 4000X/Y.06: Science and Nature in the Modern Period. HSTC 4102.03: Topics in Ancient Natural Philosophy. HSTC 4200.03: Histories and Practices of Technology I. From Techne to Technology. HSTC 4201.03: Histories and Practices of Technology II: The Questions Concerning Technology. HSTC 4300.03: Nature and Romanticism. HSTC 4400.03: Newton and Newtonianism. HSTC 4500X/Y.06: Honours Seminar in the History of Science and Technology. HSTC 4510.03: Independent Readings in History of Science and Technology. HSTC 4511.03: Independent Readings in History of Science and Technology. HSTC 4515.06: Independent Readings in History of Science and Technology. HSTC 4550X/Y.06: Honours Thesis in the History of Science and Technology.

continued on following page

Table A1. Continued

Institution	Courses
St. Thomas University	1003. Science, Technology and Society I 1503. Principles of Biology I 1513. Principles of Biology II 1613 Everyday Chemistry 1713. Science, Technology and the Earth 2103. Science, Technology and Society II 2243. Science and Technology in World History: From Pre-History to 15432253. Science and Technology in World History: From 1543 to the Present 2303. Natural Disasters 2413 - Science, Technology & Innovation 2503. History of Disease 2603. Animals: Rights, Consciousness, and Experimentation 2703. History of the Life Sciences 2803. Controversies in the Earth and Environmental Science 2903 The Politics of Science 3003 Scientific Reasoning 3063. Historical Perspectives on Science and Religion 3163. Contemporary Perspectives on Science and Religion 3203 Science, Technology and Nature 3503. Feminist Critiques of Science 3533. Science and Scientific Knowledge 3563. Philosophy of Science 3803 Space Exploration 4006 Honours Thesis 4103 Independent Study 1503. Principles of Biology I 1513. Principles of Biology II 1613. Everyday Chemistry 1713. Science, Technology and the Earth
Université du Québec à Montréal	SHM1210-STS: Atelier de méthodologie générale HIS4722-Sciences et techniques dans l'histoire des sociétés occidentales I PHI4346-Introduction à l'approche philosophique des sciences et des technologies PHY2710-L'environnement abiotique INF1052-Développements informatiques SOC4206-Méthodologie quantitative HIS4723-Sciences et techniques dans l'histoire des sociétés occidentales II SOC6209-Sociologie des sciences POL6010-Politiques scientifiques canadiennes et québécoises BIO1302-L'environnement biotique SHM1310-Atelier de méthodologie en scientométrie SOC6226-Sociologie des technologies ECO1081-Économie des technologies de l'information SHM1410-Atelier de méthodologie en évaluation des technologies en milieu de travail CHI1800-Chimie de l'environnement JUR1008-Droit, santé et environnement PHI4347 Éthique et bioéthique MOR4131 Enjeux moraux de la science
University of Alberta	STS 200 Introduction to Science, Technology and Society STS 400 Advanced Topics in Science, Technology and Society
University of Calgary	STAS 325: Technology Within Contemporary Society STAS 327: Science in Society STAS 341: New Media, Technology and Society STAS 343: Canadian Science Policy and Technology Development STAS 591: Integrative Seminar General Studies 300: Heritage I - Perspective General Studies 500: Heritage II - Integration History 477.01, 477.02 Philosophy 367

continued on following page

Table A1. Continued

Institution	Courses
University of Toronto	HPS 5001H: Fundamentals of the History of Mathematics HPS 5002H: Fundamentals of the History of Physics HPS 5004H: Fundamentals of the History of Chemistry HPS 5005H: Fundamentals of the History of Biology HPS 5006H: Fundamentals of the History of Medicine HPS 5007H: Fundamentals of the Hist. of Technology I HPS 5008H: Fundamentals of the Hist. of Tech. II HPS 5009H: Fundamentals of the History of Astronomy HPS 5010H: Fundamentals of the Philosophy of Science HPS 5011H: Fundamentals of the History and Philosophy of Science and Technology HPS 5012H: Fundamentals of the History of Psychology HPS 5013H: Fundamentals of the History and Philosophy of the Social Sciences
York University	History of Modern Science STS 2110 3.00 Revolutions in Science STS 2411 6.00 Introduction to STS STS 3170 3.00 Philosophy of Science STS 3550 6.00 Science as Practice and Culture STS 3700 6.00 History of Technology STS 3725 6.00 Science & Exploration STS 3726 3.00 Technology, Experts & Society STS 3730 6.00 Science, Technology & Modern Warfare STS 3750 6.00 Genetics, Evolution & Society STS 3755 3.00 Emergence of Cosmology as Science STS 3780 6.00 Biomedicine in Sociohistorical STS 3925 6.00 Interfaces: Technology & The Human STS 4110 3.00 Seminar in Philosophy of Science STS 5000 3.0 and 6.0 Directed Readings for M.A. Students STS 5001 3.0 Introduction to Science and Technology Studies STS 5002 0.0 M.A. Major Research Paper STS 6000 3.0 and 6.0 Directed Readings for Ph.D. Students STS 6001 3.0 Introduction to Science and Technology Studies STS 7000 0.0 Ph.D. Dissertation Research FGS/STS 6101 3.0 Rendering 'Life Itself' FGS/STS 6102 3.0 Organisms as Instruments FGS/STS 6103 3.0 Epidemics FGS/STS 6104 3.0 The Psychological Society STS 6105 3.0 - Mesmerism, Phrenology and their Influences on Twentieth Century Psychology and Psychotherapy FGS/STS 6106 3.0 Bodies and Biotechnologies in Anthropology FGS/STS 6107 3.0 - Advanced History and Theory of Psychology: Darwinian Influences on Psychology FGS/STS 6200 3.0 Bodies in Technology FGS/STS 6201 3.0 Technologies of Behavior FGS/STS 6202 3.0 Designs of War FGS/STS 6203 3.0 Modern Technology: Materiality, Identity, and Social Order FGS/STS 6300 3.0 Science and Narrative in the Nineteenth Century FGS/STS 6301 3.0 Science and Print Culture FGS/STS 6302 3.0 Race and Racism in the Human Sciences FGS/STS 6303 3.0 The Sciences in the Enlightenment FGS/STS 6304 3.0 Anticipating the Alien STS 6305 6.0 - Contexts of Victorian Science FGS/STS 6400 3.0 Earth, Time and History FGS/STS 6401 3.0 Big Science FGS/STS 6402 3.0 Understanding Oceans FGS/STS 6403 3.0 Mapping Nature: Systems and Visions STS 6404 3.0 Encountering Natural Worlds STS 4501 6.00 Seminar in Science & Technology Studies STS 4700 3.00 Independent Research in Science and Technology Studies STS 4700 6.00 Independent Research in Science and Technology Studies

continued on following page

Table A1. Continued

Institution	Courses
University of Regina	STS 100 - Science and Technology in Global Society STS 200 - The Development of Modern Science STS 230 - Science and Technology in the Ancient World STS 231 - Science and Technology in the Medieval World STS 232 - History of Astronomy STS 239 - Scientific Biography - an AA-ZZ series. STS 270 - Theories and Methods in Science and Technology Studies STS 271 - Science, Technology and Gender STS 330 - The Darwin Controversies STS 331 - Concepts of Matter, Time, Space, and Motion STS 370 - The Rationality of Science STS 371 - Philosophy of Biology STS 372 - Issues in Cognitive Science STS 373 - Technology in the Non-Western World STS 400 - Theoretical Perspectives on Science and Technology

This work was previously published in the International Journal of Technoethics, Volume 2, Issue 1, edited by Rocci Luppicini, pp. 50-63, copyright 2011 by IGI Publishing (an imprint of IGI Global).

Chapter 18
College Students, Piracy, and Ethics:
Is there a Teachable Moment?

Jeffrey Reiss
University of Central Florida, USA

Rosa Cintrón
University of Central Florida, USA

ABSTRACT

This study explores the nature of piracy prevention tools used by IT departments in the Florida State University System to determine their relative effectiveness. The study also examines the opinions of the Information Security Officer in terms of alternative piracy prevention techniques that do not involve legal action and monitoring. It was found that most institutions do not use a formal piece of software that monitors for infringing data. Furthermore, institutions agreed that students lack proper ethics and concern over the matter of copyright, but were not fully convinced that other prevention methods would be effective. The authors conclude that monitoring techniques are a short-term solution and more research must put into finding long-term solutions.

INTRODUCTION

With pressure from the Recording Industry of America (RIAA) and federal policies introduced in 2008, post-secondary institutions must implement software to monitor network activity to keep students off P2P software, dorm servers, and any other online method of transferring media il-

legally (Joachim, 2004; Worona, 2008). Though some larger universities, such as the University of Florida, had already implemented software prior to the 2008 revision of the Higher Education Act, the remaining community colleges and smaller institutions were required to follow suit. Furthermore, as the new rules currently stand, institutions will not, at the present time, suffer any penalties for failing to follow the procedures. After

DOI: 10.4018/978-1-4666-2931-8.ch018

the appropriate committees interpret the rules, it is likely, however, that institutions will lose federal financial compensation for implementation failure, making the process burdensome for smaller, financially starved institutions (Worona, 2008). Though software will always play an important role, an expensive software solution becomes useless if students find an alternate route around the software. As a result, an institution may ultimately waste money on an ineffective solution only to replace it with another solution with a limited life-span due to technology's constant evolution.

The P2P blocking software, however, is only one part of the problem. With most institutions also blocking dorm servers and other potential piracy outlets, students feel that their personal freedoms are hindered. Despite the plethora of legal uses for P2P, its adoption was stifled due to the focus on illegal use. The dorm server ban is also problematic in a historical sense. The popular search engines Yahoo! and Google, two companies that made major contributions to the overall state of the Internet, began as dorm servers. By preventing students from utilizing these resources, institutions could easily and unknowingly prevent the next major Internet innovation satisfying a large private interest (Joachim, 2004). Thus, by exploring solutions that rely on more than software and legal threats, a university could potentially eliminate digital piracy without denying freedoms to students.

The following research question guiding the present study, focused on ethics, is part of a larger study investigating the multifaceted challenges of piracy in higher education: What alternatives are being considered to discourage piracy by college students at a lower cost than monitoring software? Although throughout this paper the term 'ethics' is used, the reader should understand that we are making particular emphasis on the field of study and research of technoethics. For the purpose of this paper we have aligned our definition of technoethics with that of this journal. That is, technotethics is concerned with the "technologiclal

relationships of humans with a focus on ethical implications for human life, social norms and values, education, work, politics, law, and ecological impact" (Miah, 2010). Similar perspectives have been posited in the past by Bunge (1977) and more recently by Luppicini (2010).

REVIEW OF THE LITERATURE

According to Forester and Morrison (1994), software piracy first occurred in 1964 when Texaco was offered $5 million in stolen software. Other cases occurred over the years but were solely private corporate programs such as air-traffic control programs and CAD software. Although these instances of software piracy were a different form of stealing trade secrets, mass software piracy only surfaced with the advent of the desktop computer and Microsoft. Bill Gates created the software programming language, BASIC, as part of a package with the desktop computer kit, the Altair. While the computer was poorly constructed, the software proved more useful, and some people made copies of the program to prevent others from purchasing the entire package (Forester & Morrison, 1994).

This view combined with the more powerful viewpoint of a software package being too expensive helped fuel consumers' justification to pirate or sell counterfeit copies. With the help of the early form of the Internet, counterfeit software became easier to distribute. In 1992, a major crackdown occurred on an Internet bulletin board known as Davy Jones Locker that sold pirated versions of expensive programs such as AutoCAD, and a number of Lotus and IBM products. Considering the pirated software from this site crossed national boundaries into nations such as Iraq (Forester & Morrison, 1994), the notion of a hostile nation obtaining software that could lead to the creation of a weapon to be used against the United States or its allies could prove dangerous. Another crackdown in 1992 yielded approximately $9 million of pirated software across 10 sites. These sites not only included pirated versions of programs like MS-DOS but also possessed counterfeit versions of manuals, holograms, and even the packaging (Forester & Morrison, 1994).

In the courtroom, software piracy becomes harder to define. As Edgar (2003) noted, the term plagiarism may appear to be an appropriate term for piracy, but while a person is making a copy of another's work the copier is not claiming authorship of the file. At the same time, piracy can expand the range of one or two files or programs to the mass warehouses as described in the 1992 counterfeit software crackdown. This brings up the question as to the limits of policy (Forester & Morrison, 1994). The dilemma has been to find common ground for a policy to be applied to those who feel software should be free under any condition, and those who feel piracy is reprehensible in any situation. If a person pirates only one software program or song, it may be so minor that such an infraction, while still illegal, may be ignored. Though one infraction may be tolerable, there is the slippery slope issue as to how many infractions are permissible before the acts become intolerable.

With the prevalence of pirating and P2P, one questions the types of users on the piracy stage. Molteni and Ordanini (2003) conducted a study to identify different types of file sharers and determine potential solutions to reduce the probability that they would continue to pirate. Based on a sample of 204 individuals who answered a questionnaire on the Bocconi University website where 95% of the sample were students, the authors were able to create five different groups via cluster analysis. One group consisted of occasional downloaders who download some, but predominantly purchase, CDs. This group is essentially the control group because online strategies would least affect this group. The second group, the mass listeners, had a high dependency on P2P sites and rarely purchased CDs. Their file sharing practices are generally to obtain music for their

own entertainment. The authors suggested strategies that could provide a heightened experience or possibly a streaming service could benefit them. Third, the explorers/pioneers group used downloaded content to influence CD purchases. This group could be beneficial to new artists if free songs are provided to their consumption. Fourth, the curious group used the act of downloading as a form of entertainment. Determining a solution for these people is almost impossible. Finally, the fifth group was the duplicators who downloaded music solely for making copies. Like the curious group, finding a simple solution is difficult, but legal action may be the only solution (Molteni & Ordanini, 2003). This study provided a solid foundation for the process of identifying solutions to such a massive problem involving diverse groups of downloaders.

Cohen and Cornwell (1989) conducted one of the earliest studies that replicated prior studies of Schuster (1987) and Christoph, Forcht, and Bilbrey (1987, 1988). According to Cohen and Cornwell, both of the prior studies held no real merit alone but if combined there could be significant results. Christoph et al. (1987) examined the ethical beliefs of Information Systems students at James Madison University but could not draw any conclusions from the results because there were no expert opinions with which to compare the student responses.

IDENTIFYING ETHICS, AND MOTIVES

Taylor (2004) conducted a study to connect software piracy and ethics between standard business majors and music business majors using a five-point Likert scale questionnaire. Overall, those surveyed agreed with the standard premise that students believe piracy was a faceless crime and hurt few if any. Though music business majors were more likely to view piracy as unethical, when both majors were combined, the results were fairly neutral. Furthermore, students from either major who never pirated music thought that piracy was unethical and hurt the music industry.

Rob and Waldfogel (2006) conducted a study based upon the download habits of students and the value they placed on music they either purchased or illegally downloaded. The survey asked about whole albums either purchased or downloaded in 2003 along with a list of hit albums from 1999. By having the two lists, Rob and Waldfogel (2006) were able to explore a theory that music could be viewed as "(1) an experience good, (2) subject to depreciation as the listeners grow tired of music, or (3) both" (p. 44). Under this theory, students would download or purchase an album, enjoy the album, and grow tired of the album to the point of never listening to it again. The researchers tested this theory in their 2003 survey that was used to investigate students' habits. The list of songs from which to choose contained a list of new music and the top albums of 1999, allowing a distinction between new music a student may instinctively download and older music of which students may have become tired. The significance of this viewpoint also lends itself to a reason behind student downloading. If a student purchased an album thinking it would be good, only to realize otherwise, that student wasted money on an album. If the student downloaded the same album, the music was simply deleted. Although the purchase scenario resulted in a sale for the record label, it dissatisfied the student. The download scenario resulted in no sale for the record label, but the student's dissatisfaction was mitigated by not wasting money on a sub-par album. At the same time, if an album was generally popular via means such as the Billboard charts they would be more widely available than a less popular or more obscure track (Bhattacharjee, Gopal, Lertwachara, & Marsden, 2006). Rob and Waldfogel concluded that such an analysis tended to become obfuscated when facing individuals who both downloaded and purchased music. These individuals would contribute to a positive trend on both sides and

were equally likely to experience the same after-effects of music deprecation. Thus, downloading becomes more of a "try before you buy" scenario acting more on utilitarian methodologies to maximize the individual's happiness. Even with the justification of trying to save money, deontological and virtue ethics still support that the act is immoral (Edgar, 2003).

A study by Gupta, Gould, and Pola (2004) explored student software piracy along with ethical considerations related to the issue. Based upon some of their research, it was viewed that piracy, while immoral, has less severity and level of harm than other, more heinous acts such as breaking into a store and stealing a physical good. Their study aimed to connect the perception of software piracy with a student's habits via ethics, legality, criminality, parties harmed, and social impact. Those surveyed were 90% male, about half were under the age of 25, and the majority of the respondents were college educated. It was found that persons less concerned over ethics were more likely to pirate software. Furthermore, the social landscape of websites that encourage software piracy further strengthened the belief that piracy was acceptable. In fact, the researchers noted the strong correlation between the belief and their survey question that mentioned the ease of pirating because a person could not get caught. The legal aspect, however, did not correlate with the social one. The researchers did mention that while this result was contrary to the norm, the study had taken a more ethical focus which could have caused skewed results. Similar to Rob and Waldfogel's (2006) implications, a free demo of music would allow a potential customer to determine if an album is worth purchasing. Gupta et al. (2004) suggested that software companies put more emphasis into trial and beta versions of commercial software. Hence, free samples appear to be one outlet to help curb digital piracy. Finally, the authors have stated their beliefs that the public will not view software piracy as an ethical concern. This would indicate that there may be a need to teach ethics as early in life as possible.

SOFTWARE DETERRENTS TO PIRACY

In 2006, the Electronic Frontier Foundation (EFF) performed a study of the flaws in many of the P2P monitoring software systems. According to the EFF, all types of monitoring software or software alternatives have either contained an exploitable loophole or have restricted student rights on some level. For example, firewall software may block P2P traffic across a specific port, but the savvy student could use another port to circumvent the restriction, resulting in a massive "cat-and-mouse" scenario. Furthermore, a student may also hide the data in a port that a firewall will never block such as port 80 for standard web traffic (Nyiri, 2004). Though content-monitoring software has been considered to be the primary means to stopping piracy, file-sharing programs and crafty students can use SSL encryption to hide their tracks. Because the connection is encrypted, a monitoring program would only see nonsense due to the encryption. The EFF has also mentioned traffic-shaping programs that can alter the amount of resources for specific functions, but it also falls prey to circumvention methods. From a student rights perspective, students will not be comfortable knowing that their institution is tracking their every move online. In other cases, the measures could also generate false positives where a student is downloading a file through P2P for educational use as described through the fair use doctrine.

Considering the vast size of a campus network, it is foreseeable that there is a massive amount of traffic flowing through it. With that considered, IT professionals need to use some form of software in order to filter through the equally massive raw data. Though some software packages will do their job,

… many of the solutions generate only a slightly smaller amount of data than the raw network. The next group of companies sells a product that takes the output of the first layer and tells you

what your problems really are. A third group of companies sells yet another layer of products to finally produce actionable items (Rosenblatt, 2008, p. 9).

Thus, while IT software products may be effective, they can easily become numerous and costly (Rosenblatt, 2008). With different copyright monitoring solutions available, determining the best one is at the very least time consuming. The Common Solutions Group (2008) is a collective of IT professionals from 30 universities, 28 of which are research institutions. Common Solutions Group has further supported the EFF's findings with its own study. In this study, three major copyright-prevention software packages: CopySense, cGrid (formerly ICARUS), and Clouseau were considered. It was found that CopySense and Clouseau both relied on the vendor for blocking parameters. Students may bypass CopySense by downloading files that are not registered into the CopySense database or use SSL encryption. Clouseau sets its parameters by allowing and blocking vendor-set communication routes which operators may not modify. Though blocking whole communication channels sounds ideal, some channels may already have legitimate and legal uses through the campus. Since the network operators have no control, another piece of software may become worthless. Finally, cGrid was determined to be the most advanced of the three programs but was also the most expensive and resource intensive. cGrid differed from the others by reporting patterns rather than using outright suppression but could also be configured to act similarly to the other two programs in the ICARUS incarnation (Nyiri, 2004). This caused added network strain and more network administrators to interpret cGrid's output. It was concluded that while this type of software would improve in both quality and price in the future, the current offerings were expensive and inefficient. More importantly, the universities involved in the study concluded that software alone will not end student copyright infringement violations. They expressed the belief that constantly educating and reinforcing ethics to students would yield better results. Though the software approach appeared effective in 2004 (Spanier, 2004), the improving technology and student craftiness eventually rendered the software ineffective.

In Gopal, Sanders, Bhattacharjee, Agrawal, and Wagner's (2004) study of piracy's economic impact, the authors found that two broad types of controls exist to help combat piracy. The first control was the deterrent control that contains methods that attempt to deter users from pirating media. Legal action, educational, and media campaigns are just some of the general categories of these deterrents (Djekic & Loebbecke, 2007). Deterrent controls do not directly influence but attempt to indirectly dissuade a person with threats. At the time the article was written, the RIAA was beginning its legal campaign (EFF, 2008) which was the first and primary deterrent control for music piracy. Another deterrent control is one may that be self-inflicted by the copying community--computer viruses. Viruses, however, may only act as an effective deterrent to students if they are also bothered by legal action. Those who do not care about the threat of legal action will feel the same about viruses (Wolfe, Higgins, & Marcum, 2008).

Preventative controls operate from the viewpoint of making the process too time, labor, or financially intensive to make pirating less worthwhile (Djekic & Loebbecke, 2007). Although this has been a common practice in the software industry, the music industry has also embraced these practices. Such practices include DRM and similar fingerprinting techniques to make digital files harder to share, and software encryption to make copying files off of CDs more difficult. Like the P2P prevention programs, all preventative controls contain an exploitable weakness that a determined individual may avoid. The only difference comes from the increasing difficulty in overcoming these barriers. Considering the speed with which technology progresses, such a situation may never occur. Djekic and Loebbecke, in their study of preventative controls for computer software, provided proof that preventative controls are ineffective and that software companies would be wise to invest in deterrent controls.

ETHICS AND COMPUTERS

Rogerson (1996) recounted a speech given by professor Terrell Ward Bynum where he argued that information technology (IT) plays one of the leading roles in shaping and changing society. Compared to other societal revelations such as the Industrial Revolution, IT was thought to create a larger impact because of its ease to seep into a person's life. Furthermore, human values would exhibit the most change from IT. Bynum provided examples of IT slowly seeping into society through the ease of news travelling from person to person and the functionality IT provides for disabled people. Looking at these implications from both utilitarianism and Kantian theory, the two ethical theories share similar outcomes. From the utilitarian perspective, because IT will allow more people to take care of themselves who previously were unable, more people will find a way to achieve happiness. From the Kantian perspective, allowing IT to grant more people autonomy and control of their lives is an ethical and moral act. Regardless of the theory used, people need to consider the impacts and implications of a new form of technology (Rogerson, 1996). P2P technology and digital media easily fall into this category. Both are relatively new technologies and both have significantly impacted the way people live their lives. The impact and implications of such technologies, however, may not have been thoroughly considered, and actions may be required in light of current events.

COMPUTER ETHICS

Considering the level of harm a person could cause with a computer, having an appropriate code of ethics becomes paramount to keeping the virtual population safe. As with an offline society, unsavory characters always exist but when factoring in the global scale of the online society, strong ethics may be the only way to safely police the Internet. Forester and Morrison (1994) indicated that computer safety requires education in three principles: encouragement of more ethical behavior, better understanding of social problems brought on by computers and the digital age, and sensitivity towards computer-related moral dilemmas that will surface over the course of people's lives. Forester and Morrison viewed these principles from the perspective of a computer science program, and they believed that simply adding a required course on computer ethics to the curriculum was not sufficient to solve the problem. In actuality, they advocated that ethics should be integrated and be a recurring theme throughout all computer science courses. They recognized, however, that not all computer science professors possessed the preparedness to teach ethics in such a manner. Finally, Forester and Morrison noted that teaching ethics and morality was not a cure for all the ills of the virtual society but could significantly contribute to alleviation of many problems.

One of the most pressing issues in ethics has involved transitioning classical ethics theories into a less-defined environment:

1) it is logically argumentative, with a bias for analogical reasoning, 2) it is empirically grounded, with a bias for scenarios analysis, and 3) it endorses a problem solving approach...4) it is intrinsically decision-making oriented...5) it is based on case studies (Floridi, 1999, pp. 37-38.)

Although computer ethics follows three general guidelines of classical ethics in points one through three, it also utilizes the driving force of point four and the methodology in five. With the addition of the other two factors, it also uses the driving force of computer ethics built itself as a more applied version of ethics focusing more on real-world applications than a theoretical nature. According to Floridi (1999), this applied viewpoint of computer ethics, known as microethics, is the common viewpoint, and even though the moral implications created by computers have no non-digital equivalents little consideration goes into a theoretical component. Floridi believed that the classical theories will never fully satisfy all the conditions behind computer issues or may even risk placing computers in an anthropomorphic light. Brey (2000) supported the viewpoint that computer ethics focused heavily on application, but argued against Floridi who advocated that applied ethics could affect policies and practices if properly applied. Though applied theories require a foundation of theoretical ethics to function properly, they do not require the reliance on any particular type of ethical theory. In other words, a theoretical base used for digital piracy may not be applicable for hackers who would require the implementation of a different ethical theory. Brey also discussed applications of computer theory, noting that many seem to have ignored computer ethics' ability to solve a preexisting problem caused by a policy vacuum as well as to make the computer portion transparent so that only the ethical portion remains visible. In the case of digital piracy, this would involve removing all the computer and technological components to boil the act down to outright stealing.

STUDENT ETHICS STUDIES

In one of the first studies on computer ethics and students, Slater (1991) revealed that "information systems and business students appear to worry less about computing ethics than do today's executives" (p. 90). He reflected on a study conducted at James Madison University where over half of 300 students between the ages of 19 and 45 had admitted to using computers for some form of unethical use including software piracy. Some students had indicated they would purchase at least one copy of a program and make copies for other computers they use with the hopes of promoting the product in the future, but they could not deny that piracy took place. Slater believed that ethics must be taught from an earlier age, and the younger the better but also indicated that partnerships between IT and students would be the only way ethics will actually be applied beyond academic coursework.

Slater (1991) identified an initial problem of students using computers for unethical reasons. In a study conducted by Athey (1993), the ethical beliefs of students were compared to those of experts in the field for the purpose of determining the ethics gap between the professionals and the students and determining curricula to fill the gap. In the study students were compared by gender, income (low, middle, high), and major (computer science and computer information systems). The experts were those professionals who first examined ethical scenarios in the computing field. Of 19 different scenarios presented in the survey, females and males disagreed with seven and eight of them respectively. By major, males and females disagreed equally with 10 scenarios each, and the economic groups fared about the same. Despite all the disagreements, there were seven scenarios where all student groups agreed with the experts. Athey (1993) concluded that the scenarios should be addressed in course curricula. Although this study was not directly focused on digital or software piracy, it did provide insight into students' ethical perspectives. If students act in unethical ways due to constant exposure, stronger ethics curricula should be developed for not only computer science or computer information services students, but for the entire campus population.

Leonard and Haines (2007) conducted a study in which they attempted to determine if any differences in ethical beliefs were present when students completing a survey alone or in a group. After completing the survey alone online at computer stations, the group was divided into smaller groups of five to nine depending on the size of the group being tested at the time and allowed to chat online with group members while taking the second survey. The results of the two surveys showed that the virtual group actually strengthened the responses. If the general response to a question was considered ethical on the individual test, the group test results leaned further towards ethical and vice versa. Though the main constraint in this study was the use of students rather than IT professionals with experience, this did provide evidence regarding the influence of group behavior on decisions.

Thong and Yap (1998) conducted a study to test the ethical decision-making process theorized by Hunt and Vitell's (1986) deontological-based model. Though they did not analyze the entire model, they found that the model adequately described the ethical decision-making process in Information Systems students (Liang & Yan, 2005). Shang, Chen, and Chen (2008) followed up on Thong and Yap's work by surveying students with a similar model. They provided a scenario and alternatives of varying ethical value alongside intentions based upon a seven-point Likert scale. They found that people who pay for the use of P2P systems feel less guilty about piracy even if sharing on the P2P system breaks copyright laws. They also came to the generally shared conclusion that piracy was rationalized by students and was not considered to be a problem to them. Shang et al.'s instrument, however, was flawed in that it contained 110 items, many more than most students would want to complete.

DESIGN OF THE STUDY

The present study utilized quantitative methodology with a survey to obtain a number of descriptive statistics. The survey instrument was administered through a website on a personal, private server in order to improve safeguards against potential third-party tampering. By manually designing the webpage for the survey instrument, a correctly completed survey was guaranteed with a number of validation safeguards. The items themselves were based on (a) issues defined in the literature review and (b) an automated morality framework that questions if monitoring software could mimic human ethical behavior (Stahl, 2004). The finalized data were analyzed through SPSS software to obtain the needed results to answer the research questions.

The population consisted of the IT department's security officers for the 11 Florida State University System (SUS) institutions. Each IT department contained at least one administrator for each institution who was knowledgeable about the operation of copyright infringement detection software. Because IT departments were responsible for the deployment and maintenance of monitoring software, they were an ideal group to survey regarding the effectiveness of monitoring software.

The role of the IT department chief security officer or information security officer was relatively new at the time of the present study, but its importance has increased as universities became more connected with the digital age. Information security officers are in charge of determining university security policy and are at the forefront of the campus network's security. Their duties include but are not limited to incident management, policy development, forensics, risk assessment, and coordination with law enforcement (Goodyear, Salaway, Nelson, Petersen, & Portillo, 2009). They were the individuals who develop policies such as the University of Central Florida's (2009) three-strike policy for P2P usage and make sure the policy was enforced. Furthermore, the information security officer also possessed the authority to monitor the network to ensure that all users followed campus policy and disconnect or restrict access to offending users.

Because of the small number of information security officers in the Florida State University System, the sample consisted of the entire population, making it a census. As a result, the census was representative of the Florida State University System, but not representative of all universities in the nation. A total of 10 sample points were taken rather than 11 because two institutions share the same Internet connection. Furthermore, a 100% response rate to the survey instrument was a requirement of the researcher in order to compensate for the small population.

INSTRUMENTATION

Considering this study was the first of its type, an original survey instrument was constructed. Due to the lack of prior research in this area, the researcher relied upon the theory of automated morality as a framework for the instrument. The theory of automated morality explores computer software as a moral agent, an entity capable of making moral decisions. According to Stahl (2004), a program would be considered a moral agent if it can pass the Moral Turing Test, a test that determines whether or not software could pass for a moral being by an independent observer. In the end, the amount of trust the information security officer instills in the monitoring software reflects to what extent the software serves as a moral agent and reflects the software's overall effectiveness.

Information security officers possess the authority to monitor the campus network and remove any offenders based on policies that the officers helped develop (Goodyear et al., 2009). The Common Solutions Group (2008) looked at three programs that scour the network for data transfers that contain copyrighted material. Realistically, the information security officer cannot spend the entire day scanning the network for infringing material, so using an automated program to scan was a logical choice. The Common Solutions Group, however, based its analysis of the programs on infringements discovered and the ease of modifying the criteria. Furthermore, the software will also go so far as to restrict access to devices identified, one of the other main authorities of the information security officer. This led to the question of whether the software was capable of making ethical decisions in the same manner as an information security officer when handling network violations. The instrument consists of a web-based questionnaire. This method of surveying was chosen because web-based questionnaires provide a more enhanced method of presentation and collection than other survey methods. Although Dillman (2000) stated that web surveys may become problematic from a technological and computer penetration standpoint, the population was from a technologically-oriented profession, and such problems did not hinder the study.

The 36-item survey was divided into four sections. After the respondents logged in to the survey via credentials provided in an e-mail, they proceeded to answer questions in the first section geared towards the policy aspect of copyright monitoring software (Higher Education Act of 2008). The second, and longest, section (Items 8-20) of the instrument related to challenges to implementing monitoring software such as price, staffing/training, overall acceptance of the software, and relative effectiveness of the software. Section three (Items 21-28) explored the alternatives to monitoring software ranging from ethics, alternative programs and deals, or just stricter standards. Both sections two and three were designed using Likert scale questions. The final section inquired about demographics. As Dillman (2000) noted, demographic information is best left to the end of the questionnaire. The demographics section also contained a comment box for the respondents to add their own thoughts in order to further enhance the study.

Once the pilot data were obtained, a Cronbach's alpha test was performed to determine the reliability and validity. An alpha level of .858 was obtained, showed a strong level of reliability, but was not the most accurate value due to the small sample. Furthermore, the actual study also suffered from this problem due to a low sample size.

DATA ANALYSIS AND FINDINGS

Figure 1 shows that all of the respondents stated students would resort to P2P once they left the restrictions of the university campus. Additionally, nine of them believed that students had little concern over the consequences of their actions. As to whether ethics would help stem piracy, the respondents were divided with four agreeing, five disagreeing and one remaining neutral. In regard to its impact on monitoring software, four respondents agreed and six disagreed that such action would lessen the need for monitoring software.

According to Figure 2, the respondents were divided in regard to whether legal alternatives such as downloads would encourage students to obtain goods via legal means. Seven of the IT security directors, however, believed that having more legal means of obtaining goods would decrease the need to find them illegally. Four respondents stated that their institutions were in the process of obtaining means of providing discounted goods. Five remained neutral on the issue.

According to Figure 3, six of the respondents agreed that repeat offenders were rare occurrences. All but two agreed that students care more about their Internet connections than the act of piracy itself.

DISCUSSION

All participants responded to the items in this section completely and without use of the "Not Applicable" response, making the responses for the

Figure 1. Frequency for ethics sub-variables

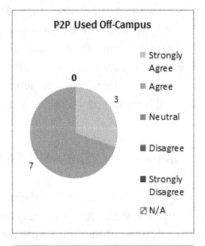

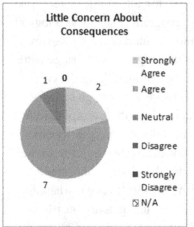

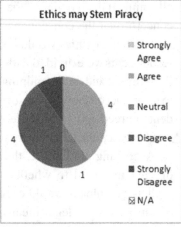

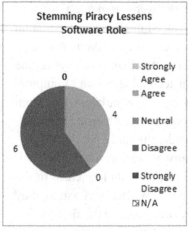

three variables more reliable. The first variable in question was the Ethics variable. It combined the opinions of a student's potential unethical behavior with opinions about ethics as a solution. The responses for the two variables related to student behavior were almost unanimously agreed upon. All of the respondents felt that students would begin using P2P once they were out of the limitations placed by the institution. Furthermore, all but one respondent agreed that students had little concern over the consequences of digital piracy. Unfortunately, this type of behavior relates directly to findings from LaRose, Lai, Lange, Love, and Wu (2005) and situations such as the Tenenbaum Trial (Anderson, 2009a, 2009b). Even with all of the security measures in place, the students viewed

it as a temporary block until they reached a site where they could download. All of the respondents had strong feelings about how students will act with and without P2P restrictions, but they were divided in regard to the effectiveness of ethics as a tool to combat piracy. Considering that researchers who had focused on student ethics advocated for it in some form (LaRose et al., 2005), this response was surprising. At the same time, ethics has not been perceived to be a stand-alone measure. It is possible that the responses acquired in the present research reflect this viewpoint. Colleges and universities have been known for teaching ethics and values to students as well as providing traditional academic curricula. Considering that digital piracy is an ethical issue, the fact that

Figure 2. Frequency for legal alternatives sub-variables

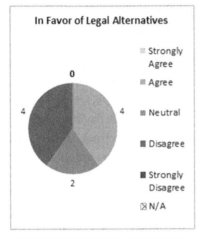

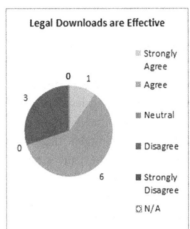

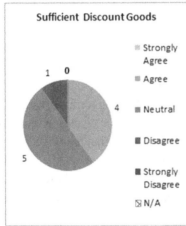

Figure 3. Frequency for legal actions sub-variables

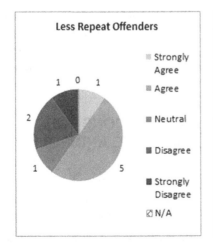

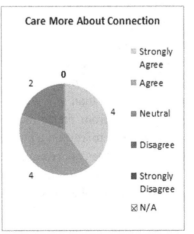

students refuse to change their opinions on the matter is somewhat alarming. Universities may need to consider treating digital piracy in a fashion similar to other popular educational topics such as drinking.

Respondents were divided in their opinions as to whether providing discounted legal goods would serve as a deterrent to piracy. Although researchers have found that some pirates illegally download in order to act out against the larger corporations, many simply pirate for the sake of pirating or for social reasons (Higgins, 2005). Since discounted legal goods may account for only one of these groups, it could explain the diversity of opinion in the data. When looking at legal downloads from a perspective of simply cutting down piracy, most of the respondents were in agreement. This could be attributed to a belief that anything to reduce piracy would be effective to some degree. With some companies providing free or heavily discounted software, students who may pirate these programs for their personal use may reconsider. Those who pirate for the sake of pirating are less likely to be impacted by such options. Finally, the respondents were divided in their opinion as to whether a sufficient number of goods were available at discounted prices would impact on decisions to pirate. This difference could be dependent on the number of goods the institutions offered. In the case of the neutral and disagree responses, this could be a signal for these institutions to look into potential arrangements with major software vendors.

Finally, the impact of Legal Actions on piracy was examined. Most of the respondents agreed with the concept that students cared more about losing their Internet connections than any legal consequences that may result from the infraction. This also linked directly to respondents' perceptions that students had little concern over the consequences of their actions. This finding provided further support for teaching students more about the ethics and consequences of digital piracy and illegal downloading. Students who were caught downloading illegally tended to avoid a subsequent offense at the majority of the institutions. Although a student not involved in a second offense is a positive notion, it is unclear if the student truly learned anything from what transpired. The student could easily have decided to go off-campus for any P2P needs, counteracting any teachings from what transpired. Since there is no way to track students after they leave the campus, finding a method to address this situation would be difficult.

SIGNIFICANT FINDINGS

The institutions were unanimous about students utilizing P2P off-campus outside of the eye of the software, and that students held little concern over the consequences about piracy. This would imply that in the long term, software has been relatively ineffective in changing the habits of students. Considering that many authors advocated for the importance of ethics, the division of opinion was unexpected. It could mean that while important, ethics has not had the intended impact that the researchers were expecting (LaRose et al., 2005). It could also mean that implementation of ethics in an institution is difficult and may be costly in the long run. Considering that most of the solutions used to combat piracy have not been directed to content monitoring, the disagreement with the role of software being lessened makes sense. There are still other security issues prevalent in a campus network such as network attacks through open ports that the current solution still combats.

IMPLICATIONS FOR PRACTICE AND POLICY

Though ethics may be one of the keys to solving the piracy problem, teaching ethics may be problematic. One institution banned servers after their numbers became problematic. There were teaching attempts, but those attempts did not get through to the students. If an ethics program is constructed, it must be able to do more than simply inform about ethics. In most cases, informational intervention tactics are the easiest to conduct but are the least effective. At the same time, ethics is neither tangible nor concrete. This forces the material to stay in an informational state and be open for interpretation. As shown in the review of the literature by Edgar (2003), ethics can take a variety of viewpoints. Even when narrowing the discussion to computer ethics, there has been considerable debate on the proper way to handle ethics. One potential solution to this conundrum involves combining both applied computer ethics and theoretical ethics. By blending the two types of ethics together, a better understanding can be achieved. Theoretical ethics provide the basis for why piracy is perceived as immoral while the applied computer ethics provide a way to show how the concepts work in a real-world situation. Considering that there are a plethora of ethical theories, by selecting the theories that best highlight the viewpoints of both sides of the piracy debate students could understand pros and cons to the reasoning behind the pirates and the corporations. This may also lead to students obtaining a better understanding of the issues behind piracy and allow them to make more educated decisions when they encounter a piracy situation.

CONCLUSION

Digital piracy is a problem that may never disappear from society. As long as people want to obtain music, movies, and software without paying for it, there will always be someone on the Internet to provide it. Furthermore, colleges and universities will always be one of the major outlets of digital piracy because of their advanced network resources. Despite digital piracy being a global phenomenon, a combination of the extensive resources and the typical college student's lack of funds makes it more lucrative for a student. Even before programs such as Napster made the act of piracy relatively easy, the type of student who would engage in piracy remained generally unaltered. Pirates have always been predominantly computer-oriented majors or students with a high level of computer knowledge.

With students abiding by the rules when they are on campus and disregarding them when they leave, it implied that future research should begin to look at more long-term solutions such as ethics in order to ensure that students will learn and accept the truth about digital piracy and uphold these lessons when they leave the institution. It was also implied that IT departments will need to be properly funded in order to ensure that the institutions can stay up-to-date on their systems since any of the current technology may be circumvented (EFF, 2006). Although digital piracy is a problem, methods to appropriately mitigate the damage do exist. Proper implementation and prevention of these methods, however, will require cooperation from all parties.

REFERENCES

Anderson, N. (2009a). *Judge rejects fair use defense as Tenenbaum P2P trial begins.* Retrieved from http://arstechnica.com/tech-policy/news/2009/07/judge-rejects-fair-use-defense-as-tenenbaum-p2p-trial-begins.ars

Anderson, N. (2009b). *Team Tenenbaum to fight on for those RIAA has screwed over.* Retrieved from http://arstechnica.com/tech-policy/news/2009/08/charlie-nesson-still-fights-for-those-riaa-has-screwed-over.ars

Athey, S. (1993). A comparison of experts' and high tech students' ethical beliefs in computer-related situations. *Journal of Business Ethics, 12*, 359–370. doi:10.1007/BF00882026

Bhattacharjee, S., Gopal, R., Lertwachara, K., & Marsden, J. R. (2006). Whatever happened to payola? An empirical analysis of online music sharing. *Decision Support Systems, 42*, 104–120. doi:10.1016/j.dss.2004.11.001

Brey, P. (2000). Method in computer ethics: Towards a multi-level interdisciplinary approach. *Ethics and Information Technology, 2*, 125–129. doi:10.1023/A:1010076000182

Bunge, M. (1977). Towards a technoethics. *The Monist, 60*(1), 96–107.

Christoph, R., Forcht, K., & Bilbrey, C. (1987/1988). The development of systems ethics: An analysis. *Journal of Computer Information Systems, 2*, 20–23.

Cohen, E., & Cornwell, L. (1989). College students believe piracy is acceptable. *CIS Educator Forum, 1*, 2–5.

Common Solutions Group. (2008). *Infringement-suppression technologies: Summary observations from a Common Solutions Group workshop.* Retrieved from http://net.educause.edu/ir/library/pdf/CSD5323.pdf

Dillman, D. A. (2000). *Mail and internet surveys* (2nd ed.). New York, NY: John Wiley & Sons.

Djekic, P., & Loebbecke, C. (2007). Preventing application software piracy: An empirical investigation of technical copy protections. *The Journal of Strategic Information Systems, 16*, 173–186. doi:10.1016/j.jsis.2007.05.005

Edgar, S. L. (2003). *Morality and machines: Perspectives on computer ethics* (2nd ed.). Sudbury, MA: Jones and Bartlett.

Electronic Frontier Foundation. (2006). *When push comes to shove: A hype-free guide to evaluating technical solutions to copyright infringement on campus networks.* Retrieved from http://www.eff.org/files/univp2p.pdf

Electronic Frontier Foundation. (2008). *RIAA v. The People: Five years later.* Retrieved from http://www.eff.org/wp/riaa-v-people-years-later

Floridi, L. (1999). Information ethics: On the philosophical foundation of computer ethics. *Ethics and Information Technology, 1*, 37–56. doi:10.1023/A:1010018611096

Forester, T., & Morrison, P. (1994). *Computer ethics: Cautionary tales and ethical dilemmas in computing* (2nd ed.). Cambridge, MA: MIT Press.

Goodyear, M., Salaway, G., Nelson, M. R., Petersen, R., & Portillo, S. (2009). *The career of the IT security officer in higher education.* Retrieved from http://www.educause.edu/library/ECP0901

Gopal, R. D., Sanders, G. L., Bhattacharjee, S., Agrawal, M., & Wagner, S. C. (2004). A behavioral model of digital music piracy. *Journal of Organizational Computing and Electronic Commerce, 14*(2), 89–105. doi:10.1207/s15327744joce1402_01

Gupta, P. B., Gould, S. J., & Pola, B. (2004). "To pirate or not to pirate": A comparative study of the ethical versus other influences on the consumer's software acquisition-mode decision. *Journal of Business Ethics, 55*, 255–274. doi:10.1007/s10551-004-0991-1

Higgins, G. E. (2005). Can low self-control help with the understanding of the software piracy problem? *Deviant Behavior, 26*, 1–24. doi:10.1080/01639620490497947

Hunt, S. D., & Vitell, S. (1986). A general theory of marketing ethics. *Journal of Macromarketing, 6*, 5–16. doi:10.1177/027614678600600103

Joachim, D. (2004). *The enforcers—The University of Florida's ICARUS P2P-blocking software has clipped students' file-sharing wings. Do its policy-enforcing capabilities go too far?* Retrieved from http://cyber.law.harvard.edu/digitalmedia/Icarus%20at%20UF.htm

LaRose, R., Lai, Y.-J., Lange, R., Love, B., & Wu, Y. (2005). Sharing or piracy? An exploration of downloading behavior. *Journal of Computer-Mediated Communication, 11*(1), 1. doi:10.1111/j.1083-6101.2006.tb00301.x

Leonard, L. N. K., & Haines, R. (2007). Computer-mediated group influence on ethical behavior. *Computers in Human Behavior, 23*, 2302–2320. doi:10.1016/j.chb.2006.03.010

Liang, Z., & Yan, Z. (2005). Software piracy among college students: A comprehensive review of contributing factors, underlying processes, and tackling strategies. *Journal of Educational Computing Research, 33*, 115–140. doi:10.2190/8M5U-HPQK-F2N5-B574

Luppicini, R. (2010). *Technoethics and the evolving knowledge society*. Hershey, PA: IGI Global. doi:10.4018/978-1-60566-952-6

Miah, A. (2010). *International Journal of Technoethics.* Retrieved from http://www.andymiah.net/2010/01/14/international-journal-of-technoethics/

Molteni, L., & Ordanini, A. (2003). Consumption patterns, digital technology and music downloading. *Long Range Planning, 36*, 389–406. doi:10.1016/S0024-6301(03)00073-6

Nyiri, J. (2004). The effects of piracy in a university setting. *Crossroads, 10*(3), 3. doi:10.1145/1027321.1027324

Rob, R., & Waldfogel, J. (2006). Piracy on the high c's: Music downloading, sales displacement, and social welfare in a sample of college students. *The Journal of Law & Economics, 49*, 29–62. doi:10.1086/430809

Rogerson, S. (1996). ETHicol: Reports on a recent public lecture on computers and human values. *IMIS Journal, 6*(5), 14–17.

Rosenblatt, J. (2008). Security metrics: A solution in search of a problem. *EDUCAUSE Quarterly, 31*(3), 8–11.

Schuster, W. V. (1987). Bootleggery, smoking guns and whistle blowing; a sad saga of academic opportunism. In *Proceedings of the Western Educational Computing Conference* (pp. 179-187).

Shang, R.-A., Chen, Y.-C., & Chen, P.-C. (2008). Ethical decisions about sharing music files in the P2P environment. *Journal of Business Ethics, 80*, 349–365. doi:10.1007/s10551-007-9424-2

Slater, D. (1991). New crop of IS pros on shaky ground. *Computerworld*, 90.

Spanier, G. B. (2004). *Peer to peer on university campuses: An update.* Retrieved from http://president.psu.edu/testimony/articles/161.html

Stahl, B. C. (2004). Information, ethics, and computers: The problem of autonomous moral agents. *Minds and Machines, 14*, 67–83. doi:10.1023/B:MIND.0000005136.61217.93

Taylor, S. L. (2004). Music piracy—Differences in the ethical perceptions of business majors and music business majors. *Journal of Education for Business, 79*, 306–310. doi:10.3200/JOEB.79.5.306-310

Thong, J. Y. L., & Yap, C.-S. (1998). Testing an ethical decision-making theory: The case of softlifting. *Journal of Management Information Systems, 15*, 213–237.

U. S. Department of Education. (2008). *Higher Education Act of 1965, Cong. §487.* Washington, DC: U.S. Department of Education.

University of Central Florida Website. (2009). *Office of student conduct.* Retrieved from http://www.osc.sdes.ucf.edu/?id=computer_misuse

Wolfe, S. E., Higgins, G. E., & Marcum, C. D. (2008). Deterrence and digital piracy: A preliminary examination of the role of viruses. *Social Science Computer Review, 26*, 317–333. doi:10.1177/0894439307309465

Worona, S. (2008). On making sausage. *EDU-CAUSE Review, 43*(6).

This work was previously published in the International Journal of Technoethics, Volume 2, Issue 3, edited by Rocci Luppicini, pp. 23-38, copyright 2011 by IGI Publishing (an imprint of IGI Global).

Chapter 19
Boys with Toys and Fearful Parents?
The Pedagogical Dimensions of the Discourse in Technology Ethics

Albrecht Fritzsche
Technical University of Darmstadt, Germany

ABSTRACT

Based on recent complaints about the neglect of the human in the philosophy of technology, this paper explores the different ways how technology ethics put the relation between the human and the technical on stage. It identifies various similarities in the treatment of the human in technology and the treatment of the child in education and compares Heidegger's concerns about the role of technology with the duplicity of childhood and adulthood in conflicts of adolescence. The findings give reason to assume that technology ethics and pedagogy are closely related. A brief review of selected topics in technology ethics illustrates exemplarily how a pedagogic interpretation of the current discussion can contribute to further progress in the field.

1. THE ROLE OF THE HUMAN FOR THE DISCUSSION IN TECHNOLOGY ETHICS

Despite all progress in the field during the last decades, it often remains the biggest challenge for studies in technology ethics to explain why they are necessary. Economic interests already exert a strong selective pressure on the development of technology that forces tools and machines to become more efficient and better adapted to the needs of their users. Ethical approaches only seem to add another dimension to the decision making process about technology. In the eyes of many engineers and managers, they are more likely to complicate the situation than to improve it. In order to make clear that the creation and use of technology requires further attention, ethical

DOI: 10.4018/978-1-4666-2931-8.ch019

approaches have to address the implications of tools and machines for human life beyond the horizon of efficiency, usability and convenience. Technology ethics, like any other kind of ethics, can therefore not be reduced to the study of a certain phenomenon. Implicitly or explicitly, it always includes a general reflection about the human condition. The anthropological aspect of technology ethics currently does not seem to attract a lot of attention, but it would be wrong to assume that it has lost its importance.

For Ernst Kapp, technology is a consequence of the fact that human beings, unlike any other animal, are not bound to a certain form of life. They use tools and machines to expand and transform their bodily functions. Similar to Aristotle, Kapp describes technical devices as a projection. They expand, enhance or replace the function of natural organs (Kapp, 1877, p. 41, cf. Micham, 1994). The hammer increases the force of a hand, the microscope improves eyesight and the telephone extends the distance in which a voice can be heard. With reference to Hegel, Kapp describes technical progress as a dialectic movement (Kapp, 1877, p. 124). The projection of natural organs to technical devices allows the recognition of the potentials of human nature. By watching technology from a distance, the consciousness realizes how powerful it is and it also gains insight into the responsibilities to use its power in the right way. If, however, the human is not considered as an unbound, but instead as a needy being, the line of conclusions changes direction. Technology does not appear as a projection of organ function, but as an aid to overcome deficiencies (Linton, 1936). Human beings rely on technical devices to survive in the world. They use clothing, weapons or fire to make up for missing fur, claws and bad adaptation to climate and vehicles, houses and other aids relieve them from too much strain (Gehlen, 1980, p. 8). With the creation and use of tools and machines, humans try to overcome their weaknesses, alienating themselves at the same time from their real nature (Freyer, 1955). Looking at

technology this way, it is essential to make sure that technological progress remains adequate to the necessities of human existence, moving the focus of technology ethics from strategic and organisational considerations of what should be done with technology to the critical analysis of its implications for human life.

Ethical reflections about the usage of tools and machines have a long tradition, going back to Plato, Aristotle and ancient mythology (cf. Hubig, 2000). Nevertheless, it was not until the late twentieth century that technology ethics – or Technoethics, as Mario Bunge suggested to call it –started to evolve into a separate field of research (cf. Bunge, 1977). The topics of technology ethics are quite diverse, ranging from surveillance and ubiquitous computing to intellectual property and artificial life. Due to the different expertise required to work on these topics, research in technology ethics is usually a quite interdisciplinary activity, involving full-time philosophers and ethicists as well as scientists and engineers (cf. Luppicini, 2009). As a result, technology ethics has become a very heterogeneous field with many separate lines of discussion. All contributors claim to be concerned with technology. Their perspectives, however, are so different that it is hard to say if they are really talking about the same thing, particularly if the public forces them to assume very specific roles in the discussion. Advocates of free access to the internet, for instance, imply that technology is beneficial to the human and everybody should have access to it, while many critics of bioengineering argue on the basis that technology poses a threat nature. This disparity in the treatment of technology can be explained by differences in the anthropological foundation of the arguments. The single lines of discussion in technology ethics, however, are rarely confronted with this disparity, because they remain within the context of one topic. For them, there does not seem to be much need to consider the role of the human element in technology. Many philosophers consequently do not include the people involved in the creation and use of technology in their research.

According to Joseph Pitt, the "failure to include the decisions and actions of the appropriate individuals results in philosophical accounts that appear isolated from the remainder of the philosophical conversation, and this is often why philosophers have been seen as having failed to provide an adequate account of technological issues" (Pitt, 2000, p. 66). The exclusion of the anthropological element from the study of technology is mistaken for an outside look at technology. In reality, it is just a neglect of the point of view from which technology is approached. Instead of helping us to understand technology, it contributes to the formation of certain attitudes towards it (Pitt, 2000, p. 70), making it practically impossible to reconcile the different positions in the debate, because the underlying conflict is not recognized to be related to the conception of the human element in technology. In order to overcome this obstacle, it is necessary to step back from the discussions in the field of technology ethics and to take a look at the setting in which the relation between the human and the technical is put on stage. The following pages explore the hypothesis that this setting actually reflects the double nature of human existence on earth: as a child and as an adult.

2. THE DUPLICITY OF CHILD AND ADULT AND THE CONFLICT OF ADOLESCENCE IN MODERN SOCIETIES

Many of the current ideas in the debate about technology can be traced back to Friedrich Hegel. Hegel never considered technology as a philosophical topic. His portrayal of the conscience in the process of becoming, however, has laid the groundwork for the studies of technical progress in the following centuries and his description of the dialectic relation of means and ends in human action is a common element in contemporary theories about the role of technology in human life (Hubig, 2006, pp. 71, 113). Hegel's writings are based on the study of an individual mind, but he did not just think of the individual as a single human being. Quite in the contrary, Hegel was rather interested in the cultural history of the world in general, which he studied as if it was one person. Narratives of technical progress tend to proceed in a very similar way. They describe the beneficial or detrimental effect of technology on human life, as if it was the life of one individual. Technology, however, affects a succession of different individuals with different expectations, convictions and abilities. Referring to all these people as one person only presents one possible way to study the implications of technical development. Every one of us is affected by technology in a different way and within a different period of time. If mankind is considered as one individual living on this planet since the Stone Age, phylogenesis is blended with ontogenesis. As individuals, we do eventually not have much choice other than to approach the world on the basis of personal experience. One can therefore say that phylogenetic studies always have to go through an ontogenetic filter. Nevertheless, it is important to understand that the treatment of humanity as if it was one person is a reduction.

With this background, it is interesting to note that treatment of the human by Kapp and Gehlen shows many parallels to the treatment of children in modern education. Since children have not finished their physical and mental development, they are not yet able to perform like adults. As compensation, childcare institutions protect children from exposure to adult life and provide an environment appropriate to their age. According to the UN convention on children's rights, countries also have to assist children in matters like family conflicts, social security and health or the representation of their interests before the law (Verhellen, 2000). The convention also states that the education of children should be directed "to the development of a child's personality, talents and mental and physical abilities to their fullest potential" (art.29). Inasmuch as childcare and education compensate for deficiencies, relieve the pressure of adult life and enable children to develop, they have the same role for the individual human being that, according to many philosophers, technology has for humanity in general. In this sense, one can therefore say that the philosophy of technology has translated pedagogical motifs to the phylogenetic level. The contrast between the approaches to technology in terms of enablement on the one hand and in terms of compensation on the other mirrors an ongoing conflict in education. While some say that it is necessary for children to learn as much as possible to lead a successful life, others emphasize that too much learning can have a detrimental effect on children and that education has to be appropriate to their individual needs.

If mature individuals are at the same time immature members of a developing society, they can be treated both as adults and as children. This paradox has always been part of human existence. The engagement of human beings in philosophy gives the best evidence for the fact that adults continue to consider themselves to be unknowing, incomplete and growing, despite their maturity. Interestingly, the opposite has also been true for many centuries. Until the eighteenth century, it was more or less understood that children participated in adult life. The sensitivity for the differences between children and adults was much lower – a fact that is, for example, illustrated by the portrayals of children in old paintings. The proportions of the body, the extremities and the face are similar to those of an adult, ignoring the changes that take place during in physical development of an individual. Children look like smaller versions of adults and there is evidence that they were also treated in this way. They had for the most part the same social responsibilities as anyone else; in particular, child labour was a common phenomenon (Cunningham, 1990). The modern perception of childhood as a separate phase

in human life in which different rules applied than in adulthood first gained popularity through the pedagogical approaches of the enlightment (cf. Nash, 1970). The increasing exploitation of children in the economies of the industrial revolution added significantly to its momentum. Ellen Key's proclamation of the century of the child in 1900 can be considered as another milestone on the way of pedagogy into public conscience (Key, 1909). In the twentieth century, the idea that children needed special treatment had become common sense.

This changing perception of childhood was accompanied by the growing awareness for adolescence as a phase of conflict (Hall, 1904). With the reduction of child labour and the expansion of the educational system, young people remained longer in a state of economic and intellectual dependency. As a consequence, they experienced the reaching of personal maturity without being able to finish the transition to adulthood. Up to this point, the term adolescence had been used as a synonym for growing up or maturing. Now it became a description of the phase in life in which young people are stuck somewhere between childhood and adulthood, belonging to both at the same time. Inasmuch as the prolongation of the economic and intellectual dependency of young people is a symptom of the increased separation between childhood and adulthood, the conflicts of adolescence can also be described as a result of change in the perception of the child. As long as the child was in general considered as an adult in transition, the conflicts of this process were already an inherent part of the child's existence. When childhood and adulthood were increasingly perceived as two exclusive phases in human life, the transition between both turned into a separate process. Childhood now actually had to be left to become an adult. The fact that adolescence became a painful and controversial experience is in many ways evidence for the fact that the exclusivity of childhood and adulthood that modern societies implied is only superficial.

3. HEIDEGGER'S PHILOSOPHY OF TECHNOLOGY AS A REFLECTION ABOUT INCOMPATIBLE POINTS OF VIEW

Approaches to the philosophy of technology that start with describing a need of the human for improvement seem to base their understanding of technology on an anthropological reflection. They assume that technology answers to human nature. However, the talk of an enhancement, expansion or replacement of natural organs by technical devices only makes sense if the organs themselves already have the characteristics of tools or machines. In order to improve human nature, it is first of all necessary to consider it improvable. More than anybody else, it was Heidegger who made clear that anthropological approaches to technology are therefore in fact technological approaches to humanity (Luckner, 2008, p. 34). Technical functioning, one can say, provides the framework for the discussion of the human condition. With the idea that technology can free the human from natural restrictions, the human becomes an object of technical reflection.

Speaking of the human being as a victim of technology is very popular in technology ethics. Critical approaches in Marxist tradition, for example, use Hegelian ideas to diagnose a dissociation of the means from the ends in modern technology and a consequential loss of authority for the individual; or they study the structural influence of technical systems on the way human thinking proceeds and the attempt to overcome it in the mode of a negative dialectic and free discourse. It is a common mistake to think that Heidegger's approach is based on the same premise. His argument, however, proceeds on a more fundamental level. From Heidegger's point of view, the danger of technology does not lie in the way it is treated. Unlike Horkheimer and Adorno, he does not think that reason has to be saved by reason (Adorno & Horkheimer, 1979; Dallmayr, 1991). Quite in the contrary, Heidegger believes that the reliance of modern societies on instrumental reason is itself the problem. In that sense, Heidegger's argument about technology is directed against Hegel. He wants to overcome the dominance of instrumental reflection through means and ends in the relation of the human being to the world. For Joseph Pitt, Heidegger's philosophy is exemplary for an approach that diverts the discussion away from the study of technological issues (Pitt, 2000, p. 67). Heidegger has moved the philosophical interest in technology to the fundamental role that it plays for the way we relate to the world and to ourselves. In a time where electronic devices have penetrated practically every human activity, his approach seems to have become more important than ever before. It remains unclear, however, how the awareness of the ubiquitous involvement of technology in human life affects decision making about technology. With Pitt and many others, we have to ask if Heidegger's concerns about technology have really contributed to progress in the field, or if they have just popularized a social criticism that makes us uneasy about technical progress (Pitt, 2000, p. 71).

One of the most crucial parts of Heidegger's argument against technology is his comparison between the work of modern technicians and the work of a silversmith in ancient Greece (Heidegger, 1977, p. 6). Heidegger uses the example of the silversmith to point out the shortcomings of the way modern societies relate to the world with technology. The reading of this example plays a critical role for the understanding of the alternative that Heidegger has in mind. Roughly speaking, Heidegger claims that the silversmith understood his work as bringing together the different causes that are responsible for something being in the world. Modern technicians, on the other hand, identify their work with the application of technical devices that carry out a determinate operation. The decision to apply these devices already anticipates the outcome of their operation. Heidegger is concerned about technology in modern societies, because it forces things to appear in a certain

285

way. Technology has become an "enframing", a structure that puts nature into the harness of our means and ends. The problem of this comparison is that the accounts of the way the silversmith proceeds and the way modern technicians proceed do not seem to fit together. According to Christoph Hubig, Heidegger confuses technology in ancient Greece with a conceptualization of technology in Aristotle's theory of action (Hubig, 2006, p. 102). What Heidegger describes as the thoughts of the silversmith at work are in fact reflections about its meaning. When it comes to the application of an instrument, there is no difference between the silversmith and a modern technician. Ever since the Neolithic Revolution, when humans started to use natural objects systematically as tools, technology has always been an enframing, a system and a challenge to nature (Hubig, 2006, pp. 37, 103). The silversmith therefore does not represent an alternative to the way modern societies relate to the world with technology. Nevertheless, Heidegger's argument can still help us to get a better understanding of what has changed in the last centuries, if we consider it as a complement to the object of his concerns. Heidegger is by far not the only one who is dissatisfied with everything we can accomplish with technology today. Strong reservations about technology are just as much a characteristic of modern societies as the penetration of human life with uncountable technical devices. What makes ancient Greece and the modern world in Heidegger's argument different is not the way humans relate to the world through technology, but the way they deal with the conflict between being involved with technology and being apart from it. While the silversmith seems to have been able to resolve the tension between the two positions, Heidegger argues against an involvement with technology that is so comprehensive that it does not leave space for such a resolution. The two positions have become fundamentally incompatible.

The previous sections of this paper have shown how the relation between the human and the technical is put on stage in the reflection of technology as an improvement to human nature. This staging has been connected to childhood as a state of growth and development. Heidegger changes the setting to a more fundamental level. The contrast between involvement in technology and being apart from it shows similarities to the contrast between childhood and adulthood as a state of completeness and saturation in which progress is not relevant. In previous centuries, human beings often found themselves in both states at the same time, able to make decisions and still subjected to a divine plan, wise and naive, powerful and exposed, free and equal but also enslaved. The natural order remained stable despite technical development. With the dawning modern age, the duplicity of a fixed state and a continuous development turned into a painful experience, embodied in adolescent conflict. In many ways, Heidegger's concerns about technology resemble the concerns of adults about the immaturity of their children. If they are mistaken for the concerns of an older generation about a younger generation, the Heideggerian approach becomes very susceptible to social criticism. Heidegger, however, does not speak about the decline of moral values or competences through technical progress. He addresses different human perspectives. With the motif of the incompatibility of childhood and adulthood in modern societies, this may become easier to understand.

4. PEDAGOGICAL INTERPRETATIONS OF THE DISCOURSE IN TECHNOLOGY ETHICS

Pedagogy is the scientific reflection of education. It studies the question how human beings should be treated to bring about a change that is beneficial to them. This question has a normative and a technical quality. It therefore does not receive a lot of attention among philosophers who engage in the study of the human condition in general. According to Pitt, however, the discussion about technology is often diverted away from the general philosophical discourse. After everything we have found out so far, there is good reason to believe that it has turned from philosophy to pedagogy. Because of the disregard for the human element in technology, this turn has so far remained unnoticed, but a look at the staging of the human and the technical in the discussion makes it easy to bring it to light. The questions that are raised in technology ethics are in many cases very similar to the question of pedagogy. They are concerned with the human as a developing being and the influence of external structures on its direction. They study different artefacts that bring about change through their effect on the human and the appropriateness of this change to the human condition. On a deeper level, both pedagogy and technology ethics are confronted with the duplicity of the developing and the saturated human being – and the exclusiveness in which modern societies seem to approach it in terms of the first (cf. Freire, 1997; Giroux, 1992). Saying that technology ethics are a form of pedagogy certainly does not do the field justice, but there can hardly be doubt that they are closely related. It therefore sounds conceivable that we can gain further insight into many discussions in technology ethics by treating them as if they were pedagogical discussions.

If, for example, the current controversy on surveillance, data mining and data profiling is considered from a pedagogical point of view, it assumes a new dynamic. The surveillance of infants and small children is hardly questioned. Quite on the contrary, adults are morally obliged to keep track of everything they do in order to minimize the possibility of an accident. Children are, in that sense, constantly under suspicion and have to earn the trust of their caretakers before they are left alone. Arguing against the permanent surveillance of children is very difficult. It is possible to assume that the surveillance has a potentially detrimental effect on them, because they never learn to be

on their own or because adults might abuse their power. These assumptions, however, only seem to address technicalities of the way the surveillance is performed and controlled. Arguments about human dignity are applicable to the surveillance of adults (Haggerty & Samatas, 2010), but when it comes to children, trading safety for dignity seems highly problematic. It is very difficult to apply the same standard to adults and children on this aspect (Korczak, 1992), and this may be the reason why surveillance is such a critical issue. Thinking about the human in terms of development and change allows tradeoffs between different rights of the individual over time and disrupts the idea of general equality. Pedagogy is constantly confronted with this problem. What we can learn from Janusz Korczak is the importance of sharing the position of those who are unequally treated to make the impact of this treatment at least partially bidirectional (Joseph, 1999).

The discussion on remote warfare using radio controlled robots and drones connects to various different fields of research in technology ethics (Singer, 2009; Royakkers & van Est, 2010). One of them is media ethics, a field that can in many respects be said to have originated in pedagogy and that often falls victim to the so-called third-person-effect (Davison, 1983). Scholars usually describe how the usage of technical devices affects others, but not themselves, creating another inequality in the treatment of the human that commonly results in cries for censorship and the abolishment of certain lines of technical development (Perloff, 1993). From a philosophical point of view, this phenomenon points once more to the incompatibility between the conceptions of childhood and adulthood in modern societies and the misinterpretation of this problem as a generation conflict that leads towards social criticism. Similar to schools that are often the last to adopt a new technology, technology ethics therefore often miss the chance to have a formative influence on the development and reduce themselves to belated critical reflections. In the particular case of remote warfare, the scholars in technology ethics therefore have to ask themselves to what extent they are willing to participate in the activities in the military sector instead of watching it from a distance.

A pedagogical interpretation of artificial intelligence can be helpful to identify the different controversies that are mixed up in the discussion. On the one hand, the creation of intelligent machines can be interpreted as an educative measure that gives us better insight into ourselves and guides us towards intellectual maturity. This point of view is contradicted by Hubert Dreyfus and others who say that research on artificial intelligence turns the development into the wrong direction and that it is inadequate to the human, or, more generally, that it is not justified to speak about human development in these terms at all (Dreyfus, 1992). On the other hand, the discussion about artificial intelligence also expresses doubts about the maturity of the people involved in it (cf. Weizenbaum, 1984). Depending on the point of view, either all scientists or the managers in contrast to the engineers are accused to be irresponsible and naive, leading to the question whether those involved in the creation and use of technology need more education. This question is in many respects essential to the whole field of technology ethics, illustrating once more its proximity to pedagogy. With the assumption that machines are in fact on the way to develop a consciousness, research in artificial intelligence gains another pedagogical dimension. The machines that engineers create can be interpreted as their children. This thought makes it possible to look at the responsibilities of engineers from a different angle and to ask if it is necessary to talk about the moral enablement of machines.

5. CONCLUSION AND OUTLOOK

In many ways, it is surprising that philosophical studies of technology focus so much on adults. Technology does not come into our lives when we reach maturity. Most of us probably got acquainted with technical operation at a much earlier age, when we played with wooden blocks, model cars and plastic hammers. The first experiences with technology take place in nurseries and children's rooms, where infants learn from their toys how tools and machines work. The presence of their parents or other adults reassures that the technology can do them little harm and encourages them to engage in it. Without much need to consider its consequences, children can treat technology in a playful manner and enjoy it; and it does not seem unreasonable to assume that these experiences shape the attitude towards technology for life. There is a general connection between the notion of using tools and machines and being childish. The playing child is a recurring motif in public debates about technology. It is used to describe the behaviour of irresponsible managers who ignore technical risks and reckless engineers who do not think of the impact that their work will have on the world. Ignorance for technical risks is often explained by immaturity and soldiers who use advanced weapons are said to treat warfare like a computer game. Another popular topic of conversation is that grown-up men behave with their cars like boys with their toys. Women show similar tendencies when it comes to what is called social technology, for example when mothers dress their children like puppets. On a more general scale, supporters and critics of technical progress mutually accuse each other of being naïve, hanging on to fairy stories instead of real fact.

References to childish behaviour are not necessarily beneficial to an adequate understanding of technological issues. They can easily be abused to discredit a point of view which is worth considering. Nevertheless, this paper has identified so many different ways how the discourse in technology ethics can be related to the pedagogic discourse that it seems quite clear that role of childhood deserves to receive more attention. Being attentive to the duplicity of the human as a child and an adult can be beneficial to technology ethics in various ways. First, it can bring the involvement of the human in technology to light; second, it can explain why it is so difficult to reconcile the positions in the discussion; third, it can refine the distinction of different points of view; and forth, it can help to find new approaches to decision making about technological issues. In short: it can help us to develop a better understanding for our own work and to keep it on the right track.

REFERENCES

Adorno, T. W., & Horkheimer, M. (1979). *Dialectic of enlightenment*. London, UK: Verso.

Bunge, M. (1977). Towards a technoethics. *The Monist, 60*(1), 96–107.

Cunningham, H. (1990). The employment and unemployment of children in England c. 1680-1851. *Past & Present, 126,* 115–150. doi:10.1093/past/126.1.115

Dallmayr, F. R. (1991). *Life-world, modernity and critique: Paths between Heidegger and the Frankfurt School*. Cambridge, UK: Polity.

Davison, W. P. (1983). The third-person effect in communication. *Public Opinion Quarterly, 47,* 1–15. doi:10.1086/268763

Dreyfus, H. L. (1992). *What computers still can't do: A critique of artificial reason*. Cambridge, MA: MIT Press.

Freire, P. (Ed.). (1997). *Mentoring the mentor: A critical dialogue with Paulo Freire (Counterpoints: Studies in the postmodern theory of education) (Vol. 60)*. Pieterlen, Switzerland: Peter Lang.

Freyer, H. (1955). *Theorie des gegenwärtigen Zeitalters*. Stuttgart, Germany: Deutsche Verlagsanstalt.

Gehlen, A. (1980). *Man in the age of technology*. New York, NY: Columbia University Press.

Giroux, H. A. (1992). *Educational leadership and the crisis of democratic culture*. University Park, VA: University Council of Educational Administration.

Haggerty, K. D., & Samatas, M. (Eds.). (2010). *Surveillance and democracy*. London, UK: Routledge.

Hall, G. S. (1904). *Adolescence: Its psychology and its relations to physiology, anthropology, sociology, sex, crime, religion, and education (Vol. 1-2)*. New York, NY: D. Appleton & Co.

Heidegger, M. (1977). *The question concerning technology, and other essays*. New York, NY: Harper & Row.

Hubig, C. (2000). Historische wurzeln der technikphilosophie. In Hubig, C., Huning, A., & Ropohl, G. (Eds.), *Nachdenken über technik. Die klassiker der technikphilosophie* (pp. 19–40). Berlin, Germany: Ed. Sigma.

Hubig, C. (2006). *Die kunst des möglichen i. technikphilosophie als reflexion der medialität*. Bielefeld, Germany: Transcript.

Joseph, S. (1999). *A voice for the child: The inspirational words of Janusz Korczak*. London, UK: Harper Collins.

Kapp, E. (1877). *Grundlinien einer philosophie der technik: Zur entstehungsgeschichte der kultur aus neuen gesichtspunkten*. Braunschweig, Germany: Westermann.

Key, E. (1909). *The century of the child*. New York, NY: Putnam.

Korczak, J. (1992). *When I am little again and the child's right to respect*. Lanham, MD: University Press.

Linton, R. (1936). *The study of man*. New York, NY: Appleton-Century.

Luckner, A. (2008). *Heidegger und das denken der technik*. Bielefeld, Germany: Transcript.

Luppicini, R. (2009). The emerging field of tech-noethics. In Luppicini, R., & Adell, R. (Eds.), *Handbook of research on technoethics* (pp. 1–19). Hershey, PA: Information Science Reference.

Mitcham, C. (1994). *Thinking through technology: The path between engineering and philosophy*. Chicago, IL: University of Chicago Press.

Nash, P. (Ed.). (1970). *History and education*. New York, NY: Random House.

Perloff, R. M. (1993). Third-person effect research 1983-1992: A review and synthesis. *International Journal of Public Opinion Research, 5*, 167–184. doi:10.1093/ijpor/5.2.167

Pitt, J. (2000). *Thinking through technology*. New York, NY: Seven Bridges.

Royakkers, L., & van Est, R. (2010). The cubicle warrior: The marionette of digitalized warfare. *Ethics and Information Technology, 12*, 289–296. doi:10.1007/s10676-010-9240-8

Singer, P. W. (2009). *Wired for war: The robotics revolution and conflict in the twenty-first century*. New York, NY: Penguin Press.

Verhellen, E. (2000). *Convention on the rights of the child: Background, motivation, strategies, main themes*. Antwerp, Belgium: Coronet.

Weizenbaum, J. (1984). *Computer power and human reason: From judgment to calculation*. London, UK: Penguin Press.

This work was previously published in the International Journal of Technoethics, Volume 2, Issue 3, edited by Rocci Luppicini, pp. 39-47, copyright 2011 by IGI Publishing (an imprint of IGI Global).

Chapter 20
Without Informed Consent

Sara Belfrage
Royal Institute of Technology, Sweden

ABSTRACT

The requirement of always obtaining participants' informed consent in research with human subjects cannot always be met, for a variety of reasons. This paper describes and categorises research situations where informed consent is unobtainable. Some of these kinds of situations, common in biomedicine and psychology, have been previously discussed, whereas others, for example, those more prevalent in infrastructure research, introduce new perspectives. The advancement of new technology may lead to an increase in research of these kinds. The paper also provides a review of methods intended to compensate for lack of consent, and their applicability and usefulness for the different categories of situations are discussed. The aim of this is to provide insights into one important aspect of the question of permitting research without informed consent, namely, how well that which informed consent is meant to safeguard can be achieved by other means.

INTRODUCTION

Research where human beings are involved as objects of investigation may be of many different kinds. There is wide agreement that no such research should take place without the practice of informed consent, in order to ensure that all human research subjects participate voluntarily. However, it is not always possible to obtain informed consent. This has been acknowledged in particular for research of some kinds, typical of biomedicine and psychology, while the absence of informed consent in other kinds of research has been largely ignored, or at least under-analysed.

The latter kinds of research are more common in fields such as infrastructure research, and may be increasingly important due to the advancement of new technology facilitating surveillance as well as the collection and handling of large amounts of data. This situates the problems investigated in this contribution within the scope of the growing field of technoethics.

DOI: 10.4018/978-1-4666-2931-8.ch020

The present paper produces an overview and categorisation of research on humans where informed consent cannot be obtained, based on the reason why informed consent is inapplicable. These reasons are that (1) providing information to participants is counter-productive to the research at hand, (2) prospective participants lack decisional capacity, (3) it is excessively costly to ask for consent, and (4) the collective nature of the study rules out voluntary participation. Again, some of these – the first two categories – have previously been extensively discussed in the literature whereas the other two, have received much less attention. These last two kinds of cases also bring problems to light that are significantly different from those of the first two.

Of course, an important question becomes whether research should be permissible without informed consent, when the necessity of consent is commonly so strongly insisted upon. At least three aspects are of relevance for answering that question, namely to what extent the lack of informed consent can be compensated for, the scientific value of the research and the risks that research participants are exposed to. This paper treats the first of these, by reviewing methods proposed as substitutes for informed consent. These methods are arranged into five groups: (A) provision of information without consent, (B) consent based on partial information, (C) advance directive, (D) proxy consent, and (E) collective decision making. It will be discussed for which types of situations each group of methods is applicable, and how well they compensate for the lack of informed consent. Furthermore, it will be pointed out what the merits are of these methods, merits that are not restricted to how well they compensate for the lack of informed consent.

SITUATIONS WHEN INFORMED CONSENT IS NOT APPLICABLE

Informed consent is commonly viewed as a necessary but insufficient condition for permissible research where human beings are involved. It is necessary in order to ensure that people do not participate in research against their will. It is insufficient because people should not participate in research projects that do not pass the scrutiny of research ethics committees (ascertaining, in addition to participants' consent, that they are not exposed to great risks of harm, do not participate in projects with low scientific value etc.) *even* if they would accept to participate in such research.

The aim of informed consent is to ensure that a person who participates in research does this *because* she wants to do so. The best way to achieve this is to make sure that she herself decides whether to participate or not. In order for such a decision to constitute valid consent, it must contain three elements: information, decisional capacity[1] and voluntariness. The information element is relatively straightforward: it means that the subject must have all relevant information available. There is, however, a tension between the view that potential research participants must in fact assimilate the information provided,[2] and the view that it suffices that the information is offered – the potential participant may or may not choose to take it all into account (Beauchamp & Childress, 2001; Hayry & Takala, 2001). The latter of these views is favoured in this paper. The element of capacity is more complex; it means that the subject must be able to understand the information, to form a will and to make decisions in line with that will. Furthermore she must understand that she is making such a decision, and *perceive* this decision as voluntary. She must not, for instance, believe there to be anyone threatening her to participate or that she is ruled by some superior force etc. Finally, in order for her capacity to decide to be effective, she must be able to communicate her decision to others. The element of voluntariness means that there must not only be perceived, but also de facto voluntariness. This includes the absence of threats or undue inducements making the subject accept participation. Furthermore, the person must not be outright forced into participation through the decisions of others or by the exclusion of all options other than to participate.

So, informed consent serves the laudable purpose of ensuring that people who participate in research do so because this is what they, and not (merely) others, want.[3] In some research situations, however, informed consent cannot be used. In this section, four categories of cases will be described where informed consent is impracticable.

Category 1: Providing Information is Counter-Productive to Research

The first category consists of situations where informed consent is impracticable because informing potential research participants renders the research project impossible, or at least meaningless. These are cases when the authentic behaviour of individuals is the object of investigation, and it is reasonable to believe that letting people know that, or in what way, they are being studied would alter this behaviour (consciously or unconsciously – the result remains distorted regardless of which). This would make the research outcome unreliable, perhaps even useless. This category includes cases where people cannot be informed about the mere fact that they are being observed, as well as cases where people know that they are being studied but not the purpose and layout of the study.

Studies of this kind are common in psychology and sociology. A famous example is Stanley Milgram's (1963) obedience study. Examples can also be found in other areas, such as traffic research. One example is a study of people's willingness to participate in car pools.[4] The researcher put in an advertisement in a newspaper, pretending to be a citizen looking for others interested in starting a car pool with him. He reasoned that if he had written in the ad that the aim was to investigate the response to such an ad, fewer – or maybe even other – people would have replied.

Other examples from the traffic field are when people's driving behaviour (e.g., stopping at red lights, not exceeding speed limits) is studied.

In one case of this type, from the medical field, patients were asked to consent to a follow-up after a stroke, but not to randomisation into two groups (Dennis, 1997). The purpose of the study was to measure the effects of contact with a "stroke family care worker", which one group was offered. The researchers suspected that if patients and their families were informed about the potential help from the care worker, and then randomised into the control group without any such contact, they would feel disappointed and worse than if they had not known about the possibility of the care worker. In other words, full information to the patients was believed to endanger the scientific quality of the study.

Category 2: Prospective Participants Lack Decisional Capacity

In the second category, informed consent is impossible because the potential participant lacks decisional capacity. People can lack decisional capacity in different ways and to different extent.[5] An infant is much further away from having this capacity than an older child or an adolescent, individuals can be more or less mentally impaired, and they can suffer from milder or more severe dementia.[6] Some lack this capacity in general, such as mentally impaired persons, whereas others only lack it in certain respects, like in cases of phobias.[7] There is also a distinction to be made between persons who possess the capacity under normal circumstances but lack it temporarily because they are unconscious, and those whose normal condition is to lack the capacity.

Furthermore, there are variants with regard to the temporal duration of the incapacity. The most important distinction is between those who have never had the capacity and those who had it at an earlier point in time. Children and many mentally impaired persons are examples of the former, while the temporarily unconscious and people with dementia are examples of the latter.

A distinction can also be made between those who are expected to eventually (re)gain the capacity, such as children, people whose incapacity depends on a treatable condition (including people who are temporarily unconscious) and those who will not have the capacity in the future, such as permanently injured or demented persons.

It is a common standard to try to avoid including persons lacking decisional capacity in research. Some research, however, cannot be performed on other participants, such as research on medical treatments for children.[8] Emergency medicine is another example. Clinical research on temporarily incapacitated patients is essential for the improvement of treatments of acute illnesses or injuries.[9]

Category 3: It Is Too Costly To Ask For Consent

The third category of cases where informed consent is impracticable consists of cases where it is too costly, or too demanding, to obtain consent from the participants. It is disproportionately costly – in terms of time, effort and money – for the researcher and/or the participants. These cases can generally be described as situations where it is not known beforehand who will come to participate, such as big research projects taking place in the public space,[10] or situations where changes and additions to what is being studied are made long after the procedure of informed consent has been carried out, such as research on personal information collected in databases or biological material in biobanks.[11] Concrete examples of the former kind could be the testing of different types of train cars on a subway line, comparative studies of two types of escalators in an airport, or investigations of alternative paving on motorways. The participants in such trials are those who just happen to pass by. They have not been actively recruited or chosen to be part of the trial – they are just those who by chance end up participating in research. The central feature of situations in this category is *not* that there is only one option available for all (that is instead a characteristic of cases in the next category) – as there can be alternatives at hand. In the examples above, both new and old train cars can be in operation, different escalators or even a lift can be available, and there can be different pavings on different lanes. A specific individual, however, ends up in one of the alternatives.

In cases like these it is very difficult to implement informed consent. Researchers would have to stop all people coming close to the train, or escalator, or road, on which a trial takes place, provide full information, adapt it in order for all persons to be able to comprehend it, and furthermore ask each individual to consent. This would obviously consume a lot of resources for the researchers as well as for the research subjects. In the examples mentioned it is reasonable to think that most prospective participants would be unwilling to spend time on an informed consent procedure. It seems to be a common feature of projects like this that many people would most likely choose not to be informed, and not to be asked to consent. [12]

In the other kind of cases in this category, the difficulty is that a long time has passed since the individuals agreed (given that they did) to give out information about themselves or to donate biological material and it may be problematic to retrace those individuals, if they are still alive. Arguments have been presented against applying the informed consent requirement in cases of this kind, in the interest of both researchers and donors (to save time and money, and to avoid being disturbed, respectively).[13]

Category 4: The Collective Nature of the Study Rules Out Voluntariness

In the fourth and last category, informed consent is impossible because the project's collective nature rules out voluntary participation of any one individual.

In cases in this category either all individuals in a certain wide group, such as those living in a particular region, participate, or none.[14] There is no way for a single individual to say no, or if we do allow one person to say no, then the project cannot be carried out at all (for example – if one person can say no to testing a new roundabout, then the roundabout has to be removed or not built, hence, nobody can try it.)

Examples of this category are (mandatory) social experiments of, for example, effects of different social security systems, where inhabitants or citizens in a certain region face one system, while others face another.[15] For a particular individual, no alternative to participation is available.[16]

It seems reasonable to include in this category both cases where there is in fact no alternative to participating in research and cases where there is such an option but it is associated with very high costs. Some social experiments, mentioned above, can belong to the latter of these two if it is for instance possible to avoid research participation by moving to a different state or region. Other examples, again, can be big traffic research projects, where the alternative to participate in a trial – for instance to drive on a new road surface, on a road with a new type of rumble strips or the like – is to choose another road which may be significantly longer/slower/more expensive, or even not to go at all. If the alternative to participating in research is to move or to drastically change one's travel habits, then we may say that there are in fact no alternatives available.

It is difficult to delimit exactly how costly the option not to participate in a collective research project has to be in order to be viewed as a non-option, or in other words, to turn the choice to participate insufficiently voluntary.[17] For our present purposes it suffices to say that there can be cases where research participation can only be avoided at costs that are so high that participating cannot reasonably be seen as voluntary.[18]

COMPENSATING FOR THE LACK OF INFORMED CONSENT

In the previous section it was established that there are indeed situations where informed consent cannot be employed, and that these situations are of different kinds depending on the reason why informed consent is inapplicable.

This section will discuss what can be done in cases where informed consent cannot be obtained. First, it should be underlined that while in most kinds of situations discussed above, informed consent is impossible to apply, it is in fact possible to use in others (those in category 3) but at a very high cost.[19] For the latter kind of cases, it could be argued that if we are serious about the requirement of informed consent, it should be enforced even if the costs are high. If we instead choose not to insist on this in these cases, we face the same alternatives as for cases where informed consent is literally impossible to use, namely either not to allow the research to take place at all, or to allow it even without informed consent – perhaps conditioned on attempts being made to compensate for this lack.

Whether there are justifiable exceptions from the requirement of informed consent seems to depend on three factors: the potential benefits of the research, the size of risks of harm, and how well the lack of informed consent can be compensated for. It is the third of these factors that will be discussed below.

A number of methods meant to compensate for the lack of informed consent can be found in the literature. These will be arranged into five groups, according to their key characteristics. Each group will be described, and then the kinds of cases in which it is applicable will be identified. It will be discussed whether the methods do well as substitutes for informed consent, and also whether they do well in other respects – if they provide something of value apart from what is related to the aim of informed consent.

Group A: Information Provision

In some cases where it is not possible to ask for participants' consent, it is still feasible to provide them with relevant information. For children or others who are not considered to have decisional capacity but who can understand some information about a research project (i.e., some, but not all, cases in category 2), this seems to be appropriate (and is indeed demanded in central research ethical guidelines).

Similarly, in cases of category 4, when people cannot decide whether or not to participate in a collective study, they can be properly informed about it. The same holds for cases of category 3 where informed consent is impracticable because it would consume too many resources. Although providing information is costly, it is generally less costly than carrying out the full informed consent procedure. Information can for instance be distributed by means of newspaper advertisements, roadside signs, etc.[20]

For cases in category 1, when providing information would be counter-productive to the research, debriefing is an option. Debriefing means that the participants are informed *after* having participated. Debriefing can also be used in some cases in category 2, for example when research has been carried out on an unconscious accident victim who has thereafter regained consciousness.

Although debriefing, as well as other methods of information provision, can at least partly compensate for the information element of informed consent, it does not make up for the lack of the third element of informed consent – it does not render research participation voluntary. In other words, *consent* is not obtained,[21] and the aim of informed consent, that people participate because they want to, is not achieved.

It seems to be valuable, however, for people to be aware of what is going on around them and what they are or have been involved in even if they cannot decide on those matters. Providing information to participants is also a requirement in research ethical guidelines that stands on its own right – it is not merely a means to achieve informed consent. Furthermore, in the case of debriefing, informing people that they have been observed may arguably to some extent compensate for the privacy intrusion that they were victims of. It could even be claimed that their dignity is to some degree restored.

Group B: Consent Based on Partial Information

Another set of methods applicable in situations where participants cannot be informed about the research project can be called "partially informed consent".[22] This means that people are asked to sign agreements in which they consent to being participants in a study without being provided with full information about the study.

The category of cases where partial consent could be a relevant option is the first one in which the detailed information about the study is the obstacle for informed consent. However, the method of partially informed consent is not always applicable. In some cases, knowledge that one is being observed (even without knowing exactly in what way or for what purpose) may affect one's behaviour and hence distort the result.[23]

In principle, partially informed consent could consist in agreeing to participate in a number of studies (as opposed to one specific), about which details are not provided. Such a method, which is sometimes labelled *blanket* or *broad*[24] consent, is of particular relevance to research on biological material collected in biobanks (which is discussed within the framework of category 3 above). In the biobank context, such unspecific consent can be justified either because there are *at present* a number of different studies to be carried out on the material to be donated, or because it is expected that *in the future* there will be a number of studies to be carried out (of which some are at present not even conceivable). In the latter case the method should perhaps better be described as an unspecific advance directive, a method to be explained in the next subsection.

Partially informed consent is of no use for cases in categories 2 and 4, since no consent at all is possible in situations of those kinds.

It is not obvious that a decision to participate in a research project does not amount to proper consent just because the information is only partial. It would, as a matter of fact, be possible for people to decline participation altogether. Furthermore, it may be argued that partially informed consent does in fact constitute valid consent, as the potential participants accept and are aware of not being fully informed. A case where a person is informed about her not being fully informed and yet accepts to participate is similar in important respects to one in which she is offered full information but declines parts of it and yet accepts to participate.

If it is indeed the case that people would prefer not to be informed in any depth, then the method of partially informed consent can rightly be said to have the benefit of not burdening unwilling people with information they wish to avoid.

Group C: Advance Directive

For many cases of category 2 and possibly also cases of category 3 regarding database research, the method of advance directive seems applicable. Advance directives are like informed consent, or partially informed consent, except for the fact that that they concern possible future research participation and are given at an earlier point in time. People fearing to fall into dementia could use such a method, as can people who wish to prepare for a situation where they end up in emergency care. Advance directives can be of two kinds: *unspecific* and *specific*. An unspecific advance directive, mentioned in the previous sub-section, is an individual's consent, based on partial information, to participate in future research projects if the opportunity arises while she is unable to provide informed consent. She does not consent to any particular project, and cannot therefore be

sufficiently informed. A specific advance directive refers to specified research projects of which information can be revealed and communicated. A form of specific advance directive, would be *randomised controlled trial cards* that people in risk groups for certain diseases or conditions have in their wallets, thereby consenting to participate in well-described studies of treatments for these diseases or conditions (Lindley, 1998). This method has been discussed under the label of *prospective consent*, which is supposed to be obtained from a large group of people, who would be included in trials were they to suffer from certain injuries.[25] Both randomised controlled trial cards, and other versions of prospective consent, can however also be of a more general nature, hence constituting unspecific rather than specific advance directives.

Even for categories 2 and 3 mentioned, advance directives are not always an option. In category 2 they require that the incompetent individuals have been competent at an earlier stage, which is not always the case. In category 3, advance directives may be relied on for research on biobank material (database research), but for other cases of type 3 – those concerning large projects in the public space for instance – advance directives do not seem much easier to apply than the procedure of informed consent itself.

In principle, unspecific advance directives could be used for cases of category 1 above. In order for such directives not to have the same negative impact on research findings as informed consent in these cases, the directives must be very unspecific and perhaps also given a long time before the research takes place in order for the person not to actively remember it and consequently letting it influence his or her behaviour.

For cases of category 4, advance directives are of no use as the question regarding research participation is still not to be settled on the level of the individual.

The more specific an advance directive is, the more does it approach the fulfillment of the aim of informed consent. Unspecific advance directives, like partially informed consent, obviously fail to ensure that peoples' decisions are informed and as long as the provision of full information is a requirement for valid consent, unspecific advance directives consequently fail to fully compensate for the lack of informed consent. The aspect of advance directives that distinguishes them from informed, or partially informed, consent is that it is given before the prospect of research participation is realised. It must therefore be asked for how long an advance directive should be recognised as valid. It seems reasonable to assume that decisions made a long time ago do not necessarily reflect the will of the person at present,[26] and it is reasonable to argue that the aim of informed consent is to ensure that it is the will of the person who is presently to participate in a research study which is to be decisive, rather than the will of the person in the past.

Group D: Proxy Consent

For persons lacking decisional capacity (i.e., category 2), proxy, or surrogate, consent is a common measure to compensate for the absence of informed consent. Proxy consent means that another person – typically a close relative or friend – is provided with adequate information and accepts or rejects the conditions of participation on the incompetent person's behalf. The proxy can either be chosen beforehand by the person to be represented, or appointed by someone else. The former method, which is often called delegated consent, is obviously preferable but requires the person to be represented to have been competent enough to choose a proxy at an earlier point in time.[27] It can be argued that a person can be competent enough to choose her own proxy, yet not competent enough to decide on her own research participation.

A relevant question in the context of proxy consent is on what grounds the proxy should make the decision. In Dworkin's words: "Ought the representative act as the principal would have acted, or as the principal should have acted, or in the interests of the principal, or in pursuit of his welfare?" (Dworkin, 1988, p. 87). There are two established ways of dealing with this issue; either the proxy is asked to decide that which the incompetent person *would have wanted* had he or she not been incompetent, or to decide that which is considered to be best for the person, in other words in his or her *best interest*.

The first alternative, the *substituted judgment standard*, aims at respecting that which the individual would decide, had she not lacked decisional capacity. For cases where the incompetent person has never, or at least not for a long time, been competent (i.e., many cases of type 2), this standard is problematic since it requires us to determine what such an individual would have wanted, had he or she possessed decisional capacity, i.e., had he or she been *different* in a significant way.[28] In practice the proxy is then likely to decide in accordance with what he or she him- or herself would have wanted or what most normal people would have wanted, if placed in the same situation but having decisional capacity. For cases where the potential research subject is only temporarily incompetent, or was recently competent, the substituted judgment standard seems more appropriate. The proxy has to determine what decision the person would have made if she had been conscious at the moment (or had not recently lost important cognitive capacities), which is a much more feasible task, in particular for someone who knows the person well. It could even be argued that such a person's will can in fact be *fully* respected, given that he or she has given the proxy the authority to provide or withhold consent on his or her behalf, i.e., when proxy consent amounts to delegated consent.

The second alternative, called the *best interest standard*, means that the proxy is to decide in accordance with what one believes to be in the decision-incapable person's best interest. This alternative is more paternalistic in nature, as the proxy's decision is to be based on what is in the interest of the individual, as opposed to what the individual would have wanted.[29] Beneficence, rather than respect for autonomously formed preferences, seems to be the guiding light behind this standard.

For cases of category 1, proxy consent could perhaps be used in order for the research output not to be distorted by participants' awareness of their being studied. In such cases, a proxy could be asked to decide, on the (uninformed) participant's behalf, whether the details of the study are acceptable.

For cases of categories 3 and 4 proxy consent does not seem to be applicable.

Group E: Collective Decision Making

For cases in categories 3 and 4, individuals' opportunities to be involved in democratic decision making regarding research participation for a collective they belong to, could perhaps to some extent compensate for the lack of informed consent. For cases of categories 1 and 2, on the other hand, such methods do not seem to be useful. In cases of category 1, participation in collective decision making requires access to the same information that would be detrimental to the research at hand in an (individual) informed consent procedure. In cases of category 2, the potential research subjects lack decisional capacity for decisions on the collective as well as the individual level.

Democratic collective decision making, can be either direct or representative. Direct democracy, in this context, can be exemplified by members in a particular group together deciding whether to take on a research project or not. This method is sometimes called community consent.[30] There seems, however, to be no accepted definitions of the concept of community consent. It is sometimes taken to mean that some representative of the community makes the decision.[31] In such cases, community consent would rather be a matter of representative democracy – given that the representative is democratically elected which is not necessarily the case. Such procedure would

perhaps better be called an ombudsman, or group proxy.[32] A system with ombudsmen for groups with special interests or views regarding research participation could potentially do rather well in protecting the will of the individuals involved, especially if the ombudsmen are democratically elected by those represented.[33] Such a group proxy could perhaps come close to using the substituted judgment standard. If instead the ombudsmen represent large and heterogeneous groups (and particularly if they are not democratically elected), their function would rather be to double the role of the research ethics committees.

Apart from the use of an ombudsman, representative democracy in this context could be limited to people's pre-existing possibilities to be engaged in the political life of their society – either on the local, regional or national level. Thereby they can influence what rules and regulations are to be followed by researchers, what kinds of research projects to be permitted and so on.

Collective democratic decision making can, in other words, mean many different things. Of relevance is the size and composition of the demos, as well as the functioning of the decision making process. For instance, how is a relevant community to be delimited (geographically, or with respect to who is likely to face a certain risk, etc)? Is the democratic structure created for the specific purpose of determining the issue of research participation at hand, or is it the political structure already in place that is used?

The advantage of democratic collective decision making is of course that people to some extent are involved in deciding matters that concern them. The extent of this influence varies between different versions of democratic participation. In some cases of community consent, there is an ambition to make decision by consensus, which means that from the point of view of the individual the method approaches the right to veto and informed consent. For the most part, however, democratic collective decision making seems to aim at achieving the common good, that which is good for society as

Table 1. *"Yes" means that the group of compensating methods can be used and would constitute an improvement as compared to no compensating methods being used; "No" means that the group of methods cannot be used and/or would not constitute an improvement*

		Categories of cases where informed consent cannot be used			
		1. Providing information is counter-productive to research	2. Prospective participants lack decisional capacity	3. It is too costly to ask for consent	4. The collective nature rules out voluntariness
Groups of methods compensating for the lack of informed consent	A. Information provision	Yes (debriefing).	Yes, for conscious individuals.	Yes, for large public space projects.	Yes.
	B. Partially informed consent	Yes, to a limited extent.	No.	Sometimes.	No.
	C. Advance directive	Yes, to a limited extent.	Yes, if the person had capacity before.	Yes, for database cases.	No.
	D. Proxy consent	Yes.	Yes.	No.	No.
	E. Collective decision making	No.	No.	Yes.	Yes.

a whole, rather than ensuring that the will of each individual is respected.

Furthermore, procedures of democratic decision making allow people to have their voices heard which seems to be valuable even if they are not permitted to decide for themselves.

A specific benefit of using ombudsmen could be that they may be better informed in some ways than the individuals to take part in research. They obviously do not know more about their wills and preferences than do the individuals themselves, but they may have a better knowledge and understanding of the subject matter at hand.

CONCLUSION

In Table 1 categories of cases when informed consent cannot be used are tabulated against the methods intended to compensate for this lack.

This paper has put focus on the relevance of examining the applicability of informed consent in contexts that have previously received fairly little attention. If we believe that people should voluntarily decide about whether or not to be involved in research in some cases, such as those in categories 1 and 2, then they should arguably do so also in cases of categories 3 and 4. It has been shown what the difficulties are in also those cases, and what methods that can be used in order to compensate for the lack of informed consent.

As pointed out above, the question whether it can be justified to allow for research when informed consent cannot be obtained is not answered in this paper. Instead the focus is on one factor of relevance for answering that question, namely how well the lack of informed consent can be compensated for. Other factors, in particular the scientific value of the research and the risk of harm faced by those participating need to be considered when deciding whether research without (full) informed consent should be performed in these cases.

It deserves mentioning that it can be questioned whether the practice of informed consent itself is a guarantee for achieving its aim – that people participate in research because this is what they want. There are many lines of argument to the contrary, focusing on for example people being incapable of understanding even basic information about standard research methodology (see for example Dawson, 2004) or the difficulty of making decisions that truly are in line with one's will (Schwab, 2006).

If there is scope for such doubts, in addition to the problems of obtaining informed consent discussed above, as well as the applicability of methods that to a varying degree compensate for there being no informed consent, this lends support to a tentative conclusion of this paper, namely that there may be reason to challenge or at least broaden the view on how and by whom decisions regarding research participation ought to be made.

REFERENCES

Beauchamp, T. L., & Childress, J. F. (2001). *Principles of biomedical ethics* (5th ed.). New York, NY: Oxford University Press.

Broström, L., Johansson, M., & Klemme Nielsen, M. (2007). 'What the patient would have decided': A fundamental problem with the substituted judgment standard. *Medicine, Health Care, and Philosophy, 10*, 265–278. doi:10.1007/s11019-006-9042-2

Brown, P. G. (1975). Informed consent in social experimentation: Some cautionary notes. In Rivlin, A. M., & Timpane, P. M. (Eds.), *Ethical and legal issues of social experimentation* (pp. 79–104). Washington, DC: The Brookings Institution.

Buchanan, A. E., & Brock, D. W. (1989). *Deciding for others: The ethics of surrogate decision making*. Cambridge, UK: Cambridge University Press.

Capron, A. (1982). The authority of others to decide about biomedical interventions with incompetents. In W. Gaylin & R. Macklin (Eds.), *Who speaks for the child?* (pp. 115-152). Hastings-on-Hudson, NY: The Hastings Center.

Dawson, A. (2004). What should we do about it? Implications of the empirical evidence in relation to comprehension and acceptability of randomisation. In Holm, S., & Jonas, M. (Eds.), *Engaging the world: The use of empirical research in bioethics and the regulation of biotechnology*. Amsterdam, The Netherlands: IOS Press.

Dennis, M. (1997). Commentary: Why we didn't ask patients for their consent. *British Medical Journal, 314*, 1077.

Dickert, N. W., & Sugarman, J. (2006). Community consultation: Not the problem--an important part of the solution. *The American Journal of Bioethics, 6*(3), 26–28. doi:10.1080/15265160600685762

Dworkin, G. (1988). *The theory and practice of autonomy*. Cambridge, UK: Cambridge University Press.

Elks, M. L. (1993). The right to participate in research studies. *The Journal of Laboratory and Clinical Medicine, 122*(2), 130–136.

Faden, R. R., Beauchamp, T. L., & King, N. M. P. (1986). *A history and theory of informed consent*. New York, NY: Oxford University Press.

Feinberg, J. (1986). *The moral limits of the criminal law*. New York, NY: Oxford University Press.

Foex, B. A. (2001). The problem of informed consent in emergency medicine research. *Emergency Medicine Journal, 18*(3), 198–204. doi:10.1136/emj.18.3.198

Fost, N. (1998). Waived consent for emergency research. *American Journal of Law & Medicine, 24*(2-3), 163–183.

Greenberg, D., Shroder, M., & Onstott, M. (1999). The social experiment market. *The Journal of Economic Perspectives, 13*(3), 157–172. doi:10.1257/jep.13.3.157

Hansson, M., Dillner, J., Bartram, C. R., Carlson, J. A., & Helgesson, G. (2006). Should donors be allowed to give broad consent to future biobank research? *The Lancet Oncology, 7,* 266–269. doi:10.1016/S1470-2045(06)70618-0

Hansson, S. O. (2004). The ethics of biobanks. *Cambridge Quarterly of Healthcare Ethics, 13,* 319–326. doi:10.1017/S0963180104134038

Hansson, S. O. (2006). Informed consent out of context *Journal of Business Ethics, 63,* 149–154. doi:10.1007/s10551-005-2584-z

Hayry, M., & Takala, T. (2001). Genetic information, rights, and autonomy. *Theoretical Medicine and Bioethics, 22*(5), 403–414. doi:10.1023/A:1013097617552

Helgesson, G., & Johnsson, L. (2005). The right to withdraw consent to research on biobank samples. *Medicine, Health Care, and Philosophy, 8,* 315–321. doi:10.1007/s11019-005-0397-6

Jones, S., Davies, K., & Jones, B. (2005). The adult patient, informed consent and the emergency care setting. *Accident and Emergency Nursing, 13*(3), 167–170. doi:10.1016/j.aaen.2005.05.005

King, N. M. P. (1999). Medical research: Using a new paradigm where the old might do. In King, N. M. P., & Henderson, G. E. (Eds.), *Beyond regulations: Ethics in human subjects research.* Chapel Hill, NC: University of North Carolina Press.

Kraybill, E. N., & Bauer, B. S. (1999). Can community consultation substitute for informed consent in emergency medicine research? In King, N. M. P., & Henderson, G. E. (Eds.), *Beyond regulations: Ethics in human subjects research.* Chapel Hill, NC: University of North Carolina Press.

Lindley, R. I. (1998). Thrombolytic treatment for acute ischaemic stroke: Consent can be ethical. *BMJ (Clinical Research Ed.), 316*(7136), 1005–1007.

Marshall, P. A., & Berg, J. W. (2006). Protecting communities in biomedical research. *The American Journal of Bioethics, 6*(3), 28–30. doi:10.1080/15265160600685770

Martin, M. W., & Schinzinger, R. (1989). *Ethics in engineering* (2nd ed.). New York, NY: McGraw-Hill.

Milgram, S. (1963). Behavioral study of obedience. *Journal of Abnormal Psychology, 67,* 371–378. doi:10.1037/h0040525

O'Neill, O. (2003). Some limits of informed consent. *Journal of Medical Ethics, 29,* 4–7. doi:10.1136/jme.29.1.4

Palmer, R. B., & Iserson, K. V. (1997). The critical patient who refuses treatment: An ethical dilemma. *Ethics of Emergency Medicine, 5*(5), 729–733.

Schultze, C. L. (1975). Social programs and social experiments. In Rivlin, A. M., & Timpane, P. M. (Eds.), *Ethical and legal issues of social experimentation* (pp. 115–125). Washington, DC: The Brookings Institution.

Schwab, A. P. (2006). Formal and effective autonomy in healthcare. *Journal of Medical Ethics, 32*(10), 575–579. doi:10.1136/jme.2005.013391

Svensson, S., & Hansson, S. O. (2007). Protecting people in research: A comparison between biomedical and traffic research. *Science and Engineering Ethics, 13,* 99–115.

United States National Bioethics Advisory Commission. (1998). *Research involving persons with mental disorders that may affect decisionmaking capacity.* Retrieved from http://bioethics.georgetown.edu/nbac/pubs.html

United States National Commission for the Protection of Human Subjects of Biomedical and Behavioral Research. (1978). *The Belmont report: Ethical principles and guidelines for the protection of human subjects of research.* Bethesda, MD: United States National Commission for the Protection of Human Subjects of Biomedical and Behavioral Research.

ENDNOTES

[1] There are several possible terms for *decisional capacity*, the main alternatives being *competence* or *decision competence*. In this paper these terms are used interchangeably.

[2] Faden et al. (1986) seem to favour this, as they propose methods of ensuring that patients or research subjects have understood the information disclosed by the physician or doctor (pp. 326-327). They do however, also argue in favour of *substantial* as opposed to *full* understanding – meaning that what matters is that the patient/subject is provided with and understands information that is material to him or her, as opposed to information that is considered relevant by others (e.g., p. 302). The testing of understanding refers, or so I gather, to that latter kind of information and not to that (which may be none!) which the patient/subject finds important.

[3] One hugely important question, of course, is whether the practice of informed consent actually succeeds in its aims. There are certainly reasons to investigate this matter, but the present paper makes no contribution in this respect as it is limited to discussing cases where the practice of informed cannot be used at all.

4 This example was provided by Peter Bonsall at The First Conference on the Ethics of Traffic and Transportation, the Royal Institute of Technology, Stockholm, Sweden, 29-30 November 2005. After receiving replies, the researcher made sure that those who answered were all connected so that they could start up a car pool.

[5] cf. Buchanan and Brock (1989, pp. 26-28) who argue that while competence (or capacity) is really a matter of degree, it is more reasonably to be viewed as a threshold concept. This means that what matters is whether the person is competent enough to decide on a certain matter.

[6] The case of mental impairments, when a person is to be considered incapable of providing informed consent, and what substitutes are appropriate in different situations are questions thoroughly discussed in the report "Research Involving Persons With Mental Disorders That May Affect Decisionmaking Capacity" by the National Bioethics Advisory Commission in the US, and the papers commissioned to the report (United States. National Bioethics Advisory Commission, 1998).

[7] Note however, that also mentally impaired persons may be capable of making many decisions in their everyday life, and yet be incapable of making difficult decisions about research participation (see for example Palmer & Iserson, 1997, p. 731; Buchanan & Brock, 1989, pp. 18-19).

[8] Note that it may be to the disadvantage of groups lacking the capacity to form autonomous choice to be excluded from research participation. Such participation may be needed to adapt treatments to their needs – see Svensson and Hansson (2007, p. 109) and Elks (1993).

[9] See for example Jones et al. (2005).

[10] Hansson (2006, pp. 151-152) discusses this problem under the label of "consenter identification problem". Martin & Schinzinger (1989, p. 67) mention the problem in relation to engineering projects, which they frame as experiments.

11 Note, however, that it is not obvious that the same informed consent requirements should be applied in cases where biological material are the objects of investigation as when individuals themselves are the subjects of research (Helgesson & Johnsson, 2005, p. 315).

12 To support this claim – see Martin and Schinzinger (1989, p. 68), who write that people in general are often unwilling to be informed of engineering projects.

13 Hansson (2004, p. 321) discusses this in relation to biobanks.

14 A related question arises in cases where (all) members of a certain group are to participate in a study, while (all) members of another group are not to participate but may still be negatively affected by it. Brown (1975, p. 95) discusses the example of certain social experiments, for instance on the effects of housing allowances, where the group not participating does not receive the benefit, and hence is made *relatively* worse off than before. Would it be appropriate to ask also members of this group for their consent, given that we could?

15 Social experiments are described by Greenberg et al. (1999) as "field studies of social programs where individuals, households or (in rare instances) firms or organisations are randomly assigned to two or more alternative policy interventions" (p. 157). They note that participation in such experiments can be either voluntary or mandatory (p. 160), and out of the cases beginning after 1982 which the authors have investigated, almost 40% of the experiments were mandatory.

16 The same problem arises of course in areas outside of research. O'Neill (2003, p. 4) mentions it in relation to the selection of public health policies.

17 cf. Feinberg's (1986, pp. 199-228) discussion on this.

18 Other typical examples of there being no acceptable alternatives are when treatments for diseases are tested, and there are no other treatments available for the patient than the one to be tested. Such cases do not belong to this category, however, because the reason for there being "no choice but to participate" is not the collective nature of the research project. Furthermore, in these cases the voluntariness of potential participants is compromised because research participation is extremely tempting, and not because it is preferable to avoid participation. This is related to the discussion on coercive offers, and to the idea of voluntariness in general. See for instance Feinberg (1986).

19 This is not the same thing as in category 4 where it may be very costly for an individual to avoid research participation. For such cases, informed consent is not present even if there may be individuals who muster the strength to decline participation, because this decision would arguably still not have been voluntary enough.

20 The method of information provision is sometimes used in Swedish traffic research. Apart from newspapers, information is also provided by means of signs beside the roads, pamphlets distributed to households etc.

21 Despite this, the method of debriefing is sometimes inappropriately called *post-debrief consent*. In some studies, however, there is an element of consent involved if people, after being informed, are asked to consent to their data remaining in the study or if it is to be withdrawn. This is not always an option, for instance when group dynamics or interactions are studied. Then the input of a particular individual cannot be singled out and removed.

22 In the Belmont Report (United States. National Commission for the Protection of Human Subjects of Biomedical and Behavioral Research., 1978, C:1), this type of method

of incomplete disclosure of information is discussed.

23   Note that there is a fundamental difference between informing potential participants that there is information that will not be disclosed, and providing false information about the study as did Milgram (1963) in his obedience study. The decision on whether to participate or not, can hardly be seen as consent if based on false information.

24   Hansson (2004, p. 322) uses the term *blanket consent*, while Hansson et al. (2006) use the term *broad consent*.

25   Fost (1998) mentions this in the context of brain injury.

26   A similar problem, but in the context of the substituted judgment standard for proxy consent, is discussed by Broström et al. (2007).

27   Foëx (2001, p. 201) argues that *only* in such cases is proxy consent at all valid. Sometimes this type of "delegated consent" is what is meant by "waived consent", but the latter term can also include the person's letting go of the consent requirement without appointing anyone specific to decide instead, also in cases where the person is not incompetent but only wishing to avoid decision.

28   Dworkin (1988) raises this issue on pp. 98-99, and concludes that "we need a theory of what it is to respect an incompetent person as a person" (p. 99).

29   Capron (1982) is using the labels of *substituted judgment doctrine* and *best interest doctrine*. Dworkin (1988, pp. 91-92) relates these categories to other categorisations of the notion of proxy representation.

30   Such procedures could also be called "community consultation", although this concepts seems to most often mean not that the community members are involved in the decision making process, but rather to denote a tool for research ethical committees to identify relevant interests and concerns of certain groups (as well as their members) who may take part or will be asked to take part in a certain study. For discussions on the appropriate use, if any, of community consultation, se for example King (1999), Dickert and Sugarman (2006), Marshall and Berg (2006), and Kraybill and Bauer (1999).

31   Schultze (1975, pp. 122-124) also discusses the legitimacy of such a model but under the label of "institutional consent", in the context of social experiments where schools and other institutions consent to research participation on behalf of all its members/clients/customers.

32   Martin and Schinzinger (1989, p. 69) propose this kind of method, which they see as a form of proxy consent.

33   Hansson (2006, p. 151) discusses this option. The method implies that there would be one person representing the interests of all participants, but the representative would still not necessarily know each and every individual involved.

This work was previously published in the International Journal of Technoethics, Volume 2, Issue 3, edited by Rocci Luppicini, pp. 48-61, copyright 2011 by IGI Publishing (an imprint of IGI Global).

Chapter 21
The Middle Ground for Nuclear Waste Management:
Social and Ethical Aspects of Shallow Storage

Alan Marshall
Asian Institute of Technology, Thailand

ABSTRACT

The 2001 terrorist attacks in the USA and the 2011 seismic events in Japan have brought into sharp relief the vulnerabilities involved in storing nuclear waste on the land's surface. Nuclear engineers and waste managers are deciding that disposing nuclear waste deep underground is the preferred management option. However, deep disposal of nuclear waste is replete with enormous technical uncertainties. A proposed solution to protect against both the technical vagaries of deep disposal and the dangers of surface events is to store the nuclear waste at shallow depths underground. This paper explores social and ethical issues that are relevant to such shallow storage, including security motivations, intergenerational equity, nuclear stigma, and community acceptance. One of the main ethical questions to emerge is whether it is right for the present generation to burden local communities and future generations with these problems since neither local peoples nor future people have sanctioned the industrial and military processes that have produced the waste in the first place.

INTRODUCTION TO SHALLOW STORAGE

Nuclear waste comes in a variety of forms, ranging from low level radioactively-contaminated materials, such as radiated clothing and equipment, to high level radioactive waste such as spent nuclear fuel rods and plutonium. The high-level waste is incredibly dangerous; if a human were to ingest just micrograms, Fetter and von Hippel (1990) point out, it would probably be a fatal dose. The low-level waste is also very dangerous if not contained properly since its escape into the human environment could cause large increases in cancer rates (Perlic, 1990).

DOI: 10.4018/978-1-4666-2931-8.ch021

Technical approaches about how to store or dispose of nuclear waste stress the importance of the following question: 'where do we put it?' Most of the time, the proposed answer to this question involves discussions about how to isolate the waste from the human environment. Suggestions in this matter are numerous and varied (NRC, 2001; Alexander & Mckinley, 2007) from burying it deep below the Earth (usually two to four kilometres), to dispersing the waste far away into open ocean spaces, or even launching it into extraterrestrial space.

All of the above are problematic for technical reasons, political reasons, managerial reasons and ethical reasons, and all the above should be approached within the scope of technoethics, since they involve problems about a) technologically created pollution, b) unevenly distributed environmental risk, and c) democratically-dubious technical decision-making. For instance, the decision to bury nuclear waste more than a few hundred metres below the surface will probably subject future peoples and the local environment to the vagaries of untried technologies, uncertain geological knowledge and probable hydrological return of radioactive liquids. Dispersing waste into the oceans would undoubtedly contaminate marine life and probably devastate some ocean resources. Blasting nuclear waste into space makes the human environment vulnerable to a Challenger or Columbia-type accident that would create radioactive fall-out in hemispheric proportions.

Given the problems noted above, an alternative answer to the question 'where do we put it?' might rely less on final disposal--which can never fulfill its promise to permanently isolate radioactive waste from the human and natural environment--but to leave it on the surface where it can be watched and taken care of. This solution keeps the waste visible and actively-managed instead of hidden and buried away where it may do unpredictable things in the near or long-term future. This 'surface storage' idea was often-supported waste management option until a) the 2001 ter-

rorist attacks in the USA, and b) the 2011 seismic events in eastern Japan. These events showed up the vulnerabilities of nuclear facilities located on the surface. Before these events, though, the real reason that surface stores were popular and supported usually revolved around their relative ease of construction and lower cost rather than some higher ideals to do with good management of environmental risk.

It should probably be considered quite strange that surface stores have become seen as 'very risky' only in the past 10 years since, throughout the history of the 'Atomic Age', numerous accidents have occurred at surface storage sites. A very short list could include the following events:

- The 1957 Kyshtym nuclear waste accident, Russia (in which a cooling system failure in a waste facility failed, resulting in massive and widespread radioactive contamination),
- The 1966 US Air Force nuclear accident, Spain (in which nuclear material was spread about the Spanish coastal countryside as a result of two planes colliding, before being buried—both physically and politically),
- The 1967 Mayak nuclear waste accident, Russia (in which unsecured radioactive dust blew off of a containment site before contaminating large populated areas).
- The 1972 West Valley nuclear waste accident, USA (in which 600,000 gallons of high-level wastes leaked into the environment, including Lakes Ontario and Erie).
- The 1973 Hanford nuclear waste accident, USA (in which thousands of cubic meters of radioactive waste flowed out of a nuclear weapons complex),
- The 1981 Muroroa nuclear waste accident, French Territory (in which a tornado washed nuclear waste from a French nuclear site into a Pacific lagoon).

- The 1985 Karlsruhe nuclear waste accident, Germany (in which a barrel of nuclear waste went up in flames),
- The 2010 Hunterston nuclear waste accident, Scotland (in which radioactive liquid leaked into the Firth of Clyde).

So, if not on the surface (and if not deep underground) then yet another option must emerge as we ask 'where do we put nuclear waste?' An up and coming suggestion is 'we should put it somewhere in the middle'; somewhere between surface storage and deep disposal. This is the 'shallow storage' option. A shallow store would be situated between 20 to 100 metres underground; deep enough to evade most terrorist attacks and tsunamis, it is said, but shallow enough to keep an eye on.

Shallow storage goes by various names, the technical purists at the United Nations' International Atomic Energy Agency (IAEA) would call it the 'near surface storage below ground level' option (EKRA, 2000) but this label hardly reads very easily so I have chosen to use the term 'shallow storage' in this paper.

Options for nuclear waste management can be divided into two broad categories; those that require on-going care and maintenance and those which are supposed to be safe without ongoing management and therefore do not require human intervention. This option, shallow storage, fits into the former category rather than the latter. The level of care and management required would depend on the exact design, of course, but as they are construed as 'stores' and not 'disposal sites' shallow stores have been proposed as being operationally open (Smith, 1998); i.e. to require monitoring and to allow for retrievability of the stored waste sometime in the future.

SHALLOW STORES AND BURDENS FOR FUTURE GENERATIONS

There is shared agreement within many nuclear waste organisations around the world that the waste producing generation should work to minimise the burdens of long-lived nuclear waste on future generations. What this actually means in practice is the subject of intense debate, however. Both surface storage and deep disposal are advocated by their respective proponents as being better systems to avoid passing on burdens to future generations. Deep disposal, for instance is fashioned as a permanent solution that doesn't burden future generations with ongoing management hassles. Surface storage, however, is a way to avoid burdening future generations with the limitations of early 21st century tools and techniques by allowing future generations to implement their own nuclear waste management plans. It is also a retrievable option. If something goes wrong with the waste or its containers, as may well happen over the future history of the waste, it can be far more easily retrieved from a well-maintained shallow store than an abandoned deep disposal site.

Because the waste *has been* produced, however, and because this waste will be dangerous for so long, there will be burdens no matter which system of management is planned. Deep disposal burdens future generations with an ever-degrading system of containment that will eventually breakdown and let radioactivity back into the environment. It also locks the future generation into inheriting an unchangeable system of disposal with an irretrievable waste. Surface storage on the other hand, means that future peoples will be burdened with physically and financially managing the waste forevermore. If they don't manage it, then the system of containment will degrade and radioactivity is likely to be released into their environment. The more optimistic nuclear waste professionals feel that the burdens can be minimized in one or other of the systems through good science, good management and good engineering. Not all nuclear waste professionals are optimistic, however.

The shallow store option, as the name implies, is a storage system, and so the burdens it will pass on to future generations revolve most closely around the issues of on-going management. Shallow stores are designed to be continuously monitored and managed, and will probably need to be as refurbished every few decades, and, if used over a period of more than a hundred years, it is probable, according to Ellis (2004) they will also need to be re-sited due to degradation of the subterranean environment. If shallow stores are envisaged as indefinite solutions, this ongoing attention will have to be projected far into the future. For instance, for shallow stores to last for the entire lifetime of nuclear waste then ongoing attention will have to be maintained for up to one million years.

Whether utilised over a hundred year period or for the full million years, the management burden associated with shallow stores will come in various forms:

- **Administration/Record Keeping Burdens:** This includes the problem of information extinction. It is probable that changing regimes of information storage and retrieval may result in the loss of information about the quantity, characteristics and location of the waste, and the best way to manage it technically. Add to this the myriad of external problems that a record system may face over a long period of time (fires, floods, theft, vandalism etc) and it seems likely that some, if not most, of the current information about the waste is sure to be lost over its lifetime.

- **Financial Burdens:** Someone will have to pay for a) ongoing technical maintenance, b) record-keeping, c) for any necessary research and development, and also for d) compensation to injured parties. Surely, the producing generation should do this but given the vagaries of future circumstances and given the various opinions over whether the waste is likely to be entirely safely contained or not, any calculation of the exact amount left in a trust fund to enable all of the above is likely to be widely divergent from the amount needed.

- **Environmental and Health Risk Burdens:** Future generations and their environments will have increased chances of exposure to radiological and non-radiological hazards.

Although individual shallow stores are expected to last for a lifetime of only 50 to 100 years, over this time, significant changes in society and policy may affect the operation of the facility. Some changes may enhance the operation of a shallow store (new skills, better understanding of risks, greater financial security, better environmental management, increased scientific and technological development) but some may degrade operations (funding cuts, extinction of information, military and terrorist interference). Nuclear waste storage facilities may be more susceptible to social and political change than nuclear waste disposal facilities but a store some 50-100 meters below the surface has a greater degree of geological protection than a surface store (so that if abandonment or neglect comes about due to social or political change, then the waste may be more likely to be contained for a longer period). Problems start to arise however if you begin to rely on this geological protection to do a job it was not intended to do; to isolate the waste from the biosphere permanently.

Related to the problem of social and political change and its impact on future nuclear waste management is the changing skill-base in nuclear operations. Refurbishment of the stores, replacement of the stores, and the need to repackage the waste periodically will require expertise in nuclear engineering. In nations like the United Kingdom and Germany there is already concern that the nuclear workforce is aging and that few young people are joining the industry. The shortage of new recruits is something felt already in the nuclear industry in Europe and this could cause problems with decommissioning and waste management in the next few years.

Looking into the far future, when the nuclear industry will likely have collapsed for whatever reason, this may be an even greater problem. This not meant to be an argument for the retention of nuclear power and nuclear weapons, however. The need to keep an actively trained nuclear waste workforce can probably be done via the environmental remediation and environmental engineering professions rather than through nuclear operations. Also, if nuclear operations are phased out, this obviously ceases the production of ever-increasing volumes of nuclear waste that has to be dealt with.

The ethical question that emerges from considering the long-term nature of nuclear waste is whether it is right for the present generation to burden future generations with the problems listed above since the future generation has not directly benefited from, nor has it sanctioned, the industrial and military processes that have produced the waste. Even if we dare to suggest that future generations should be thankful for the greater economic and technological legacy we have left them due to our investment in nuclear energy and nuclear weapons, we should still feel both regret and responsibility that these nuclear operations will leave a waste that we know is going to be extremely hazardous and enormously difficult to manage for local communities for many generations to come. Be that as it may, the waste is here and we have to plan what to do with it. Shallow storage requires ongoing management and it is an open question whether we can trust in such management being available as we project our plans into the future. However, shallow storage also enables retrievability if something goes wrong, and since there is a strong possibility of something going wrong in the future, this option may serve future generations better than a supposed final solution like deep disposal.

FUTURE UNCERTAINTIES

Discussion about future burdens, and the best way to manage them, must also involve a discussion of future uncertainties. With regards to shallow stores there are a large number of uncertainties that affect our appraisal of their suitability to deal with nuclear waste burdens.

Firstly, we have to acknowledge political uncertainties over the lifetime of an initial shallow store (i.e.: over a 50 to 100 year period). On the local community scale, when a host community is being asked to consent to a shallow store we have to ask the question: what exactly is the host community consenting to? How risky is it that the nature of the store will change over the initial 50 to 100 year period into a facility which transgresses its original design and function? Despite lengthy lists of promises, plans, design specifications and political deals, how confident should anyone in this generation be that any provision to monitor, refurbish and re-site a store (and to retrieve the waste if something goes wrong) will not be changed after 50 or so years to effect a more convenient (or less expensive) solution in the future? Here in the early 21st century, we can confidently state that our shallow store designs will be perfectly safe and manageable just as long as the technical operations we lay out are followed over the next century. But should we really be confident that such technical operations *will* be followed? This cautionary question suggests that when we ask a community to consent to a safe and temporary shallow store we may be asking them to do a lot more than that.

Shallow stores also possess a number of technical and scientific uncertainties. Most of these are common to all land-based nuclear waste management options. For example:

- **Engineering Uncertainties:** For example, the waste containers may corrode faster than we think, the ventilation of the subterranean environment may be more complicated than we have figured, the emplacement and removal of the waste may be more difficult and expensive than we have estimated, and monitoring the waste may prove both hazardous and haphazard.

- **Geological Uncertainties:** For example, the rocks surrounding the store may not be as strong and uniform as we estimate, the underground water-flow may not be behave as we believe, the landforms around the site may not be as stable as we guess. The impact of earthquakes and tsunamis on underground stores may be underestimated.
- **Chemical Uncertainties:** For example, the interaction of chemicals within the waste containers and the interaction of the chemistry of the surrounding geological environment with the container may produce unforeseeable events: increased container degradation for example or excess localized heat or pressure.
- **Biological Uncertainties:** For example, microbial activity may degrade the store environment or it may alter the chemistry of the containers to degrade the entire system. Pests may invade the shallow store and re-emerge upon the surface in a radioactive form.

This rather condensed list of uncertainties shows, in a small way, that estimating both the reliability and expected burdens of shallow stores is fraught with unknown factors.

SECURITY IN SHALLOW STORAGE

As introduced above, shallow storage can be interpreted as a balance between two driving forces in the current social and political climate; the desire to have a monitored facility, so the waste can be observed, versus the desire to protect the facility and its contents from near-term malicious human interference. This balance has become of interest in the post 9/11 world where a lot of attention (Allison, 2003; Medalia, 2003) has been focused on the way nuclear facilities, including nuclear waste facilities, may be subject to two types of intentional human interference:

1. Direct attack (via aircraft, missiles, or bombs) or
2. Theft of materials.

The second of these, theft, can only be mitigated by some use of active security. For instance, according to the NRC Committee on Disposition of High-Level Radioactive Waste Through Geological Isolation:

The security of surface or near-surface storage can be achieved by restricting access of individuals and groups that might divert fissionable materials for weapons use or use radioactive material for acts of terrorism. (NRC, 2001, p. 115)

Active security necessitates ongoing management. Shallow stores will require ongoing active security throughout the lifetime of the facility and future generations will have to have the resources available to enable this.

Direct attack, on the other hand can be countered by the passive security of the waste lying 50 to 100m meters underground. Direct attack might still affect damage to surface buildings, however; offering a path to the human environment for gaseous radionuclides or radioactive particulate matter that may have escaped from their containers. While some have commented that such 'nuclear terrorism' takes advantage more of the fear it will induce within the public than the physical damage it may cause (UN Wire, 2003), others point out that a direct attack on some nuclear waste storage facilities will result in both large-scale loss of life and massive regional environmental contamination (Pugwash Council, 2003).

The UN IAEA has declared that:

Putting hazardous materials underground increases the security of the materials. (IAEA, 2003.

If security is a major driver for nuclear waste to be located underground, it may be of little advantage if:

A. There is continued operation of other vulnerable nuclear operations such as nuclear power plants.

B. A shallow store system will involve a lot of transportation.

Both 'a' and 'b', above, are, of course, quite likely. Near surface stores may indeed decrease the risk of nuclear terrorism but unless all nuclear power plants are decommissioned, the risk of nuclear terrorism will still be rather high.

With regards to 'b', the extra physical security that a shallow store offers over surface stores may mean nothing if the waste has to be transported upon the backs of easily targeted trucks and trains for large distances. The transportation phase of nuclear waste management is an extremely vulnerable phase no matter what storage or disposal option is chosen (Abt, 2003).

NUCLEAR STIGMA SURROUNDING NEAR SURFACE STORAGE

If we acknowledge that public acceptability is a difficult matter whatever type of nuclear waste facility is planned then it is likely that the problem of nuclear stigma will emerge during the planning and execution of a shallow store. Nuclear stigma is a term that acknowledges the sensitivity of nuclear host community members to negative stereotyping emerging externally about their community. This concern for what outsiders think may then be internalized so that local members stigmatize themselves above and beyond external attitudes.

In the UK, nuclear stigma has been identified by the UK-based Gosforth Parish Action group as having identifiably negative economic consequences. The following comments, which were made to the House of Lords Select Committee on Science and Technology (1999) refer to the thoughts of this group:

New industry was reluctant to set up nearby and one food firm had to close because customers elsewhere in the UK refused to buy products coming from near Sellafield.

This pattern has been repeated in elsewhere in Europe (Sundqvist, 2006) and also in the North America (Nuclear Energy Agency, 2003).

Nuclear stigma in the case of a shallow storage facility may be slightly different in nature and intensity compared to other options like surface storage and deep disposal. For example, because it is a storage option, the national consciousness within a nation may not imagine the community surrounding the store as being forever tainted by the waste.

Maybe, however, it will do the opposite; a surface store may be perceived by potential host community members as a double nuisance involving disruption of the community in two distinct stages; during the initial build and then reconstruction or re-siting at the end of the 50 to 100 year lifetime of the first store. If the initial construction is marred by public inconvenience, safety problems, time and cost over-runs, then this may be an unsettling prospect for a community committed to doing it all again in the future.

The issue of refurbishing or re-siting a shallow store may also be regarded in a positive way by people within the community who perceive that their family members, or their community, will benefit from the ongoing (re)construction. This benefit is off-set into the future but people concerned about the continued economic prosperity of their community may perceive future industrial nuclear activity as a positive thing. For example, McCutcheon (2002) recounts that the community leaders raised this as a pro-facility argument when they briefed both their community members and a congressional committee during the early stages of a New Mexican nuclear waste facility.

OUT OF SIGHT-OUT OF MIND

The underground aspect of a shallow store may also affect the intensity of community stigma. Some community members may be impressed by the extra geological barriers guarding against radiological escape. This confidence in the safety aspect of the project may well salve their worries. Other community members, though, will fear that a certain degree control over the waste is forfeited if a store goes underground; thus raising fears of future contamination and hidden health risks.

The pre-facility land use patterns may affect the intensity of community stigma of underground options as well. Stigma may be less likely in a site that has either good knowledge of nuclear facilities or has experienced safe economic use of underground industries (i.e.: mining). Certainly, though, this is not always the case since some of the most vehemently-expressed safety worries are often raised by people working within these industries (Martin, 1995).

However, in a region that depends heavily on untainted land (such as agriculture and tourism) then waste hidden within that land may be regarded with more suspicion than waste stored at the surface.

One conceptual issue associated with shallow stores which might interact with issues of nuclear stigma is the 'Out of Sight--Out of Mind' attitude that some environmentalists believe pervade all underground options. Storing dangerous waste underground for the anticipated length of time that shallow storage calls for, raises the problem that some members of the public will consider shallow storage as 'waste burial'. While shallow storage calls for continuous monitoring and, by definition, is not a form disposal, it may well be interpreted as an attempt to hide the waste and forget the all about it.

The Nuclear Guardianship Project (NGP, n. d.) founded by ecophilosopher Joanna Macy, advocates citizen-initiated, monitorable retrievable storage of nuclear waste, envisioning that such

storage will keep the fires of community care and awareness burning. According to the NGP, indefinite surface stores will encourage what NGP members refer to as 'Guardianship', the continued retelling of stories regarding the dangers and meaning of nuclear waste, and the continued citizen involvement with managing that waste.

It is not entirely clear whether stores located at 50 to 100 metres underground would fulfill the requirements of 'guardianship'. On the face of it, shallow storage is a concept that enables monitorability and retreivability so it would seem to be a Guardianship-like concept. On the other hand, the Guardianship status of shallow stores would depend a lot on three things:

1. The confidence that the store would not turn into a *de facto* disposal facility.
2. The confidence that monitoring and retrievbility will be up to the same standard as surface stores.
3. The dependence on visibility of a store to foster citizen-initiated care. Given that stores will have surface buildings, there will be a visible component. But as this building contains no actual waste, it could be interpreted as just a proxy, no more real to the citizenry than radioactive waste management files located in far-away offices. The Guardianship proponent Thea Bauriedl (a psychoanalyst) says, for instance:

By storing the waste where we can keep an eye on it, we also keep the danger, and the guilt it generates, from being suppressed (Bauriedl, 1991).

From the perspective of the Bauriedl comment, shallow storage may not be a Guardianship concept since we "can't keep an eye on it". Perhaps this is too literal an interpretation of Guardianship (but it is an interpretation that acknowledges that we live in an increasingly visual culture, one which privileges the eye over other senses in all aspects of social life). If we don't take Bauriedl so literally, shallow storage might still be problematic since it involves the delivery of the waste to a point where humans would not normally consider living and working; the waste may therefore be taken out of the community care.

If there is an 'out of sight--out of mind' issue with regards to shallow storage--yet it is decided that it must go ahead to satisfy security concerns--then, to satisfy Guardianship values, there has be a way of keeping the stored wastes 'in mind'; encouraging stories about underground waste in the popular and political culture. A store is, by definition, a 'work in progress'; with all the positive and negative connotations that that may imply. Optimistically, this would suggest stores are adjustable, on-going and involving, but, negatively, they are also unfinished and may be interpretable as being a dirty and dangerous procrastination.

AESTHETIC ASPECTS OF SHALLOW STORES

The aesthetics of large engineering projects concern not just the way they may look, but the way they reflect, deepen or challenge people's values and images of the sensual features of their landscape and community. A shallow storage facility imposes a certain functional relationship with the landscape. There needs to be a vault in the ground in which to put the waste and there needs to be a building above this vault in order to facilitate operations.

Some have expressed a concern that an unimaginative and intensive devotion to function in the design of such a store will not do anything to garner public acceptance. Professor Eric Van Hove, of the University of Antwerp, for instance, suggests that creative design has to be included in any nuclear waste facility plan, such that the facility fosters a positive response within the community (Van Hove, 2003). There are three broad approaches to this issue:

- To blend the shallow storage site into the surroundings (proceeds on the basis that a facility should not subtract from the aesthetic value of the landscape),
- To valorise the site and celebrate the place it has in culture (proceeds on the basis that a facility should add value to the landscape),
- To demonise the site and warn people away (proceeds on the basis that aesthetics should be devoted to ensuring safety and security).

Any one of these approaches may be appropriate for a particular shallow storage project but unlike a disposal option they will be have to be workable within a 50 to 100 year framework which:

- Acknowledges the possibility of disturbance or wholesale destruction of the local landscape if re-siting is needed.
- Allows the landscape aesthetics to be reassessed according to each succeeding generation.
- Allows communities to adjust their relationship with the site if safety and security issues change.

One aesthetic difference between stores and final repositories is that the latter may tend to presuppose the 'closedness' of the aesthetic design of a facility, forging an assumption or hope that the aesthetic components will remain intact for some time approximating the radioactive lifetime of the waste, for instance, the symbolic sculptures surrounding the low-level waste radioactive facility known as 'WIPP' in New Mexico have been designed in this way (SNL, 1992). Stores on the other hand, may encourage re-assemblage of the aesthetic components on a rolling basis in an institutionalised way. This may lead to advantages by encouraging each successive generation to be involved with the cultural management of the waste (something which will lead to better physical management of the waste).

Another point that has to be raised as far as aesthetics is concerned is that the shallow store option involves two elements. As well as the buildings or landforms that may cover the shaft, an essential functional part of a shallow store is the subsurface part. This part must be subjected to aesthetic investigation as well since human cultural relationships to the subterranean environment are likely to change over the million year lifetime of the waste. A deep disposal option could not adjust to this as readily as a shallow store.

SHALLOW STORES AND THE DEVELOPMENT OF FUTURE SCIENCE AND TECHNOLOGY

One of the arguments for storage options of all kinds is the fact that some form of delaying tactics to final disposal will give the opportunity for nations, societies and researchers to develop better technological solutions to the nuclear waste problem. This argument reflects a belief in techno-evolutionary optimism which is imbued within much of modern culture; especially the culture of the professional classes. According this point of view, given that society is always advancing and progressing in a scientific, technological and economic way, radioactive waste managers shouldn't be in too much of a hurry to invent a solution because any generation that comes after us will provide a better one. This is not an argument for shallow stores so much as it is an argument for all types of stores.

Arguments against this techno-optimistic waiting game include the following:

- The development of technology doesn't progress in a linear way (skills and technological capability can decrease as well as increase over time)
- Economic, demographic, and resource uncertainty may make it more difficult

for future generations to solve these problems despite advanced technological development,

- Just because future generations MAY be able to solve the problem better than present generations, this does not mean they SHOULD solve the problem (Shrader-Frechette, 1993)

PUBLIC PARTICIPATION VIA COMMUNITY SITE RESTORATION OF A SHALLOW STORAGE FACILITY

To increase social and political acceptability of waste management projects, many organisations involved in hazardous waste management have acknowledged the importance of public participation. By this reckoning, in order to foster project success and to foster ongoing management over the long lifetime of a shallow storage project, the facility has to be thoroughly embedded in the community context.

Further to this, it has been advanced by a number of researchers (Light, 2003) intimate that to get people really involved in the management of a long term environmental project in their area you have to involve the community in three important aspects of the project:

A. Pre-operational decision-making in siting and management strategy
B. Design of the facility
C. Active involvement in caring for the waste so as to keep it from harming the environment.

This last aspect may be deemed community site restoration. This aspect in itself would involve three factors:

1. Information upkeep and administration
2. Upkeep of cultural aspects of the environment.
3. Physical upkeep of the restored site.

Generally, the goals of participatory site restoration is to encourage those members of a community affected by a degraded or contaminated site to democratically agree about a design for how the site should be utilised in the future, and to work out how to get to this desired point, and then become involved in the physical work to do this. When it comes to hazardous waste sites, including shallow nuclear waste stores, important safety issues may preclude certain options (such as green field restoration) but participatory site restoration presupposes that the safety issues are worked out via close collaboration between the community participants, design experts and safety experts (to ascertain what is and is not possible according to site safety criteria).

From the perspective of garnering community support, site restoration will involve communities in adding value to their landscape, repelling stigmatisation, maintaining community stories of care and management (a cultural project that Guardianship concept tries to foster) and integrating the facility positively into community identity.

A shallow nuclear waste store site restoration project will also enable direct involvement in the aesthetics of the site, enabling the community to democratically contribute to the landscape it lives within.

Using site restoration as a vehicle for public engagement in a shallow storage project is not without problems however. Nuclear waste stores generally require high levels of active security. This may be especially important in the case of shallow stores since this particular option has security against terrorism as an important driver. Such a level of security precludes active physical involvement of the host community in restoring the site. This, in turn, may foster a lack of acceptability within the community.

A related issue is transparency. Transparency and openness are a stated aim for many national radioactive waste management organisations (CoRWM, 2004; Mohanty & Sagar, 2002). In some quarters, security is often laid out as imping-ing upon transparency (Chang, 2004). If security against terrorism is a driving force for shallow stores, many of the physical and engineering details of the facility may have to be kept secret in order not to compromise the reason for choosing this option in the first place. This will obviously negate the chances of community site restoration, but it will also foster distrust about the whole project since lack of access to technical details will give rise to suspicion that the project is unsafe.

Site restoration, in itself, may be regarded as somewhat of an oxymoron by community members since restoration projects rely on the presence of an historically degraded site. This point acknowledges the danger that a restoration project could be used by proponents to advertise the benefits of accepting a waste management facility. There is a clear distinction between a corporate site restoration and a community site restoration. Whereas the former may be regarded with suspicion, an attempt to hide the problem, the former involves the community becoming intimately involved in caring for and making decisions about the site. This point also acknowledges Elliot's (1994) two-type categorisation of site restoration:

1. **Malicious/Pernicious Restoration:** Which tends to make people believe that all environmental misdeeds are ultimately undoable by a mix of good science and good management. Some would say this is the likeliest restoration response of the nuclear industry.
2. **Benevolent Restoration:** which is instigated to remedy past human harms on the environment without justifying such past harms and without believing that human science and technology can manipulate environments with total control.

CONCLUSION

Shallow stores seek to be a middle ground solution for nuclear waste management; a compromise between deep disposal of nuclear waste and its indefinite storage on the surface. Shallow stores offer some degree of geological protection against surface events, whether manmade or natural, and they also offer some degree of monitorability and retrievability. The compromise lies in the fact that shallow stores are not as easy to monitor and retrieve as surface stores, and the geological protection they offer is not as great as deep disposal. The degree of confidence that shallow stores elicit from both engineers and the public will probably depend on the respective weights given to the following factors:

• How concerned you are that nuclear waste is a terrorist target?
• How suspicious you are in geological knowledge professed by government scientists?
• How confident you are that future governments will stick to policies decided today?
• How confident you are that any community would want to host such a project?
• How frightened you are that natural disasters may wreak havoc on human-made surface structures?

However these concerns may be balanced against each other by various players a large degree of uncertainty will remain about how suitable shallow stores are in addressing them. For example, if security is a major force to prompt the burial of nuclear waste underground, then it seems logical to question the ongoing use of nuclear reactors upon the surface. Similarly, if you trust governments to implement promised crises-prevention policies then the nuclear waste issue should have been solved thirty or forty years ago when it was first declared a crisis situation.

Given that other storage and disposal options possess similar uncertainties and present similar burdens it seems too harsh to pronounce shallow storage as an unsuitable option for nuclear waste management. Perhaps shallow storage cannot be

relied upon to do all the things of it that we ask. But I would venture to say that deep disposal and surface storage are similarly cursed. All options possess compromises and weaknesses and none can be said to be able to neutralise nuclear waste in any way. The next stage should be to summon not only expert advice on choosing where to put the waste, but on assessing the public and community response to both the options, and the future of an un-neutralizable waste.

REFERENCES

Alexander, W. R., & Mckinley, L. (Eds.). (2007). *Deep geological disposal of radioactive waste* (*Vol. 9*). Amsterdam, The Netherlands: Elsevier.

Allison, G. (2003). Nuclear terrorism poses the greatest threat today. *Wall Street Journal*, A14.

Bauriedl, T. (1991). On guilt, grief, responsibility and mythology-a psychoanalysts view of nuclear guardianship project. *NUX Forum for Responsible Scientists, 69-70.*

Chang, D. (2004, March 8-10). *The facets of technology convergence in telecommunications.* Paper presented at the Wireless and Communications Conference, Taipei, Taiwan.

Clark, C. (2003). *The economic impact if nuclear terrorist attacks on freight transport systems in age of seaport vulnerability* (Tech. Rep. No. DTRS57-03-P-80130). Cambridge, MA: US DOT National Transportation Systems Center.

Committee on Radioactive Waste Management. (CoRWM). (2004). *Annual report.* Retrieved from http://corwm.decc.gov.uk/

EKRA. (2000). *Disposal concepts for radioactive waste: Final report.* Bern, Switzerland: Federal Office of Energy.

Elliot, R. (1994). *Faking nature: The ethics of environmental restoration.* London, UK: Routledge.

Ellis, T. (2004). *Near surface storage below ground level* (Tech. Rep. No. 477002). Oxford, UK: Nirex Limited.

Fetter, S., & von Hippel, F. (1990). The hazard from plutonium dispersal by nuclear-warhead accidents. *Science & Global Security*, *2*(1), 21–41. doi:10.1080/08929889008426345

House of Lords Select Committee on Science and Technology. (1999). *Management of nuclear waste* (p. 43). London, UK: The Stationery Office.

IAEA. (2003). *The long-term storage of radioactive waste: Safety and sustainability*. Vienna, Austria: IAEA Publications.

Light, A. (2003). Urban ecological citizenship. *Journal of Social Philosophy*, *34*(1), 44–63. doi:10.1111/1467-9833.00164

Martin, S. (1995). *Protecting environmental and nuclear whistleblowers: A litigation manual*. Washington, DC: Nuclear Information and Resource Service.

McCutcheon, C. (2002). *Nuclear reactions: The politics of opening a radioactive waste disposal site*. Santa Fe, NM: University of New Mexico Press.

Medalia, J. E. (Ed.). (2003). *Nuclear terrorism*. Hauppauge, NY: Nova Science.

Mohanty, S., & Sagar, B. (2002). Importance of transparency and traceability in building a safety case for high-level nuclear waste repositories. *Risk Analysis*, *22*(1), 7–15. doi:10.1111/0272-4332.t01-1-00005

NRC. (2001). *Disposition of high level waste and spent nuclear fuel: The continuing societal and technical challenges*. Atlanta, GA: National Academy Press.

Nuclear Energy Agency. (2003). *Forum on stakeholder confidence*. Paris, France: OECD.

Nuclear Guardianship Project (NGP). (n. d.). *An ethic of nuclear of guardianship—values to guide decision-making on the management of radioactive materials*. Retrieved from http://www.nonukes.org/r02ethic.htm

Perlic, T. (1995). Low-level waste-high risk, getting closer state of the states. *The Planet*, *2*(9), 11–12.

Pugwash Council. (2002). Nuclear terrorism: The danger of highly enriched uranium. *Pugwash Newsletter*, *2*(1), 5–6.

Shrader-Frechette, K. S. (1993). *Burying uncertainty: Risk and the case against geological disposal of waste*. Berkeley, CA: University of California Press.

SNL. (1992). *Expert judgment on markers to deter inadvertent human intrusion into the waste isolation pilot plant* (Tech. Rep. No. SAND92-1382/UC-721). Livermore, CA: Sandia National Laboratories.

Van Hove, E. (2003, November 18-21). *Valorisation of a repository in an added value project*. Paper presented at the Plenary Lecture to NEA Forum on Stakeholder Confidence, Brussels, Belgium.

UN Wire. (2003, July 21). IAEA says terrorism necessitates deep waste burial. *UN Wire*.

This work was previously published in the International Journal of Technoethics, Volume 2, Issue 2, edited by Rocci Luppicini, pp. 1-13, copyright 2011 by IGI Publishing (an imprint of IGI Global).

Compilation of References

Abbate, J. (2000). *Inventing the Internet*. Cambridge, MA: MIT Press.

Abernathy, W., & Tien, L. (2003). *Biometrics: Who's Watching You?* Retrieved from http://www.eff.org/wp/biometrics-whos-watching-you, 2003

Acton, L. (1917). Letter to Bishop Creighton, 1887. In *Acton-Creighton Correspondence*. London: Figgis and Laurence.

Adam, A. (2008). Ethics for things. *Ethics and Information Technology*, *10*, 149–154. doi:10.1007/s10676-008-9169-3

Adorno, T. W., & Horkheimer, M. (1979). *Dialectic of enlightenment*. London, UK: Verso.

Aiello, J. R., & Kolb, K. J. (1995). Electronic performance monitoring and social context: Impact on productivity and stress. *The Journal of Applied Psychology*, *80*, 339–353. doi:10.1037/0021-9010.80.3.339

Alder, G. S., Schminke, M., & Noel, T. W. (2007). The impact of individual ethics on reactions to potentially invasive HR practices. *Journal of Business Ethics*, *75*, 201–214. doi:10.1007/s10551-006-9247-6

Alexander, W. R., & Mckinley, L. (Eds.). (2007). *Deep geological disposal of radioactive waste (Vol. 9)*. Amsterdam, The Netherlands: Elsevier.

Allen, C., Smit, I., & Wallach, W. (2005). Artificial morality: top-down, bottom-up, and hybrid approaches. *Ethics and Information Technology*, *7*, 149–155. doi:10.1007/s10676-006-0004-4

Allen, C., Varner, G., & Zinser, J. (2000). Prolegomena to any future artificial moral agent. *Journal of Experimental & Theoretical Artificial Intelligence*, *12*(3), 251–261. doi:10.1080/09528130050111428

Allison, G. (2003). Nuclear terrorism poses the greatest threat today. *Wall Street Journal*, A14.

Allport, G. W., & Postman, L. (1947). *The psychology of rumor*. New York: Henry Holt.

Alvarez, R. (2006). Controlling democracy: The principal-agent problems in election administration. *Policy Studies Journal: the Journal of the Policy Studies Organization*, *34*, 491–520. doi:10.1111/j.1541-0072.2006.00188.x

Amichai-Hamburger, Y., McKenna, K., & Tal, S. (2008). E-empowerment: Empowerment by the Internet. *Computers in Human Behavior*, *24*(5), 1776–1789. doi:10.1016/j.chb.2008.02.002

Anderson, N. (2009a). *Judge rejects fair use defense as Tenenbaum P2P trial begins*. Retrieved from http://arstechnica.com/tech-policy/news/2009/07/judge-rejects-fair-use-defense-as-tenenbaum-p2p-trial-begins.ars

Anderson, N. (2009b). *Team Tenenbaum to fight on for those RIAA has screwed over*. Retrieved from http://arstechnica.com/tech-policy/news/2009/08/charlie-nesson-still-fights-for-those-riaa-has-screwed-over.ars

Anderson, E. (1999). *Code of the street: Decency, violence and the moral life of the inner city*. New York, NY: W.W. Norton & Co.

Arias, A. V. (2008). Life, liberty, and the pursuit of sword and armor: Regulating the theft of virtual goods. *Emory Law Journal*, *57*, 1301–1345.

Aristotle,. (1920). *Poetics (I. Bywater trans.)*. Oxford, UK: Clarendon Press.

Aristotle,. (1995). *Aristotle: Selections* (Irwin, T., & Fine, G. (Trans. Eds.)). Indianapolis, IN: Hackett.

Aristotle,. (1998). *Nicomachean ethics* (Ross, D., Ackrill, J. L., & Urmson, J. O., Trans.). Oxford, UK: Oxford University Press.

Armin, J. (2009). *Hacker deploys cloud to smash passwords.* Retrieved from http://news.hostexploit.com/cyber-security-news/4739-hacker-deploys-cloud-to-smash-passwords.html

ARPA (DARPA). (2004). *Velocity guide.* Retrieved from http://www.velocityguide.com/internet-history/arpa-darpa.html

Arthur, C., & Kiss, J. (2010). Facebook: The club with 500m 'friends'. *The Guardian.*

Association of Canadian Community Colleges. (2007). *Programs Database.* Retrieved July 20, 2007, from http://www.aucc.ca/can_uni/search/Index_e.html

Athey, S. (1993). A comparison of experts' and high tech students' ethical beliefs in computer-related situations. *Journal of Business Ethics, 12,* 359–370. doi:10.1007/BF00882026

Au, W. J. (2008). *The making of Second Life: Notes from the new world.* New York, NY: HarperCollins.

Axelrod, R. (1984). *The evolution of cooperation.* New York, NY: Basic Books.

Ayotte, B. J., Bosworth, H. B., Potter, G. G., Williams, H. T., & Steffens, D. C. (2009). The moderating role of personality factors in the relationship between depression and neuropsychological functioning among older adults. *International Journal of Geriatric Psychiatry, 24*(9), 1010–1019. doi:10.1002/gps.2213

Bäck, A. (2009). Thinking clearly about violence. In Bufacchi, V. (Ed.), *Violence: A philosophical anthology* (pp. 365–374). Basingstoke, UK: Palgrave Macmillan.

Balkin, J. (2004). Virtual liberty: Freedom to design and freedom to play in virtual worlds. *Virginia Law Review, 90*(8), 2043–2098. doi:10.2307/1515641

Baran, P. (1964). *On distributed communications series.* Santa Monica, CA: RAND.

Bardone, E. (2011). *Seeking chance: From biased rationality to distributed cognition.* Berlin, Germany: Springer-Verlag.

Barkham, P. (2008, July 17). The Question: What is the knowledge economy? *The Guardian.*

Barnett, P. W., & Littlejohn, S. W. (1997). *Moral conflict: When social worlds collide.* London, UK: Sage.

Barr, J. R., Bowyer, K. W., & Flynn, P. J. (2011, January 5-7). Detecting questionable observers using face track clustering. In *Proceedings of the IEEE Workshop on Applications of Computer Vision* (pp. 182-189).

Barrett, D. (2009). *Supernormal stimuli: How primal urges overran their evolutionary purpose.* New York, NY: W.W. Norton.

Barro, R. (1973). The control of politicians: an economic model. *Public Choice, 14,* 19–42. doi:10.1007/BF01718440

Battelle, J. (2005). *The birth of Google.* Retrieved from http://www.wired.com/wired/archive/13.08/battelle.html?pg=2&topic=battelle&topic_set=

Bauer, H. (1990). Barriers Against Interdisciplinarity: Implications for Studies of Science, Technology, and Society (STS). *Science, Technology & Human Values, 15*(1), 105–119. doi:10.1177/016224399001500110

Baumeister, R. (1997). *Evil: Inside human violence and cruelty.* New York, NY: Freeman/Holt.

Bauriedl, T. (1991). On guilt, grief, responsibility and mythology-a psychoanalysts view of nuclear guardianship project. *NUX Forum for Responsible Scientists, 69-70.*

Beardon, C. (1992). The ethics of virtual reality. *Intelligent Tutoring Media, 3,* 23–28.

Beauchamp, T. L., & Childress, J. F. (2001). *Principles of biomedical ethics* (5th ed.). New York, NY: Oxford University Press.

Beckett, D. (2000). Internet technology. In Langford, D. (Ed.), *Internet ethics.* New York, NY: St. Martin's Press.

Bedau, M. (2003). Artificial life: organization, adaptation and complexity from the bottom up. *Trends in Cognitive Sciences, 7*(11), 505–511. doi:10.1016/j.tics.2003.09.012

Bedau, M. (2004). Artificial life. In Floridi, L. (Ed.), *Philosophy of computing and information* (pp. 97–211). Oxford, UK: Blackwell Press.

Bem, D. J. (1967). Self-perception: An alternative interpretation of cognitive dissonance phenomena. *Psychological Review*, *74*(3), 183–200. doi:10.1037/h0024835

Bernays, E. (2005). *Propaganda*. Brooklyn, NY: Ig Publishing.

Bertolotti, T., Bardone, E., & Magnani, L. (2011). Perverting activism: Cyberactivism and its potential failures in enhancing democratic institutions. *International Journal of Technoethics*, *2*(2), 19–29. doi:10.4018/jte.2011040102

Bhattacharjee, S., Gopal, R., Lertwachara, K., & Marsden, J. R. (2006). Whatever happened to payola? An empirical analysis of online music sharing. *Decision Support Systems*, *42*, 104–120. doi:10.1016/j.dss.2004.11.001

Bijker, W. (1995). *Of Bicycles, Bakelite, and Bulbs: Toward a Theory of Sociotechnical Change*. Cambridge, MA: MIT Press.

Bimbot, F., Magrin-Chagnolleau, I., & Mathan, L. (1995). Second-Order Statistical Measures for Text-Independent Speaker Identification. *Speech Communication*, *17*(1-2), 177–192. doi:10.1016/0167-6393(95)00013-E

BioSecure. (2005, August 5). In *Proceedings of the BioSecure Residential Workshop*, Paris. Retrieved from http://biosecure.it-sudparis.eu/AB/index.php?option=com_content&view=article&id=23&Itemid=23

BioTech. (2002). *Biometric Technical Assessment*. Retrieved from http://www.bioconsulting.com/Bio_Tech_Assessment.html, 2002

BIS. Department of Business Innovation and Skill. (2010). *P2P Consultation Responses January 2009*. Retrieved August 8, 2010, from http://www.bis.gov.uk/policies/business-sectors/digital-content/p2p-consultation-responses

Blackburn, S. (2002). *Being Good: A short introduction to ethics*. Oxford, UK: Oxford University Press.

Blanchard, A. L., & Henle, C. A. (2008). Correlates of different forms of cyberloafing: The role of norms and external locus of control. *Computers in Human Behavior*, *24*, 1067–1084. doi:10.1016/j.chb.2007.03.008

Block, N. (1981). Psychologism and behaviorism. *The Philosophical Review*, *90*, 5–43. doi:10.2307/2184371

Bocij, P. (2006). *The dark side of the Internet*. Westport, CT: Greenwood Publishing.

Bock, G.-W., & Ho, S. L. (2009). Non-work related computing (NWRC). *Communications of the ACM*, *52*(4), 124–128. doi:10.1145/1498765.1498799

Bock, G.-W., Shin, Y., Liu, P., & Sun, H. (2010). The role of task characteristics and organizational culture in non-work-related computing: A fit perspective. *ACM SIGMIS Database*, *41*(2), 132–151. doi:10.1145/1795377.1795385

Boddy, C. R. (2010). Corporate psychopaths, bullying and unfair supervision in the workplace. *Journal of Business Ethics*, *100*(3), 367–379. doi:10.1007/s10551-010-0689-5

Boddy, C. R., Ladyshewsky, R. K., & Galvin, P. (2010). The influence of corporate psychopaths on corporate social responsibility and organizational commitment to employees. *Journal of Business Ethics*, *97*, 1–19. doi:10.1007/s10551-010-0492-3

Boellstorff, T. (2008). *Coming of age in Second Life: An anthropologist explores the virtually human*. Princeton, NJ: Princeton University Press.

Borgmann, A. (1987). *Technology and the Character of Contemporary Life: A philosophical inquiry*. Chicago: University of Chicago Press.

Boylan, M. (1995). *Ethical Issues in Business*. London: Harcourt Brace.

Brady, E. N. (1985). A Janus-headed model of ethical theory: Looking two ways at business/society issues. *Academy of Management Review*, *10*, 568–576.

Brenner, S. W. (2008). Fantasy crime: The role of criminal law in virtual worlds. *Vanderbilt Journal of Entertainment and Technology Law*, *11*(1).

Brey, A., Innerarity, D., & Mayos, G. (2009). *The ignorance society and other essays*. Barcelona, Spain: Infonomia.

Brey, P. (1999). The ethics of representation and action in virtual reality. *Ethics and Information Technology*, *1*, 5–14. doi:10.1023/A:1010069907461

Brey, P. (2000). Method in computer ethics: Towards a multi-level interdisciplinary approach. *Ethics and Information Technology*, *2*, 125–129. doi:10.1023/A:1010076000182

Brey, P. (2006). Freedom and privacy in AmI. *Ethics and Information Technology*, *7*(3), 157–166. doi:10.1007/s10676-006-0005-3

Brey, P. (2008). Virtual reality and computer simulation. In Himma, K., & Tavani, H. (Eds.), *Handbook of information and computer ethics*. New York, NY: John Wiley & Sons. doi:10.1002/9780470281819.ch15

Bringsjord, S. (2004). On building robot persons: A response to Zlatev. *Minds and Machines*, *14*(3), 381–385. doi:10.1023/B:MIND.0000035477.39773.21

Broadbrand Suppliers. (2010). *Recap the Internet history*. Retrieved from http://www.broadbandsuppliers.co.uk/uk-isp/recap-the-history-of-Internet/

Brookman, F. (2005). *Understanding homicide*. London, UK: Sage.

Brooks, M. J. (2010). New biometric modalities using internal physical characteristics, Biometric Technology for Human Identification VII. In V. Kumar et al. (Eds.), *Proceedings of the SPIE, Volume 7667* (pp. 76670N-76670N-12).

Broström, L., Johansson, M., & Klemme Nielsen, M. (2007). 'What the patient would have decided': A fundamental problem with the substituted judgment standard. *Medicine, Health Care, and Philosophy*, *10*, 265–278. doi:10.1007/s11019-006-9042-2

Brown, P. G. (1975). Informed consent in social experimentation: Some cautionary notes. In Rivlin, A. M., & Timpane, P. M. (Eds.), *Ethical and legal issues of social experimentation* (pp. 79–104). Washington, DC: The Brookings Institution.

Brunton, F., & Nissenbaum, H. (in press). Vernacular resistance to data collection and analysis: A political theory of obfuscation. In Hildebrandt, M., & de Vries, E. (Eds.), *Privacy, due process and the computational turn: Philosophers of law meet philosophers of technology*. London, UK: Routledge.

Buchanan, A. (2009). Moral status and human enhancement. *Philosophy & Public Affairs*, *37*(4), 346–381. doi:10.1111/j.1088-4963.2009.01166.x

Buchanan, A. E., & Brock, D. W. (1989). *Deciding for others: The ethics of surrogate decision making*. Cambridge, UK: Cambridge University Press.

Buffardi, L. E., & Campbell, W. K. (2008). Narcissism and social networking web sites. *Personality and Social Psychology Bulletin*, *34*(10), 1303–1314. doi:10.1177/0146167208320061

Bunge, M. (1977). Towards a technoethics. *The Monist*, *60*(1), 96–107. doi:10.5840/monist197760134

Bush, G. W. (2001). *Address to a joint session of Congress and the American People*. Retrieved from http://georgewbush-whitehouse.archives.gov/news/releases/2001/09/20010920-8.html

Bynum, T. W., & Rogerson, S. (Eds.). (2004). *Computer ethics and professional responsibility*. Malden, MA: Blackwell.

Callon, M. (1986). The sociology of an actor-network: The case of the electric vehicle. In Callon, M., Law, J., & Rip, A. (Eds.), *Mapping the dynamics of science and technology* (pp. 19–34). Basingstoke, UK: Macmillan Press.

Calluzzo, V. J., & Cante, C. J. (2004). Ethics in information technology and software use. *Journal of Business Ethics*, *51*(3), 301–312. doi:10.1023/B:BUSI.0000032658.12032.4e

Campbell, R. (1974). The sorites paradox. *Philosophical Studies*, *26*(3), 175–191. doi:10.1007/BF00398877

Capron, A. (1982). The authority of others to decide about biomedical interventions with incompetents. In W. Gaylin & R. Macklin (Eds.), *Who speaks for the child?* (pp. 115-152). Hastings-on-Hudson, NY: The Hastings Center.

Capurro, R. (2008). On Floridi's metaphysical foundation of information ecology. *Ethics and Information Technology*, *10*(2-3), 167–173. Retrieved from http://www.capurro.de/floridi.htmldoi:10.1007/s10676-008-9162-x

Caramore, M. B. (2008). Help! My intellectual property is trapped: Second Life, conflicting ownership claims and the problem of access. *Richmond Journal of Law and Technology*, *15*(1), 1–20.

Cargile, J. (1969). The sorites paradox. *The British Journal for the Philosophy of Science*, 193–202. doi:10.1093/bjps/20.3.193

Carhart, R. (2010). Pacifism and virtue ethics. *The Lyceum*, *11*(2).

Carleton University Technology, Society, and Environmental Studies. (2010). Retrieved March 1, 2010, from http://www.carleton.ca/tse/

Carlson, N. (2010, March 5). *At last-the full story of how Facebook was founded.* Retrieved from http://www.businessinsider.com/how-facebook-was-founded-2010-3#we-can-talk-about-that-after-i-get-all-the-basic-functionality-up-tomorrow-night-1

Carpenter, B. (1996). *The architectural principles of the Internet.* Retrieved from http://www.ietf.org/rfc/rfc1958.txt

Carroll, W., & Hackett, R. (2006). Democratic media activism through the lens of social movement theory. *Media Culture & Society, 28*(1), 83–104. doi:10.1177/0163443706059289

Castells, M. (2001). *The Internet galaxy: Reflections on the Internet, business, and society.* Oxford, UK: Oxford University Press.

Castronova, E. (2005). *Synthetic worlds: The business and culture of online games.* Chicago, IL: University of Chicago Press.

Castronova, E. (2007). *Exodus to the virtual world: How online fun is changing reality.* New York, NY: Palgrave Macmillan.

Catley, B. (2003). Philosophy – the luxurious supplement of violence? In *Proceedings of the Third International Critical Management Studies Conference on Critique and Inclusivity: Opening the Agenda,* Lancaster, UK.

Cerf, V. G. (1995). *Computer networking: Global infrastructure for the 21st century.* Retrieved from http://www.cs.washington.edu/homes/lazowska/cra/networks.html

Cerf, V. G. (1998). *I remember IANA.* Retrieved from http://www.rfc-editor.org/rfc/rfc2468.txt

Cerf, V. G. (2008). The scope of Internet governance. In Doria, A., & Kleinwachter, W. (Eds.), *Internet governance forum (IGF): The first two years* (pp. 51–56). Geneva, Switzerland: IGF Office.

Cerf, V. G., & Kahn, R. (1974). A protocol for packet network interconnection. *IEEE Transactions on Communications, 22*(5). doi:10.1109/TCOM.1974.1092259

Chang, D. (2004, March 8-10). *The facets of technology convergence in telecommunications.* Paper presented at the Wireless and Communications Conference, Taipei, Taiwan.

Chick Net. (n. d.). *Pioneers of the net.* Retrieved from http://www.chick.net/wizards/pioneers.html

Chollet, G., Dorizzi, B., & Petrovska, D. (2009). Introduction—About the Need of an Evaluation Framework in Biometrics. In *Guide to Biometric Reference Systems and Performance Evaluation* (pp. 1–10). London: Springer Verlag. doi:10.1007/978-1-84800-292-0_1

Chomsky, N. (2002). *Understanding power.* New York, NY: New Press.

Christoff, K., Gordon, A. M., Smallwood, J., Smith, R., & Schooler, J. (2009). Experience sampling during fMRI reveals default network and executive system contributions to mind wandering. *Proceedings of the National Academy of Sciences of the United States of America, 106*(21), 8719–8724. doi:10.1073/pnas.0900234106

Christoph, R., Forcht, K., & Bilbrey, C. (1987/1988). The development of systems ethics: An analysis. *Journal of Computer Information Systems, 2,* 20–23.

Chulov, M. (2010). *Research links rise in Falluja birth defects and cancers to US assault.* Retrieved from http://www.guardian.co.uk/world/2010/dec/30/faulluja-birth-defects-iraq

Chun, Z. Y., & Bock, G.-W. (2006). Why employees do non-work-related computing: An investigation of factors affecting NWRC in a workplace. In *Proceedings of the Tenth Pacific Asia Conference on Information Systems* (pp. 1259-1273).

Clark, C. (2003). *The economic impact if nuclear terrorist attacks on freight transport systems in age of seaport vulnerability* (Tech. Rep. No. DTRS57-03-P-80130). Cambridge, MA: US DOT National Transportation Systems Center.

Clark, D., Field, F., & Richards, M. (2010). *Computer networks and the Internet: A brief history of predicting their future.* Retrieved from http://groups.csail.mit.edu/ana/People/DDC/Working%20Papers.html

Clark, A. (2003). *Natural-born cyborgs: Minds, technologies, and the future of human intelligence*. Oxford, UK: Oxford University Press.

Clarke, R. (1999). Internet privacy concerns confirm the case for intervention. *Communications of the ACM*, 60–67. doi:10.1145/293411.293475

Coeckelberg, M. (2010). Robot rights? Towards a social-relational justification of moral consideration. *Ethics and Information Technology*, *12*, 209–221. doi:10.1007/s10676-010-9235-5

Coghlin, T. (2008, July 7). *Afghan inquiry into American bombing of 'wedding party'*. Retrieved September 21, 2011, from The Sunday Times website: http://www.timesonline.co.uk/tol/news/world/asia/article4281078.ece

Cohen, D. (1998). *Remembering Jonathan B. Postel.* Retrieved from http://www.postel.org/remembrances/cohen-story.html

Cohen, E., & Cornwell, L. (1989). College students believe piracy is acceptable. *CIS Educator Forum*, *1*, 2–5.

Committee on Radioactive Waste Management. (CoRWM). (2004). *Annual report*. Retrieved from http://corwm.decc.gov.uk/

Common Solutions Group. (2008). *Infringement-suppression technologies: Summary observations from a Common Solutions Group workshop*. Retrieved from http://net.educause.edu/ir/library/pdf/CSD5323.pdf

Computer Professionals for Social Responsibility. (2005). *CPSR history.* Retrieved from http://cpsr.org/about/history/

Conn, K. (2002). *The Internet and the law: What educators need to know*. Alexandria, VA: Association for Supervision and Curriculum Development.

Cooper, H. (2010). *Research synthesis & meta-analysis: A step-by-step approach*. Thousand Oaks, CA: Sage.

Cornish, P. (2009). *Cyber security and politically, socially and religiously motivated cyber attacks*. London, UK: European Parliament.

Corrigan, M., & Sprehe, T. (2010). Cleaning up your information wasteland. *Information & Management*, *443*, 27–31.

Cosmides, L., & Tooby, J. (2000). Consider the source: the evolution of adaptations for decoupling and metarepresentation. In Sperber, D. (Ed.), *Metarepresentations: A Multidisciplinary Perspective* (pp. 53–115). Oxford, UK: Oxford University Press.

Council of Europe. (n. d.). *Convention on cybercrime: Preamble.* Retrieved from http://cis-sacp.government.bg/sacp/CIS/content_en/law/item06.htm

Crabb, P. B. (1996). Video camcorders and civil inattention. *Journal of Social Behavior and Personality*, *11*, 805–816.

Crabb, P. B. (1999). The user of answering machines and caller ID to regulate home privacy. *Environment and Behavior*, *31*, 657–670. doi:10.1177/00139169921972281

Crabb, P. B., & Stern, S. E. (2010). Technology traps: Who is responsible? *International Journal of Technoethics*, *1*(2), 19–26.

Creswell, J. (2007). *Qualitative inquiry & research design: Choosing among five approaches*. Thousand Oaks, CA: Sage.

Creswell, J. (2009). *Research design, qualitative, quantitative, and mixed methods approaches* (3rd ed.). Thousand Oaks, CA: Sage.

Csikszentmihalyi, M. (1996). *Creativity: Flow and the psychology of discovery and invention*. New York, NY: Harper Perennial.

Cunningham, H. (1990). The employment and unemployment of children in England c. 1680-1851. *Past & Present*, *126*, 115–150. doi:10.1093/past/126.1.115

Curran, J., & Seaton, J. (2009). *Power without responsibility*. London, UK: Routledge.

Cutcliffe, S. (1989). The emergence of STS as and academic field. *Research in Philosophy and Technology*, *9*, 287–301.

Cutcliffe, S. (1990). The STS Curriculum: What Have We Learned in Twenty Years? *Science, Technology & Human Values*, *15*(3), 360–372. doi:10.1177/016224399001500305

Dalhousie University History of Science and Technology. (2010). Retrieved March 1, 2010, from http://ug.cal.dal.ca/HSTC.htm#2

Dallmayr, F. R. (1991). *Life-world, modernity and critique: Paths between Heidegger and the Frankfurt School.* Cambridge, UK: Polity.

Damasio, A. R. (1999). *The feeling of what happens: Body and emotion in the making of consciousness.* New York, NY: Harcourt Brace.

Dartnell, M. (2003). Weapons of mass instruction: Web activism and the transformation of global security. *Millennium: Journal of International Studies,* 477-499.

Daugman, J. J. (2006). Probing the uniqueness and randomness of IrisCodes: Results from 200 billion iris pair comparisons. *Proceedings of the IEEE, 94*(11), 1927–1935. doi:10.1109/JPROC.2006.884092

Davis, N. D., & Splichal, S. L. (2000). *Access denied: Freedom of information in the information age.* Ames, IA: Iowa State University Press.

Davison, P. (2009). *Sion chicken.* Retrieved from http://www.nextnature.net/2009/12/sion-chicken

Davison, P. (n. d.). *On chickens: Sion and Second Life* [Weblog]. Retrieved from http://00pd.com/post.php?id=2

Davison, W. P. (1983). The third-person effect in communication. *Public Opinion Quarterly, 47,* 1–15. doi:10.1086/268763

Dawson, A. (2004). What should we do about it? Implications of the empirical evidence in relation to comprehension and acceptability of randomisation. In Holm, S., & Jonas, M. (Eds.), *Engaging the world: The use of empirical research in bioethics and the regulation of biotechnology.* Amsterdam, The Netherlands: IOS Press.

de Vries, M. J. (2009). A multi-disciplinary approach to Technoethics. In Luppicini, R., & Adell, R. (Eds.), *Handbook of research on technoethics* (pp. 20–31). Hersey, PA: IGI Global.

Debatin, B., Lovejoy, J. P., Horn, A. K., & Hughes, B. N. (2009). Facebook and online privacy: Attitudes, behaviors, and unintended consequences. *Journal of Computer-Mediated Communication, 15*(1), 83–108. doi:10.1111/j.1083-6101.2009.01494.x

Dennett, D. (1997). When HAL kills, who's to blame? Computer ethics. In Stork, D. G. (Ed.), *HAL's Legacy: 2001's Computer as Dream and Reality.* Cambridge, MA: MIT Press.

Dennis, M. (1997). Commentary: Why we didn't ask patients for their consent. *British Medical Journal, 314,* 1077.

Dessalles, J. (2000). Language and hominid politics. In Knight, J. H. C., & Suddert-Kennedy, M. (Eds.), *The evolutionary emergence of language: Social function and the origin of linguistic form* (pp. 62–79). Cambridge, UK: Cambridge University Press. doi:10.1017/CBO9780511606441.005

Dickert, N. W., & Sugarman, J. (2006). Community consultation: Not the problem--an important part of the solution. *The American Journal of Bioethics, 6*(3), 26–28. doi:10.1080/15265160600685762

Digital Economy Act. (2010). *The UK Law Database.* Retrieved August 8, 2010, from http://www.stautelaw.gov.uk/content.aspx?activeTextDocId=3699621

Dillman, D. A. (2000). *Mail and internet surveys* (2nd ed.). New York, NY: John Wiley & Sons.

Djekic, P., & Loebbecke, C. (2007). Preventing application software piracy: An empirical investigation of technical copy protections. *The Journal of Strategic Information Systems, 16,* 173–186. doi:10.1016/j.jsis.2007.05.005

Dodd, J. (2009). *Violence and phenomenology.* New York, NY: Routledge.

Dogan, M. K. (2010). Transparency and political moral hazard. *Public Choice, 142,* 215–235. doi:10.1007/s11127-009-9485-0

Dorsey, J., & Enge, E. (2007). *Jack Dorsey and Eric Enge talk about Twitter.* Retrieved from http://www.stonetemple.com/articles/interview-jack-dorsey.shtml

Dowding, K. (2005). Is it rational to vote? Five types of answer and a suggestion. *British Journal of Politics and International Relations, 7*(3), 442–459. doi:10.1111/j.1467-856X.2005.00188.x

Dowd, K. (2009). Moral hazard and the financial crisis. *The Cato Journal, 29*(1), 141–166.

Dreamscape, N. (2010). *The saga of Sion: SL chicken blog.* Retrieved from http://slchickenblog.wordpress.com/2010/04/01/the-saga-of-sion/

Dreyfus, H. L. (1992). *What computers still can't do: A critique of artificial reason.* Cambridge, MA: MIT Press.

Dubow, C. (2005). The Internet: An overview. In Mur, C. (Ed.), *Does the Internet benefit society?* Farmington Hills, MI: Greenhaven.

Dunbar, R. (2004). Gossip in an evolutionary perspective. *Review of General Psychology*, *8*, 100–110. doi:10.1037/1089-2680.8.2.100

Duncan, L. (1999). Motivation for collective action: Group consciousness as mediator of personality, life experiences, and women's rights activism. *Political Psychology*, *20*(3), 611–635. doi:10.1111/0162-895X.00159

Dunstone, T., & Yager, N. (2009). *Biometrics system and data analysis*. New York, NY: Springer. doi:10.1007/978-0-387-77627-9

Dutt, K. S., & Aditya, A. V. (n.d.). *Biometrics*. Retrieved from http://issuu.com/sohildutt/docs/4_biometrics

Dworkin, G. (1988). *The theory and practice of autonomy*. Cambridge, UK: Cambridge University Press.

Eaton, K. (2010, December 13). *Senate Approves Bill Requiring Silent EVs Like Nissan's Leaf to Make Noise*. Retrieved September 20, 2011, from FastCompany.com website: http://www.fastcompany.com/1709381/senate-approves-bill-making-noisy-electric-cars-mandatory-as-nissan-ships-first-leaf-ev

Eckstrand, N., & Yates, C. S. (Eds.). (2011). *Philosophy and the return of violence*. New York, NY: Continuum.

Edgar, S. L. (2003). *Morality and machines: Perspectives on computer ethics* (2nd ed.). Sudbury, MA: Jones and Bartlett.

EKRA. (2000). *Disposal concepts for radioactive waste: Final report*. Bern, Switzerland: Federal Office of Energy.

Electronic Frontier Foundation. (2006). *When push comes to shove: A hype-free guide to evaluating technical solutions to copyright infringement on campus networks*. Retrieved from http://www.eff.org/files/univp2p.pdf

Electronic Frontier Foundation. (2008). *RIAA v. The People: Five years later*. Retrieved from http://www.eff.org/wp/riaa-v-people-years-later

Elks, M. L. (1993). The right to participate in research studies. *The Journal of Laboratory and Clinical Medicine*, *122*(2), 130–136.

Elliot, R. (1994). *Faking nature: The ethics of environmental restoration*. London, UK: Routledge.

Ellis, T. (2004). *Near surface storage below ground level* (Tech. Rep. No. 477002). Oxford, UK: Nirex Limited.

Ellul, J. (1964). *The Technological Society*. New York: Vintage Books.

Eron, L. D. (1963). Relationship of television viewing habits and aggressive behavior in children. *Journal of Abnormal and Social Psychology*, *67*, 193–196. doi:10.1037/h0043794

European Commission. (2005). *Biometric at the frontiers*. Brussels, Belgium: Joint Research Centre.

Eusebi, C., Gliga, C., John, D., & Maisonave, A. (2008). A Data Mining Study of Mouse Movement, Stylometry, and Keystroke Biometric Data. In *Proceedings of the Student-Faculty Research Day (CSIS)*, Pace University, New York (pp. B1.1-B1.6).

Faden, R. R., Beauchamp, T. L., & King, N. M. P. (1986). *A history and theory of informed consent*. New York, NY: Oxford University Press.

Fearon, F. (1998). Deliberation as discussion. In Elster, J. (Ed.), *Deliberative democracy* (pp. 123–140). Cambridge, UK: Cambridge University Press.

Feenberg, A. (1991). *Critical Theory of Technology*. New York: Oxford University Press.

Feinberg, J. (1986). *The moral limits of the criminal law*. New York, NY: Oxford University Press.

Ferguson, C. J. (2010). The modern hunter-gatherer hunts aliens and gathers power-ups: The evolutionary appeal of violent video games and how they can be beneficial. In Koch, N. (Ed.), *Evolutionary psychology and information systems research* (pp. 329–342). New York, NY: Springer. doi:10.1007/978-1-4419-6139-6_15

Ferguson, C. J., & Rueda, S. M. (2010). The hitman study: Violent video game exposure effects on aggressive behavior, hostile feelings and depression. *European Psychologist*, *15*(2), 99–108. doi:10.1027/1016-9040/a000010

Fetter, S., & von Hippel, F. (1990). The hazard from plutonium dispersal by nuclear-warhead accidents. *Science & Global Security*, *2*(1), 21–41. doi:10.1080/08929889008426345

Finn, E. (2011, June 14). *'Smart cars' that are actually, well, smart*. Retrieved June 17, 2011, from MIT News website: http://web.mit.edu/newsoffice/2011/smart-cars-0614.html

Flinn, M., Geary, D., & Ward, C. (2005). Ecological dominance, social competition, and coalitionary arms races. Why humans evolved extraordinary intelligence. *Evolution and Human Behavior*, *26*(1), 10–46. doi:10.1016/j.evolhumbehav.2004.08.005

Floridi, L. (2011a.) *The fourth technological revolution*. Retrieved from http://www.youtube.com/watch?v=c-kJsyU8tgI

Floridi, L., & Sanders, J. W. (1999). Entropy as Evil in Information Ethics. *Etica & Politica, 1*(2). Retrieved from http://www2.units.it/etica/1999_2/floridi/node17.html

Floridi, L. (1999). Information ethics: On the philosophical foundation of computer ethics. *Ethics and Information Technology*, *1*, 37–56. doi:10.1023/A:1010018611096

Floridi, L. (2001). *Philosophy and Computing*. London: Routledge.

Floridi, L. (2002). On the intrinsic value of information objects and the infosphere. *Ethics and Information Technology*, *4*(4), 287–304. doi:10.1023/A:1021342422699

Floridi, L. (2004). Open problems in the philosophy of information. *Metaphilosophy*, *35*(4), 554–582. doi:10.1111/j.1467-9973.2004.00336.x

Floridi, L. (2006). Information ethics, its nature and scope. *SIGCAS Computers and Society*, *36*(3), 21–36. doi:10.1145/1195716.1195719

Floridi, L. (2008). Information ethics: A reappraisal. *Ethics and Information Technology*, *10*(2-3), 189–204. doi:10.1007/s10676-008-9176-4

Floridi, L. (2009). The information society and its philosophy: Introduction to the special issue on the philosophy of information, its nature and future developments. *The Information Society*, *25*(3). doi:10.1080/01972240902848583

Floridi, L. (2010). *Information – a very short introduction*. Oxford, UK: Oxford University Press.

Floridi, L. (2010). *The Cambridge handbook of information and computer ethics*. Cambridge, UK: Cambridge University Press.

Floridi, L. (2011b). The informational nature of personal identity. *Minds and Machines*, *21*(4), 549–566. doi:10.1007/s11023-011-9259-6

Floridi, L., & Sanford, J. W. (2001). Artificial Evil and the Foundation of Computer Ethics. *Ethics and Information Technology*, *3*(1), 55–66. doi:10.1023/A:1011440125207

Foex, B. A. (2001). The problem of informed consent in emergency medicine research. *Emergency Medicine Journal*, *18*(3), 198–204. doi:10.1136/emj.18.3.198

Foot, P. (1967). Abortion and the doctrine of the double effect. *Oxford Review*, *5*, 5–15.

Ford, P. (2001). A further analysis of the ethics of representation in virtual reality: Multi-user environments. *Ethics and Information Technology*, *3*, 113–121. doi:10.1023/A:1011846009390

Forester, T., & Morrison, P. (1994). *Computer ethics: Cautionary tales and ethical dilemmas in computing* (2nd ed.). Cambridge, MA: MIT Press.

Fost, N. (1998). Waived consent for emergency research. *American Journal of Law & Medicine*, *24*(2-3), 163–183.

Frankfurt, H. (2005). *On bullshit*. Princeton, NJ: Princeton University Press.

Franks, T. (2011). *Iraq body count*. Retrieved from http://www.iraqbodycount.org/

Freire, P. (Ed.). (1997). *Mentoring the mentor: A critical dialogue with Paulo Freire (Counterpoints: Studies in the postmodern theory of education)* (*Vol. 60*). Pieterlen, Switzerland: Peter Lang.

Freyer, H. (1955). *Theorie des gegenwärtigen Zeitalters*. Stuttgart, Germany: Deutsche Verlagsanstalt.

Friess, S. (2006, June 22). *Tales from packaging hell*. Retrieved September 20, 2011, from Wired.com website: http://www.wired.com/science/discoveries/news/2006/05/70874

F-Secure. (2003). *Iraq War and information security.* Retrieved from http://www.fsecure.com/virus-info/iraq.shtml

Fuller, S. (2005). *The Philosophy of STS.* New York: Routledge.

Furedi, F. (2002). *Culture of fear.* London, UK: Continuum.

Furnell, S., & Ward, J. (2006). Malware: An evolving threat. In Kanellis, P., Kiountouzis, E., Kolokotronis, N., & Martakos, D. (Eds.), *Digital crime and forensic science in cyberspace* (p. 357). Hershey, PA: IGI Global. doi:10.4018/978-1-59140-872-7.ch002

Gackenbach, J. (1998). *Psychology and the Internet.* San Diego: Academic Press.

Gafurov, D., Helkala, K., & Sondrol, T. (2006). Biometric Gait Authentication Using Accelerometer Sensor. *Journal of computers, 1*(7), 51-59.

Garrett, R. K., & Danziger, J. N. (2008). Disaffection or expected outcome: Understanding personal Internet use during work. *Journal of Computer-Mediated Communication, 13*, 937–958. doi:10.1111/j.1083-6101.2008.00425.x

Gehlen, A. (1980). *Man in the age of technology.* New York, NY: Columbia University Press.

Gibson, J. J. (1979). *The ecological approach to visual perception.* Boston, MA: Houghton Mifflin.

Gillies, J., & Cailliau, R. (2000). *How the Web was born: The story of the World Wide Web.* Oxford, UK: Oxford University Press.

Gillin, P. (2007). *The new influencers: A marketer's guide to the new social media.* New York, NY: Quill Driver Books.

Girard, R. (1977). *Violence and the sacred.* Baltimore, MD: Johns Hopkins University Press. (Original work published 1972)

Girard, R. (1982). *Le bouc émissaire.* Paris, France: Grasset.

Girard, R. (1986). *The scapegoat.* Baltimore, MD: Johns Hopkins University Press.

Giroux, H. A. (1992). *Educational leadership and the crisis of democratic culture.* University Park, VA: University Council of Educational Administration.

Gluckman, M. (1963). Papers in Honor of Melville J. Herskovits: Gossip and scandal. *The American Economic Review, 4*(3), 307–316.

Glushko, B. (2007). Tales of the (virtual) city: Governing property disputes in virtual worlds. *Berkeley Technology Law Journal, 22*, 507–531.

Godwin, M. (1998). *Cyber rights: Defending free speech in the digital age.* Toronto, ON, Canada: Random House.

Goffman, E. (1961). *Encounters: Two studies in the sociology of interaction.* Indianapolis, IN: Bobbs-Merrill.

Goffman, E. (1986). *Frame analysis: an essay on the organization of experience.* Boston, MA: Northeastern University Press.

Goldman, A. (1992). *Liaisons: Philosophy meets the cognitive and social sciences.* Cambridge, MA: MIT Press.

Goldsmith, J., & Wu, T. (2006). *Who controls the Internet? Illusions of a borderless world.* New York, NY: Oxford University Press.

Goodyear, M., Salaway, G., Nelson, M. R., Petersen, R., & Portillo, S. (2009). *The career of the IT security officer in higher education.* Retrieved from http://www.educause.edu/library/ECP0901

Gopal, R. D., Sanders, G. L., Bhattacharjee, S., Agrawal, M., & Wagner, S. C. (2004). A behavioral model of digital music piracy. *Journal of Organizational Computing and Electronic Commerce, 14*(2), 89–105. doi:10.1207/s15327744joce1402_01

Gotterbarn, D. (2000). Virtual information and the software engineering code of ethics. In Langford, D. (Ed.), *Internet ethics* (pp. 200–219). New York, NY: Macmillan.

Greenberg, D., Shroder, M., & Onstott, M. (1999). The social experiment market. *The Journal of Economic Perspectives, 13*(3), 157–172. doi:10.1257/jep.13.3.157

Grice, H. (1975). Logic and conversation. In Sternberg, R., & Kaufman, K. (Eds.), *Syntax and semantics 3: Speech acts.* New York, NY: Academic Press.

Griere, R. (1993). STS: Prospects for an Enlightened Postmodern Synthesis. *Science, Technology & Human Values*, *18*(1), 102–112. doi:10.1177/016224399301800106

Griffiths, R. T. (2002). *Search engines*. Retrieved from http://www.let.leidenuniv.nl/history/ivh/chap4.htm

Grimes, J. M., Fleischmann, K. R., & Jaegger, P. T. (2010). Emerging ethical issues of life in virtual worlds. In Wankel, C., & Malleck, S. (Eds.), *Emerging ethical issues of life in virtual worlds*. Charlotte, NC: Information Age Publishing.

Gromov, G. R. (1996). *The roads and crossroads of Internet history*. Retrieved from http://www.netvalley.com/intvalnext.html

Gupta, P. B., Gould, S. J., & Pola, B. (2004). "To pirate or not to pirate": A comparative study of the ethical versus other influences on the consumer's software acquisition-mode decision. *Journal of Business Ethics*, *55*, 255–274. doi:10.1007/s10551-004-0991-1

Gurian-Sherman, D. (2008). *CAFOs uncovered: The untold costs of confined animal feeding operations*. Retrieved from http://www.ucsusa.org/food_and_environment/sustainable_food/cafos-uncovered.html

Gusterson, H. (2007). Anthropology and militarism. *Annual Review of Anthropology*, *16*, 155–175. doi:10.1146/annurev.anthro.36.081406.094302

H'obbes' Zakon. R. (2010). *H'obbes' Internet timeline 10.1*. Retrieved from http://www.zakon.org/robert/Internet/timeline/

Hackett, E., Amsterdamska, O., Lynch, M., & Wajcman, J. (Eds.). (2007). *The Handbook of STS*. Boston: MIT Press.

Hafner, A., & Lyon, M. (1998). *Where wizards stay up late: The origins of the Internet*. New York, NY: Simon & Schuster.

Haggerty, K. D., & Samatas, M. (Eds.). (2010). *Surveillance and democracy*. London, UK: Routledge.

Hall, G. S. (1904). *Adolescence: Its psychology and its relations to physiology, anthropology, sociology, sex, crime, religion, and education* (*Vol. 1-2*). New York, NY: D. Appleton & Co.

Hammond, K. R. (1996). *Human judgment and social policy: Irreducible uncertainty, inevitable error, unavoidable injustice*. New York, NY: Oxford University Press.

Hample, D., Han, B., & Payne, D. (2010). The aggressiveness of playful arguments. *Argumentation*, *24*(4), 405–421. doi:10.1007/s10503-009-9173-8

Hansson, M., Dillner, J., Bartram, C. R., Carlson, J. A., & Helgesson, G. (2006). Should donors be allowed to give broad consent to future biobank research? *The Lancet Oncology*, *7*, 266–269. doi:10.1016/S1470-2045(06)70618-0

Hansson, S. O. (2004). The ethics of biobanks. *Cambridge Quarterly of Healthcare Ethics*, *13*, 319–326. doi:10.1017/S0963180104134038

Hansson, S. O. (2006). Informed consent out of context. *Journal of Business Ethics*, *63*, 149–154. doi:10.1007/s10551-005-2584-z

Haraway, D. (1991). *A Cyborg Manifesto, Science Technology and Socialist-Feminism in the Late Twentieth Century, The reinvention of nature*. New York: Routledge.

Hardin, G. (1968). The tragedy of the commons. *Science*, *162*, 1243–1248. doi:10.1126/science.162.3859.1243

Hare, R. D. (2003). *The Psychopathy Checklist—Revised*. Toronto, Canada: Multi-Health Systems.

Hassan, G. (2005). *Living conditions in Iraq: A criminal tragedy*. Retrieved from http://www.globalresearch.ca/articles/HAS506A.html

Hauben, R. (1998). *Lessons from the early MsgGroup mailing list as a foundation for identifying the principles for Internet governance: Abstract*. Retrieved from http://www.columbia.edu/~rh120/other/talk_governance.txt

Hauben, M., & Hauben, R. (1997). *Netizens: On the history and impact of UseNet and the Internet*. Washington, DC: IEEE Computer Society.

Hayry, M., & Takala, T. (2001). Genetic information, rights, and autonomy. *Theoretical Medicine and Bioethics*, *22*(5), 403–414. doi:10.1023/A:1013097617552

Heidegger, M. (1975). *Poetry, Language, Thought (A. Hofstadter trans.)*. New York: Harper and Rowe.

Heidegger, M. (1977). *The Question Concerning Technology and Other Essays (W. Lovitt trans.)*. New York: Harper and Rowe.

Heidegger, M. (2005). *Being and Time (J. Macquarrie & E. Robinson trans.)*. London: Blackwell.

Helgesson, G., & Johnsson, L. (2005). The right to withdraw consent to research on biobank samples. *Medicine, Health Care, and Philosophy, 8*, 315–321. doi:10.1007/s11019-005-0397-6

Henle, C. A., & Blanchard, A. L. (2008). The interaction of work stressors and organizational sanctions on cyberloafing. *Journal of Managerial Issues, 20*(3), 383–400.

Hickman, L. (2001). *Philosophical Tools for Technological Culture, Putting Pragmatism to Work*. Indianapolis, IN: Indiana University Press.

Higgins, G. E. (2005). Can low self-control help with the understanding of the software piracy problem? *Deviant Behavior, 26*, 1–24. doi:10.1080/01639620490497947

Hildebrandt, M. (2008). A vision of ambient law. In Brownsword, R., & Yeung, K. (Eds.), *Regulating Technologies* (pp. 175–191). Oxford, UK: Hart.

Hildebrandt, M. (2008). Ambient intelligence, criminal liability and democracy. *Criminal Law and Philosophy, 2*(2), 163–180. doi:10.1007/s11572-007-9042-1

Hildebrandt, M. (2008). Profiling and the rule of law. *Identity in the Information Society, 1*(1), 55–70. doi:10.1007/s12394-008-0003-1

Hildebrandt, M. (2009). Technology and the end of law. In Keirsbilck, B., Devroe, W. W., & Claes, E. (Eds.), *Facing the limits of the law* (pp. 1–22). Heidelberg, Germany: Springer-Verlag. doi:10.1007/978-3-540-79856-9_23

Hildebrandt, M. (2009b). Profiling and AmI. In Rannenberg, K., Royer, D., & Deuker, A. (Eds.), *The future of identity in the information society* (pp. 273–310). Heidelberg, Germany: Springer-Verlag. doi:10.1007/978-3-642-01820-6_7

Hildebrandt, M., & Gutwirth, S. (2008). *Profiling the European Citizen*. New York, NY: Springer. doi:10.1007/978-1-4020-6914-7

Hill, R., & Dunbar, R. (2003). Social network size in humans. *Human Nature (Hawthorne, N.Y.), 14*(1), 53–72. doi:10.1007/s12110-003-1016-y

Himma, K. E. (2004). There's something about Mary: the moral value of things *qua* information objects. *Ethics and Information Technology, 6*, 145–159. doi:10.1007/s10676-004-3804-4

History of Computing Project. (2001). *Donald W. Davies CBE, FRS*. Retrieved from http://www.thocp.net/biographies/davies_donald.htm

Hobbes, T. (1985). *The Leviathan*. London, UK: Penguin Classics. (Original work published 1651)

Hogan, L. (2008, May 20). Cyberbullying 'widespread' but parents kept in the dark. *The Irish Independent*.

Holvast, J. (2009). *The Future of Identity in the Information Society*. Boston: Springer.

Hoppenbrouwers, S. S., Farzan, F., Barr, M. S., Voineskos, A. N., Schutter, D. J. L. G., Fitzgerald, P. B., & Daskalakis, Z. J. (2010). Personality goes a long a way: An interhemispheric connectivity study. *Frontiers in Psychiatry, 1*(140), 1–6.

Horowitz, S. J. (2007). Competing Lockean claims to virtual property. *Harvard Journal of Law & Technology, 20*.

Houellebecq, M. (2006). *The possibility of an island*. New York, NY: Knopf.

House of Lords Select Committee on Science and Technology. (1999). *Management of nuclear waste* (p. 43). London, UK: The Stationery Office.

Hoven, V. D. J. (2000). The Internet and varieties of moral wrongdoing. In Langford, D. (Ed.), *Internet ethics* (p. 257). New York, NY: St. Martin's Press.

Hubig, C. (2000). Historische wurzeln der technikphilosophie. In Hubig, C., Huning, A., & Ropohl, G. (Eds.), *Nachdenken über technik. Die klassiker der technikphilosophie* (pp. 19–40). Berlin, Germany: Ed. Sigma.

Hubig, C. (2006). *Die kunst des möglichen i. technikphilosophie als reflexion der medialität*. Bielefeld, Germany: Transcript.

Hunter, V. (1990). Gossip and the politics of reputation in classical Athens. *Phoenix, 44*(4), 299–325.

Hunt, S. D., & Vitell, S. (1986). A general theory of marketing ethics. *Journal of Macromarketing, 6,* 5–16. doi:10.1177/027614678600600103

Hutchins, E. (1995). *Cognition in the wild.* Cambridge, MA: MIT Press.

IAEA. (2003). *The long-term storage of radioactive waste: Safety and sustainability.* Vienna, Austria: IAEA Publications.

Iaria, G., Petrides, M., Dagher, A., Pike, B., & Bohbot, V. (2003). Cognitive strategies dependent on the hippocampus and caudate nucleus in human navigation: Variability and change with practice. *The Journal of Neuroscience, 23*(13), 5945–5952.

ibiblio. (n. d.). *Internet pioneers.* Retrieved from http://www.ibiblio.org/pioneers/

ICANN. (2010). *About ICANN.* Retrieved from http://www.icann.org/en/about/

Illia, L. (2003). Passage to cyberactivism: How dynamics of activism change. *Journal of Public Affairs, 3*(4), 326–337. doi:10.1002/pa.161

Information Warfare Monitor & Shadowserver Foundation. (2010). *Shadows in the cloud: Investigating cyber espionage 2.0.* Retrieved from: http://shadows-in-the-cloud.net

Information Warfare Monitor. (2009). *Tracking GhostNet: Investigating a cyber espionage network.* Toronto, ON, Canada: Information Warfare Monitor.

International Atomic Energy Agency. (n. d.). *Depleted uranium.* Retrieved from http://www.iaea.org/newscenter/features/du/du_qaa.shtml

Internet World Stats. (2010). *Internet growth statistics.* Retrieved from http://www.Internetworldstats.com/emarketing.htm

Ito, K., Iitsuka, S., & Aoki, T. (2009). A palmprint recognition algorithm using phase-based correspondence matching. In. *Proceedings of the IEEE International Conference on Image Processing, 2009,* 1977–1980.

Jacobs, D. (2005). Internet activism and the democratic emergency in the US. *Ephemera: Theory & Politics in Organization, 5*(1), 68–77.

Jain, A. K., & Ross, A. (2007). Introduction to Biometrics. In *Handbook of Biometrics* (pp. 1–22). New York: Springer.

Jain, A. K., Ross, A., & Pankanti, S. (2006). Biometrics: a tool for information security. *Transactions on Information Forensics and Security, 1*(2), 125–143. doi:10.1109/TIFS.2006.873653

JANET. (n. d.). *About JANET.* Retrieved from http://www.ja.net/company/about.html

Jankowich, A. E. (2005). Property and democracy in virtual worlds. *Boston University Journal of Science and Technology Law, 11,* 173–220.

Jenkins, H. (2006). *Fans, bloggers, and gamers: Exploring participatory culture.* New York, NY: New York University Press.

Jenkins, H. (2007). Afterword: The future of fandom. In Gray, J., Jones, J. P., & Thompson, E. (Eds.), *Fandom: Identities and communities in a mediated world.* New York, NY: New York University Press.

Jenkins, P. (2001). *Beyond tolerance: Child pornography on the Internet.* New York, NY: New York University Press.

Joachim, D. (2004). *The enforcers—The University of Florida's ICARUS P2P-blocking software has clipped students' file-sharing wings. Do its policy-enforcing capabilities go too far?* Retrieved from http://cyber.law.harvard.edu/digitalmedia/Icarus%20at%20UF.htm

Johnson, D. (2001). *Computer Ethics.* Upper Saddle River, NJ: Prentice Hall.

Johnson, D., & Post, D. (1997). The rise of law on the global network. In Kahin, B., & Nesson, C. (Eds.), *Borders in cyberspace.* Cambridge, MA: MIT Press.

Jones, S., Davies, K., & Jones, B. (2005). The adult patient, informed consent and the emergency care setting. *Accident and Emergency Nursing, 13*(3), 167–170. doi:10.1016/j.aaen.2005.05.005

Joseph, S. (1999). *A voice for the child: The inspirational words of Janusz Korczak.* London, UK: Harper Collins.

Just, M. A., Keller, T., & Cynkar, J. (2008). A decrease in brain activation associated with driving when listening to someone speak. *Brain Research, 1205*, 70–80. doi:10.1016/j.brainres.2007.12.075

Kant, I., Abbott, T. K., & Liberty Fund. (1898). *Kant's critique of practical reason and other works on the theory of ethics*. London, UK: Longmans, Green. Retrieved from http://files.libertyfund.org/files/360/Kant_0212_EBk_v6.0.pdf

Kapp, E. (1877). *Grundlinien einer philosophie der technik: Zur entstehungsgeschichte der kultur aus neuen gesichtspunkten*. Braunschweig, Germany: Westermann.

Karagiozakis, M. (2009). Counting excess civilian casualties of the Iraq War: Science or Politics? *The Journal of Humanitarian Assistance*. Retrieved from http://sites.tufts.edu/jha/archives/559

Kayser, J. J. (2007). The new-world: Virtual property and the end user license agreement. *Loyola of Los Angeles Entertainment Law Review, 27*, 59–85.

Keen, A. (2007). *The cult of amateur. How today's Internet is killing our culture and assaulting our economy*. London, UK: Nicholas Brealey Publishing.

Kellner, D. (n. d.). *The media and social problems*. Retrieved from http://gseis.ucla.edu/faculty/kellner/essays/medsocialproblems.pdf

Kennedy, R. (2008). Virtual rights? Property in on-line game objects and characters. *Information & Communications Technology Law, 17*(2), 95–106. doi:10.1080/13600830802204195

Kern, M. L., & Friedman, H. S. (2008). Do conscientious individuals live longer? A quantitative review. *Health Psychology, 27*(5), 505–512. doi:10.1037/0278-6133.27.5.505

Key, E. (1909). *The century of the child*. New York, NY: Putnam.

Kiesler, S. (1997). *Culture of the Internet*. Mahwah, NJ: Lawrence Erlbaum Associates.

Killingsworth, M., & Gilbert, D. (2010). A wandering mind is an unhappy mind. *Science, 330*, 932. doi:10.1126/science.1192439

Kincaid, P. (2011). *Occupy Wall Street protestors demands the abolishing of the Federal Reserve*. Retrieved from http://presscore.ca/2011/?p=4637

King, M. L., Jr. (1963). *Letter from the Birmingham Jail*. Retrieved from http://mlk-kpp01.stanford.edu/index.php/resources/article/annotated_letter_from_birmingham

King, M. L. Jr. (2011). *The trumpet of conscience*. Boston, MA: Beacon Press.

King, N. M. P. (1999). Medical research: Using a new paradigm where the old might do. In King, N. M. P., & Henderson, G. E. (Eds.), *Beyond regulations: Ethics in human subjects research*. Chapel Hill, NC: University of North Carolina Press.

Kipnis, D. (1976). *The powerholders*. Chicago: University of Chicago Press.

Kipnis, D. (1990). *Technology and power*. New York: Springer Verlag.

Kipnis, D. (1991). The technological perspective. *Psychological Science, 2*, 62–69. doi:10.1111/j.1467-9280.1991.tb00101.x

Kirkpatrick, D. (2010). *The Facebook effect*. New York, NY: Simon and Schuster.

Kitchin, R., & Dodge, M. (2006). Software and the mundane management of air travel. *First Monday, 7*.

Kleinrock, L. (1961). *Information flow in large communication nets*. Unpublished doctoral dissertation, MIT, Cambridge, MA.

Kleinrock, L. (1996). *Personal history/biography: The birth of the Internet*. Retrieved from http://www.lk.cs.ucla.edu/LK/Inet/birth.html

Kleinrock, L. (1973). *Communication nets: Stochastic message flow and design*. New York, NY: Dover Publications.

Kleinrock, L. (2002). Creating a mathematical theory of computer networks. *Operations Research, 50*(1), 125–131. doi:10.1287/opre.50.1.125.17772

Kleinrock, L. (2008). History of the Internet and its flexible future. *IEEE Wireless Communications, 15*(1), 8–18. doi:10.1109/MWC.2008.4454699

Kleinrock, L. (2010). An early history of the Internet. *IEEE Communications Magazine*, *48*(8), 26–36. doi:10.1109/MCOM.2010.5534584

Klempous, R. (2009). Biometric Motion Identification Based on Motion Capture. In *Computational Intelligence* (pp. 335–348). Berlin: Springer.

Korczak, J. (1992). *When I am little again and the child's right to respect*. Lanham, MD: University Press.

Kornberg, A., & Brehm, M. (1971). Ideology, institutional identification, and campus activism. *Social Forces*, *49*(3), 445–459. doi:10.2307/3005736

Kouzes, R. T., Myers, J. D., & Wulf, W. (1996). Collaboratories: Doing science on the Internet. *Computer*, *29*(8), 40–46. doi:10.1109/2.532044

Kraybill, E. N., & Bauer, B. S. (1999). Can community consultation substitute for informed consent in emergency medicine research? In King, N. M. P., & Henderson, G. E. (Eds.), *Beyond regulations: Ethics in human subjects research*. Chapel Hill, NC: University of North Carolina Press.

Kuhn, T. (1962). *The Structure of Scientific Revolutions*. Chicago: University of Chicago Press.

Kuran, T. (1998). Moral overload and its alleviation. In Ben-Ner, A., & Putterman, L. (Eds.), *Economics, Values, Organization* (pp. 231–266). Cambridge, UK: Cambridge University Press.

Kurokawa, T. (2007). Ontology for cross-organizational communication. *The Quarterly Review*, 13–20.

Laland, K., & Brown, G. (2006). Niche construction, human behavior, and the adaptive-lag hypothesis. *Evolutionary Anthropology*, *15*, 95–104. doi:10.1002/evan.20093

Langford, D. (2000). *Internet ethics*. New York, NY: St. Martin's Press.

Langford, D. (Ed.). (2000). *Internet ethics*. New York, NY: Macmillan.

LaRose, R., Lai, Y.-J., Lange, R., Love, B., & Wu, Y. (2005). Sharing or piracy? An exploration of downloading behavior. *Journal of Computer-Mediated Communication*, *11*(1), 1. doi:10.1111/j.1083-6101.2006.tb00301.x

Lastowka, G., & Hunter, D. (2004). Law of virtual worlds. *California Law Review*, *92*(1), 3–73. doi:10.2307/3481444

Latour, B. (1991). Technology is society made durable. In Law, J. (Ed.), *A sociology of monsters: Essays on power, technology and domination* (pp. 196–233). London, UK: Routledge.

Latour, B. (2005). *Reassembling the social: an introduction to actor-network-theory*. Oxford, UK: Oxford University Press.

Law, J., & Law, J. (1999). After ANT: Complexity, naming and topology. *The Sociological Review*, *46*, 1–14. doi:10.1111/1467-954X.46.s.1

Lederman, L. (2007). Stranger than fiction: Taxing virtual worlds. *New York University Law Review*, *82*, 1620–1672.

Lee, Y., Lee, Z., & Kim, Y. (2007). Understanding personal Web usage in organizations. *Journal of Organizational Computing and Electronic Commerce*, *17*(1), 75–99.

Leiner, B. M., Cerf, V. D., Clark, D. D., Kahn, R. E., Kleinrock, L., Lynch, D. C. et al. (2009). A brief history of the Internet. *ACM SIGCOMM Communications Review*, *39*(5).

Leiner, B. M., Cerf, V. G., Clark, D. D., Kahn, R. E., Kleinrock, L., & Lynch, D. C. (1997). The past and future history of the Internet. *Communications of the ACM*, *40*(2), 103. doi:10.1145/253671.253741

Leonard, L. N. K., & Haines, R. (2007). Computer-mediated group influence on ethical behavior. *Computers in Human Behavior*, *23*, 2302–2320. doi:10.1016/j.chb.2006.03.010

Leopold, A., Schwartz, C. W., & Leopold, A. (1970). *A Sand County Almanac: with essays on conservation from Round River*. New York, NY: Ballantine Books.

Leung, L., & Wei, R. (2000). More than just talk on the move: Uses and gratifications of the cellular phone. *Journalism & Mass Communication Quarterly*, *77*, 308–322.

Levmore, S. (2010). The Internet's anonymity problem. In Levmore, S., & Nussbaum, M. C. (Eds.), *The offensive Internet: Speech, privacy, and reputation* (pp. 50–67). Cambridge, MA: Harvard University Press.

Levy, S. (2010). *Hackers: Heroes of the Computer Revolution* (25th anniversary ed.). Sebastopol, CA: O'Reilly Media Inc.

Li, H., Ma, B., Lee, K., Sun, H., Zhu, D., Sim, K. C., et al. (2009). The I4U system in NIST 2008 speaker recognition evaluation. In *Proceedings of the 2009 IEEE International Conference on Acoustics, Speech and Signal Processing,* Taipei, Taiwan (pp. 4201-4204).

Liang, Z., & Yan, Z. (2005). Software piracy among college students: A comprehensive review of contributing factors, underlying processes, and tackling strategies. *Journal of Educational Computing Research, 33,* 115–140. doi:10.2190/8M5U-HPQK-F2N5-B574

Libin, A., & Libin, E. (2004). Person–robot interactions from the robopsychologists point of view: the robotic psychology and robotherapy approach. *Proceedings of the IEEE, 92*(11), 1789–1803. doi:10.1109/JPROC.2004.835366

Lidsky, D. (2010, February 1). *The brief but impactful history of YouTube.* Retrieved from http://www.fastcompany.com/magazine/142/it-had-to-be-you.html

Light, A. (2003). Urban ecological citizenship. *Journal of Social Philosophy, 34*(1), 44–63. doi:10.1111/1467-9833.00164

Linden Lab. (2009). *XML statistics.* Retrieved from http://secondlife.com/xmlhttp/secondlife.php

Lindley, R. I. (1998). Thrombolytic treatment for acute ischaemic stroke: Consent can be ethical. *BMJ (Clinical Research Ed.), 316*(7136), 1005–1007.

Ling, R., & Yttri, B. (2006). Control, emancipation, and status: The mobile telephone in teen's parental and peer relationships. In Kraut, R., Brynin, M., & Kiesler, S. (Eds.), *Computers, phones, and the internet* (pp. 219–234). New York: Oxford University Press.

Linton, R. (1936). *The study of man.* New York, NY: Appleton-Century.

Lippmann, W. (1997). *Public opinion.* London, UK: Free Press.

Li, S. Z., & Jain, A. K. (2009). *Encyclopedia of Biometrics* (1st ed.). New York: Springer. doi:10.1007/978-0-387-73003-5

Living Internet. (2000). *DARPA/ARPA Defense: Advanced research project agency.* Retrieved from http://www.livingInternet.com/i/ii_darpa.htm

Locke, J. (1980). *Second treatise of government.* Indianapolis, IN: Hackett Publishing.

Locke, J. (2003). First treatise of government. In Shapiro, I. (Ed.), *Two treatises of government and a letter concerning toleration.* New Haven, CT: Yale University Press.

Lowensohn, J. (2010, December 10). *Microsoft made $15bn bid to buy Facebook, says exec.* Retrieved from http://www.zdnet.co.uk/news/mergers-and-acquisitions/2010/12/10/microsoft-made-15bn-bid-to-buy-facebook-says-exec-40091124/

Luckner, A. (2008). *Heidegger und das denken der technik.* Bielefeld, Germany: Transcript.

Luppicini, R. (2009). Technoethical inquiry: From technological systems to society. *Global Media Journal, 2*(1), 5–21.

Luppicini, R. (2010). *Technoethics and the evolving knowledge society: Ethical issues in technological design, research, development and innovation.* Hershey, PA: IGI Global.

Luppicini, R., & Adell, R. (2009). *Handbook of research in technoethics.* Hershey, PA: Information Science Reference.

Lu, X., Wang, Y., & Jain, A. K. (2003). Combining classifiers for face recognition. In *Proceedings of the 2003 International Conference on Multimedia and Expo (ICME '03),* Baltimore. *MD Medical Newsmagazine, 3,* 13–16.

Lyotard, J.-F. (1979). *The postmodern condition: A report on knowledge.* Minneapolis, MN: University of Minnesota Press.

MacKinnon, R. (1997). Virtual rape. *Journal of Computer-Mediated Communication, 2*(4).

Madigan, D., Genkin, A., Lewis, D. D., Argamon, S., Fradkin, D., & Ye, L. (2005). Author identification on the large scale. In *Proceedings of the Joint Annual Meeting of the Interface and the Classification Society of North America (CSNA).*

Magnani, L. (in press). Abducing personal data, destroying privacy. Diagnosing profiles through artifactual mediators. In M. Hildebrandt & E. Vries de (Eds.), *Privacy, Due Process and the Computational Turn: Philosophers of Law Meet Philosophers of Technology*. London, UK: Routledge.

Magnani, L. (in press). *Understanding violence. morality, religion, and violence intertwined: A philosophical stance*.

Magnani, L., & Bardone, E. (2011). From epistemic luck to chance-seeking. The role of cognitive niche construction. In A. König, A. Dengel, K. Hinkelmann, K. Kise, R. Howlett, & L. Jain (Eds.), *Proceedings of the 15ᵗʰ International Conference on Knowledge-Based and Intelligent Information and Engineering Systems, Part II* (LNCS 6882, pp. 486-494).

Magnani, L., & Bertolotti, T. (2011). Cognitive bubbles and firewalls: Epistemic immunizations in human reasoning. In *Proceedings of the 33rd Annual Conference of the Cognitive Science Society* (pp. 3370-3375). Austin, TX: Cognitive Science Society.

Magnani, L. (2005). An abductive theory of scientific reasoning. *Semiotica, 2005*, 261–286. doi:10.1515/semi.2005.2005.153-1-4.261

Magnani, L. (2007). *Morality in a technological world: Knowledge as duty*. Cambridge, UK: Cambridge University Press. doi:10.1017/CBO9780511498657

Magnani, L. (2009), *Abductive cognition: The episte mological and eco-cognitive dimensions of hypothetical reasoning*. Heidelberg, Germany: Springer-Verlag.

Magnani, L. (2009). Beyond Mind: How brains make up artificial cognitive systems. *Minds and Machines, 19*(4), 477–493. doi:10.1007/s11023-009-9171-5

Magnani, L. (2010). Is knowledge a duty? Yes, it is, and we also have to "respect people as things", at least in our technological world: Response to Bernd Carsten Stahl's Review of Morality in a Technological World: Knowledge as Duty. *Minds and Machines, 20*(1), 161–164. doi:10.1007/s11023-010-9179-x

Magnani, L. (2011). *Understanding violence: The intertwining of morality, religion, and violence: A philosophical stance*. Heidelberg, Germany: Springer-Verlag.

Magnani, L. (in press). Abducing personal data, destroying privacy. Diagnosing profiles through artifactual mediators. In Hildebrandt, M., & de Vries, E. (Eds.), *Privacy, due process and the computational turn: Philosophers of law meet philosophers of technology*. London, UK: Routledge.

Magnani, L. (in press). Structural and technology-mediated violence. Profiling and the urgent need of new tutelary technoknowledge. *International Journal of Technoethics*.

Magnani, L., & Bardone, E. (2008). Distributed morality: externalizing ethical knowledge in technological artifacts. *Foundations of Science, 13*(1), 99–108. doi:10.1007/s10699-007-9116-5

Magnani, L., & Bardone, E. (2008). Sharing representations and creating chances through cognitive niche construction. The role of affordances and abduction. In Iwata, S., Oshawa, Y., Tsumoto, S., Zhong, N., Shi, Y., & Magnani, L. (Eds.), *Communications and Discoveries from Multidisciplinary Data* (pp. 3–40). Heidelberg, Germany: Springer-Verlag. doi:10.1007/978-3-540-78733-4_1

Magnani, L., & Nersessian, N. J. (2002). *Model-Based Reasoning: Science, Technology, Values*. Boston, MA: Kluwer Academic. doi:10.1007/978-1-4615-0605-8

MaHring, M., Holmstrom, J., Keil, M., & Montealerge, R. (2004). Trojan actor-networks and swift translation. *Information Technology & People, 17*(2), 210–238. doi:10.1108/09593840410542510

Malaby, T. (2009). *Making virtual worlds: Linden Lab and Second Life*. Ithaca, NY: Cornell University Press.

Malik, O. (2009). *A brief history of Twitter*. Retrieved from http://gigaom.com/2009/02/01/a-brief-history-of-twitter/

Malinowski, C. (2006). Training the cyber investigator. In Kanellis, P., Kiountouzis, E., Kolokotronis, N., & Martakos, D. (Eds.), *Digital crime and forensic science in cyberspace* (pp. 311–333). Hersey, PA: IGI Global. doi:10.4018/978-1-59140-872-7.ch014

Malyutov, M. B. (2006). Authorship Attribution of Texts: A Review. In *General Theory of Information Transfer and Combinatorics* (pp. 362–380). Berlin: Springer. doi:10.1007/11889342_20

Maple, C., Short, E., & Brown, A. (2011). *Cyberstalking in the United Kingdom: An analysis of the echo pilot survey*. Bedfordshire, UK: University of Bedfordshire National Centre for Cyberstalking Research.

Markoff, J. (2010, October 9). *Google Cars Drive themselves, in Traffic*. Retrieved June 17, 2011, from The New York Times website: http://www.nytimes.com/2010/10/10/science/10google.html?_r=1

Marks, R. (2003). *EverQuest companion: The inside lore of a game world*. New York, NY: McGraw-Hill.

Marshall, K. P. (1999). Has technology introduced new ethical problems? *Journal of Business Ethics*, *19*(1), 81–90. doi:10.1023/A:1006154023743

Marshall, P. A., & Berg, J. W. (2006). Protecting communities in biomedical research. *The American Journal of Bioethics*, *6*(3), 28–30. doi:10.1080/15265160600685770

Martin, M. W., & Schinzinger, R. (1989). *Ethics in engineering* (2nd ed.). New York, NY: McGraw-Hill.

Martin, S. (1995). *Protecting environmental and nuclear whistleblowers: A litigation manual*. Washington, DC: Nuclear Information and Resource Service.

Masnick, M. (2009, September 22). *Nissan to Add Futuristic Sound Effects to its Electric Car to keep it from Hitting Unaware Pedestrians*. Retrieved September 23, 2011, from TechDirt website: http://www.techdirt.com/articles/20090921/0141396258.shtml

Mason, R. O. (1986). Four ethical issues of the information age. *Management Information Systems Quarterly*, *10*(1), 5–12. doi:10.2307/248873

Mastrangelo, P. M., Everton, W., & Jolton, J. A. (2006). Personal use of work computers: Distraction versus destruction. *Cyberpsychology & Behavior*, *9*(6), 730–741. doi:10.1089/cpb.2006.9.730

Mawhood, J., & Tysver, D. (2000). Law and internet. In Langford, D. (Ed.), *Internet ethics* (pp. 96–126). New York, NY: Macmillan.

McCracken, H. (2010, May 24). *A history of AOL, as told in its own old press releases*. Retrieved from http://technologizer.com/2010/05/24/aol-anniversary/

McCumber, J. (2011). Philosophy after 9/11. In Eckstrand, N., & Yates, C. S. (Eds.), *Philosophy and the return of violence* (pp. 17–30). New York, NY: Continuum.

McCutcheon, C. (2002). *Nuclear reactions: The politics of opening a radioactive waste disposal site*. Santa Fe, NM: University of New Mexico Press.

McFadden, C., & Fulginiti, M. (2008, March 24). Searching for justice: Online harassment. *ABC News Transcript*.

Meadows, M. S. (2008). *I, Avatar: The culture and consequences of having a Second Life*. Berkeley, CA: New Riders.

Medalia, J. E. (Ed.). (2003). *Nuclear terrorism*. Hauppauge, NY: Nova Science.

Meehan, M. (2006). Virtual property: Protecting bits in context. *Richmond Journal of Law and Technology*, *13*(2), 1–48.

Meikle, G. (2010). Intercreativity: Mapping online activism. *The International Handbook of Internet Research*, 363-377.

Meikle, G. (2002). *Future active: Media activism and the Internet*. London, UK: Routledge.

Miah, A. (2010). *International Journal of Technoethics*. Retrieved from http://www.andymiah.net/2010/01/14/international-journal-of-technoethics/

Micro, T. (2009). *Data-stealing malware on the rise—solutions to keep businesses and consumers safe*. Retrived from http://us.trendmicro.com/imperia/md/content/us/pdf/threats/securitylibrary/data_stealing_malware_focus_report_-_june_2009.pdf

Milgram, S. (1963). Behavioral study of obedience. *Journal of Abnormal Psychology*, *67*, 371–378. doi:10.1037/h0040525

Mill, J. S. (1957). *Utilitarianism*. Indianapolis, IN: The Liberal Arts Press.

Mill, J. S. (1966). On liberty. In Mill, J. S. (Ed.), *On liberty, representative government, the subjection of women* (12th ed., pp. 5–141). Oxford, UK: Oxford University Press. (Original work published 1859)

Mills, E. (2010). *Report unearths targeted attacks on oil firms*. Retrieved from http://news.cnet.com/security/?keyword=espionage

Mistral, P. (2009, April 7). *Bio-warfare attacks on second life chicken farms*. Retrieved from http://alphavilleherald.com/2009/07/biowarfare-attack-on-second-life-chicken-farms.html

Mitcham, C. (1994). *Thinking through technology: The path between engineering and philosophy*. Chicago, IL: University of Chicago Press.

Mohanty, S., & Sagar, B. (2002). Importance of transparency and traceability in building a safety case for high-level nuclear waste repositories. *Risk Analysis, 22*(1), 7–15. doi:10.1111/0272-4332.t01-1-00005

Molteni, L., & Ordanini, A. (2003). Consumption patterns, digital technology and music downloading. *Long Range Planning, 36*, 389–406. doi:10.1016/S0024-6301(03)00073-6

Moor, J. H. (2005). Why we need better ethics for emerging technologies. *Ethics and Information Technology, 7*(3), 111–119. doi:10.1007/s10676-006-0008-0

Mosaic Communications Corporation. (1994). *Who are we: Our mission*. Retrieved from http://home.mcom.com/MCOM/mcom_docs/backgrounder_docs/mission.html

Moseley, A. (2009). *Just war theory*. Retrieved from http://www.iep.utm.edu/justwar

Mossoff, A. (2002). Locke's labor lost. *University of Chicago Law School Roundtable, 9*(155).

Mumford, L. (1963). *Technics and civilization*. New York: Harcourt, Brace & World.

Nagaraja, S., & Anderson, R. (2009). *Snooping dragon: Social malware surveillance of the Tibetan movement* (Tech. Rep. No. 746). Cambridge, UK: University of Cambridge.

Nakashima, H., Aghajan, H., & Augusto, J. C. (2010). *Handbook of ambient intelligence and smart environments*. New York, NY: Springer. doi:10.1007/978-0-387-93808-0

Nanavati, S., Thieme, M., & Nanavati, R. (2002). *Biometrics*. New York, NY: John Wiley & Sons.

Nash, P. (Ed.). (1970). *History and education*. New York, NY: Random House.

National Academy of Engineering. (2011). *Timeline*. Retrieved from http://www.greatachievements.org/Default.aspx?id=2984

National Academy of Engineering. (n. d.). *Internet history*. Retrieved from http://www.greatachievements.org/?id=3747

National Association for Shoplifting Prevention. (2006). *Shoplifting statistics*. Retrieved September 20, 2011, from Shopliftingprevention.org website: http://www.shopliftingprevention.org/WhatNASPOffers/NRC/PublicEducStats.htm

Needleman, S. E. (2010, May 20). *A Facebook-free workplace? Curbing cyberslacking*. Retrieved from http://online.wsj.com/

Nelson, M. R. (2010). A response to responsibility of and trust in ISPs by Raphael Cohen-Almagor. *Knowledge, Technology and Policy, 23*(3).

Neumayer, C., & Raffl, C. (2008). Facebook for global protest: The potential and limits of social software for grassroots activism. In *Proceedings of the 5th Prato Community Informatics & Development Informatics Conference*.

Newbould, R., & Collingridge, R. (2003). Profiling - Technology. *BT Technology Journal, 21*(1), 44–55. doi:10.1023/A:1022400226864

Nino, T. (2006). *Under the grid - Other causes of lag*. Retrieved from http://www.secondlifeinsider.com/2006/12/26/under-the-grid-other-causes-of-lag

Norman, D. A. (n.d.). *The design of everyday things*. New York: Doubleday.

Norman, D. (1999). Affordance, conventions and design. *Interaction, 6*(3), 38–43. doi:10.1145/301153.301168

NRC. (2001). *Disposition of high level waste and spent nuclear fuel: The continuing societal and technical challenges*. Atlanta, GA: National Academy Press.

NSTC. (2006). National Sciences and Technology Council NSTC. *Biometrics history*. Retrieved from http://www.biometrics.gov/Documents/BioHistory.pdf

Nuclear Energy Agency. (2003). *Forum on stakeholder confidence*. Paris, France: OECD.

Nuclear Guardianship Project (NGP). (n. d.). *An ethic of nuclear of guardianship—values to guide decision-making on the management of radioactive materials*. Retrieved from http://www.nonukes.org/r02ethic.htm

Nyiri, J. (2004). The effects of piracy in a university setting. *Crossroads, 10*(3), 3. doi:10.1145/1027321.1027324

Odling-Smee, F., Laland, K., & Feldman, M. (2003). *Niche Construction: A Neglected Process in Evolution*. Princeton, NJ: Princeton University Press.

O'Neill, O. (2003). Some limits of informed consent. *Journal of Medical Ethics, 29*, 4–7. doi:10.1136/jme.29.1.4

Ouamour, S., Guerti, M., & Sayoud, H. (2008). A New Relativistic Vision in Speaker Discrimination. *Canadian Acoustics Journal, 36*(4), 24–34.

Ouamour, S., Sayoud, H., & Guerti, M. (2009). Optimal Spectral Resolution in Speaker Authentication, Application in noisy environment and Telephony. *International Journal of Mobile Computing and Multimedia Communications*, 36–47.

Palmer, R. B., & Iserson, K. V. (1997). The critical patient who refuses treatment: An ethical dilemma. *Ethics of Emergency Medicine, 5*(5), 729–733.

Paré, D. (2005). The digital divide: Why the 'the' is misleading. In Klang, M., & Murray, A. (Eds.), *Human rights in the digital age*. London, UK: GlassHouse.

Parker, Uncle B. (1962, August). The Amazing Spider-man. *Amazing Stories*, 15.

Pauzie, A., & Amditis, A. (2011). Intelligent driver support system functions in cars and their potential consequences for safety. In Barnard, Y., Risser, R., & Krems, J. (Eds.), *The safety of intelligent driver support systems: Design, evaluation and social perspectives* (pp. 7–26). Farnham, UK: Ashgate.

Pee, L. G., Woon, I. M. Y., & Kankanhalli, A. (2008). Explaining non-work-related computing in the workplace: A comparison of alternative models. *Information & Management, 45*(2), 120–130. doi:10.1016/j.im.2008.01.004

Peñalosa, E. (2003). Parks for livable cities: Lessons from a radical mayor. *Places, 15*(3), 30.

Perlic, T. (1995). Low-level waste-high risk, getting closer state of the states. *The Planet, 2*(9), 11–12.

Perloff, R. M. (1993). Third-person effect research 1983-1992: A review and synthesis. *International Journal of Public Opinion Research, 5*, 167–184. doi:10.1093/ijpor/5.2.167

Peterson, D. K. (2002). Computer ethics: The influence of guidelines and universal moral beliefs. *Information Technology & People, 15*(4), 346–361. doi:10.1108/09593840210453124

Phillips, S. (2007, July 25). *A brief history of Facebook*. Retrieved from http://www.guardian.co.uk/technology/2007/jul/25/media.newmedia

Pickard, V. (2008). Cooptation and cooperation: Institutional exemplars of democratic internet technology. *New Media & Society, 10*(4), 625–645. doi:10.1177/1461444808093734

Pitt, J. (2000). *Thinking through technology*. New York, NY: Seven Bridges.

Plato,. (1997). *Complete works* (Cooper, J. M., & Hutchinson, D. S. (Trans. Eds.)). Indianapolis, IN: Hackett.

Popper, K. R. (1945). *Open society and its enemies*. London, UK: Routledge.

Postel, J. (1998, October 19). *'God of the Internet' is dead*. Retrieved from http://news.bbc.co.uk/1/hi/sci/tech/196487.stm

Postill, J. (2008). Localizing the Internet beyond communities and networks. *New Media & Society, 10*(3), 413–431. doi:10.1177/1461444808089416

Potthast, M., Eiselt, A., & Barron-Cedeno, A. (2009). Uncovering Plagiarism, Authorship and Social Software Misuse. In *Proceedings of the 1st International Competition on Plagiarism Detection (SEPLN'09)*. Retrieved from www.uni-weimar.de/medien/webis/research/ workshop-series/pan-09/competition.html

Prigg, M. (2010, February 25). *Satnav is to blame, says driver rescued from swollen river*. Retrieved September 23, 2011, from London Evening Standard website: http://www.thisislondon.co.uk/standard/article-23809822-satnav-is-to-blame-says-driver-rescued-from-swollen-river.do

Pugwash Council. (2002). Nuclear terrorism: The danger of highly enriched uranium. *Pugwash Newsletter*, *2*(1), 5–6.

Putnam, R. (2000). *Bowling alone*. New York, NY: Simon & Schuster.

Qian, Y., Liang, J., & Dang, C. (2008). Knowledge structure, knowledge granulation and knowledge distance in a knowledge base. *International Journal of Approximate Reasoning*, *50*, 174–188. doi:10.1016/j.ijar.2008.08.004

RAND Corporation. (1994). *Paul Baran and the origins of the Internet*. Retrieved from http://www.rand.org/about/history/baran.html

Ratha, N. K., & Govindaraju, V. (2008). *Advances in biometrics*. New York, NY: Springer. doi:10.1007/978-1-84628-921-7

Rawls, J. (2007). *Lectures on the history of political philosophy*. Cambridge, MA: Harvard University Press.

Redelmeier, D. A., & Thbshirani, R. A. (1997). Association between cellular-telephone calls and motor vehicle collisions. *The New England Journal of Medicine*, *336*, 453–458. doi:10.1056/NEJM199702133360701

Regan, T. (1983). *The case for animal rights*. Berkeley, CA: UCLA Press.

Reiman, J. H. (1976). Privacy, intimacy, personhood. *Philosophy & Public Affairs*, *6*(1).

Reiman, J. H. (1995). Driving to the Panopticon: a philosophical exploration of the risks to privacy posed by the highway technology of the future. *Computer and High Technology Law Journal*, *11*(1), 27–44.

Reuters. (2010). *Facebook is worth $52 billion, and that's not a good thing*. Retrieved from http://blogs.reuters.com/mediafile/2010/12/14/facebook-is-worth-52-billion-and-thats-not-a-good-thing/

Reuveni, E. (2007). On virtual worlds: Copyright and contract law at the dawn of the virtual age. *Indiana Law Journal (Indianapolis, Ind.)*, 82.

Revett, K. (2008). *Behavioral biometrics*. New York, NY: John Wiley & Sons. doi:10.1002/9780470997949

Reymers, K. (2004). Communitarianism on the internet: An ethnographic analysis of the Usenet newsgroup tpy2k, 1996-2004 (Doctoral dissertation, Morrisville State College). *WorldCat Dissertations* (OCLC: 57415279).

Roberts, L. G. (1999). *Internet chronology*. Retrieved from http://www.ziplink.net/users/lroberts/InternetChronology.html

Rob, R., & Waldfogel, J. (2006). Piracy on the high c's: Music downloading, sales displacement, and social welfare in a sample of college students. *The Journal of Law & Economics*, *49*, 29–62. doi:10.1086/430809

Rogerson, S. (1996). ETHicol: Reports on a recent public lecture on computers and human values. *IMIS Journal*, *6*(5), 14–17.

Rogoff, K. (1990). Equilibrium political budget cycles. *The American Economic Review*, *80*, 21–36.

Rosen, C. (2007). Virtual friendship and the new narcissism. *The New Atlantis: A Journal of Technology and Society*, 15-31.

Rosen, J. (2010, July 19). *The Web means the end of forgetting*. Retrieved from http://www.nytimes.com/2010/07/25/magazine/25privacy-t2.html

Rosenblatt, J. (2008). Security metrics: A solution in search of a problem. *EDUCAUSE Quarterly*, *31*(3), 8–11.

Rowlands, M. (1999). *The body in mind*. Cambridge, UK: Cambridge University Press. doi:10.1017/CBO9780511583261

Royakkers, L., & van Est, R. (2010). The cubicle warrior: The marionette of digitalized warfare. *Ethics and Information Technology*, *12*, 289–296. doi:10.1007/s10676-010-9240-8

Rubinstein, I. (2009, June 4-6). *Anonymity reconsidered*. Paper presented at the Second Annual Berkeley-GW Privacy Law Scholars Conference, Berkeley, CA.

Ruggles, T. (2002). *Comparison of biometric techniques*. Retrieved from www.bioconsulting.com/bio.htm

Russell, A., Ito, M., Richmond, T., & Tuters, M. (2008). Culture: Media convergence and networked participation. In Varnelis, K. (Ed.), *Networked publics*. Cambridge, MA: MIT Press.

Ryan, J. (2011). *The essence of the 'net': A history of the protocols that hold the network together.* Retrieved from http://arstechnica.com/tech-policy/news/2011/03/the-essence-of-the-net.ars/

Sagan, C. (1998). *Billions and billions: Thoughts on life and death at the brink of the millennium.* New York, NY: Ballantine Books.

Salus, P. H. (1995). *Casting the net: From ARPANET to Internet and beyond.* Reading, MA: Addison-Wesley.

Sandel, M. (2009). *Justice: What's the right thing to do?* New York, NY: Farrar, Straus and Giroux.

Sandhana, L. (2011, January 24). *Crime-Fighting Technology Spots Lingering Arsonists in the Crowd.* Retrieved September 20, 2011, from FastCompany website: http://www.fastcompany.com/1719341/questionable-observer-detector-spots-lingering-arsonists-at-the-scene-of-the-crime

Sandvoss, C. (2007). The death of the reader. In Gray, C. S. J. A., & Harrington, C. L. (Eds.), *Fandom: Identities and communities in a mediated world.* New York, NY: New York University Press.

Scanlon, T. (1998). *What we owe to each other.* Cambridge, MA: The Belknap Press of Harvard University Press.

Schiller, H. J. (2000). The global information highway. In Winston, M. E., & Edelbach, R. D. (Eds.), *Society, ethics and technology* (pp. 171–181). Belmont, CA: Wadsworth.

Schneider, G. P., & Evans, J. (2007). *New perspectives on the Internet: Comprehensive* (6th ed.). Boston, MA: Thomson.

Schuler, D., & Day, P. (2004). *Shaping the network society.* Cambridge, MA: MIT Press.

Schulkind, M. (2000). Perceptual interference decays over short unfilled intervals. *Memory & Cognition, 28,* 949–956. doi:10.3758/BF03209342

Schultze, C. L. (1975). Social programs and social experiments. In Rivlin, A. M., & Timpane, P. M. (Eds.), *Ethical and legal issues of social experimentation* (pp. 115–125). Washington, DC: The Brookings Institution.

Schuster, W. V. (1987). Bootleggery, smoking guns and whistle blowing; a sad saga of academic opportunism. In *Proceedings of the Western Educational Computing Conference* (pp. 179-187).

Schwab, A. P. (2006). Formal and effective autonomy in healthcare. *Journal of Medical Ethics, 32*(10), 575–579. doi:10.1136/jme.2005.013391

Schwarz, G., & Cohen, T. (2011). *Rumsfeld: WMD issue was 'the big one' in Iraq invasion.* Retrieved from http://politicalticker.blogs.cnn.com/2011/02/20/rumsfeld-wmds-werent-only-reason-for-war-in-iraq/

Science, technologie et société. (2010). *Accueil.* Retrieved March 1, 2010, from http://www.sts.uqam.ca/accueil/accueil.as

Searle, J. (1980). Minds, brains and programs. *The Behavioral and Brain Sciences, 3,* 417–457. doi:10.1017/S0140525X00005756

Searle, J. R. (1995). *The construction of social reality.* New York, NY: Simon & Schuster.

Searle, J. R. (2006). Social ontology: Some basic principle. *Anthropological Theory, 6*(1), 12–29. doi:10.1177/1463499606061731

Searle, J. R. (2010). *Making the social world: The structure of human civilization.* New York, NY: Oxford University Press.

Second Life Wiki. (2010). *Griefing.* Retrieved from http://wiki.secondlife.com/wiki/Griefing

Selby, N., Henderson, D., & Tayyabkahn, T. (2011). *Shotspotter gunshot location system efficacy report.* Retrieved November 10, 2011, from http://csganalysis.files.wordpress.com/2011/08/shotspotter_efficacystudy_gls8_45p_let_2011-07-08_en.pdf

Seto, T. P. (2009). When is a game only a game? The taxation of virtual worlds. *University of Cincinnati Law Review, 77,* 1027.

Shang, R.-A., Chen, Y.-C., & Chen, P.-C. (2008). Ethical decisions about sharing music files in the P2P environment. *Journal of Business Ethics, 80,* 349–365. doi:10.1007/s10551-007-9424-2

Sharkey, N. (2008). Grounds for discrimination: Autonomous robot weapons. *RUSI Defense Systems, 11*(2), 86–89.

Sharp, E. S., Pedersen, N. L., Gatz, M., & Reynolds, C. A. (2010). Cognitive engagement and cognitive aging: Is openness protective? *Psychology and Aging, 25*(1), 60–73. doi:10.1037/a0018748

Shavell, S. (1979). On moral hazard and insurance. *The Quarterly Journal of Economics, 93*, 541–562. doi:10.2307/1884469

Shelley, C. (in press). Fairness and regulation of violence in technological design. *International Journal of Technoethics.*

Shelley, C. (in press). Fairness in technological design. *Science and Engineering Ethics.*

Shepardson, D. (2011, June 13). *NHTSA chief skeptical of Google's driverless vehicles.* Retrieved June 17, 2011, from The Detroit News website: http://www.detnews.com/article/20110613/AUTO01/106130418/1361/NHTSA-chief-skeptical-of-Google-s-driverless-vehicles

Shoeman, F. (Ed.). (1984). *Philosophical dimensions of privacy: an anthology.* Cambridge, UK: Cambridge University Press. doi:10.1017/CBO9780511625138

Shrader-Frechette, K. S. (1993). *Burying uncertainty: Risk and the case against geological disposal of waste.* Berkeley, CA: University of California Press.

Shulman, A. N. (2011). GPS and the end of the road. *The New Atlantis: A Journal of Technology and Society.* Retrieved September 16, 2011, from http://www.thenewatlantis.com/publications/gps-and-the-end-of-the-road

Sicart, M. (2009). *The ethics of computer games.* Cambridge, MA: MIT Press.

Simon, H. (1983). *Reason in human affairs.* Stanford, CA: Stanford University Press.

Simon, H. (1993). Altruism and economics. *The American Economic Review, 83*(2), 156–161.

Simon, L. (Ed.). (2002). *Democracy and the Internet: Allies or adversaries?* Washington, DC: Woodrow.

Singer, P. W. (2009). *Robots at War: The New Battlefield.* Retrieved September 21, 2011, from The Wilson Quarterly website: http://www.wilsonquarterly.com/article.cfm?aid=1313

Singer, P. (1993). *Practical ethics.* Cambridge, UK: Cambridge University Press.

Singer, P. (1994). *Ethics.* Oxford, UK: Oxford University Press.

Singer, P. W. (2009). *Wired for war: The robotics revolution and conflict in the twenty-first century.* New York, NY: Penguin Press.

Slater, D. (1991). New crop of IS pros on shaky ground. *Computerworld*, 90.

Slevin, J. (2000). *The Internet and society.* Oxford, UK: Polity Press.

Smallwood, J., McSpadden, M., & Schooler, J. (2008). When attention matters: The curious incident of the wandering mind. *Memory & Cognition, 36*, 1144–1150. doi:10.3758/MC.36.6.1144

Smith Holt, S., Loucks, N., & Adler, J. R. (2009). Religion, culture, and killing. In Rothbart, D., & Korostelina, K. V. (Eds.), *Why we kill: Understanding violence across cultures and disciplines* (pp. 1–6). London, UK: Middlesex University Press.

Smith, D. B. (2011, May 1). *The university has no clothes.* Retrieved May 27, 2011, from New York Magazine website: http://nymag.com/news/features/college-education-2011-5/

SNL. (1992). *Expert judgment on markers to deter inadvertent human intrusion into the waste isolation pilot plant* (Tech. Rep. No. SAND92-1382 / UC-721). Livermore, CA: Sandia National Laboratories.

Social Bakers. (n. d.). *Facebook page statistics.* Retrieved from http://www.socialbakers.com/facebook-pages/

Solove, D. (2007). *The future of reputation.* New Haven, CT: Yale University Press.

Sommerfeld, R., Krambeck, H., & Milinski, M. (2008). Multiple gossip statements and their effect on reputation and trustworthiness. *Proceedings. Biological Sciences, 275*(1650), 2529–2536. doi:10.1098/rspb.2008.0762

Spanier, G. B. (2004). *Peer to peer on university campuses: An update.* Retrieved from http://president.psu.edu/testimony/articles/161.html

Sperber, D., & Wilson, D. (1995). *Relevance: communication and cognition.* Oxford, UK: Blackwell.

Spinello, A. R. (2002). *Regulating cyberspace: The policies and technologies of control*. Westport, CT: Quorum Books.

Spinello, R. A. (2000). *Cyberethics: Morality and law in cyberspace*. Sudbury, MA: Jones and Bartlett.

St. Thomas University Science and Technology Studies. (2010). Retrieved March 1, 2010, from http://w3.stu.ca/stu/sites/sts/whatis.htm

Stahl, B. C. (2004). Information, ethics, and computers: The problem of autonomous moral agents. *Minds and Machines*, *14*, 67–83. doi:10.1023/B:MIND.0000005136.61217.93

Stair, R. M. (1996). *Principles of information systems: A managerial approach*. Florence, KY: Thomson Publishing.

Stallman, R. (1991). Why software should be free. In Stallman, R. (Ed.), *Free software, free society: The selected essays of Richard M. Stallman*. Boston, MA: Free Software Foundation.

Steinberg, A. B. (2009). For sale - One level 5 barbarian for 94,800 Won: The international effects of virtual property and the legality of its ownership. *Georgia Journal of International and Comparative Law*, *37*, 381–420.

Steinhart, E. (1999). Emergent values for automations: ethical problems of life in the generalized internet. *Journal of Ethics and Information Technology*, *1*(2), 155–160. doi:10.1023/A:1010078223411

Stephens, K., & McKee, L. (2010). *Cyber espionage: Is the United States getting more than its giving?* Smithfield, VA: National Security Cyberspace Institute.

Stern, S. E. (1999). Addiction to technologies: A social psychological perspective of Internet addiction. *Cyberpsychology & Behavior*, *2*, 419–424. doi:10.1089/cpb.1999.2.419

Stern, S. E., & Handel, A. D. (2001). Sexuality and mass media: The historical context of psychology's reaction to sexuality on the Internet. *Journal of Sex Research*, *38*, 283–291. doi:10.1080/00224490109552099

Stevens, S. (2010). *Google studies*. Retrieved from http://www.discoverdecaturil.com/

Strader, T. J., Simpon, L. A., & Clayton, S. R. (2010). Using computer resources for personal activities at work: Employee perceptions of acceptable behavior. *Journal of International Technology and Information Management*, *18*(3-4), 465–476.

Strayer, D. L., & Johnston, W. A. (2001). Driven to distraction: Dual-task studies of simulated driving and conversing on a cellular telephone. *Psychological Science*, *12*, 462–466. doi:10.1111/1467-9280.00386

Strickland, J. (2010). *How ARPANET works*. Retrieved from http://www.howstuffworks.com/arpanet.htm/printable

Strickland, L. H. (1958). Surveillance and trust. *Journal of Personality*, *26*, 201–215. doi:10.1111/j.1467-6494.1958.tb01580.x

Suler, J. (2004). The online disinhibition effect. *Cyberpsychology & Behavior*, *7*(3), 321–326. doi:10.1089/1094931041291295

Sutcliffe, A. (2003). Symbiosis and synergy? Scenarios, task analysis and reuse of HCI knowledge. *Interacting with Computers*, *15*, 245–263. doi:10.1016/S0953-5438(03)00002-X

Svensson, S., & Hansson, S. O. (2007). Protecting people in research: A comparison between biomedical and traffic research. *Science and Engineering Ethics*, *13*, 99–115.

Tavani, H. (2003). *Ethics and Technology: Ethical Issues in an Age of Information and Communication Technology*. London: Wiley.

Taylor, H. C., & Russell, J. T. (1939). The relationship of validity coefficients to the practical effectiveness of tests in selection: Discussion and tables. *The Journal of Applied Psychology*, *23*, 565–578. doi:10.1037/h0057079

Taylor, J. A., Lips, M., & Organ, J. (2009). Identification practices in government: citizen surveillance and the quest for public service improvement. *Identity in the Information Society*, *1*(1), 135–154. doi:10.1007/s12394-009-0007-5

Taylor, S. L. (2004). Music piracy—Differences in the ethical perceptions of business majors and music business majors. *Journal of Education for Business*, *79*, 306–310. doi:10.3200/JOEB.79.5.306-310

Taylor, T. L. (2002). Living digitally: Embodiment in virtual worlds. In Schroeder, R. (Ed.), *The social life of avatars: Presence and interaction in shared virtual environments* (pp. 40–62). New York, NY: Springer.

Taylor, T. L. (2006). *Play between worlds: Exploring online game culture*. Cambridge, MA: MIT Press.

The Economist. (2008, October 23). *Surveillance technology: if looks could kill.* Retrieved from The Economist website: http://www.economist.com/node/12465303?story_id=12465303

The London Speaker Bureau. (n. d.).*Sir Tim Berners-Lee.* Retrieved from http://www.londonspeakerbureau.co.uk/sir_tim_berners_lee.aspx

Thompson, P. (2007). Deception as a semantic attack. In Kott, A., & McEneaney, W. (Eds.), *Adversarial reasoning: Computational approaches to reading the opponent's mind* (pp. 125–144). London, UK: Chapman & Hall/CRC.

Thom, R. (1980). *Modèles mathématiques de la morphogenèse* (Brookes, W. M., & Rand, D., Trans.). Paris, France: Christian Bourgois.

Thong, J. Y. L., & Yap, C.-S. (1998). Testing an ethical decision-making theory: The case of softlifting. *Journal of Management Information Systems, 15,* 213–237.

Tolstoy, L., & Garnett, C. (1917). Ivan the Fool. In *The Harvard Classics Shelf of Fiction (Vol. 17)*. New York, NY: P.F. Collier & Son.

Tombs, S., & White, D. (2009). A deadly consensus. *The British Journal of Criminology, 20,* 1–20.

Tremlett, G. (2010, October 4). *GPS directs driver to death in Spain's largest reservoir.* Retrieved September 23, 2011, from The Guardian website: http://www.guardian.co.uk/world/2010/oct/04/gps-driver-death-spanish-reservoir

Trillin, C. (2007, January, 26). *Park, He Said.* Retrieved June 17, 2011, from The New York Times website: http://www.nytimes.com/2007/01/26/opinion/26trillin.html?_r=1

Tufekci, Z. (2008). Grooming, gossip, Facebook and Myspace. *Information Communication and Society, 11*(4), 544–564. doi:10.1080/13691180801999050

Tully, J. (1983). *A discourse on property: John Locke and his adversaries.* Cambridge, UK: Cambridge University Press.

Turilli, M., & Floridi, L. (2010). The ethics of information transparency. *Ethics and Information Technology, 11*(2), 105–112. doi:10.1007/s10676-009-9187-9

Turing, A. (1950). Computing machinery and intelligence. *Mind, 59*(236), 433–460. doi:10.1093/mind/LIX.236.433

Turkle, S. (1997). *Life on the screen: Identity in the age of the Internet.* New York, NY: Simon & Schuster.

Turkle, S. (2011). *Alone Together: Why we expect more from technology and less from each other.* New York, NY: Basic Books.

Turner, F. (2006). *From counterculture to cyberculture: Stewart Brand, the whole earth network, and the rise of digital utopianism.* Chicago, IL: University of Chicago Press.

U. S. Department of Education. (2008). *Higher Education Act of 1965, Cong. §487.* Washington, DC: U.S. Department of Education.

U.S. National Science and Technology Council. (2006). *Privacy & biometrics building a conceptual foundation.* Washington, DC: Author.

Ugrin, J. C., Pearson, J. M., & Odom, M. D. (2007). Profiling cyber-slackers in the workplace: Demographic, cultural, and workplace factors. *Journal of Internet Commerce, 6*(3), 75–89. doi:10.1300/J179v06n03_04

UN Wire. (2003, July 21). IAEA says terrorism necessitates deep waste burial. *UN Wire.*

United States National Bioethics Advisory Commission. (1998). *Research involving persons with mental disorders that may affect decisionmaking capacity.* Retrieved from http://bioethics.georgetown.edu/nbac/pubs.html

United States National Commission for the Protection of Human Subjects of Biomedical and Behavioral Research. (1978). *The Belmont report: Ethical principles and guidelines for the protection of human subjects of research.* Bethesda, MD: United States National Commission for the Protection of Human Subjects of Biomedical and Behavioral Research.

University of Alberta Science Technology and Society. (2010). Retrieved March 1, 2010, from http://www.ois.ualberta.ca/sts.cfm

University of Calgary Science, Technology and Society Program. (2010). *Society.* Retrieved March 1, 2010, from http://www.ucalgary.ca/comcul/scitech

University of Central Florida Website. (2009). *Office of student conduct.* Retrieved from http://www.osc.sdes.ucf.edu/?id=computer_misuse

University of Regina BA in STS. (2010). Retrieved March 1, 2010, from http://pager.uregina.ca/gencal/ugcal/facultyofArts/ugcal_223.shtml

University of Surrey. (2000). *UNIX introduction.* Retrieved from http://www.ee.surrey.ac.uk/Teaching/Unix/unixintro.html

University of Toronto IHPST. (2010). *Institute for the History and Philosophy of Science and Technology.* Retrieved March 1, 2010, from http://www.hps.utoronto.ca/

Urry, J. (2006). Inhabiting the car. *The Sociological Review, 54,* 17–31. doi:10.1111/j.1467-954X.2006.00635.x

US Federal Highway Administration. (2005). *Safety Evaluation of Red-Light Cameras.* Retrieved September 23, 2011 from Federal Highway Administration website: http://www.fhwa.dot.gov/publications/research/safety/05049/

UseNet. (2009). *What is UseNet? - User network.* Retrieved from http://www.usenet.com/usenet.html

Van den Hove, J., Lokhorst, G.-J., & Van Poel, I. (in press). Engineering and the problem of moral overload. *Science and Engineering Ethics.*

Van den Hoven, J. (2000). The Internet and varieties of moral wrongdoing. In Langford, D. (Ed.), *Internet ethics* (pp. 127–157). New York, NY: Macmillan.

Van Hove, E. (2003, November 18-21). *Valorisation of a repository in an added value project.* Paper presented at the Plenary Lecture to NEA Forum on Stakeholder Confidence, Brussels, Belgium.

Verhellen, E. (2000). *Convention on the rights of the child: Background, motivation, strategies, main themes.* Antwerp, Belgium: Coronet.

Versanyi, L. (1974). Can robots be moral? *Ethics, 84*(3), 248–259. doi:10.1086/291922

Vitanza, V. (2005). *Cyberreader* (Abridged ed.). New York, NY: Pearson Education.

W3C. (2003). *Who's who at the World Wide Web consortium.* Retrieved from http://www.w3.org/People/all#timbl

Waldron, J. (1988). *The right to private property.* New York, NY: Clarendon Press.

Waldron, J. (2002). *God, Locke, and equality: Christian foundations in Locke's political thought.* Cambridge, UK: Cambridge University Press. doi:10.1017/CBO9780511613920

Wallace, R. J. (1994). *Responsibility and the moral sentiments.* Cambridge, MA: Harvard University Press.

Wall, D. (2003). *Cyberspace crime.* London, UK: Dartmouth Publishing.

Walton, D. (2004). *Relevance in argumentation.* Mahwah, NJ: Lawrence Erlbaum.

Wang, L., & Geng, X. (2010). *Behavioral biometrics for human identification.* Hershey, PA: IGI Global.

Warwick, K. (2003). Cyborg morals, cyborg values, cyborg ethics. *Ethics and Information Technology, 5,* 131–137. doi:10.1023/B:ETIN.0000006870.65865.cf

Watkins, C., Mazerolle, L. G., & Rogan, D. F. (2002). Technological approaches to controlling random gunfire: Results of a gunshot detection system field test. *Policing: An International Journal of Police Strategies and Management, 25*(2), 345–370. doi:10.1108/13639510210429400

Watters, E. (2007). *ShotSpotter.* Retrieved June 24, 2011, from Wired Magazine website: http://www.wired.com/wired/archive/15.04/shotspotter.html

Wei, G., & Li, D. (2005). Biometrics: Applications, Challenges and the Future. In *Privacy and Technologies of Identity* (pp. 135–149). New York: Springer.

Weinberger, D. (2002). *Small pieces loosely joined: A unified theory of the web.* New York, NY: Perseus Publishing.

Weisberg, J. (2010, October 16). *Hyper-libertarian Facebook billionaire Peter Thiel's appalling plan to pay students to quit college.* Retrieved September 20, 2011 from Slate website: http://www.slate.com/id/2271265/

Weizenbaum, J. (1984). *Computer power and human reason: From judgment to calculation.* London, UK: Penguin Press.

Wells, A. J. (2002). Gibson's affordances and Turing's theory of computation. *Ecological Psychology, 14*(3), 141–180. doi:10.1207/S15326969ECO1403_3

Westbrook, J. (2006). Owned: Finding a place for virtual world property rights. *Michigan State Law Review*, 779-812.

White, J. (in press). *The mechanism of morality* [Monograph].

White, A. E. (2006). *Virtually obscene: The case for an uncensored Internet.* Jefferson, NC: McFarland & Company.

White, B. (in press). Infosphere to ethosphere. Moral mediators in the nonviolent transformation of self and world. *International Journal of Technoethics.*

White, J. (2006). *Conscience: Toward the mechanism of morality.* Columbia, MO: University of Missouri-Columbia.

White, J. (2010). Understanding and augmenting human morality: An introduction to the ACTWith model of conscience. *Studies in Computational Intelligence, 314*, 607–621. doi:10.1007/978-3-642-15223-8_34

White, J. (in press). An information processing model of psychopathy and anti-social personality disorders integrating neural and psychological accounts towards the assay of social implications of psychopathic agents. In *Moral Psychology.* New York, NY: Nova.

White, J. (in press). Manufacturing morality: A general theory of moral agency grounding computational implementations: the ACTWith model. In *Computational Intelligence.* New York, NY: Nova.

White, T. (2007). *In defense of dolphins: the new moral frontier.* Oxford, UK: Blackwell. doi:10.1002/9780470694152

Wieviorka, W. (2009). *Violence: A new approach.* London, UK: Sage.

Wikipedia. (2010). *Wikipedia. World population.* Retrieved from http://en.wikipedia.org/wiki/World_population

Wikipedia. (n. d.). *Wikipedia.* Retrieved from http://en.wikipedia.org/wiki/Wikipedia

Williams, M. (2005). Forget fingerprints and eye scans; the latest in biometrics is in vein. In *Computerworld.* Retrieved from www.computerworld.com/s/article/102861/Forget_fingerprints_and_eye_scans_the_latest_in_biometrics_is_in_vein

Williams, P., Shimeall, T., & Dunlevy, C. (2002). Intelligence analysis for Internet security. *Contemporary Security Policy, 23*(2), 1–38. doi:10.1080/713999739

Wilson, C. (2009). Botnets, cybercrime, and cyberterrorism: Vulnerabilities and policy issues for Congress. In Jacobson, G. V. (Ed.), *Cybersecurity, botnets, and cyberterrorism* (p. 71). New York, NY: Nova Science Publishers.

Wilson, D., Wilczynski, C., Wells, A., & Weiser, L. (2002). Gossip and other aspects of language as group-level adptations. In Hayes, C., & Huber, L. (Eds.), *The evolution of cognition* (pp. 347–365). Cambridge, MA: MIT Press.

Windsor, W. L. (2004). An ecological approach to semiotics. *Journal for the Theory of Social Behaviour, 34*(2), 179–198. doi:10.1111/j.0021-8308.2004.00242.x

Winner, L. (1993). Upon opening the black box and finding it empty: Social Constructivism and the Philosophy of Technology. *Science, Technology & Human Values, 18*, 362–378. doi:10.1177/016224399301800306

Winston, M. E., & Edelbach, R. D. (Eds.). (2000). *Society, ethics, and technology.* Belmont, CA: Wadsworth/Thomson Learning.

Winter, S. J., Stylianou, A. C., & Giacalone, R. A. (2004). Individual differences in the acceptability of unethical information technology practices: The case of Machiavellianism and ethical ideology. *Journal of Business Ethics, 54*(3), 275–296. doi:10.1007/s10551-004-1772-6

Wolfendale, J. (2007). My Avatar, My Self: Virtual harm and attachment. *Ethics and Information Technology, 9*(2), 111–119. doi:10.1007/s10676-006-9125-z

Wolfe, S. E., Higgins, G. E., & Marcum, C. D. (2008). Deterrence and digital piracy: A preliminary examination of the role of viruses. *Social Science Computer Review, 26*, 317–333. doi:10.1177/0894439307309465

Woolgar, S. (1991). The turn to technology in social studies of science. *Science, Technology, & Human Values, 17*, 20-50.

Worona, S. (2008). On making sausage. *EDUCAUSE Review, 43*(6).

Yahoo. (n. d.). *Birth of the Internet – timeline*. Retrieved from http://smithsonian.yahoo.com/timeline.html

Yahoo. Media Relations. (2005). *The history of Yahoo! -How it all started*. Retrieved from http://docs.yahoo.com/info/misc/history.html

Yampolskiy, R. V. (2008). Behavioral modeling: An overview. *American Journal of Applied Sciences, 5*(5), 496–503. doi:10.3844/ajassp.2008.496.503

Yankelovich, D. (1991). *Coming to public judgment: Making democracy work in a complex world*. Syracuse, NY: Syracuse University Press.

Yee, N. (2006). The labor of fun: How video games blur the boundaries of work and play. *Games and Culture, 1*(1), 68–71. doi:10.1177/1555412005281819

Yerkovich, S. (1977). Gossiping as a way of speaking. *The Journal of Communication, 27*, 192–196. doi:10.1111/j.1460-2466.1977.tb01817.x

Yin, R. K. (2003). *Case study resarch: Design and methods* (2nd ed.). Thousand Oaks, CA: Sage.

Young, J. (2010). *Woodbury U. banned from Second Life, again*. Retrieved from http://chronicle.com/blogs/wiredcampus/woodbury-u-banned-from-second-life-again/23352

Young, K. (2010). Killer surf issues: Crafting an organizational model to combat employee Internet abuse. *Information & Management*, 34–38.

Young, K. S. (1998). *Caught in the net: How to recognize the signs of Internet addiction and a winning strategy for recovery*. New York: John Wiley and Sons.

Zhang, D., & Jain, A. K. (2006). *Handbook of multibiometrics*. New York, NY: Springer.

Zhang, D., Song, F., Xu, Y., & Liang, Z. (2009). *Advanced pattern recognition technologies with applications to biometrics*. Hershey, PA: IGI Global. doi:10.4018/978-1-60566-200-8

Zhang, Y., Li, Q., & You, J. (2007). Palm Vein Extraction and Matching for Personal Authentication. In Bhattacharya, P. (Ed.), *Advances in Visual Information Systems*. Berlin: Springer. doi:10.1007/978-3-540-76414-4_16

Zimmerman, P. R. (1996). *The official PGP user's guide*. Boston, MA: MIT Press.

Žižek, S. (2009). *Violence*. London, UK: Profile Books.

About the Contributors

Rocci Luppicini is an Associate Professor in the Department of Communication and affiliate of the Institute for Science, Society, and Policy (ISSP) at the University of Ottawa (Canada) and acts as the Editor-in-Chief for the *International Journal of Technoethics*. He is a leading expert in technology studies (TS) and technoethics. He has published over 25 peer reviewed articles and has authored and edited several books including, *Online Learning Communities in Education* (IAP, 2007), the *Handbook of Conversation Design for Instructional Applications* (IGI, 2008), *Trends in Canadian Educational Technology and Distance Education* (VSM, 2008), *the Handbook of Research on Technoethics: Volume I & II* (with R. Adell) (IGI, 2008, 2009), *Technoethics and the Evolving Knowledge Society: Ethical Issues in Technological Design, Research, Development, and Innovation* (2010), *Cases on Digital Technologies in Higher Education: Issues and Challenges* (with A. Haghi) (IGI, 2010), *Education for a Digital World: Present Realities and Future Possibilities* (AAP, in press). His most recent edited work, the *Handbook of Research on Technoself: Identity in a Technological Society: Vol I & II* (IGI, 2012), provides the first comprehensive reference work in the English language on human enhancement and identity within an evolving technological society.

* * *

Emanuele Bardone earned his PhD in Philosophy at the University of Pavia (2009). He currently teaches Philosophy of Cognition at the University of Pavia (Italy) and he is Marie Curie Fellow at the University of Tallinn (Estonia). He is currently working on the distributed aspects of hypothesis generation. His publications include the book *Seeking Chances: From Biased Rationality to Distributed Cognition* (2011, Springer).

Sara Belfrage is at the Divison of Philosophy of the Royal Institute of Technology in Stockholm, Sweden. Besides philosophy, she has a background in political science and human rights. Her research focuses on ethical issues arising when people are involved in research as objects of investigation. She has for instance published a paper together with Professor Sven Ove Hansson involving a comparison between how research subjects are protected in and affected by biomedical and traffic research. Currently, she deals with questions concerning voluntariness, coercion and exploitation that have bearing beyond the scope of research.

Tommaso Bertolotti is a PhD student at the Department of Philosophy, University of Pavia and a member of the Computational Philosophy Laboratory at the same university. He graduated with a dissertation concerning the abductive origin of belief in supernatural agents (2010). Under the supervision of Lorenzo Magnani, with whom he co-authored several articles, he is currently working on a series of topics concerning some characteristic aspects of human cognition: inferences regulating religious thinking, nature and use of models in science with respect to other rational domains, inferential modeling of biological phenomena persisting in human environments (e.g. camouflage) and ethical consequences of technological development.

Daniele Cantore got his MA degree in Philosophy in 2010, with a thesis titled *An Epistemological Approach to Biometrics. Identification, Verification, and Identity Issue*s. He works with Lorenzo Magnani's group at the Computational Philosophy Laboratory of the Department of Philosophy (University of Pavia, Italy). He is interested in ethical and epistemological problems raised by contemporary technological development.

Rosa Cintrón is an Associate Professor at the University of Central Florida, Department of Educational and Human Sciences in the College of Education. Her first career was in the mental health field working as a Bilingual Psychotherapist in Puerto Rico, Connecticut, and New York. Her academic career started in the early 1980s in SUNY/College at Old Westbury. Since then she has occupied various positions as staff, administrator and faculty in the states of Illinois and Oklahoma. She is the past chair of the NASPA Faculty Fellows and holds various other leadership positions in professional associations. Her latest book, *College Student Death: Guidance for a Caring Campus* (co-authored with Erin Taylor and Katherine Garlough) has been listed among the most important academic resources dealing with crisis intervention. Her expertise is in the area of access, retention and issues of social justice in American colleges and universities.

Suzanne R. Clayton is Assistant Professor of Practice in Information Systems at the Drake University College of Business and Public Administration. Professor Clayton received her MBA at Drake in 1989. Ms. Clayton began her career over 25 years ago working in the consulting division of Arthur Andersen & Co (now Accenture) and has also worked as an IT consultant, programming manager, and project leader. She taught for ten years at the American Institute of Business and eight years at Iowa State University prior to coming to Drake. Professor Clayton has received teaching and mentoring awards and is a past Founding President of the Association of Information Technology and Management.

Raphael Cohen-Almagor (D. Phil., Oxon) is an educator, researcher, human rights activist and Chair in Politics, University of Hull. He has published extensively in the fields of political science, philosophy, law, media ethics, medical ethics, sociology, history and education. He was Visiting Professor at UCLA and Johns Hopkins, Fellow at the Woodrow Wilson International Center for Scholars, Founder and Director of the Center for Democratic Studies, University of Haifa, and Member of The Israel Press Council. Among his recent books are *Speech, Media and Ethics* (2005), *The Scope of Tolerance* (2006), *The Democratic Catch* (2007), and his second poetry book *Voyages* (2007). His sixteenth book is scheduled to be published in late 2011, dealing with public responsibility in Israel. Further information http://www.hull.ac.uk/rca and http://almagor.blogspot.com.

J. Royce Fichtner is an Assistant Professor of Business Law in the College of Business and Public Administration at Drake University where he teaches in the areas of business law, accounting law, and company law. He received his Bachelor of Arts in Business Administration from the University of Northern Iowa and his Juris Doctorate from Drake University. He is a former law clerk for Justice Michael Streit of the Iowa Supreme Court and a former staff attorney for the Honorable Terry Huitink of the Iowa Court of Appeals and the Honorable Robert Mahan of the Iowa Court of Appeals. He has published his research in the *Journal of Intellectual Property Law and Practice*, the *International Journal of Disclosure and Governance*, and the *Drake Law Review*. His major research interests are in the areas of company law, corporate governance, and information technology law.

Albrecht Fritzsche graduated from Albert-Ludwigs-University in Freiburg with degrees in Mathematics (Diplom-Mathematiker) and Educational Science & Philosophy (Magister Artium). He worked for many years as a technology consultant in the automotive industry. He received a doctoral degree in economics from Hohenheim University, Stuttgart, and another one in philosophy from the Technical University of Darmstadt. His doctoral dissertation in economics studies heuristic methods of decision support for industrial planning activities in complex networks. His doctoral dissertation in philosophy is concerned with indeterminacy in technical systems. His current research interests include intellectual property rights, knowledge management, narratives of progress and the cultural foundations of technical expertise. In addition to his academic activities, he continues to work as a consultant in Germany and abroad.

Benjamin Grounds is a student at the Penn State College of Medicine. In 2010 he earned a BA in sociology and BS in psychology from the University of Pittsburgh at Johnstown. As an undergraduate researcher, he actively pursued research on earwitness identification as well as stereotypes of people with disabilities and perceptions of computer synthesized speech. Prior to attending college he served in the United States Marine Corp for over six years.

Miles Kennedy was born in Bellingham Washington and lives in Galway Ireland. He teaches philosophy and IT ethics for the National University of Ireland Galway. He also teaches preparatory courses in philosophy on several outreach programmes that prepare socially and economically disadvantaged students for entry into university. His first book *Home: A Bachelardian Concrete Metaphysics* is currently being prepared for publication with Peter Lang (Oxford). Miles is married with two children and a dog.

Xue Lin completed her Master's Degree at the University of Ottawa specializing in Organizational Communication and Technology. She is interested in IT security, IT policy issues, and other technology issues within the workplace.

Lorenzo Magnani, philosopher and cognitive scientist, is a professor at the University of Pavia, Italy, and the director of its Computational Philosophy Laboratory. He is visiting professor at the Sun Yat-sen University, Canton (Guangzhou), China. He has taught at the Georgia Institute of Technology and at The City University of New York and currently directs international research programs in the EU, USA, and China. His book *Abduction, Reason, and Science* (New York, 2001) has become a well-respected work in the field of human cognition. The recent book *Morality in a Technological World* (Cambridge, 2007) develops a philosophical and cognitive theory of the relationships between ethics and technology in a naturalistic perspective. The book *Abductive Cognition. The Epistemological and Eco-Cognitive Dimensions of Hypothetical reasoning* has been published by Springer, Berlin/Heidelberg/New York (2009). The last book *Understanding Violence. Morality, Religion, and Violence Intertwined: A Philosophical Stance* has also been published by Springer in 2011, In 1998 he started the series of International Conferences on Model-Based Reasoning (MBR). Since 2011 he is the editor of the Book Series *Studies in Applied Philosophy, Epistemology and Rational Ethics* (SAPERE), Springer, Heidelberg/Berlin.

Alan Marshall is currently a postdoctoral fellow at the School of Management, Asian Institute of Technology, Thailand. His specialty subject is the social aspects of technology and his research within this field has been conducted at a number of research institutes worldwide, including the Institute for Advanced Studies in Austria; Masaryk University in the Czech Republic; Curtin University of Technology in Australia, and also at the Nuclear Industry Radioactive Waste Executive in England. His books include: 'Dangerous Dawn: The New Nuclear Age' (FoE: Melbourne), 'Wild Design' (North Atlantic Books: Berkeley), Lancewood (Indra, Melbourne) and 'The Unity of Nature' (Imperial College Press: London).

Jeffrey Reiss is a Systems Analyst at the University of Central Florida (UCF). He holds a BS in Statistics, an MS in Statistical Computing and Data Mining, and an EdD in Higher Educational Leadership, all from UCF. Dr. Reiss also serves as an adjunct professor of Statistics at Seminole State College and works as a private statistical consultant, assisting students and organizations in need of methodology and survey support in education and the social sciences. His research interests lie within the realm of technology in higher education, whether related to prevalence, usage issues, or its effects on students and faculty.

Kurt Reymers is a cybersociologist studying the social construction of virtual worlds. In virtual reality, he can be reached as Kurt Karsin in Second Life and has a laboratory at the Nature.com island, Elucian Omega. In face-to-face reality, he is a professor of science and technology studies, sociology and anthropology at Morrisville State College of the State University of New York.

Halim Sayoud is an Associate Professor at the FEI-USTHB: College of Electronics and Computer Engineering (www.usthb.dz). He received his MSc in 1994, and his PhD (in Automatic Speaker Recognition) in 2003 from the USTHB University in collaboration with the LIA laboratory of Avignon (in France). He was a research visitor at several universities in France and Greece: IRIT of Toulouse, LIA of Avignon, ENST of Paris and ILSP of Athens. His research works focus on the area of speaker identification, biometrics and text mining. He has been investigating the effects of different types of reduced features in speaker characterization and recognition. He is also interested in developing novel research and methods for text mining, author verification and biometrics.

Marcus Schulzke is a PhD candidate in political science at the State University of New York at Albany. His primary research interests are political theory and comparative politics, with special attention to contemporary political theory, moral theory, and political violence. He is currently working on a dissertation about how soldiers make moral decision in combat.

Cameron Shelley is a lecturer with the Centre for Society, Technology, and Values at the University of Waterloo. There, he teaches courses in Design and Society, Biotechnology and Society, and Information Technology and Society. He earned his Ph.D. in Philosophy from the University of Waterloo in 1999. His research interests include analogical cognition, model-based reasoning, fairness in the design of technology, and motivational influences in design. His recent publications include "Fairness in technological design" (in press), *Science and Engineering Ethics* and "Why study animals to treat humans?" (2010), *Studies in History and Philosophy of Biological and Biomedical Sciences*. He is also the author of *Multiple analogies in science and philosophy* (2003), published by John Benjamins.

LouAnn Simpson is a Professor of Business Law in the College of Business and Public Administration at Drake University where she teaches the introductory course in business law along with advanced courses in property and employment law. She received her Bachelor of Science in Business Administration as well as her Juris Doctorate from Drake University. After practicing law for a few years in Des Moines (private practice in Des Moines and City Prosecutor for the city of West Des Moines), she joined the faculty at Drake. Her major research interests are in the areas of employment law and information technology. She frequently reviews Business Law textbooks.

Steven Stern received his PhD in Social and Organizational Psychology from Temple University in 1995. He is Professor of Psychology and Chair of the Division of Natural Sciences at the University of Pittsburgh at Johnstown. For nearly two decades he has been studying the impact of technology on behavior and social functioning. His scholarly work has included research on the use of computer synthesized speech by people with speech disabilities, studies on how automation can affect perceptions of performance, the use of the Internet as a tool for collecting experimental data, and an examination of why the term "Internet Addiction" might not be appropriate to describe compulsive Internet usage. Along with John Mullennix, he has recently co-authored *Computer Synthesized Speech Technologies: Tools for Aiding Impairment*, published by IGI Global.

Troy J. Strader is Professor of Information Systems in the Drake University College of Business and Public Administration. Dr. Strader received his Ph.D. in Business Administration (Information Systems) from the University of Illinois at Urbana-Champaign in 1997. His research interests include information technology ethics, digital product management, online consumer behavior, information technology adoption, and the impact of the Internet and e-business on initial public offerings. Dr. Strader has published in the *International Journal of E-Commerce, Communications of the ACM*, the *European Journal of Information Systems*, the *Journal of the Association of Information Systems*, *Decision Support Systems,* and other academic and practitioner journals and books. He has edited three books, *Digital Product Management, Technology and Practice: Interdisciplinary Perspectives*, the *Handbook on Electronic Commerce*, and *Mobile Commerce: Technology, Theory and Applications*. His work experience is in software development and information systems analysis.

Christopher Wareham is engaged in an interdisciplinary PhD programme in the foundations and ethics of the life sciences conducted by the European School of Molecular Medicine and the University of Milan. His research interests include philosophy of science, political theory and the ethics of emerging biotechnologies.

Jeffrey Benjamin White trained in chemistry and bio/medical ethics with emphases on philosophy of chemistry and complex systems before taking the PhD in Philosophy in 2006 from the University of Missouri – Columbia where his primary mentors were John Kultgen, Alexander VonSchoenborn, and Ron Sun. His focal interests are philosophy of mind and cognition at the intersection of the cognitive sciences and computational models, evolutionary ethics, and phenomenology, and he has most recently published in robot ethics and moral psychology. He maintains an abiding interest in ancient Greek philosophy, especially Socratic ethics, as well as in the triad of Kant, Hegel, and Heidegger. His forthcoming monograph, *The Mechanism of Morality* is an integration of these interests around the ACTWith model of moral cognition, addressing the enduring philosophical questions: What is the life worth living and what is the world worth living in?

Index